Glacier-Influenced Sedimentation on High-Latitude Continental Margins

Special Publications reviewing procedures

The Society makes every effort to ensure that the scientific and production quality of its books matches that of its journals. Since 1997, all book proposals have been refereed by specialist reviewers as well as by the Society's Books Editorial Committee. If the referees identify weaknesses in the proposal, these must be addressed before the proposal is accepted.

Once the book is accepted, the Society has a team of Book Editors (listed above) who ensure that the volume editors follow strict guidelines on refereeing and quality control. We insist that individual papers can only be accepted after satisfactory review by two independent referees. The questions on the review forms are similar to those for *Journal of the Geological Society*. The referees' forms and comments must be available to the Society's Book Editors on request.

Although many of the books result from meetings, the editors are expected to commission papers that were not presented at the meeting to ensure that the book provides a balanced coverage of the subject. Being accepted for presentation at the meeting does not guarantee inclusion in the book.

Geological Society Special Publications are included in the ISI Science Citation Index, but they do not have an impact factor, the latter being applicable only to journals.

More information about submitting a proposal and producing a Special Publication can be found on the Society's web site: www.geolsoc.org.uk.

It is recommended that reference to all or part of this book should be made in one of the following ways:

DOWDESWELL, J. A. & Ó COFAIGH, C. (eds) 2002. *Glacier-Influenced Sedimentation on High-Latitude Continental Margins.* Geological Society, London, Special Publications, **203.**

BARRIE, J. V. & CONWAY, K. W. 2002. Contrasting glacial sedimentation processes and sea-level changes in two adjacent basins on the Pacific margin of Canada. *In:* DOWDESWELL, J. A. & Ó COFAIGH, C. (eds) *Glacier-Influenced Sedimentation on High-Latitude Continental Margins.* Geological Society, London, Special Publications, **203,** 181–194.

GEOLOGICAL SOCIETY SPECIAL PUBLICATION NO. 203

Glacier-Influenced Sedimentation on High-Latitude Continental Margins

EDITED BY

J. A. DOWDESWELL & C. Ó COFAIGH
Scott Polar Research Institute, University of Cambridge, UK

2002
Published by
The Geological Society
London

THE GEOLOGICAL SOCIETY

The Geological Society of London (GSL) was founded in 1807. It is the oldest national geological society in the world and the largest in Europe. It was incorporated under Royal Charter in 1825 and is Registered Charity 210161.

The Society is the UK national learned and professional society for geology with a worldwide Fellowship (FGS) of 9000. The Society has the power to confer Chartered status on suitably qualified Fellows, and about 2000 of the Fellowship carry the title (CGeol). Chartered Geologists may also obtain the equivalent European title, European Geologist (EurGeol). One fifth of the Society's fellowship resides outside the UK. To find out more about the Society, log on to www.geolsoc.org.uk.

The Geological Society Publishing House (Bath, UK) produces the Society's international journals and books, and acts as European distributor for selected publications of the American Association of Petroleum Geologists (AAPG), the American Geological Institute (AGI), the Indonesian Petroleum Association (IPA), the Geological Society of America (GSA), the Society for Sedimentary Geology (SEPM) and the Geologists' Association (GA). Joint marketing agreements ensure that GSL Fellows may purchase these societies' publications at a discount. The Society's online bookshop (accessible from www.geolsoc.org.uk) offers secure book purchasing with your credit or debit card.

To find out about joining the Society and benefiting from substantial discounts on publications of GSL and other societies worldwide, consult www.geolsoc.org.uk, or contact the Fellowship Department at: The Geological Society, Burlington House, Piccadilly, London W1J 0BG: Tel. +44 (0)20 7434 9944; Fax +44 (0)20 7439 8975; Email: enquiries@geolsoc.org.uk.

For information about the Society's meetings, consult *Events* on www.geolsoc.org.uk. To find out more about the Society's Corporate Affiliates Scheme, write to enquiries@geolsoc.org.uk.

Published by The Geological Society from:
The Geological Society Publishing House
Unit 7, Brassmill Enterprise Centre
Brassmill Lane
Bath BA1 3JN,
UK

(*Orders*: Tel. +44 (0)1225 445046
Fax +44 (0)1225 442836)
Online bookshop: http://bookshop.geolsoc.org.uk

The publishers make no representation, express or implied, with regard to the accuracy of the information contained in this book and cannot accept any legal responsibility for any errors or omissions that may be made.

British Library Cataloguing in Publication Data
A catalogue record for this book is available from the British Library.

ISBN 1–86239–120–3
ISSN 0305–8719

Distributors

USA
AAPG Bookstore
PO Box 979
Tulsa
OK 74101–0979
USA
Orders: Tel. +1 918 584-2555
Fax +1 918 560-2652
E-mail *bookstore@aapg.org*

India
Affiliated East–West Press PVT Ltd
G-1/16 Ansari Road, Daryaganj,
New Delhi 110 002
India
Orders: Tel. +91 11 327-9113
Fax +91 11 326-0538
E-mail *affiliat@nda.vsnl.net.in*

Japan
Kanda Book Trading Co.
Cityhouse Tama 204
Tsurumaki 1-3-10
Tama-shi
Tokyo 206–0034
Japan
Orders: Tel. +81 (0)423 57-7650
Fax +81 (0)423 57-7651

Typeset by Type Study, Scarborough, UK
Printed by Cromwell Press, Trowbridge, UK

Contents

Referees

The Editors are grateful to the following people for their assistance with the reviewing of papers submitted to this Special Publication.

J. Anderson
J. Andrews
P. Barker
H. Bauch
C. Clark
E. Cowan
E. Domack
A. Elverhøi
D. Evans
J. Evans
R. Gilbert
M. Hambrey
R. Hiscott
J. Howe
J. Jaeger
H. Josenhans
A. Kuijpers
J. Laberg
J. Landvik
R. Larter
J. Mienert
D. Praeg
C. Pudsey
D. Roberts
J. Scourse
M. Seigert
A. Solheim
M. Stoker
O. Suitcliffe
J. Syvitski
P. Talling
J. Taylor
M. Tranter
T. Vorren
P. Weaver

Glacier-influenced sedimentation on high-latitude continental margins: introduction and overview

JULIAN A. DOWDESWELL & COLM Ó COFAIGH

Scott Polar Research Institute & Department of Geography, University of Cambridge, Cambridge CB2 1ER, UK

(e-mail: jd16@cam.ac.uk; co232@cam.ac.uk)

The papers in this volume are the outcome of a Geological Society (London) meeting concerning glacier-influenced sedimentation on high-latitude continental margins, held at the University of Bristol, UK, in 2001. Since the publication of the Geological Society Special Publication *Glacimarine Environments: Processes and Sediments* (Dowdeswell & Scourse 1990), the intervening 11 years have seen a significant increase in research into the processes and patterns of glacimarine sedimentation, and the relationship between ice dynamics and the sediments and landforms preserved in the glacimarine environment. In this introduction, we outline the papers that make up the volume and we highlight the main findings and significance of each. First, however, we highlight three areas relating to sedimentation in glacimarine environments where we believe significant progress has been made in the last 11 years. These are: (1) glacially-influenced continental slope sedimentation; (2) iceberg-rafting processes and events; and (3) reconstructing the dynamics of former marine-terminating ice sheets from glacial geomorphological evidence on continental shelves. Each of these areas is represented by papers in this volume.

Glacier-influenced continental slope sedimentation

Over the last decade, an increasing amount of research has focused on the record of glacier-influenced sedimentation on high-latitude continental slopes (e.g. Kuvaas & Kristoffersen 1991; Aksu & Hiscott 1992; Vorren & Laberg 1997; Anderson 1999; Kleiber *et al*. 2000). Some of the most detailed work on this topic has been conducted around the continental margins of the Polar North Atlantic, where extensive geophysical and geological data-sets have been used to investigate the morphology and sedimentary architecture of glacier-influenced continental slopes (e.g. Stoker 1990; Laberg & Vorren 1995, 1996, 2000; Dowdeswell *et al*. 1996,

1997, 1998; Vorren *et al*. 1998). This work demonstrates that during glacial maxima, ice sheets expanded across continental shelves to the shelf edge. Furthermore, zones of streaming flow have been identified within the regional ice sheet cover using numerical modelling and geophysical investigations of sea-floor morphology (e.g. Dowdeswell & Siegert 1999; Siegert & Dowdeswell 2002; **Ottesen *et al*.**). Associated sediment flux along these ice-sheet margins was non-uniform, with the highest rates of sedimentation focused at ice-stream termini.

Marine geophysical and geological research efforts have focused on the morphology and sedimentary architecture of the continental slope in front of these ice-stream termini. This research indicates that these locations were characterized by significant sediment progradation resulting in the formation of trough-mouth fans. Much of the sediment within these submarine fans comprise glacigenic debris-flow deposits. These debris flows originated by sediment release along the ice-stream front at the shelf break, and subsequent downslope remobilization by debris-flow processes (e.g. Laberg & Vorren 1995, 1996, 2000; Dowdeswell *et al*. 1996, 1997; Elverhøi *et al*. 1997; King *et al*. 1998; Taylor 1999). Major periods of fan progradation are associated with relatively short intervals when ice streams terminated at the shelf break during glacial maxima. These fans therefore contain an archive of past ice-sheet expansion and recession. By contrast, in areas between ice streams where slower moving ice reached the shelf break, fans are not present and the continental slope is characterized by large-scale sediment slides and canyon systems (Dowdeswell *et al*. 1996; Van Weering *et al*. 1998; Vorren *et al*. 1998; Taylor *et al*. 2000). This reflects slower ice-sheet velocities and correspondingly lower rates of sediment delivery to the shelf edge.

More recent research is extending this work by continued investigation into the sedimentary processes operating on trough-mouth fans. This includes attempts to model and explain the long

From: DOWDESWELL, J. A. & Ó COFAIGH, C. (eds) 2002. *Glacier-Influenced Sedimentation on High-Latitude Continental Margins.* Geological Society, London, Special Publications, **203**, 1–9. 0305-8719/02/$15.00
© The Geological Society of London 2002.

run-out distances of glacigenic debris flows (Laberg & Vorren 2000; **Elverhøi et al.**; **Talling et al.**) and assessment of the role of subglacial meltwater and contour-current activity in trough-mouth fan formation (**Taylor et al.**). On-going investigations of the morphology and sedimentary architecture of continental slopes in front of ice-stream termini also demonstrates considerable variation from that of the trough-mouth fan model.

Investigations of ice-rafted debris and iceberg-rafting events

Iceberg rafting is the principal mechanism by which glacial debris is transported from marine-terminating glaciers and ice sheets into the deep ocean. Over the last decade, intense research interest has centred around the investigation of ice-rafted debris (IRD) layers in ocean cores and the utilization of these data to reconstruct palaeoceanographic and palaeoclimatic change, especially ice-sheet fluctuations. The identification of six sand-rich layers deposited by iceberg rafting in cores from the North Atlantic (i.e. the 'Heinrich layers' (Heinrich 1988)) and the recognition that these were related to massive discharges of icebergs from the Laurentide Ice Sheet (e.g. Andrews & Tedesco 1992; Bond et al. 1992; Dowdeswell et al. 1995), provided the impetus for an increased focus on the use of IRD as a proxy by the palaeoceanographic community. Subsequent research has identified IRD layers in marine cores that are related to the other major Northern hemisphere late Quaternary ice sheets (e.g. Bond & Lotti 1995; Andrews et al. 1998; Dowdeswell et al. 1999; Scourse et al. 2000; Bischof 2000).

A key question that is central to many of these studies is what mechanism drives the collapse of these ice-sheet drainage basins? Is it some internal glaciological mechanism or is it externally forced by climate (cf. MacAyeal 1993; Bond & Lotti 1995)? In part, resolution of this debate hinges on the correlation of IRD records between cores, sometimes hundreds of kilometres apart, and thus the accurate dating of these events is critical. This is particularly important given the increasing focus on high-resolution records where the aim is the investigation of ice-sheet history and palaeoclimatic change at millennial and even sub-millennial scales (e.g. Dokken & Jansen 1999; Knutz et al. 2001). Attempts have also been made to correlate IRD layers with other palaeoenvironmental records such as ice cores (e.g. Bond et al. 1997).

More recently, it has been recognized that the sedimentology of the IRD layers themselves is variable. For example, the work of Hesse and others from the Labrador Sea has shown that, in addition to iceberg rafting, the depositional processes responsible for Heinrich layer formation in this region also include suspension sedimentation from meltwater plumes and turbidity-current activity (Wang & Hesse 1996; Hesse & Khodabakhsh 1998; Andrews et al. 1998). Several workers (e.g. Andrews 1998, 2000; **Andrews & Principato**) have also highlighted the issue of how IRD is defined in palaeoceanographic studies with respect to its grain size, and they point out that that IRD events identified using one grain size cut-off may be indistinguishable using another (see also Dowdeswell et al. 2001).

Current research efforts in this area are increasingly focused on fingerprinting the various IRD sources to individual ice-sheet drainage basins in order to identify iceberg drift trajectories and to allow examination of ice-sheet interaction (e.g. synchronous v. asynchronous iceberg-rafting events; Barber et al. 1995; Hemming et al. 1998; Bischof 2001). In addition, the continued search for millennial and finer-resolution records using IRD in marine sediment cores is currently a major avenue of research. In conjunction with the efforts of glacimarine sedimentologists to define the nature of IRD layers in terms of sedimentology and depositional processes, and the efforts of the modelling community (both glaciological and palaeoceanographic) to better understand the physical relationships between glacier dynamics, iceberg calving and debris release, research in this area is likely to continue to remain central for some time to come.

Reconstructions of former ice-sheet dynamics on continental shelves

Recent improvements in geophysical techniques, in particular the application of multibeam swath bathymetry, has enabled the extensive mapping of glacial and glacimarine landforms and sediments on high-latitude continental shelves and has allowed reconstruction of former ice-sheet flow paths, geometry and dynamics. One of the most exciting developments in this regard has been the identification of the locations of former ice streams on continental shelves. Around the Polar North Atlantic and Antarctic margins, streamlined subglacial bedforms located within cross-shelf bathymetric troughs record the former presence of ice streams which drained towards the continental shelf edge during glacial maxima (e.g. Shipp *et al.* 1999; Canals *et al.* 2000, 2002; Wellner *et al.* 2001; Ó Cofaigh *et al.* 2002; **Ottesen *et al.*; Shipp *et al.**). This work follows on from developments in the field of terrestrial glacial geomorphology (e.g. Clark 1993, 1994; Punkari 1995; Stokes & Clark 1999; Clark & Stokes 2001), and a range of criteria have now been developed by which zones of former streaming flow within a regional ice-sheet cover can be identified. These include: mega-scale glacial lineations that are confined to, and orientated parallel with, the trough long axis; an absence of such lineations on intervening shallow banks; lateral shear moraines which demarcate the edge of prominent zones of lineations; and the presence of deformation tills within the troughs.

Given the importance of ice streams and fast-flowing outlet glaciers to ice-sheet mass-balance and stability, and the current interest in the dynamics of the Antarctic Ice Sheet, identification of former ice-stream beds is important. Such identification also facilitates investigations of a range of questions related to ice-stream dynamics and subglacial processes that, in the case of contemporary ice streams, are constrained by the relative inaccessibility of the subglacial environment. For example, do marine-terminating ice streams deform their beds and is this deformation pervasive or localized? Does streaming flow operate independently of bedrock lithology and if so by what mechanism(s)? What is the form of the subglacial hydrological system under ice streams and what is its role in fast-flow? These are key questions which are actively being pursued on contemporary ice streams in Antarctica (e.g. Engelhardt & Kamb 1998; Tulacyzk *et al.* 2000; Studinger *et al.* 2001), but which can also be addressed using the palaeo-ice stream record on formerly glacier-covered continental shelves.

Overview

This volume contains 18 papers which deal with various aspects of glacier-influenced sedimentation on high-latitude continental margins ranging from fjords through continental shelf and slope to the deep sea. Many of the papers relate specifically to one or more of the three topics discussed above and provide excellent examples of field-based investigations or more general discussions of these issues that draw on empirical data from various high-latitude, geographical locations.

The first two papers consider the large-scale sediment architecture of Arctic continental margins and the controls thereon. **O'Grady & Syvitski** examine the relationship between glacial processes and the morphology of Arctic continental slopes at a regional scale. They apply a GIS approach to their analysis of a bathymetric grid of the Arctic Ocean in order to compare slope angle and macro-roughness, and they relate these results to ice-sheet dynamics. The authors find that in areas of convergent flow across continental shelves (that is where ice streams flow in cross-shelf bathymetric troughs), the continental slopes beyond the trough mouths have a gentler gradient than those slopes in areas characterized by divergent (slower) ice flow. The authors also highlight the fact that the morphology of trough-mouth fans is directly related to the width of the adjacent continental shelf such that the slope angle of the fan has an inverse relationship to shelf width. This emphasizes the control of sediment supply on slope morphology.

The paper by **Dowdeswell *et al.*** focuses on the sedimentary architecture of high-latitude continental margins, specifically the Polar North Atlantic (PNA). These authors discuss the contrasting nature of sedimentation around the PNA margins and relate these data to variations in past ice-sheet dynamics. Significantly, they present new geophysical evidence in the form of swath bathymetric and side-scan sonar data from the Greenland Basin, which investigates the morphology of a large submarine channel system and associated sediment wave fields. This work builds on previous research utilizing lower resolution GLORIA side-scan sonar records from this region (Mienert *et al.* 1993; Dowdeswell *et al.* 1996). The authors relate the development of these features to the intermittent downslope flow of dense water and turbidity currents occasioned by sea-ice formation and brine rejection on the adjacent continental shelf and upper slope. This may have fundamental implications for the rate of bottom-water production in the Greenland Sea and, therefore, in the thermohaline circulation of the PNA. They conclude by presenting a simple conceptual model of glacier-influenced sedimentation on high-latitude continental margins, which summarizes the influence of ice sheets and sea ice on the nature and rate of sediment build-up.

The following three papers each focus on the themes of depositional processes on trough-mouth fans and the mechanics of glacigenic submarine debris flows. The paper by **Taylor *et al.*** investigates Late Quaternary depositional processes on the two largest trough-mouth fans in the Polar North Atlantic, the Bear Island and North Sea fans. This is an important contribution because as well as discussing the nature and origin of glacigenic debris flows, which are commonly emphasized in models of trough-mouth fan formation (e.g. Laberg & Vorren 1995, 2000; Dowdeswell et al. 1996; Vorren & Laberg 1997; King et al. 1998), it also highlights the role of meltwater sedimentation in fan formation. Using geophysical and geological data from the Bear Island Fan, the authors demonstrate that considerable areas of trough-mouth fans may be dominated by deposition of suspension sediments associated with meltwater release from a warm-based ice sheet, as well as contour-current activity. The presence of abundant meltwater is also proposed as a means of facilitating the long run-out distances characteristic of glacigenic debris flows (e.g. Elverhøi et al. 1997; Vorren et al. 1998; Laberg & Vorren 2000; **Elverhøi et al.**). Other studies have also recognized the importance of hemipelagic–meltwater sedimentation to trough-mouth fan development during periods when debris flow activity has ceased (e.g. **Davison & Stoker**). However, the work of **Taylor et al.** is significant because it shows that meltwater sedimentation operates concurrently with the delivery of glacigenic debris flows to other areas of the fan during glacial maxima. This demonstrates spatial variability in depositional processes on trough-mouth fans during times of maximum ice sheet extent and fan development.

The theme of glacigenically-derived, submarine debris flows on high-latitude continental slopes is further developed in the paper by **Elverhøi et al.** The authors examine recent progress in the modelling of these muddy flows and the slide events that may trigger them, through the various stages of release, break-up, flow and final deposition. They argue that the long run-out distances that are characteristic of glacigenic debris flows on low gradient slopes are due to hydroplaning, but that this mechanism requires large volumes (>1 km³) of sediment in order to be initiated. By contrast, for small slides and debris flows that travel only short distances, a visco-plastic model is applicable in which flow behaviour is determined by the sediment rheology.

The final paper of this group, by **Talling et al.**, explores the subject of shear mixing and the transformation of submarine debris flows into turbidity currents. This is directly relevant to the sedimentary processes on glacier-influenced, high-latitude continental margins, in that glacigenic debris flows, which are a basic building-block of trough-mouth fans, can travel hundreds of kilometres along the sea floor without undergoing significant dilution (e.g. Elverhøi et al. 1997; **Elverhøi et al.**; Laberg & Vorren 2000). By contrast, in other locations, submarine slope failures can transform completely into turbidity currents before exiting the continental slope. The authors utilize experimental data to predict rates of shear mixing for beds of cohesive sediment overlain by a turbulent flow. They demonstrate that shear mixing can initiate at relatively low velocities but that the addition of small amounts of mud (c. 3%) can significantly decrease shear mixing rates. This paper illustrates the potential for highly variable mixing rates in submarine mass-flow events

The next four papers in the volume focus on the glacimarine sedimentary processes and deposits on high-latitude continental margins from a range of spatially and temporally contrasting settings, and they explore the wider implications of these results for reconstructions of glaciology and climate from these regions. **Hambrey et al.** present the results of a sedimentological investigation of late Oligocene to early Miocene strata obtained during recent offshore drilling in the southwestern Ross Sea, Antarctica. Sediment facies recovered in these drill cores indicate considerable variation in former glacier dynamics with alternating episodes of ice-proximal and ice-distal sedimentation. The lithofacies documented in these cores are indicative of a climate that was significantly warmer than that which characterizes Antarctica today, and one in which wet-based glaciers produced abundant meltwater.

Davison & Stoker present the results of a sedimentological investigation of a Late Pleistocene succession recovered from a deep-water borehole at the base of the West Shetland continental slope off the northwestern United Kingdom. The core data are integrated with high-resolution seismic records and sea-bed imagery to provide a comprehensive picture of deep-water glacigenic sedimentation. These data indicate that the glacigenic part of the sequence is dominated by stacked glacigenic debris flows. The paper also highlights the role of meltwater sedimentation in the deposition of hemipelagic muds contemporaneous with debris-flow delivery (cf. **Taylor et al.**). The data suggest past extension of ice sheets to the shelf edge in this region during glacial maxima and the delivery of large volumes of subglacial sediment to the continental slope. On the basis of their investigation from the Shetland slope, and the integration of these data with onshore outcrop data from mainland Scotland, the authors present a depositional model for glacially-influenced, deep-marine environments.

Glacimarine sedimentation along a fjord-shelf-slope transect on the East Greenland continental margin forms the subject of the contribution by **Evans *et al.*** This paper provides geophysical and geological data relating to glacimarine processes and sediments from this region and links this information to the extent of the Late Weichselian glaciation on the East Greenland continental shelf. It demonstrates the first clear evidence from the marine record that the Late Weichselian ice sheet terminated on the middle continental shelf in this region. The paper also highlights the significance of meltwater processes in Holocene glacimarine sedimentation in the East Greenland fjords and, in conjunction with related work on this topic from fjords further to the south (Smith & Andrews 2000; Ó Cofaigh *et al.* 2001), indicates that this is a regional phenomenon.

The theme of ice-proximal glacimarine sedimentation and controls thereon, and the relationship with glacier dynamics, is continued in the paper by **Barrie & Conway** which examines geophysical and geological records of Late Wisconsinan glacial sedimentation and sea-level change in two adjacent basins on the Pacific margin of Canada. By comparing the glacial and sea-level histories of the two regions, this study demonstrates that glacioisostatically-induced variations in sea level can directly influence sedimentary processes on high-latitude continental shelves.

The paper by **Jaeger** explores methods of obtaining high-resolution chronologies in glacimarine sediments, based on research in SE Alaskan fjords. In glacimarine environments, sedimentation rates are highest in ice-proximal locations such as near the grounding-lines of ice-shelves or within a few kilometres of tide-water–glacier termini. However, although these areas provide the highest resolution records, they are correspondingly the most difficult to access. Jaeger describes a variety of geochronological methods that utilize the short half-life of ^{234}Th to measure sedimentation rates in ice-proximal settings. These methods are then applied to a series of progressively more ice-distal cores of glacimarine sediments from Icy Bay, Alaska and a high-resolution record is obtained for the proximal cores. The techniques described in this paper are applicable to most glacimarine environments and, hence, are of significance in the evaluation of seasonal and short-term sedimentary processes operating in this setting.

The following set of four papers deal with glacial and glacimarine processes and sediments on high-latitude shelves and their relationship to ice-sheet dynamics. The first paper, by **Powell & Cooper,** develops a sequence–stratigraphic model for temperate glaciated continental shelves. The model is based on the integration of a range of geophysical and sedimentological data from Pleistocene continental-shelf successions in southern Alaska, contemporary Alaskan fjords, the stratigraphic record of the late Miocene to recent Yakataga Formation from the same region, and facies sequences from the Oligocene–Miocene Cape Roberts cores from McMurdo Sound, Antarctica. The authors define four glacial-systems tracts related to glacial minimum, advance, maximum and retreat, which can be evaluated in relation to changes in other external variables such as relative sea level. These systems tracts may also include glacial-erosion surfaces or their equivalent conformities. The authors detail the various sediment types associated with these glacial-systems tracts and provide a model by which glacigenic sediment sequences on continental shelves may be evaluated.

The three subsequent papers describe geophysical investigations of sea-floor morphology on the glaciated continental shelves of Norway and Antarctica and discuss the implications of these data for ice-sheet history and dynamics. **Ottesen *et al.*** describe a series of spectacular streamlined subglacial bedforms, including mega-scale glacial lineations, that are aligned along bathymetric troughs on the mid-Norwegian continental shelf. Bedforms are confined to the troughs and are absent from the intervening shallow banks. These subglacial bedforms record the former flow of ice streams through these troughs to the continental-shelf edge during the last glacial maximum (LGM). The glacial geomorphological data are then used to test existing numerical-model predictions of fast-flowing ice on the Norwegian continental shelf during the LGM (Dowdeswell & Siegert 1999). This paper adds to the rapidly increasing data on the geomorphological imprint of palaeo-ice streams and illustrates the potential for integrating field-based and modelling-based studies in order to reconstruct former ice-sheet extent and spatial variations in glacier dynamics.

Rafaelsen *et al.* provide the first 3-dimensional seismic investigation of multiple levels of subglacial lineations on the SW Barents Sea shelf. They identify several generations of lineations on four distinct palaeo-surfaces and interpret these as reflecting former ice-sheet flow patterns. The lineations demonstrate conclusively that warm-based, grounded glacier ice reached the southern part of the Bear Island Trough on at least four occasions during the last 0.8 Ma. The predominant trend of the lineations is N–S, suggesting that ice-flow was directed northwards across the Barents Sea shelf. This contrasts with previous reconstructions from this region which propose westward-directed ice flow through the Bear Island Trough (Vorren *et al.* 1990; Vorren & Laberg 1996; Landvik *et al.* 1998). **Rafaelsen *et al.*** discuss several possible interpretations of these flow-sets, and suggest that the northward-oriented lineations record the latest flow events and were formed during deglaciation. This investigation demonstrates the utility of 3-dimensional industrial seismic data in the mapping of glacial landforms and reconstructing former ice-sheet dynamics.

The final paper from this group is that of **Shipp *et al.*** from the Ross Sea, Antarctica. This contribution details a geophysical and geological investigation of a glacially-eroded trough on the Ross Sea continental shelf that contained an ice stream. The ice stream acted as a 'conveyor belt', eroding and advecting sediment from the inner to the outer shelf where the material was then deposited. On the basis of sediment stratigraphy and glacial geomorphology, these authors subdivide the former ice-stream flow path into six zones recording a progressive transition from a subglacial environment characterized by net erosion to one of net deposition. Mega-scale glacial lineations formed in sediment are inferred to record the highest velocities associated with low basal shear stresses over a saturated soft bed. A range of recessional landforms, including corrugation moraines and back-stepping grounding-zone wedges, record progressive positions of the ice margin during deglaciation and indicate that the ice margin remained grounded during retreat. On a regional scale, this contrasts with reconstructions of ice-stream break-up from elsewhere on the Antarctic Peninsula shelf (Ó Cofaigh *et al.* 2002) and indicates spatial variation in Pleistocene Antarctic ice-stream dynamics during deglaciation.

Three papers deal with various aspects of glacimarine and glacially-related sedimentation in the deep-sea. **Andrews & Principato** address the question of the characteristic grain-size distribution of glacial and glacimarine sediments. This paper discusses several issues that link the research findings and interests of workers in the glacimarine environment with those of the deep-sea palaeoceanographic community. The underlying rationale for this is the increasing emphasis on the use of specific sand-size fractions as indicators of iceberg-rafting by the palaeoceanographic community. The paper addresses a series of questions concerning the grain-size distribution of glacimarine sediments using data-sets from the Labrador Sea, East Greenland, North Iceland and the Ross Sea, Antarctica. These questions include: what is the total grain-size distribution of glacimarine sediments and what is their matrix grain-size distribution? In addition, the paper explores the problem of distinguishing glacial sources using grain-size spectra and differences in the grain size of glacimarine sediments deposited on continental margins v. the deep sea. Based on analysis of down-core variations of different sand-size fractions and the correlation between these fractions, the authors demonstrate that there is no consistent relationship between the sand fractions. A key implication of this is that different interpretations of iceberg-rafting events might be determined from the same set of samples purely on the basis of the choice of the sand-size cut off. Furthermore, **Andrews & Principato** point out that although the focus of many palaeoceanographic investigations is on the sand-size fraction, in fact the silt and clay fraction dominates in the total ice-proximal sediment spectra. The paper discusses many key questions relating to the recognition of ice-rafted debris and the identification of iceberg-rafting events in deep-sea sediments and contains much of relevance to both the glacimarine and palaeoceanographic communities.

The paper by **Ó Cofaigh *et al.*** details the contrasting styles of sediment reworking around the continental margins of the Norwegian–Greenland Sea. It is based on detailed lithofacies analysis of sediment cores, placed in context by accompanying geophysical records. Sediment reworking in the form of mass flow, bottom- and contour-current activity and bioturbation is widespread across the region, but is spatially variable in style. This spatial variability reflects variations in past ice-sheet dynamics. The authors highlight the implications of the widespread nature of sediment reworking for palaeoceanographic investigations in the Norwegian–Greenland Sea where <7% of sediment delivery since the last glaciation is derived from pelagic and hemipelagic sources. Areas of high sedimentation in this region are characterized by extensive reworking. Hence, sediment reworking is likely to be a significant problem where continuous, high-resolution records of hemipelagic and pelagic sedimentation are required and attempts are made to correlate with other high-resolution proxies such as ice cores at sub-millennial scales.

Wilson & Austin provide an investigation of millennial and sub-millenial scale variability in sediment colour in marine sediment cores from the Barra Fan, off NW Scotland, and they discuss the implications of these data for Late Pleistocene British Ice-Sheet dynamics. Sediment colour has not commonly been used in investigations of high-latitude, continental margin records to date. This technique uses changes in lightness and reflectance to provide quantitative estimates of down-core variations in calcium carbonate and clay content. Interstadials are carbonate-rich/clay-poor (higher lightness and reflectivity) while stadials are carbonate-poor/clay-rich (lower lightness and reflectivity). The combined use of sediment colour with other proxy data indicates a shift in the response of the British Ice Sheet to external climate forcing before and after 30 ka BP. Prior to this time, strong Dansgaard/Oeschger cyclicity dominates the record, while after 30 ka BP lower amplitude but higher frequency variability characterizes the record as the ice sheet reached its maximum extent. These authors highlight the potential difficulties of correlating millennial-scale IRD events when IRD is derived from different ice sheets.

The final paper in the volume by **Woodward et al.** describes the results of an investigation of glacier-surge periodicity in East Greenland using trace metal analysis from marine sediment cores. Two peaks in Molybdenum in sediment cores recovered from the Noret Inlet in the Mesters Vig area, East Greenland, are interpreted to record surge-termination events. Mo is not found in the geology of the Noret Inlet catchment area, however, it is found in the neighbouring Mesters Vig Inlet. The Mo peaks are inferred to be the result of the release of sediment-rich meltwater from Östre Gletscher, a large surge-type glacier that drains into Mesters Vig Inlet, at surge-termination. The meltwater plumes subsequently entered Noret Inlet and deposited Mo-rich sediment. A recurrence interval of approximately 400 years for surge activity is suggested by these data.

The editors would like to thank the participants at the Geological Society of London meeting on 'Glacier-influenced sedimentation on high-latitude continental margins: modern and ancient' held in the School of Geographical Sciences, University of Bristol, in March 2001. We also thank the staff of the Geological Society Publishing House for their assistance in producing this volume. We are grateful to the referees who gave their time to review the papers published here. Finally, we acknowledge the support of the conference sponsors, LASMO and BP AMOCO.

References

AKSU, A. E. & HISCOTT, R. N. 1992. Shingled Quaternary debris flow lenses on the north-east Newfoundland Slope. *Sedimentology*, **39**, 193–206.

ANDERSON, J. B. 1999. *Antarctic Marine Geology*. Cambridge University Press, Cambridge.

ANDREWS, J. T. 1998. Abrupt changes (Heinrich events) in late Quaternary North Atlantic marine environments. *Journal of Quaternary Science*, **13**, 3–16.

ANDREWS, J. T. 2000. Icebergs and iceberg-rafted detritus (IRD) in the North Atlantic: facts and assumptions. *Oceanography*, **13**, 100–108.

ANDREWS, J. T. & TEDESCO, K. 1992. Detrital carbonate-rich sediments, northwestern Labrador Sea: implications for ice-sheet dynamics and iceberg-rafting (Heinrich) events in the North Atlantic. *Geology*, **20**, 1087–1090.

ANDREWS, J. T., KIRBY, M., JENNINGS, A. E. & BARBER, D. C. 1998. Late Quaternary stratigraphy, chronology, and depositional processes on the slope of S.E. Baffin Island, detrital carbonate and Heinrich Events: Implications for onshore glacial history. *Géographie Physique et Quaternaire*, **52**, 1–15.

BARBER, D. C., FARMER, L., ANDREWS, J. T. & KIRBY, M. E., 1995. Mineralogic and isotopic tracers in late Quaternary sediments in the Labrador Sea: implications for iceberg-sources during Heinrich events. *EOS, Transactions of the American Geophysical Union*, **76**, F296.

BISCHOF, J. 2000. *Ice Drift, Ocean Circulation and Climate Change*. Springer-Praxis, London.

BOND, G. & LOTTI, R. 1995. Iceberg discharges into the North Atlantic on millennial timescales during the last glaciation. *Science*, **267**, 1005–1010.

BOND, G., HEINRICH, H., BROECKER, W., ET AL. 1992. Evidence for massive discharges of icebergs into the North Atlantic during the last glacial period. *Nature*, **360**, 245–249.

BOND, G., SHOWERS, W., CHESEBY, M., ET AL. 1997. A pervasive millennial-scale cycle in North Atlantic Holocene and glacial climates. *Science*, **278**, 1257–1266.

CANALS, M., URGELES, R. & CALAFAT, A. M. 2000. Deep sea-floor evidence of past ice streams off the Antarctic Peninsula. *Geology*, **28**, 31–34.

CANALS, M., CASAMOR, J. L., URGELES, R., CALAFAT, A. M., DOMACK, E. W., BARAZA, J., FARRAN, M. & DE BATIST, M. 2002. Sea-floor evidence of a subglacial sedimentary system off the northern Antarctic Peninsula. *Geology*, **30**, 603–606.

CLARK, C. D. 1993. Mega-scale glacial lineations and cross-cutting ice-flow landforms. *Earth Surface Processes and Landforms*, **18**, 1–19.

CLARK, C. D. 1994. Large-scale ice-moulding: a discussion of genesis and glaciological significance. *Sedimentary Geology*, **91**, 253–268.

CLARK, C. D. & STOKES, C. R. 2001. Extent and basal characteristics of the M'Clintock Channel Ice Stream. *Quaternary International*, **86**, 81–101.

DOKKEN, T. M. & JANSEN, E. 1999. Rapid changes in

the mechanism of ocean convection during the last glacial period. *Nature*, **401**, 458–461.

DOWDESWELL, J. A. & SCOURSE, J. D. (eds) 1990. *Glacimarine Environments: Processes and Sediments*. Geological Society, London, Special Publications **53**.

DOWDESWELL, J. A. & SIEGERT, M. J. 1999. Ice-sheet numerical modelling and marine geophysical measurements of glacier-derived sedimentation on the Eurasian Arctic continental margins. *Geological Society of America Bulletin*, **111**, 1080–1097.

DOWDESWELL, J. A., MASLIN, M. A., ANDREWS, J. T. & McCAVE, I. N. 1995. Iceberg production, debris rafting, and the extent and thickness of Heinrich layers (H-1, H-2) in North Atlantic sediments. *Geology*, **23**, 301–304.

DOWDESWELL, J. A., KENYON, N., ELVERHØI, A., LABERG, J. S., MIENERT, J. & SIEGERT, M. J. 1996. Large-scale sedimentation on the glacier-influenced Polar North Atlantic margins: long-range side-scan sonar evidence. *Geophysical Research Letters*, **23**, 3535–3538.

DOWDESWELL, J. A., KENYON, N. H. & LABERG, J. S. 1997. The glacier-influenced Scoresby Sund Fan, East Greenland continental margin: evidence from GLORIA and 3.5 kHz records. *Marine Geology*, **143**, 207–221.

DOWDESWELL, J. A., ELVERHØI, A. & SPIELHAGEN, R. 1998. Glacimarine sedimentary processes and facies on the Polar North Atlantic margins. *Quaternary Science Reviews*, **17**, 243–272.

DOWDESWELL, J. A., ELVERHØI, A., ANDREWS, J. T. & HEBBELN, D. 1999. Asynchronous deposition of ice-rafted layers in the Nordic seas and North Atlantic Ocean. *Nature*, **400**, 348–351.

DOWDESWELL, J. A., Ó COFAIGH, C., ANDREWS, J. T. & SCOURSE, J. D. 2001. Workshop explores debris transported by icebergs and its paleo-environmental implications. *EOS, Transactions of the American Geophysical Union*, **82** (35), 382 & 386.

ELVERHØI, A., NOREM, H., ANDERSEN, E. S., *ET AL.* 1997. On the origin and flow behaviour of submarine slides on deep-sea fans along the Norwegian-Barents Sea continental margin. *Geo-Marine Letters*, **17**, 119–125.

ENGELHARDT, H. & KAMB, B. 1998. Basal sliding of Ice Stream B, West Antarctica. *Journal of Glaciology*, **44**, 223–230.

HEINRICH, H. 1988. Origin and consequences of cyclic ice rafting in the Northeast Atlantic Ocean during the past 130,000 years. *Quaternary Research*, **29**, 143–152.

HEMMING, S. R., BROECKER, W. S., SHARP, W. D., BOND, G. C., GWIAZDA, R. H., McMANUS J. F., KLAS, M. &. HAJDAS, I. 1998. Provenance of Heinrich layers in core V28–82, northeastern Atlantic: $^{40}Ar/^{39}Ar$ ages of ice-rafted hornblende, Pb isotopes in feldspar grains and Nd–Sr–Pb isotopes in the fine sediment fraction. *Earth and Planetary Science Letters*, **164**, 317–333.

HESSE, R. & KHODABAKHSH, S. 1998. Depositional

facies of late Pleistocene Heinrich Events in the Labrador Sea. *Geology*, **26**, 103–106.

KING, E. L., HAFLIDASON, H., SEJRUP, H. P. & LØVLIE, R. 1998. Glacigenic debris flows on the North Sea Trough Mouth Fan during ice-stream maxima. *Marine Geology*, **152**, 217–246.

KLEIBER, H. P., KNIES, J. & NIESSEN, F. 2000. The Late Weichselian glaciation of the Franz Victoria Trough, northern Barents Sea: ice sheet extent and timing. *Marine Geology*, **168**, 25–44.

KNUTZ, P. C., AUSTIN, W. E. N. & JONES, E. J. W. 2001. Millennial-scale depositional cycles related to British Ice Sheet variability and North Atlantic paleocirculation since 45 kyr B.P., Barra Fan, U.K. margin. *Paleoceanography*, **16**, 53–64.

KUVAAS, B. & KRISTOFFERSEN, Y. 1991. The Crary Fan: a trough–mouth fan on the Weddell Sea continental margin, Antarctica. *Marine Geology*, **97**, 345–362.

LABERG, J. S. & VORREN, T. O. 1995. Late Weichselian submarine debris flow deposits on the Bear Island Trough Mouth Fan. *Marine Geology*, **127**, 45–72.

LABERG, J. S. & VORREN, T. O. 1996. The glacier-fed fan at the mouth of Storfjorden Trough, western Barents Sea: a comparative study. *Geologische Rundschau*, **85**, 338–349.

LABERG, J. S. & VORREN, T. O. 2000. Flow behaviour of the submarine glacigenic debris flows on the Bear Island Trough Mouth Fan, western Barents Sea. *Sedimentology*, **47**, 1105–1117.

LANDVIK, J. Y., BONDEVIK, S., ELVERHØI, A., FJELDSKAAR, W., MANGERUD, J., SALVIGSEN, O., SIEGERT, M. J., SVENDSEN, J. I. & VORREN, T. O. 1998. The last glacial maximum of Svalbard and the Barents Sea area: ice sheet extent and configuration. *Quaternary Science Reviews*, **17**, 43–75.

MACAYEAL, D. R. 1993. Binge/purge oscillations of the Laurentide Ice Sheet as a cause of the North Atlantic Heinrich Events. *Paleoceanography*, **8**, 775–784.

MIENERT, J., KENYON, N. H., THIEDE, J. & HOLLENDER, F.-J. 1993. Polar continental margins: studies off East Greenland. *EOS, Transactions of the American Geophysical Union*, **74**, 225–236.

Ó COFAIGH, C., DOWDESWELL J. A. & GROBE, H. 2001. Holocene glacimarine sedimentation, inner Scoresby Sund, East Greenland: the influence of fast-flowing ice-sheet outlet glaciers. *Marine Geology*, **175**, 103–129.

Ó COFAIGH, C., PUDSEY, C. J., DOWDESWELL, J. A. & MORRIS, P. 2002. Evolution of subglacial bedforms along a paleo-ice stream, Antarctic Peninsula continental shelf. *Geophysical Research Letters*, **29** (8), 10.1029/2001GL014488.

PUNKARI, M. 1995. Glacial flow systems in the zone of confluence between the Scandinavian and Novaya Zemlya ice sheets. *Quaternary Science Reviews*, **14**, 589–603.

SCOURSE, J. D., HALL, I. R., McCAVE, I. N., YOUNG, J. R. & SUGDON, C. 2000. The origin of Heinrich layers: evidence from H2 for European precursor events. *Earth and Planetary Science Letters*, **182**, 187–195.

SHIPP, S., ANDERSON, J. B. & DOMACK, E. W. 1999.

Late Pleistocene-Holocene retreat of the West Antarctic Ice-Sheet system in the Ross Sea: Part 1 – Geophysical results. *Geological Society of America Bulletin*, **111**, 1486–1516.

SIEGERT M. J. & DOWDESWELL, J. A. 2002. Late Weichselian iceberg, meltwater and sediment production from the Eurasian Ice Sheet: results from numerical ice-sheet modelling. *Marine Geology*, **188**, 109–127.

SMITH, L. M. & ANDREWS, J. T. 2000. Sediment characteristics in iceberg dominated fjords, Kangerlussuaq region, East Greenland. *Sedimentary Geology*, **130**, 11–25.

STOKER, M. S. 1990. Glacially-influenced sedimentation on the Hebridean slope, northwest United Kingdom continental margin. *In*: Dowdeswell, J.A. & Scourse, J.D. (eds) *Glacimarine Environments: Processes and Sediments*. Geological Society, London, Special Publications, **53**, 349–362.

STOKES, C. R. & CLARK, C. D. 1999. Geomorphological criteria for identifying Pleistocene ice streams. *Annals of Glaciology*, **28**, 67–74.

STUDINGER, M., BELL, R. E., BLANKENSHIP, D. D., FINN, C. A., ARKO, R. A., MORSE, D. L. & JOUGHIN, I. 2001. Subglacial sediments: a regional geological template for ice flow in West Antarctica. *Geophysical Research Letters*, **28**, 3493–3496.

TAYLOR, J. 1999. *Large-scale sedimentation and ice sheet dynamics in the Polar North Atlantic*. PhD thesis, University of Bristol.

TAYLOR, J., DOWDESWELL, J. A. & KENYON, N. H. 2000. Canyons and Late Quaternary sedimentation on the North Norwegian margin. *Marine Geology*, **166**, 1–9.

TULACYZK, S. M., KAMB, B. & ENGELHARDT, H. F. 2000. Basal mechanics of Ice Stream B, West Antarctica. I. Till mechanics. *Journal of Geophysical Research*, **105** (B), 483–494.

VAN WEERING, T. C. E., NIELSEN, T., KENYON, N. H., AVENTIEVA, K. & KUIPERS, A. H. 1998. Sediments and sedimentation at the NE Faeroes continental margin: contourites and large-scale sliding. *Marine Geology*, **152**, 159–176.

VORREN, T. O. & LABERG, J. S. 1996. Late glacial air temperature, oceanographic and ice sheet interactions in the southern Barents Sea region. *In*: ANDREWS, J. T., AUSTIN, W. E. N., BERGSTEN, H. & JENNINGS, A. E. (eds) *Late Quaternary Palaeoceanography of the North Atlantic Margins*. Geological Society, London, Special Publications, **111**, 303–321.

VORREN, T. O. & LABERG, J. S. 1997. Trough mouth fans – palaeoclimate and ice-sheet monitors. *Quaternary Science Reviews*, **16**, 865–881.

VORREN, T. O., LEBESBYE, E. & LARSEN, K. B., 1990. Geometry and genesis of the glacigenic sediments in the southern Barents Sea. *In*: DOWDESWELL, J. A. & SCOURSE, J. D. (eds) *Glacimarine Environments: Processes and Sediments*. Geological Society, London, Special Publications, **53**, 269–288.

VORREN, T. O., LABERG, J. S., BLAUME, F., DOWDESWELL, J. A., KENYON, N. H., MIENERT, J., RUMOHR, J. & WERNER, F. 1998. The Norwegian–Greenland Sea continental margins: morphology and late Quaternary sedimentary processes and environment. *Quaternary Science Reviews*, **17**, 273–302.

WANG, D. & HESSE, R. 1996. Continental slope sedimentation adjacent to an ice-margin. II. Glaciomarine depositional facies on the Labrador Slope and glacial cycles. *Marine Geology*, **135**, 65–96.

WELLNER, J. S., LOWE, A. L., SHIPP, S. S. & ANDERSON, J. B. 2001. Distribution of glacial geomorphic features on the Antarctic continental shelf and correlation with substrate: implications for ice stream behaviour. *Journal of Glaciology*, **47**, 397–411.

Large-scale morphology of Arctic continental slopes: the influence of sediment delivery on slope form

DAMIAN B. O'GRADY* & JAMES P. M. SYVITSKI

*Institute of Arctic and Alpine Research, University of Colorado, Boulder, Colorado,
80309-0450, USA (e-mail: damian.ogrady@colorado.edu)*
*Present address: Exxon Mobil Upstream Research Company,
P.O. Box 2189, Houston TX 77252 (e-mail: damian.b.ogrady@exxonmobil.com)*

Abstract: The continental slopes of the pan-Arctic region exhibit a range of morphological expression inherently related to different styles of glaciation and sediment delivery to the slope. This study examines the basic associations between glacial processes and morphology at the regional scale. A Geographical Information System (GIS) is applied to a new bathymetric grid of the Arctic Ocean in order to compare the slope angle and sea-floor roughness of 70% of Arctic continental slopes. We also subdivide the circum-Arctic continental slope with respect to parameters likely to influence sediment delivery to the slope. These include proximity to Late Quaternary glacial advance, convergent versus divergent ice termini, and the presence or absence of glacial shelf troughs. Comparison shows that those continental slopes that experience higher sediment input dip more gently than slopes with less sediment input. On glaciated margins where the expanding ice sheet would have produced convergent (faster) ice flow, continental slopes have mean slopes of 1.3° on average. This is in contrast with margins that have experienced divergent (slower) ice flow which tend to have steeper slopes (mean of 2.2°). There is a direct relationship between morphological variability among trough-mouth fans and the width of the adjacent continental shelf (i.e. size of the trough). Longer troughs give rise to slope fans with a more gentle profile geometry. This finding, though simple, suggests a dependency of slope morphology on sediment input to the slope through basal ice sheet erosion. Published values for sediment discharge to several trough mouth fans support this concept. A model is proposed for fan development that relates a fan's geomorphic state to its stage of stratigraphic development.

Introduction

Recent studies show that the morphology of high-latitude continental slopes can vary considerably between regions of recent glacial advance (Vorren and Laberg 1997; Dowdeswell et al. 1998; Taylor et al. 2000). Such variations seem to reflect diversity in the mechanisms and rates of sediment delivery to the margin by glacial termini (Fig. 1; Dowdeswell et al. 1996). Understanding the link between sedimentary processes and morphology of continental slopes begins with recognizing diversity in sedimentary environments and the resultant morphology. Our current understanding of this link for the Arctic glacial system stems from a small number of well-studied continental margins for which ample data on morphology and sedimentary processes exist. Examples include Arctic trough mouth fans (Laberg & Vorren 1996a); the canyon-dissected northern Norway margin (Taylor et al. 2000); and the turbidite-dominated southern Greenland margin (Hesse et al. 1999).

Such examples highlight the associations between glacial process and slope morphology because they represent ends-of-the-spectrum in terms of glacial margin sedimentary systems. For example, high sediment delivery from trough–mouth systems (Elverhøi et al. 1998) commonly leads to a low-angle, debris-flow dominated fan. In contrast, sediment starved regions may be more prone to canyon development such as in the Norwegian Andøya margin (Taylor et al. 2000). It is unclear whether such associations are relevant only to these end-member systems or whether they may be applied to all high-latitude slopes.

The purpose of this study is to understand whether a systematic relationship exists between glacial processes and continental slope morphology for the entire range of northern high-latitude margins. To address this problem we compare the large-scale continental slope morphology for several regions of the Arctic covering a broad spectrum of glacial to non-glacial sedimentary environments.

From: DOWDESWELL, J. A. & Ó COFAIGH, C. (eds) 2002. *Glacier-Influenced Sedimentation on High-Latitude Continental Margins.* Geological Society, London, Special Publications, **203**, 11–31. 0305-8719/02/$15.00
© The Geological Society of London 2002.

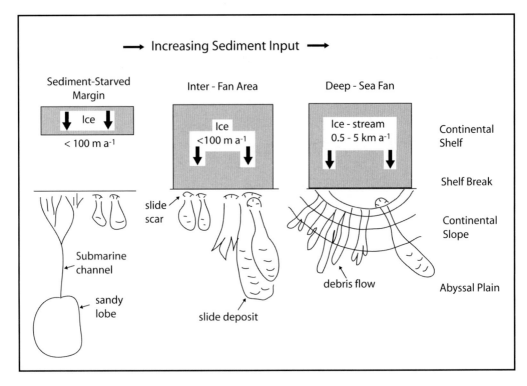

Fig. 1. Conceptual model showing the change in the style of slope deposition in relation to ice proximity and rate of glacial sedimentation (Dowdeswell *et al.* 1996).

Arctic bathymetry

Due to a lack of suitable bathymetric data, regional comparisons of Arctic slope morphology have lagged behind those of more temperate regions (i.e. O'Grady *et al.* 2000; Adams & Schlager 2000). Consequently, most previous comparisons of Arctic slope morphology are limited to only the best studied margins (Laberg & Vorren 1996a). In this study we use the recent International Bathymetric Chart of the Arctic Ocean (IBCAO; Jakobsson *et al.* 2000); a compilation and fusion of digitized soundings and contours from a variety of contributing sources. The grid is 6000 × 6000 km with 2.5 km cell spacing in a Cartesian coordinate system and a Polar stereographic projection (true scale at 75°N). The IBCAO covers all of the Arctic Ocean and adjacent shelf seas as well as most of the Labrador, Norwegian, and Greenland–Iceland Seas. In addition to bathymetry, land elevation is also included (Fig. 2). The IBCAO grid is an improvement on other regional bathymetric charts available for the Arctic Ocean. However, it contains notable errors and processing artefacts. In particular, the errors consist of artificial terracing that results from a processing overemphasis of digitized contours, as well as miscalibrated ship track soundings that create visible 'streaks' of erroneous data. In addition, limited spatial resolution (6.25 km²) permits only large-scale analysis and does not warrant the analysis of many smaller slope features.

The new grid is used in this study to characterize the large-scale morphology of a span of Arctic and other high-latitude continental slopes by measuring slope angle and bathymetric roughness with a raster GIS package. These simple parameters are then compared between recently ice-dominated and ice-free regions, as well as differing glacial sedimentary environments to assess possible links between sedimentary processes and margin form.

Data preparation and analysis

Previous GIS-based studies of the continental slope have used high-resolution multibeam bathymetry to assess such problems as slope

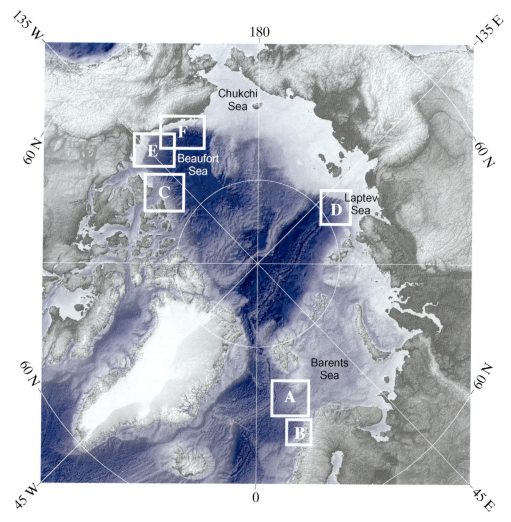

Fig. 2. International Bathymetric Chart of the Arctic Ocean (IBCAO) digital bathymetric model (Jakobsson *et al.* 2000). The six rectangles show the locations of margins discussed in the text and shown in Figure 6.

stability and the geomorphology of mass movements (Lee *et al.* 1999; McAdoo 1999; McAdoo *et al.* 2000). This study attempts to capture a regional perspective on continental slope morphology from data with a much lower spatial resolution. Because of the limited resolution, our morphological analysis is generalized in its treatment of most slope features. For example, only deep-sea features larger than several pixels (50 km²) can be resolved clearly by the IBCAO data, therefore, large deep-sea canyons may not be clearly defined. Often such large features are present and visible but their

dimensions are imprecise. Smaller features, such as slope gullies and rills, are not resolved at all. This aspect of the data limits the quality of spatial analysis that can be performed. By default, the focus of our analysis is on large-scale morphological parameters, particularly general slope gradient and macro roughness. For lower-latitude margins, this type of large-scale, limited-resolution approach has been successful in illustrating the morphological diversity of global continental slopes and its relation to a margin's general sedimentary environment (O'Grady *et al.* 2000).

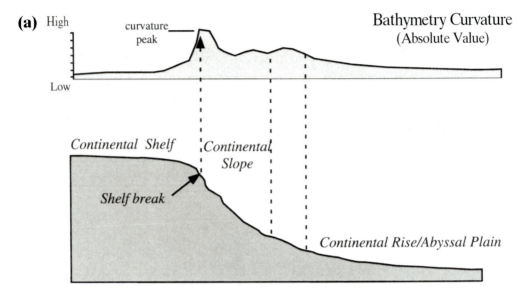

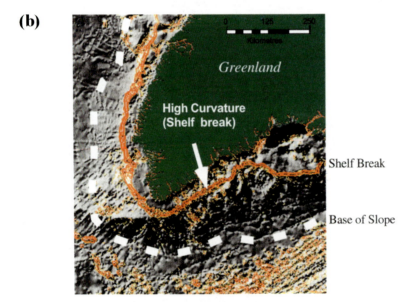

Fig. 3. Definition of the continental slope in terms of bathymetric curvature, or second derivative. (**a**) Relationship between curvature and a hypothetical bathymetric profile. (**b**) An example from the margin of southern Greenland where shelf-break curvature values are much higher than those at the base of the continental slope.

Defining the continental slope

A GIS mask defining the boundaries of the continental slope for the Arctic Ocean was produced to constrain the spatial analysis to the slope. The upper and lower boundaries of the continental slope are defined primarily by bathymetric curvature (the rate of change of sea-floor gradient; Fig. 3a). The upper boundary, the shelf–slope break, is marked as the region just

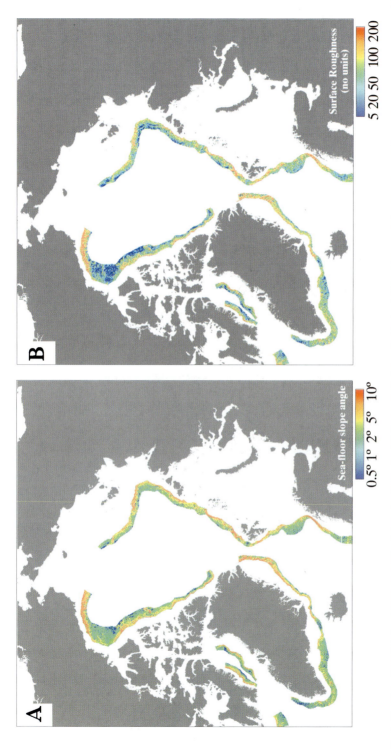

Fig. 4. Results of spatial algorithms on masked slope bathymetry. (**a**) Slope angle for each pixel calculated by averaging the slopes within a 3 × 3 moving window. (**b**) The residual result of a high-pass filter applied to bathymetry. Parameter in (**b**) is used to characterize bathymetric 'roughness' of the continental slope.

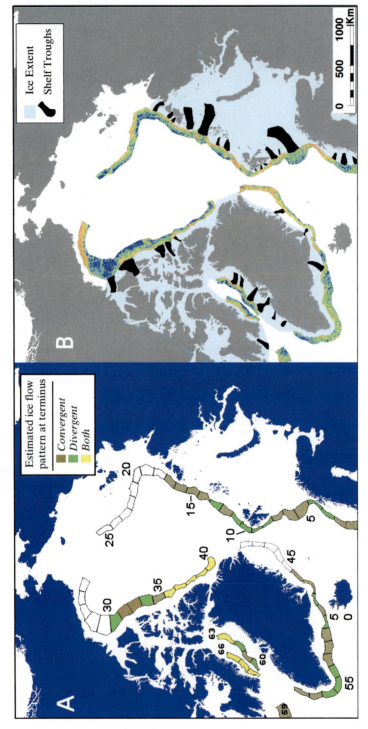

Fig. 5. (**a**) Arbitrary polygons used in the analysis. Numbers correspond to those in Appendix 1. (**b**) Limits of maximum ice extent for the last 130 ka, used in this study. Compiled from several sources described in text.

below the continental shelf with the highest curvature values. The lower boundary is defined as the zone where a significant decrease in slope angle occurs. This boundary is more difficult to define because the changes in slope angle are less dramatic than for the shelf break and thus curvature values are lower and less indicative of its location. The nature of the lower transition is also inconsistent along the margin. A generalized method, such as picking a cut-off slope angle to define the transition from slope to rise (i.e. 1.43° as defined by Heezen 1956) was not used because it is too rigid and not widely applicable to the diverse morphologies in the study. Instead, the base of the slope is defined on a case-by-case basis, using shaded relief, curvature, slope, and general appearance of the surrounding bathymetry as a guide (Fig. 3b). The continental slope mask avoids seamounts and other volcanic or tectonic features. Excluded from the mask are the Chukchi Borderland and the margin of northern Greenland. In addition, the continental slopes of Iceland are left out because of the dominant influence of tectonics on slope morphology.

The slope angle of each pixel was calculated using the average slope of a 3 pixel by 3 pixel scanning window (Fig. 4a). Surface roughness of the bathymetry was analysed by applying a high-pass filter that accentuates high-frequency changes in sea-floor elevation (Fig 4b; Appendix 1). The roughness parameter assists in estimating the presence of slope canyons and down-slope sediment-drainage features. Many smaller canyons are not resolved by the grid, but instead appear in the data as undulating or rough surfaces. We assume that areas with greater density of canyons are likely to have higher surface roughness.

Subdivisional polygons

To simplify the spatial analysis, our span of Arctic continental slopes is divided into 66 polygons, equally spaced at 200 km along margin strike (Fig. 5a). Averages and/or variance statistics within each polygon are used for comparison of slopes and roughness among the different regions. Mean slope angle of a polygon is calculated by averaging the slope of all pixels within the polygon boundaries. Average surface roughness of a polygon is estimated by first calculating the difference between the results of a high-pass filter and the original bathymetric grid. The standard deviation of the residuals for the entire polygon is then calculated. This provides an 'average' roughness parameter to compare between polygons. The arbitrary polygon width

of 200 km is used to avoid over-selection of prominent margins and to keep polygon areas relatively equal. The average area of the polygons is 25×10^3 km^2. Spatial statistics are summarized in Appendix 1.

Ice extent

We use previously published studies to define regions of the Arctic that have experienced significant continental-shelf glaciation in the Late Quaternary. Delineating ice extent for this period is difficult, due to existing controversy regarding the Last Glacial Maximum (LGM, c. 18 ka) and prior ice advances (Dyke & Prest 1987; England 1998; Grosswald & Hughes 1999; Svendsen et al. 1999; Jakobsson et al. 1999; Brigham-Grette 2001). However, reasonable confidence exists for many regions of the Arctic (Funder & Hansen 1996; Landvik et al. 1998).

Because modern slope morphology is a product of past deposition and erosion as well as recent and modern processes, we chose a span of geological time to focus our study. The time frame considered begins at the last interglacial from isotope substage 5e (c. 130 ka) and ends after the LGM. This time interval covers several different ice advances that have brought glacial termini (and river mouths) to the shelf edge, thus influencing slope sedimentation. By encompassing several ice advances, the chosen time frame also allows enough flexibility for diachronous limits of maximum ice extent since glacial ice advance to the shelf edge was episodic and spatially variable. The ice limit shown in Figure 5b is a non-contemporaneous maximum for the Late Quaternary based on several studies (Dyke & Prest 1987; Funder & Hansen 1996; England 1998; Landvik et al. 1998; Svendsen et al. 1999). For the purpose of our study, we assume that glacial conditions over this time period were important in modifying the shape and character of the continental slope so that their effects are observed in the modern morphology of the margin. This assumption may not be entirely valid since the rates of erosion and deposition are highly variable for the few continental slopes where these parameters have been measured (Faleide et al. 1996; Elverhøi et al. 1998; Nam & Stein 1999; Andrews 2000). However, generalization of the last glacial cycle provides a baseline from which to analyse process/morphology associations.

Character of ice flow

Using the methods of Syvitski & Praeg (1989) and Syvitski (1993) and maps of Denton &

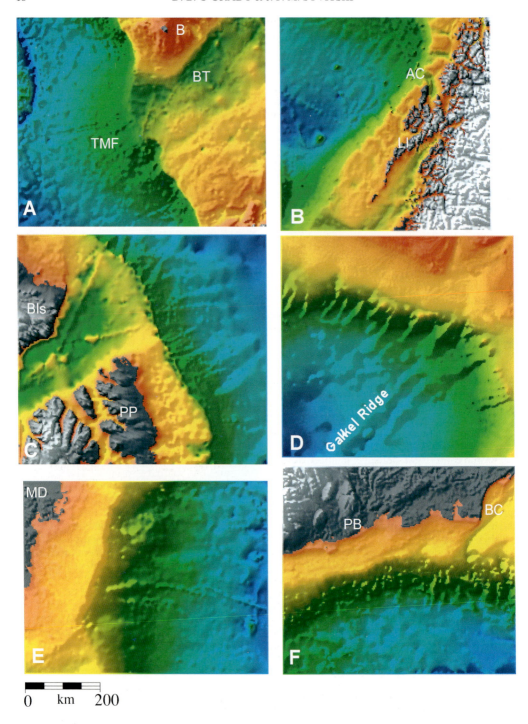

Fig. 6. Six shaded images of continental margins from the IBCAO. In all images red indicates shallow water, blue indicates deeper water. (**a**) Bear Island Fan: B, Bear Island; BT, Bear Island Trough; TMF, trough mouth fan. (**b**) Northern Norway: LI, Lofoten Islands; AC, Andøya Canyon. (**c**) Canadian Archipelago: BIs, Banks Island; PP, Prince Patrick Island. (**d**) Laptev Sea near Lena River. (**e**) Mackenzie Delta Margin: MD, Mackenzie Delta. (**f**) North Slope Alaska: BC, Barrow Canyon; PB, Prudhoe Bay.

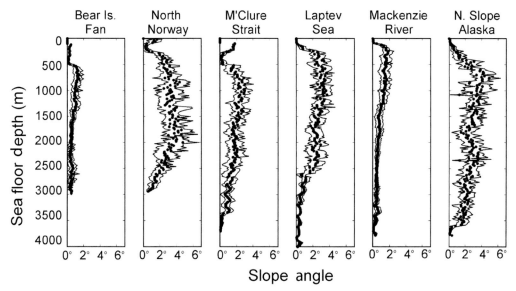

Fig. 7. Slope angle of the sea floor as a function of depth for the margins in Figure 6. Mean slope angle at 20 m depth intervals is represented by a dot and bracketed by lines indicating ± one standard deviation. The sharp increase in slope near the top of each profile represents the shelf–slope break.

Hughes (1981) and Stokes & Clark (2001), we characterize the sedimentary environment at the palaeo-ice sheet terminus in terms of the flow conditions of the glacial ice (Fig. 5a). Ice sheet isopach maps representing the LGM (Syvitski 1993 and references therein) along with modern bathymetry are used to estimate where ice sheet flow onto continental shelves would have experienced either convergent or divergent lines of flow. Glacial troughs on continental shelves provide strong evidence for convergent flow lines (Stokes & Clark 2001), yet convergence is also interpreted for regions that do not have visible troughs. Conversely, inter-trough areas were typically denoted as locations of divergent flow under slow-moving ice conditions. In several instances, the 200 km width of a polygon contains regions of both divergent and convergent palaeo-ice flow. In these cases, an estimate of the dominant flow type is made based on proportions of each flow type. In 9 of 66 polygons we interpret equally convergent and divergent conditions with neither condition having dominance (Figure 5a; Appendix 1).

The convergence or divergence of glacial flow is directly linked to ice velocity and sediment transport to the glacial terminus (Boulton 1990; Dowdeswell & Siegert 1999). In regions of convergence, fast-moving ice delivers relatively greater amounts of sediment in comparison to regions where slow-moving, divergent flow occurs.

Representative Arctic margins

We use six Arctic margins to illustrate various aspects of slope morphology and its relevance to the recent styles of margin sedimentation (Figs 2 & 6): (A) convergent ice flow (ice stream) margin (Bear Island Fan); (B) divergent ice flow (ice-sheet) margin (northern Norway); (C) ice shelf margin (M'Clure Strait); (D) non-glaciated margin with a wide continental shelf (Laptev Sea); (E) non-glaciated, high sediment supply margin (Canadian Beaufort Sea) and (F) non-glaciated, alpine river-dominated margin (North Slope of Alaska). These examples represent a variety of sedimentary environments and exhibit a wide range of morphologies. They also provide a means to evaluate the clarity and quality of the IBCAO bathymetric data used in our analyses (Fig. 6). Plots of sea-floor slope versus sea-floor depth for each locality (Fig. 7) assist in the evaluation of slope morphology. A brief and qualitative summary of each margin is included in Table 1.

Convergent-flow ice-stream margin, Bear Island Fan

Ice streams form as a result of an ice sheet experiencing strongly convergent ice flow. The resultant high flow velocities promote high basal shear stress along with significant concomitant basal sediment transport (Boulton 1990; Alley

Table 1. *Qualitative description of six representative Arctic continental slopes (see Figs 2, 6 & 7)*

Margin	Sedimentary environment	Slope	Shelf width	Shelf break depth	Canyons
Bear Island Fan	Convergent flow ice stream margin,	Gentle	Wide	Deep	Low
Northern Norway	Divergent flow (ice-sheet) margin	Steep	Narrow	Medium	High
M'Clure Strait, Can. Arch.	Ice shelf margin	Medium	Wide*	Deep	High
Laptev Sea	Non-glaciated, large shelf margin	Steep	Very wide	Shallow	High
Beaufort Sea	Non-glaciated, high sediment-supply margin	Gentle	Narrow–medium	Shallow	Low
North Slope of Alaska	Non-glaciated, alpine river-dominated margin	Steep	Narrow–medium	Shallow	High

*Includes M'Clure Strait itself.

et al. 1996). Trough mouth fans, such as the Bear Island Fan (BIF; Fig 6a), are products of high sediment delivery to the shelf edge by ice streams. Sea floor gradients on the BIF are very low, rarely exceeding 1° (Fig. 7; Laberg & Vorren 1996*b*). The fan is composed of successions of mass-flow deposits from repetitive debris flows. The debris flows were generated during times of glacial advance when an ice stream delivered large quantities of basal debris to the shelf edge (e.g. Alley *et al.* 1996; Dowdeswell & Siegert 1999; Dimakis *et al.* 2000). Often the episodic advances remain at the shelf edge only briefly, for a few thousand years at a time (Vorren & Laberg 1997). The persistent recurrence of shelf trough erosion and slope depositional processes, during repeated glaciations, has contributed to the immense size of the Bear Island Fan.

The IBCAO image of the fan shows a smooth and gently sloping feature with few to no visible evidence for canyons or other channels (Fig. 6a). One prominent feature is the large slide scar that breached the shelf break in the late Pleistocene (Laberg & Vorren 1993).

Divergent-flow (ice-sheet) margin, Northern Norway

Slow-flowing areas of ice sheets are often associated with divergent flow line gradients, little basal erosion, and low sediment supply to the continental slope (Boulton 1990). The continental margin of northern Norway (Figs 2 & 6b) is an example of a recently-glaciated shelf located adjacent to a steep continental slope with numerous canyons (Taylor *et al.* 2000). The glaciers in this region during the LGM were not

as significant producers and transporters of sediment as the shelf-eroding ice streams to the north and south (Dowdeswell & Siegert 1999; Stokes & Clark 2001). Instead, this portion of the Scandinavian Ice Sheet was slow moving (divergent ice flow) and probably relatively clean, transporting little sediment to the slope, even though the terminus probably reached the shelf edge (Dowdeswell *et al.* 1996). The Lofoten Islands may also have acted as a barrier to ice and sediment flux to the shelf edge (Laberg *et al.* 1999). As a consequence, sediment deposition was not able to fill in the canyons or prograde the slope and contourite deposition was prominent (Laberg *et al.* 1999). In addition to sedimentary processes, the steepness of the margin and perhaps initiation of the canyons themselves may have been influenced by uplift of the region in the Cenozoic (Henriksen & Vorren 1996).

The IBCAO data are not able to show individual canyons aside from perhaps Andøya Canyon (cf. Laberg *et al.* 2000). Several other canyons to the south of Andøya Canyon that have been observed with independent, higher-resolution data (Taylor *et al.* 2000), are represented in the IBCAO only as an irregular sea floor (Fig. 6b).

Ice-shelf margin, M'Clure Strait, Canadian Archipelago

An ice shelf is the seaward margin of an ice sheet where a significant portion of the terminus is not grounded but floating above the sea floor. As such, ice shelves are unable to transport sediment as a subglacial deforming layer to their seaward edge, rather only to their grounding line. Ice shelves are restricted to high-polar environments (Anderson *et al.* 1991), where ablation occurs from iceberg calving and basal melting. Ice shelves often require high ice velocities, and sometimes the aid of geometric constraints such as valley walls and pinning points to grow and survive (Syvitski 1993). Although the palaeo-distribution of ice shelves has been highly controversial (Syvitski 1993), as is the ice extent in the LGM in the NW Canadian Archipelago (England 1998), M'Clure Strait may have been occupied by an ice shelf (Dyke & Prest 1987) (Figs 2, 6c). Bathymetry within the strait and on the adjacent shelf shows a possible glacial trough and suggests the presence of a pre-LGM ice stream (e.g. Denton & Hughes 1981). The continental slope seaward of M'Clure Strait exhibits large canyon-like features in the IBCAO, although no direct studies confirm this.

Non-glaciated large shelf margins, Laptev Sea

The Laptev Sea shelf extends 300 km from the coast to the shelf-slope break and was not glaciated during the LGM (Svendsen *et al.* 1999; Fig. 6d). During that time, the Laptev Sea shelf was exposed due to the lowered sea level (*c.* 120 m; Fairbanks 1989), and several studies reveal that there was no large LGM ice sheet in this area (e.g. Dunayev & Pavlidis 1988; Hahne & Melles 1997; Kleiber & Niessen 1999; Svendsen *et al.* 1999). Instead, the surrounding land masses of the Laptev Sea contained local glaciers (Arkhipov *et al.* 1986; Velichko *et al.* 1997; Kleiber & Niessen 1999). During the LGM, lowstand river runoff continued through four valleys on the exposed Laptev Sea shelf (Kleiber & Niessen 2000). Late Weichselian shelf sediments are generally very fine grained (up to 70% clay), and sand and gravel are almost absent indicating that sediment transport by icebergs was of minor importance (Müller 1999).

The modern environment of the Laptev Sea is influenced strongly by a large volume of fresh water delivered by five major river systems (Khatanga, Anabar, Olenek, Lena, Yana). The Lena River alone contributes approximately 520 km^3 a^{-1} of fresh water (Aargaard & Carmack 1989). Riverine sediment discharge into the Laptev Sea contributes 24 Mt a^{-1}, of which more than 70% comes from the Lena River (Rachold *et al.* 2000; Gordeev *et al.* 1996). Sediment input by coastal erosion contributes another 30 to 58 Mt a^{-1} (Rachold *et al.* 2000;

Müller-Lupp *et al.* 2000). Most of the modern sediment deposition is confined to the inner shelf. The depositional history of the outer Laptev Sea shelf during the Holocene is, therefore, coupled strongly with the postglacial sea-level rise (Müller-Lupp *et al.* 2000).

The continental slope of much of the Laptev Sea appears to be dissected with canyons that are very large and regularly spaced. It is unknown whether these canyons relate to sediment erosion by turbidity currents during periods of low sea-level stands, or whether they relate to density currents created by seasonal sea-ice formation. The loss of buoyancy of the shelf waters caused by convective cooling and brine input due to sea-ice formation may lead to convective downward motion on the shelves and the draining of shelf water via density plumes into the central Arctic Ocean basin (Stein 1998).

Non-glaciated, high sediment-supply margin, Canadian Beaufort Sea

The continental slope seaward of the Mackenzie Delta is broad and relatively featureless as imaged by the IBCAO data (Figs 2 & 6e). A weak sign of a modern canyon in the data is more likely an erroneous ship track since seismic data in the region provide no such evidence (Dixon *et al.* 1992). Unlike the Laptev margin, the Beaufort Sea shelf is small, less than 100 km wide, and at least some fraction of sediment discharged by the modern Mackenzie River bypasses the shelf and is deposited on the continental slope (Giovando & Herlinveaux 1981). The Laurentide Ice Sheet has periodically covered the Mackenzie Delta, sometimes affecting the direction of the Mackenzie drainage (Duk-Rodkin & Hughes 1994). However, there is little evidence that the ice sheet reached the shelf–slope break during the LGM (Lemmen *et al.* 1994). The Canadian Beaufort Sea receives a riverine sediment discharge of 64 Mt a^{-1}, mainly from the Mackenzie River, which is the largest single source of sediment delivery in the Arctic (Rachold *et al.* 2000). The low-angle and uniform swath of continental slope is comparable to that of the Bear Island Fan (Fig. 7). In both cases, high sediment flux is a common factor.

Non-glaciated, alpine river-dominated margin, North Slope of Alaska

This section of continental slope is noticeably irregular in texture (Figs 2 & 6f). The large

Average polygon slope

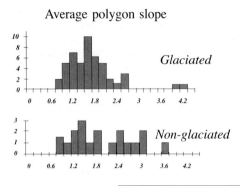

	Mean	Stan. Dev.	N
Glaciated	1.58	0.67	48
Non-glaciated	1.87	0.85	18

Average polygon surface roughness

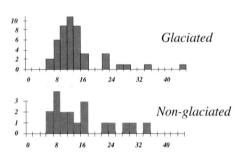

	Mean	Stan. Dev.	N
Glaciated	12.9	7.07	48
Non-glaciated	13.8	8.66	18

Fig. 8. Histograms and statistics of polygon slope angle and roughness for glaciated and non-glaciated polygons.

Barrow Canyon, which incises the shelf on its way to the continental slope, was carved by drainage of the Chukchi Sea during Quaternary sea-level fall (Garrison & Becker 1976). The shelf of the Alaskan North Slope is commonly interpreted as having been ice-free during the LGM (Dyke & Prest 1987), although extensive ice cover has been proposed (Hughes & Hughes 1994).

The Barrow Canyon is resolved adequately by the IBCAO but no canyons to the east are resolved. The rugged surface texture probably reflects massive slope failure caused by decomposing gas hydrates (Kayen & Lee 1991). In seismic reflection images, the margin is characterized by large, deep-seated faults that have

contributed to slope failure at the sea floor (Grantz *et al.* 1990).

The North Slope margin of Alaska receives direct sediment input from high-yield alpine rivers, but the local drainage area is small by Siberian and Mackenzie scales, and this limits the sediment delivery to the margin. The largest river is the Colville River which delivers 6 Mt a^{-1} (Milliman & Syvitski 1992), but even that system does not discharge its sediment across the continental shelf (Barnes & Reimnitz 1974; Walker 1974). Much of the sediment on the continental slope is believed to have been delivered during periods of low sea level.

Pan-Arctic comparison of slope morphology

The results of the comparative analyses of Arctic slopes are discussed by addressing two hypotheses:

Hypothesis 1: Continental slope morphology for high-latitude margins is related to different methods of sediment delivery to the slope.

Of the 66 polygons, 48 are adjacent to shelves that were glaciated in the Late Quaternary, with the remaining 18 being predominantly ice-free (Appendix 1). Average slope angle of the continental slope for glaciated margins ranges from 0.61° to 4.0° with a mean of 1.58°. For the non-glaciated margins the range is from 0.75° to 3.5° with a mean of 1.87° (Fig. 8). Results of a Kolmogorov–Smirnov (K–S) test, a non-parametric statistical test that determines whether two datasets differ significantly (Davis 1986), suggests that values of slope angle and roughness between glaciated and non-glaciated polygons are statistically distinct. In the case of the two distributions in Figure 8, the K–S statistic is 0.260 and the two-tailed critical K–S value at 95% confidence is 0.247. The null hypothesis, which states that the two distributions are the same, is subsequently rejected. This result reinforces an intuitive concept that slope transport driven by glacial activity is different from that controlled by fluvial sources. However, in both environments the range in slope and slope roughness is large and their ranges overlap significantly. The range of slopes shown in this analysis of pan-Arctic margins, both glacial and otherwise, generally falls within the range of mean slopes calculated for most low-latitude, river-dominated margins (O'Grady *et al.* 2000).

A closer look at glacial environments shows

Average polygon slope

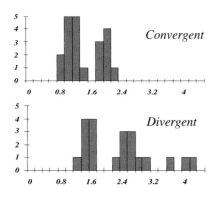

	Mean	Stan. Dev.	N
Convergent	1.29	0.46	21
Divergent	2.17	0.90	21

Average polygon surface roughness

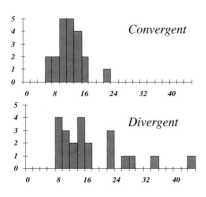

	Mean	Stan. Dev.	N
Convergent	10.8	3.7	21
Divergent	16.5	9.4	21

Fig. 9. Histograms of slope angle and roughness for polygons adjacent to convergent and divergent ice margins.

that margins that experienced a predominance of convergent ice flow have gentler and less rough continental slopes when compared to margins that have experienced predominantly divergent ice flow conditions (Fig. 9). Many of the divergent margins are quite steep. Divergent ice margins are likely to have received less sediment input during the previous glacial cycles leading to a dominance of erosional

processes on the continental slope and steeper slope angles (O'Grady *et al.* 2000; Taylor *et al.* 2000). A K–S test between these two distributions also suggests a statistical distinction.

Given our generalized subdivisions, we can safely accept Hypothesis 1 as reasonable. However, we cannot discern the influence of specific sedimentary processes on the shape of the continental slope.

Hypothesis 2: The continental slope gradient of trough-mouth fans is a direct function of long term sediment delivery from fast-flowing ice streams.

Glacial troughs are assumed to indicate erosion of the continental shelf and sediment transport to the slope by fast-moving ice streams (Alley 1996; Ottesen *et al.* 2002). At the glacier terminus (commonly the shelf–slope break) deposition occurs rapidly, leading to sea-floor failure, debris-flow accumulation, and fan development (Vorren *et al.* 1998; Kuvaas & Kristoffersen 1991). Investigated trough-mouth fans tend to show similar morphologies, having low-angle profiles and few canyons (Faleide *et al.* 1996), suggesting that a distinction from the morphology of other margins is likely. We find that trough-fed polygons exhibit a range in mean slope from 0.82° to 2.57° with a mean of 1.59°.

Twenty-seven glacial trough features have been identified on the continental shelf in our analysis of the IBCAO. These features correspond to 20 of the glaciated polygons (Fig. 5b; Appendix 1). All of these troughs are assumed to be at least partially modified by fast-flowing ice. Glacially carved troughs are distinguished from other fluvial-derived channels by their dimensions and morphology. Many small troughs are overlooked because of the spatial resolution of the bathymetric data. In addition, ice-stream marginal width is likely to be smaller than the polygon width of 200 km. Therefore the mean slope gradient of a polygon includes the fans and, if the fan is small, portions of the adjacent continental slope as well.

To help explain the variation in slope gradient of trough-fed polygons, the average distance of a polygon's shelf edge border to the nearest coastline was calculated. This parameter is essentially a rough calculation of the width of the continental shelf adjacent to each polygon (Appendix 1). We suggest that shelf width may also be considered a proxy for the amount of sediment delivered to the slope by an adjacent ice stream. Anderson (2001) demonstrates that

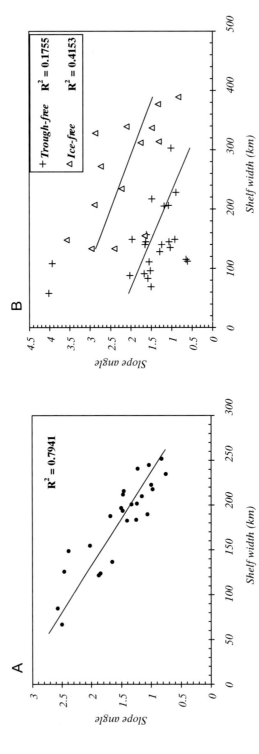

Fig. 10. Scatter plots of approximate shelf width *v.* mean polygon slope angle. (**a**) Only trough-fed polygons are represented. (**b**) Polygons without adjacent troughs, both glaciated and ice-free groups.

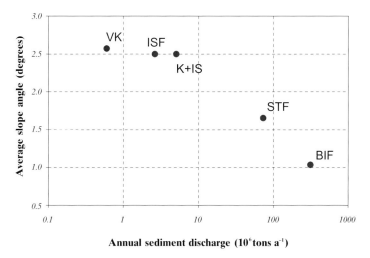

Fig. 11. Sediment discharge data for several glacial margins and trough-mouth fans (Elverhøi *et al.* 1998) plotted against average slope gradient of the adjacent polygon. Three measurements are from fans: BIF, Bear Island Fan; Polygon 5–6. STF, Storfjorden Fan; Polygon 7. ISF, Isforden Fan; Polygon 9. Two measurements are from fjords: K+IS, Kongsforden and Isforden; Polygon 9. VK, Van Keulenfjorden; Polygon 8.

continental shelf sediments are more easily eroded by fast-moving ice than is crystalline bedrock, suggesting that a wide swath of shelf is apt to contribute more to an ice-stream's sediment load than hinterland erosion.

A simple plot of mean slope angle versus approximate shelf width for trough-fed margins (Fig. 10a) shows a roughly linear relationship between the two parameters. This relationship is not exhibited by trough-absent, or ice-free margins (Fig. 10b). Near one end of the linear correlation is the Bear Island Fan which has a very large shelf width (245 km) and a low average slope (1.04°). At the other end (2.6°) is the fan adjacent to Kongsfjorden in western Spitsbergen (Landvik *et al.* 1998).

Estimates of sediment flux to a few trough mouth fans presented in the literature support this finding (Elverhøi *et al.* 1998). The flux estimates come from both volumetric estimates on the fans themselves over portions of the Cenozoic or in the adjacent fjords during the Holocene (Elverhøi *et al.* 1998). Those fans with the highest estimates of sediment input have most gentle gradients with a systematic trend towards steeper slopes and waning supply (Fig. 11). These results suggest the acceptance of Hypothesis 2.

Model of fan geomorphic development

The association between slope angle and sediment input to trough-mouth fans may be related to the stratigraphic development of a fan. As a glacial trough develops by erosion and sediment delivery across the shelf, initial accumulation of debris on the continental slope by mass flow occurs mostly at the base of the steep bedrock-controlled continental slope (Fig. 12). Deposition at the base of a steep slope is typical for subaqueous debris flows, observed in experiments (Mohrig *et al.* 1999), field data (Hampton *et al.* 1996), and numerical models (O'Grady & Syvitski 2001). A feedback occurs whereby base-of-slope deposition builds up the lower slope and reduces its gradient, in turn inducing more deposition on the slope. The consequence is that the slope angle of a developing fan becomes more gentle and more prone to on-slope deposition by mass flows through time and less prone to bypass of flows to the basin floor. However, many large slides have occurred on very shallow slopes (Hampton *et al.* 1996).

A geomorphically 'mature' fan is one that has built up the slope to the point where subsequent deposition perpetuates a similar low-gradient morphology (e.g. the Bear Island Fan). Maturity is likely to be a function of time and/or the rate of sediment delivery to the fan. Fans with high gradients (about 4°) are closer to a geomorphically 'young' developmental stage because they have received less sediment over their history than 'mature', low gradient fans. The terms 'youthful' or 'mature' refer to stages of geomorphic development and do not imply any

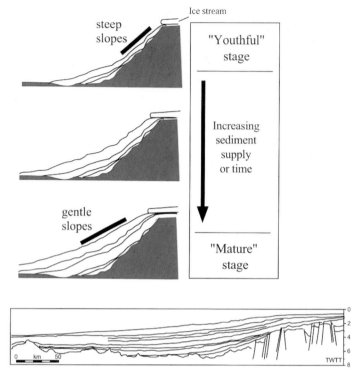

Fig. 12. The relationship between slope angle of trough-mouth fans and the stratigraphic development of a fan. Below is an interpreted multi-channel seismic record of the Bear Island Fan (from Faleide *et al.* 1996). Stratigraphy spans more than 50 Ma; however, most deposition occurred over the last 2.5 Ma with the onset of widespread glaciation.

respective age. For example, the margins of northern Svalbard were recipients of glacial sediment approximately 200 000 years before the Bear Island Fan (Solheim *et al.* 1998). However, the Bear Island Fan has matured faster due to the high rate of sediment influx.

Examples from multi-channel seismic reflection data for Bear Island Fan (Faleide *et al.* 1996) suggest that the proposed evolution model might be relevant for this margin. Sediments at the base of the stratigraphic section, although probably not glacial in origin (Vorren *et al.* 1991; Solheim *et al.* 1998), have accumulated initially at what seems to be the base of the continent–ocean crust boundary slope. Over time, the continental slope has built outward and upward, decreasing its overall gradient (Fig. 12). Debris flow deposition of the last glacial advance is entirely on the slope and is fairly uniformly distributed (Vorren *et al.* 1998). This latest pattern of accumulation has lead to a perpetuation of surface morphology as observed in the repetition of sub-parallel

reflections in shallow seismic reflection data (Laberg & Vorren 1996*b*).

Conclusions

The method of sediment delivery to the shelf edge and continental slope appears to play an important role in determining the basic shape and surface morphology of the continental slope. The means of sediment delivery influences the rate and patterns of sediment accumulation on the slope, the types of sediment dispersal mechanisms that are present (e.g. buoyant plumes, icebergs, basal transport), and ultimately the origin of down-slope processes such as slope failures, debris flows, slumps, slides, and turbidity currents. Though the actual sculpting of the sea floor is carried out by these specific processes, the characteristics of sediment delivery (its rate, amount, and location) govern their distribution and activity.

In this study, we have subdivided delivery mechanisms into fluvial (ice-free), convergent ice termini, and divergent ice termini. With simple statistics and morphological analyses we show that margins fed by probable fluvial sources are steeper and exhibit greater surface roughness than those fed from glacial sources. We observe that divergent ice termini associate with steeper and rougher slopes than convergent termini. Glacial trough-mouth fans show a wider range of mean slope angle than expected from previous studies. This range in slope is linked closely with the width of the adjacent continental shelf, and the rate of sediment discharge to the slope. We suggest that the range in the mean slope angle for modern fans may be related to a fan's stage of geomorphic development with respect to the amount of sediment delivered to the margin over its history.

Research support by the Office of Naval Research STRATAFORM project under the leadership of J. Kravitz and C. Nittrouer. An earlier version of the manuscript was greatly improved through the comments of reviewers T. Vorren, A. Elverhøi, and J. A. Dowdeswell. Thanks to S. Principato and W. Manley for helpful insight and suggestions.

References

AAGAARD, K. & CARMACK, E. C. 1989. The role of sea ice and other fresh water in the Arctic Circulation. *Journal of Geophysical Research*, **94**, 485–498.

ADAMS, E. W. & SCHLAGER, W. 2000. Basic types of submarine slope curvature. *Journal of Sedimentary Research*, **70**, 814–828.

ALLEY, R. B., ANANDAKRISHNAN, S. & CUFFEY, K. M. 1996. Subglacial sediment transport and ice-stream behavior. *Antarctic Journal of the United States*, **31**, 81–82.

ANDERSON, J. B. 2001. Geomorphic and sedimentologic records of grounded ice sheets and paleo-ice streams. *Glacier-influenced sedimentation on high-latitude continental margins; a meeting of the Marine Studies Group of the Geological Society, Programme and Abstracts*, 4.

ANDERSON, J. B., KENNEDY, D. S., SMITH, M. J. & DOMACK, E. W. 1991. Sedimentary facies associated with Antarctica's floating ice masses. *In:* ANDERSON J. B. & ASHLEY, G. M. (eds) *Glacial marine sedimentation; paleoclimatic significance.* Geological Society of America, Special Publications, **261**, 1–25.

ANDREWS, J. T. 2000. Explained and unexplained spatial and temporal variability in rates of marine sediment accumulation along the northeast margin of the Laurentide ice sheet < or = 14 Ka. *Journal of Sedimentary Research*, **70**, 782–787.

ARKHIPOV, S. A., BESPALY, V. G., FAUSTOVA, M. A., GLUSHKOVA, O., ISAYEVA, L. L. & VELICHKO, A. A. 1986. Ice sheet reconstructions. *Quaternary Science Reviews*, **5**, 475–483.

BARNES, P. W. & REIMNITZ, E. 1974. Sedimentary processes on arctic shelves off the northern coast of Alaska. *In:* REED, J. C. & SATER, J. E. (eds) *The coast and shelf of the Beaufort Sea.* Arctic Institute of North America, Arlington, 439–476.

BOULTON, G. S. 1990. Sedimentary and sea level changes during glacial cycles and their control on glacimarine facies architecture. *In:* DOWDESWELL, J. A. & SCOURSE, J. D. (eds) *Glacimarine environments: processes and sediments.* Geological Society, London, Special Publications, **53**, 15–52.

BRIGHAM-GRETTE, J. 2001. New perspectives on Beringian Quaternary paleogeography, stratigraphy, and glacial history. *Quaternary Science Reviews*, **20**, 15–24.

DAVIS, J. C. 1986. *Statistics and data analysis in Geology.* Wiley, New York.

DENTON, G. H. & HUGHES, T. J. 1981. The Arctic Ice Sheet: An outrageous hypothesis. *In:* DENTON, G. H. & HUGHES, T. J. (eds) *The last great ice sheets.* Wiley, New York.

DIMAKIS, P., ELVERHØI, A., HOEG, K., SOLHEIM, A., HARBITZ, C., LABERG, J. S., VORREN, T. O. & MARR, J. 2000. Submarine slope stability on high-latitude glaciated Svalbard-Barents Sea Margin. *Marine Geology*, **162**, 303–316.

DIXON, J., DIETRICH, J., SNOWDON, L., MORRELL, G. & MCNEIL, D. H. 1992. Geology and petroleum potential of Upper Cretaceous and Tertiary strata, Beaufort-Mackenzie area, Northwest Canada. *Bulletin of American Association of Petroleum Geologists*, **76**, 927–947.

DOWDESWELL, J. A. & SIEGERT, M. J. 1999. Ice-sheet numerical modeling and marine geophysical measurements of glacier-derived sedimentation on the Eurasian Arctic continental margins. *Bulletin of the Geological Society of America*, **111**, 1080–1097.

DOWDESWELL, J. A., KENYON, N. H., ELVERHØI, A., LABERG, J. S., HOLLENDER, F. J., MIENERT, J. & SIEGERT, M. J. 1996. Large-scale sedimentation on the glacier-influenced polar North Atlantic margins; long-range side-scan sonar evidence. *Geophysical Research Letters*, **23**, 3535–3538.

DOWDESWELL, J. A., ELVERHØI, A. & SPEILHAGEN, R. 1998. Glacimarine sedimentary processes and facies on the Polar North Atlantic margins. *Quaternary Science Reviews*, **17**, 243–272.

DUK-RODKIN, A. & HUGHES, O. L. 1994. Tertiary-Quaternary drainage of the pre-glacial Mackenzie Basin. *Quaternary International*, **22–23**, 221–241.

DUNAYEV, N. N. & PAVLIDIS, J. A. 1988. A model of the Late Pleistocene glaciation of Eurasiatic Arctic Shelf. *In:* KOTLYAKOV, V. M. & SOKOLOV, V. E. (eds) *Arctic research: advances and prospects 2.* Proceedings of the Conference of Arctic and Nordic Countries on Coordination of Research in the Arctic. Academy of Science of the USSR, 70–72.

DYKE, A. S. & PREST, V. K. 1987. Late Wisconsinan and Holocene history of the Laurentide ice sheet. *Geographie Physique et Quaternaire*, **61**, 237–263.

ELVERHØI, A., HOOKE, R. L. & SOLHEIM, A. 1998. Late Cenozoic erosion and sediment yield from the Svalbard-Barents Sea region; implications for understanding erosion of glacierized basins. *Quaternary Science Reviews*, **17**, 209–241.

ENGLAND, J. 1998. Support for the Innuitian ice sheet in the Canadian High Arctic during the last glacial maximum. *Journal of Quaternary Science*, **13**, 275–280.

FAIRBANKS, R. G. 1989. A 17,000-year glacio-eustatic sea level record: influence of glacial melting rates on the Younger Dryas event and deep-ocean circulation. *Nature*, **342**, 637–642.

FALEIDE, J. I., SOLHEIM, A., FIEDLER, A., HJELSTUEN, B. O., ANDERSEN, E. S. & VANNESTE, K. 1996. Late Cenozoic evolution of the western Barents Sea-Svalbard continental margin. *Global and Planetary Change*, **12**, 53–74.

FUNDER S. & HANSEN, L. 1996. The Greenland ice sheet; a model for its culmination and decay during and after the last glacial maximum. *Bulletin of the Geological Society of Denmark*, **42**, 137–152.

GARRISON, G. R. & BECKER, P. 1976. The Barrow submarine canyon; a drain for the Chukchi Sea. *Journal of Geophysical Research*, **81**, 4445–4453.

GIOVANDO, L. F. & HERLINVEAUX, R. H. 1981. A discussion of factors influencing dispersion of pollutants in the Beaufort Sea. *Pacific Marine Science Report*, **81**, 198 pp.

GORDEEV, V. V., MARTIN, J. M., SIDOROV, I. S. & SIDOROVA, M. V. 1996. A reassessment of the Eurasian River Input of water, sediment, major elements, and nutrients to the Arctic Ocean. *American Journal Science*, **296**, 664–691.

GRANTZ, A., MAY, S. D. & HART, P. E. 1990. Geology of the Arctic continental margin of Alaska. *In:* GRANTZ, A., JOHNSON, L. & SWEENEY, J. F. (eds) *The Arctic Ocean Region*. Geological Society of America, Boulder, 257–288.

GROSSWALD, M. G. & HUGHES, T. J. 1999. The case for an ice shelf in the Pleistocene Arctic Ocean. *Polar Geography*, **23**, 23–54.

HAHNE, J. & MELLES, M. 1997. Late- and post-glacial vegetation and climate history of the south-western Taymyr Peninsula, central Siberia, as revealed by pollen analysis of a core from Lama Lake. *Vegetation History Archaeobotany*, **6**, 1–8.

HAMPTON, M. A., LEE, H. J. & LOCAT, J. 1996. Submarine landslides. *Reviews of Geophysics*, **34**, 33–59.

HEEZEN, B. C. 1956. Outline of north Atlantic deep-sea geomorphology. *Geological Society of America Bulletin*, **67**, 1703–1704.

HENRIKSEN, S. & VORREN, T. O. 1996. Late Cenozoic sedimentation and uplift history on the mid-Norwegian continental shelf. *Global and Planetary Change*, **12**, 171–199.

HESSE, R., KLAUCK, I., KHODABAKHSH, S. & PIPER, D. 1999. Continental slope sedimentation adjacent to an ice margin. III. The upper Labrador Slope. *Marine Geology*, **155**, 249–276.

HUGHES, B. A. & HUGHES, T. J. 1994. Transgressions: rethinking Beringian glaciation. *Palaeogeogra-*

phy, Palaeoclimatology, Palaeoecology, **110**, 275–294.

JAKOBSSON, M., BACKMAN, J. & KNUTSEN, J. O. 1999. Late Pleistocene ice grounding on the Lomonosov Ridge? *In: European Union of Geosciences conference abstracts*; EUG 10 Journal of Conference Abstracts, **4**, 758.

JAKOBSSON, M., CHERKIS, N., WOODWARD, J., MACNAB, R. & COAKLEY, B. 2000. New grid of Arctic bathymetry aids scientists and mapmakers. *Eos, Transactions of the American Geophysical Union*, **81**, 89, 93, 96.

KAYEN, R. E. & LEE, H. J. 1991. Pleistocene slope instability of gas hydrate-laden sediment on the Beaufort Sea margin. *Marine Geotechnology*, **10**, 125–141.

KLEIBER, H. P. & NIESSEN, F. 1999. Late Pleistocene paleoriver channels on the Laptev Sea shelf: implications from sub-bottom profiling. *In:* KASSENS, H., BAUCH, H., DMITRENKO, I., EICKEN, H., HUBBERTEN, H. W., MELLES, M., THIEDE, J. & TIMOKHOV, L. (eds) *Land-ocean systems in the Siberian Arctic: dynamics and history*. Springer, Berlin-Heidelberg-New York, 657–666.

KLEIBER, H. P. & NIESSEN, F. 2000. Variations of continental discharge pattern in space and time: Implications from the Laptev Sea continental margin, Arctic Siberia. *International Journal of Earth Science*, **89**, 605–616.

KUVAAS, B. & KRISTOFFERSEN, Y. 1991. The Crary Fan; a trough-mouth fan on the Weddell Sea continental margin, Antarctica. *Marine Geology*, **97**, 345–362.

LABERG, J. S. & VORREN, T. O. 1993. A late Pleistocene submarine slide on the Bear Island trough mouth fan. *Geo-Marine Letters*, **13**, 227–234.

LABERG, J. S. & VORREN, T. O. 1996a. The glacier-fed fan at the mouth of Storfjorden trough, western Barents Sea: a comparative study. *Geologische Rundschau*, **85**, 338–349.

LABERG, J. S. & VORREN, T. O. 1996b. The Middle and Late Pleistocene evolution of the Bear Island Trough Mouth Fan. *Global and Planetary Change*, **12**, 309–330.

LABERG, J. S., VORREN, T. O. & KNUTSEN, S. M. 1999. The Lofoten contourite drift off Norway. *Marine Geology*, **159**, 1–6.

LABERG, J. S., VORREN, T. O., DOWDESWELL, J. A., KENYON, N. H. & TAYLOR, J. 2000. The Andøya slide and the Andøya canyon, north-eastern Norwegian–Greenland Sea. *Marine Geology*, **162**, 259–275.

LANDVIK, J. Y., BONDEVIK, S., ELVERHØI, A., FJELD-SKAAR, W., MANGERUD, J., SIEGERT, M. J., SVENDSEN, J. I., & VORREN, T. O. 1998. The last glacial maximum of Svalbard and the Barents Sea area; ice sheet extent and configuration. *Quaternary Science Reviews*, **17**, 43–75.

LEE, H., LOCAT, J., DARTNELL, P., ISRAEL, K. & WONG, F. L. 1999. Regional variability of slope stability; application to the Eel margin, California. *Marine Geology*, **154**, 305–321.

LEMMEN, D. S., DUK-RODKIN, A. & BEDNARSKI, J. 1994. Late glacial drainage systems along the northwestern margin of the Laurentide Ice Sheet. *Quaternary Science Reviews*, **13**, 805–828.

McAdoo, B. G. 1999. Mapping submarine slope failures, *In:* Wright, D. & Bartlett D. (eds) *Marine and coastal geographic information systems*. Taylor and Francis, London.

McAdoo, B. G., Pratson, L. F. & Orange, D. L. 2000. Submarine landslide geomorphology, US continental slope. *Marine Geology*, **169**, 103–136.

Milliman, J. D. & Syvitski, J. P. M. 1992. Geomorphic/tectonic control of sediment discharge to the ocean: the importance of small mountainous rivers. *Journal of Geology*, **100**, 525–544.

Mohrig, D., Elverhøi, A. & Parker, G. 1999. Experiments on the relative mobility of muddy subaqueous and subaerial debris flows, and their capacity to remobilize antecedent deposits. *Marine Geology*, **154**, 117–129.

Müller, C. 1999. Rekonstruktion der Paläo-Umweltbedingungen am Laptev-See-Kontinentalrand während der letzten beiden Interglazial/Glazial-Zyklen anhand sedimentologischer und mineralogischer Untersuchungen. *Berichte zur Polarforsch*, **328**, 1–146.

Müller-Lupp, T., Bauch, H., Erlenkeuser, H., Hefter, J., Kassens, H. & Thiede, J. 2000. Input of terrestrial organic matter into the Laptev Sea during Holocene: Evidence from stable carbon isotopes. *International Journal of Earth Science*, **89**, 563–568.

Nam, S. I. & Stein, R. 1999. Late Quaternary variations in sediment accumulation rates and their paleoenvironmental implications; a case study from the East Greenland continental margin. *In:* Bruns, P. & Hass, H. C. (eds) *On the determination of sediment accumulation rates*. Geo-Research Forum, **5**, 223–240.

O'Grady, D. B. & Syvitski, J. P. M. 2001. Predicting profile geometry of continental slopes with a multiprocess sedimentation model. *In:* Merriam, D. & Davis, J. (eds) *Geologic modeling and simulation: sedimentary systems*. Kluwer Academic/Plenum, New York, 99–117.

O'Grady, D. B., Syvitski, J. P. M., Pratson, L. F. & Sarg, J. F. 2000. Categorizing the morphologic variability of siliciclastic passive continental margins, *Geology*, **28**, 207–210.

Ottesen, D., Dowdeswell, J. A., Rise, L., Rokoengen, K. & Henriksen, S. 2002. Large-scale morphological evidence for past ice-stream flow on the mid-Norwegian continental margin. *In*: Dowdeswell, J. A. & Ó Cofaigh, C. (eds) *Glacier-Influenced Sedimentation on High-Latitude Continental Margins*. Geological Society, London, Special Publication, **203**, 245–258.

Rachold, V., Grigoryev, M. N., Are, F., Solomon, S., Reimnitz, E., Kassens, H. & Antonov, M. 2000. Coastal erosion vs. riverine sediment discharge in the Arctic shelf seas. *International Journal of Earth Science*, **89**, 450–460.

Solheim, A., Faleide, J. I., Andersen, E. S., Elverhøi, A., Forsberg, C. F., Vanneste, K., Uenzelmann, N. G. & Channell, J. 1998. Late Cenozoic seismic stratigraphy and glacial geological development of the East Greenland and Svalbard–Barents Sea continental margins. *Quaternary Science Reviews*, **17**, 155–184.

Stein, R. 1998. *Arctic paleo-river discharge.* Berichte zur Polarforschung, **279**.

Stokes, C. R. & Clark, C. D. 2001. Palaeo-ice streams. *Quaternary Science Reviews*, **20**, 1437–1457.

Svendsen, J. I., Astakhov, V. I., Bolshiyanov, D. Y., et al. 1999. Maximum extent of the Eurasian ice sheets in the Barents and Kara Sea region during the Weichselian. *Boreas*, **28**, 234–242.

Syvitski, J. P. M. 1993. Glacimarine environments in Canada: An overview. *Canadian Journal of Earth Sciences*, **30**, 354–371.

Syvitski, J. P. M. & Praeg, D. B. 1989. Quaternary sedimentation in the St. Lawrence Estuary and adjoining areas: an overview based on high resolution seismo-stratigraphy. *Geographie Physique et Quaternaire*, **43**, 291–310.

Taylor, J., Dowdeswell, J. A. & Kenyon, N. H. 2000. Canyons and late Quaternary sedimentation on the North Norwegian margin. *Marine Geology*, **166**, 1–9.

Velichko, A. A., Zelikson, E. M., Morozova, T. D., Nechaev, V. P., Porozhnyakova, O. M. & Chichagova, O. A. 1997. Paleogeographic conditions in the central Russian Plain during the Atlantic period of the Holocene; evidence from data on fossil soil study. *Transactions of the Russian Academy of Sciences*, **6**, 897–900.

Vorren, T. O. & Laberg, J. S. 1997. Trough mouth fans; palaeoclimate and ice-sheet monitors. *Quaternary Science Reviews*, **16**, 865–881.

Vorren, T. O., Richardsen, G., Knutsen, S. M. & Henriksen, E. 1991. Cenozoic erosion and sedimentation in the western Barents Sea. *Marine and Petroleum Geology*, **8**, 317–340.

Vorren, T. O., Laberg, J. S., Blaume, F., Dowdeswell, J. A., Kenyon, N. H., Mienert, J., Rumohr, J. & Werner, F. 1998. The Norwegian–Greenland Sea continental margins: morphology and Late Quaternary sedimentary processes and environment. *Quaternary Science Reviews*, **17**, 273–302.

Walker, H. J. 1974. The Colville River and the Beaufort Sea: Some Interactions. *In:* Reed, J. C. & Sater, J. E. (eds) *The Coast and Shelf of the Beaufort Sea*. Arctic Institute of North America, Arlington, 513–540.

Appendix 1. *Arctic continental slopes polygon statistics*

Polygon no.	Flow regime	Late Quat. shelf edge ice	Mean slope	Max. slope	St. Dev. slope	Roughness (unitless)	Troughs	Shelf width (km)
1	C	Yes	0.97°	10.8°	0.73°	7.7	1	218
2	C	Yes	0.82°	2.9°	0.46°	5.1	1	252
3	D	Yes	2.46°	9.6°	1.34°	20.6	1	126
4	D	Yes	4.02°	23.3°	3.12°	42.5	–	58
5	C	Yes	1.04°	7.4°	0.76°	9.0	1	245
6	C	Yes	1.00°	5.7°	0.79°	9.0	1	223
7	C	Yes	1.65°	9.6°	0.93°	15.4	2	137
8	D	Yes	2.57°	9.1°	1.18°	25.2	1	85
9	D	Yes	2.50°	10.8°	1.36°	21.7	2	67
10	D	Yes	1.58°	10.2°	1.47°	12.6	–	83
11	D	Yes	3.94°	31.0°	3.52°	33.5	–	108
12	C	Yes	1.80°	13.5°	1.49°	13.4	1*	224
13	D	Yes	2.22°	9.6°	1.72°	13.4	1*	206
14	C	Yes	2.02°	14.6°	1.74°	13.9	1	155
15	C	Yes	1.69°	8.0°	0.94°	12.7	2	188
16	C	Yes	1.73°	6.8°	1.29°	10.6	1*	118
17	C	Yes	1.85°	6.3°	1.17°	10.6	2	124
18	?	Yes	1.66°	6.8°	1.22°	9.3	–	155
19	–	Ice Free	1.22°	4.0°	0.80°	6.7	2	241
20	–	Ice Free	1.76°	7.4°	1.34°	10.6	–	312
21	–	Ice Free	2.10°	8.0°	1.50°	14.9	–	339
22	–	Ice Free	1.34°	4.6°	0.77°	8.4	–	377
23	–	Ice Free	1.48°	5.1°	0.97°	11.5	–	337
24	–	Ice Free	1.32°	4.6°	0.91°	7.3	–	314
25	–	Ice Free	0.84°	4.6°	0.65°	4.8	–	389
26	–	Ice Free	2.89°	19.8°	2.06°	32.2	–	207
27	–	Ice Free	2.96°	14.6°	1.85°	29.9	–	133
28	–	Ice Free	2.41°	9.6°	1.90°	23.1	–	133
29	–	Ice Free	1.06°	10.2°	1.04°	6.4	1	190
30	–	Ice Free	0.75°	4.6°	0.46°	5.8	1	235
31	D	Yes	1.24°	10.8°	1.23°	10.8	2	202
32	Ice Shelf	Yes	1.46°	6.3°	1.04°	7.9	1	216
33	?	Yes	1.18°	7.4°	1.04°	9.8	–	205
34	D	Yes	1.48°	9.1°	1.29°	9.1	–	217
35	C	Yes	1.33°	5.7°	0.68°	11.4	2	201
36	C/D	Yes	1.50°	8.5°	0.89°	12.6	2	197
37	C/D	Yes	1.64°	6.8°	1.13°	15.1	–	145
38	C/D	Yes	1.55°	8.5°	1.55°	10.8	–	111
39	C/D	Yes	1.61°	7.4°	1.29°	10.6	–	157
40	C/D	Yes	1.65°	6.8°	1.36°	26.5	–	140
41	–	Ice Free	3.57°	11.9°	2.36°	21.6	–	148
42	–	Ice Free	2.74°	12.4°	2.14°	27.8	–	272
43	–	Ice Free	2.88°	11.3°	2.04°	15.7	–	328
44	–	Ice Free	2.23°	11.9°	1.28°	15.3	–	234
45	–	Ice Free	2.39°	9.6°	1.56°	13.5	1	149
46	C	Yes	1.97°	7.4°	1.36°	15.2	–	149
47	C	Yes	1.88°	6.3°	1.10°	20.7	1	122
48	D	Yes	2.02°	6.8°	1.48°	9.4	–	88
49	C	Yes	0.92°	8.0°	1.05°	10.8	–	149
50	D	Yes	1.47°	5.1°	0.84°	7.1	1	212
51	C	Yes	1.08°	11.3°	0.91°	7.6	–	206
52	C	Yes	0.90°	9.1°	0.80°	9.5	–	228
53	C	Yes	1.07°	4.0°	0.62°	13.2	–	145
54	D	Yes	1.29°	7.4°	0.86°	9.4	–	128
55	D	Yes	1.03°	4.0°	0.50°	12.2	–	135
56	D	Yes	1.24°	9.6°	1.02°	7.1	–	140

Appendix 1. *continued.*

Polygon no.	Flow regime	Late Quat. shelf edge ice	Mean slope	Max. slope	St. Dev. slope	Roughness (unitless)	Troughs	Shelf width (km)
57	C	Yes	0.65°	9.1°	0.60°	4.4	–	115
58	C	Yes	0.61°	6.3°	0.84°	8.7	–	112
59	C	Yes	1.02°	8.0°	0.87°	9.1	–	303
60	D	Yes	1.25°	5.7°	1.11°	7.1	1	184
61	C	Yes	1.15°	5.1°	0.97°	10.5	1	210
62	D	Yes	1.48°	7.4°	1.09°	11.4	1	194
63	C/D	Yes	1.40°	8.5°	1.45°	9.5	1	183
64	C/D	Yes	1.50°	7.4°	1.17°	10.6	–	69
65	C/D	Yes	1.52°	8.5°	1.39°	13.5	–	96
66	C/D	Yes	1.67°	10.2°	1.42°		–	91

Polygon no. refers to the numbering scheme in Figure 5.

Flow regime refers to the dominance of convergent, C, or divergent, D, ice flow at the shelf edge terminus.

Late Quat shelf edge ice lists whether or not we assume glaciation to the shelf edge at any time during the Late Quaternary (last 130 ka).

Mean, Max and Standard deviation slope are statistics for the entire polygon.

Roughness is the value for surface roughness of the sea floor within the polygon using a high-pass filter (see below and text).

Troughs refers to the number of troughs that are observed to be adjacent to the polygon.

Shelf width is the average distance to land from the polygon edge.

* Not used in analysis due to questionable quality of bathymetry data.

High-Pass Filter definition used for roughness estimation:

$$\frac{1}{9} \times \left(\frac{-S\left(x_i \times w_i\right)}{n} \right)$$

where,

x_i = cells in the 3×3 window

w_i = weight assigned to cells in window

n = number of cells in window (9)

The 3×3 kernel for cell weights consists of a central cell with a value of 17 and a value of –1 for the eight surrounding cells.

On the architecture of high-latitude continental margins: the influence of ice-sheet and sea-ice processes in the Polar North Atlantic

J. A. DOWDESWELL[1], C. Ó COFAIGH[1], J. TAYLOR[2], N. H. KENYON[3],
J. MIENERT[4] & M. WILKEN[4]

[1]Scott Polar Research Institute and Department of Geography, University of Cambridge, Lensfield Road, Cambridge CB2 1ER, UK (e-mail: jd16@cam.ac.uk)
[2]Bristol Glaciology Centre, School of Geographical Sciences, University of Bristol, Bristol BS8 1SS, UK
[3]Southampton Oceanography Centre, Empress Way, Southampton SO14 3ZH, UK
[4]Department of Geology, University of Tromsø, N-9037 Tromsø, Norway

Abstract: The presence of ice during the Late Cenozoic distinguishes the nature and rates of processes on high-latitude margins from those elsewhere. Ice sheets terminating in marine waters deliver icebergs, meltwater and debris to high-latitude seas. Sea ice influences ocean salinity structure and downslope water and sediment transfer, and also transports fine-grained sediments over long distances. These cryospheric processes have led to the development of a distinctive sedimentary architecture on modern high-latitude continental margins. Large submarine fans made up almost entirely of stacked debris flows are present around the Norwegian–Greenland Sea. Large slides are located in a variety of settings relative to rates of sediment delivery from Quaternary ice-sheet margins, but no large slides have been mapped on the East Greenland margin. However, extensive channel systems and sediment-wave fields are present in the Greenland Basin, probably related to intermittent downslope flow of dense water and turbidity currents. The extensive NE Greenland shelf was not innundated by ice-sheet advance during recent full-glacial conditions, allowing sea-ice and deep-water production during both interglacials and full-glacials. Changes in the nature and rate of sedimentation within the Greenland Basin should provide clues on the rate of dense-water production, with implications for thermohaline circulation in the North Atlantic. Other erosional and depositional features on the Norwegian–Greenland Sea margins include canyons and contourite drifts. High-relief tectonic features influence sediment reworking by turbidity currents at abyssal depths. A simple conceptual model for glacier-influenced marine sedimentation summarizes the role of cryospheric processes in high-latitude margin sedimentary environments.

Introduction

Continental margins in the polar regions are subject to many of the same processes which operate in lower-latitude marine environments; for example, the action of ocean currents, mass-wasting processes and the rise and fall of sea level over the past few million years. However, it is the presence of ice during the Late Cenozoic that distinguishes both the nature and rates of operation of processes on high-latitude margins from those elsewhere. The ice takes two forms. First, glaciers and ice sheets terminating in marine waters deliver icebergs, melt-water and associated debris to high-latitude seas. During full-glacial periods, delivery is often focused on the continental shelf break and upper slope. In interglacials, by contrast, many

northern hemisphere glaciers terminate at fjord heads, or have disappeared, and the shelf and slope therefore become more distal environments (e.g. Dowdeswell et al. 1998). Secondly, sea ice, formed from the freezing of the sea-surface rather than the build-up of snow and ice on adjacent land, exerts a significant influence on ocean salinity structure and downslope water and sediment transfer (e.g. Midttum 1985; Honjo et al. 1988; Wadhams 2000), and also incorporates and transports predominantly fine-grained sediments over long distances (e.g. Nürnberg et al. 1994).

The operation of these cryospheric processes, over the past few million years in the Arctic, and the last few tens of millions of years in Antarctica, has led to the development of a distinctive sedimentary architecture on modern

From: DOWDESWELL, J. A. & Ó COFAIGH, C. (eds) 2002. *Glacier-Influenced Sedimentation on High-Latitude Continental Margins.* Geological Society, London, Special Publications, **203**, 33–54. 0305-8719/02/$15.00
© The Geological Society of London 2002.

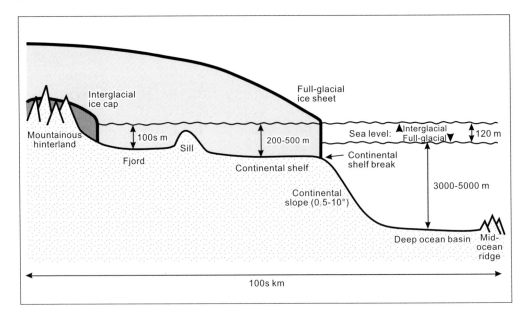

Fig. 1. Idealized high-latitude continental margin, with the locations of interglacial and full-glacial ice sheets shown. Order-of-magnitude dimensions for an idealized high-latitude margin are shown.

high-latitude continental margins that, in many cases, varies from that observed at lower latitudes, where fluvial delivery mechanisms and oceanic processes dominate (Weaver *et al.* 2000). The operation of cryospheric processes during past, pre-Cenozoic ice ages has also contributed to the formation of distinctive glacimarine depositional suites in a number of ancient sedimentary basins that have avoided destruction at plate boundaries through geological time (e.g. Eyles *et al.* 1985; Eyles 1993).

About 10% of the global ocean is influenced by ice sheets and icebergs today, and this proportion approximately doubled under Late Cenozoic full-glacial conditions. In modern winters, sea ice covers about 16 million km^2 of the Arctic seas and 19 million km^2 around Antarctica (Wadhams 2000). During summer, these values are reduced to about 9 and 4 million km^2, respectively. In this paper, we use a series of geophysical and geological datasets from the Norwegian–Greenland Sea to discuss and illustrate the nature of ice-influenced sedimentation on high-latitude continental margins.

Glaciological background

Ice-sheet behaviour

Ice sheets have advanced and retreated across high-latitude continental margins in a series of

climate-related glacial–interglacial, or stadial–interstadial, cycles (Fig. 1). Polar continental shelves have also been affected strongly by sea-level fluctuations, linked to both eustatic changes in global ice volume, and to regional isostatic loading and unloading effects (e.g. Boulton 1990). In the present interglacial, there are no areas in either the Arctic or the Antarctic where glacier ice is grounded at the continental shelf edge. Arctic glaciers today typically terminate at the heads of fjords, with the consequence that sediments are delivered mainly to inner-fjord locations and relatively little glacier-derived debris reaches the shelf and slope (Fig. 1). In Antarctica, much of the ice-sheet margin is currently located on the continental shelf, either as grounded ice walls or as extensive areas of floating ice shelves (Drewry *et al.* 1982). The continental slope and often the outer continental shelf is, therefore, distant from glacier-ice termini during interglacials, and iceberg-rafting is the principal means by which glacial debris is transported to more distal locations.

By contrast, during full-glacial periods, ice sheets in both the northern and southern hemispheres advanced across the continental shelf (e.g. Elverhøi *et al.* 1998*a*; Anderson 1999) (Fig. 1). In many high-latitude areas, and as far south as the Laurentian Channel and the NW British continental margin in the northern hemisphere

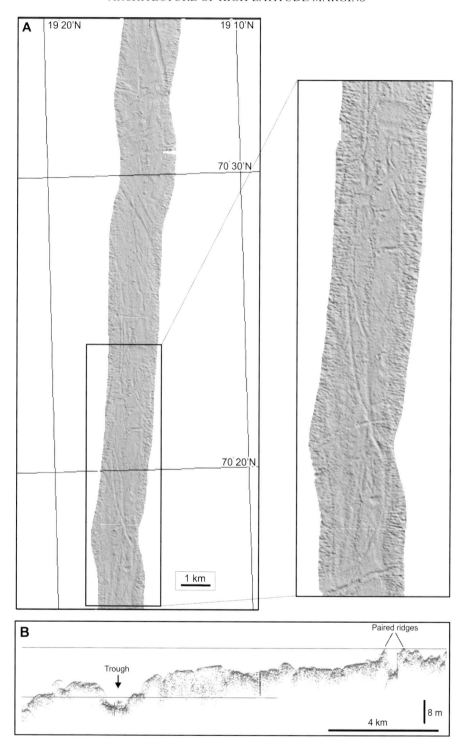

Fig. 2. Iceberg scours on East Greenland continental shelf. (**a**) 12 kHz swath bathymetry of iceberg scours in water depths of between 350 and 480 m outside Scoresby Sund. (**b**) TOPAS sub-bottom profiler record across scours, showing paired ridges or berms on either side of scours produced by iceberg keels.

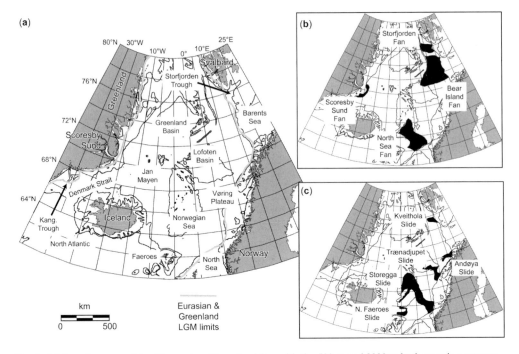

Fig. 3. (a) Location map of the Norwegian–Greenland Sea with the 500 m and 3000 m bathymetric contours shown. **(b)** Main submarine fans (shaded black). **(c)** Main slides (shaded black).

(e.g. Piper & Normark 1982; Stoker 1990), glacier ice terminated at the shelf break, where increased mass-loss by iceberg calving into deepening water prevented further expansion. This ice advance affected both the nature of the sedimentary record on high-latitude shelves and greatly increased the rate of sediment delivery to the upper continental slope during glacial periods (e.g. Elverhøi *et al.* 1995*a*; Dowdeswell & Siegert 1999). On many shelves, ice-sheet advance produced erosion and hiatuses within the sedimentary record (e.g. Cadman 1996). Further sediment reworking continues even during interglacials, through both the ploughing action of iceberg keels impinging on the sea floor and from bottom current activity (Fig. 2) (e.g. Dowdeswell *et al.* 1993, 1998; Syvitski *et al.* 2001).

Ice-sheet dynamics were not uniform over space, however. Instead, fast-flowing ice streams and outlet glaciers, punctuating slower-moving ice, drained interior ice-sheet drainage basins of the order of 10^4 to 10^6 km^2. Rapid flow was due to the presence of lubricating water at the bed or within saturated sub-glacial sediments, whereas in slower-flowing areas ice was frozen to the substrate. Both geophysical

observations of large-scale streamlined, ice-moulded landforms crossing high-latitude shelves (e.g. Canals *et al.* 2000; Ó Cofaigh *et al.* 2002; Ottesen *et al.* 2002; Shipp *et al.* 2002), and numerical-model predictions of rapid glacier flow in topographic troughs (e.g. Marshall & Clarke 1996; Dowdeswell & Siegert 1999), support the view that the delivery of ice and sediments to the continental shelf edge during full-glacials was non-uniform along the ice margin. Ice and sediment flux was focused mainly at fast-flowing ice stream termini, ranging from about 10 to 100 kilometres in width. This pattern of flow-partitioning into fast- and slow-flowing areas appears to be typical of both Quaternary and modern ice sheets and ice caps (e.g. Bentley 1987; Dowdeswell *et al.* 1999*a*; Stokes & Clark 2001).

Ice sheets and sea ice around the Norwegian–Greenland Sea

The continental margins of the Norwegian–Greenland Sea (Fig. 3) are used here to illustrate the sedimentary architecture of a high-latitude ice-influenced margin, for several reasons. First, a large body of geophysical and geological observations exists for this area (e.g. Vorren *et al.* 1989, 1998; Mienert *et al.* 1993; Dowdeswell *et al.* 1996; Solheim *et al.* 1998). Secondly, the time-dependent behaviour of the Eurasian Ice Sheet, and the associated flux of sediments, meltwater and icebergs, has been modelled numerically for the last, Late Weichselian, full-glacial period (Dowdeswell & Siegert 1999; Siegert *et al.* 1999; Siegert & Dowdeswell 2002). Thirdly, the margins of the Norwegian–Greenland Sea have experienced marked spatial variations in past ice-sheet dynamics, in terms of both fast- and slow-flowing regions, and in terms of the distance ice extended onto the continental shelves even under full-glacial conditions (Funder *et al.* 1998; Landvik *et al.* 1998; Evans *et al.* 2002).

Where glacier ice failed to advance significantly across the continental shelf, sea-ice and related oceanographic processes provided the dominant control on margin sedimentary architecture. In the Greenland Sea during both glacials and interglacials, in the Barents Sea during interglacials, and on the Norwegian margin during glacials (Dokken & Jansen 1999), cold and dense water has been produced by a combination of two processes operating mainly during winter: cooling of the ocean-surface layer and brine rejection during sea-water freezing (e.g. Davis 2000). It is the variability in full-glacial ice-sheet growth between the NE Greenland margin and the Norwegian margin that allows us to illustrate and discuss the operation of both ice-sheet and sea-ice processes around this deep-ocean basin, and the nature of the resulting sedimentary record in each case.

Glacier ice began to build up in high northern latitudes between 5 and 10 million years ago. However, a significant increase in ice-rafted debris (IRD) on the Greenland margin from about 7 Ma ago and the Norwegian margin from about 2.6 Ma ago indicates that large ice masses reached sea level at about these times and began to input large volumes of icebergs and debris to the continental margins of the Norwegian–Greenland Sea (e.g. Jansen & Sjøholm 1991; Larsen *et al.* 1994; Butt *et al.* 2000). The dated geological record of the most recent glacial stage demonstrates that ice advanced across the eastern margin of the Norwegian–Greenland Sea during the Late Weichselian, to cover the entire continental shelf from Norway to Svalbard, together with the epicontinental Barents Sea (e.g. Mangerud & Svendsen 1992; Elverhøi *et al.* 1995*b*; Andersen *et al.* 1996; Landvik *et al.* 1998). Proxy micro-palaeontological indicators of sea-ice extent during the Late Weichselian suggest

that there were significant periods of relatively high biogenic productivity and, hence, of seasonally open-water conditions, even around the time of full glaciation (Hebbeln *et al.* 1994, 1998; Siegert *et al.* 2001). This provided a moisture source to nourish the growth of the western margins of the Eurasian Ice Sheet.

By contrast, on the East Greenland margin, ice sheets appear to have advanced to the shelf break during the Late Weichselian only south of Scoresby Sund (69 °N) (e.g. Funder & Hansen 1996; Stein & Syvitski 1997; Andrews *et al.* 1998; Funder *et al.* 1998). North of this, there is little evidence that Weichselian glaciers reached outward from the mountain-rimmed fjord systems as far as the shelf break, which is up to about 250 km from the coast (Evans *et al.* 2002). This is probably because the NE sector of the Greenland Ice Sheet is an area of low precipitation even today and, although full-glacial temperatures would have been significantly colder, ice-sheet growth in this area was limited by very low snow accumulation. This was, in turn, linked to a lack of open water associated with high concentrations of sea ice in the cold East Greenland Current. Further south in East Greenland, Atlantic low-pressure cells provided a precipitation source even during glacial periods, allowing glacier expansion to proceed.

An ice-sheet numerical model, which accounts for subglacial deformation and transport of water-saturated sediment, has also been used to calculate the Late Weichselian rates of sediment and iceberg delivery to the continental margins of the eastern Norwegian–Greenland Sea (Dowdeswell & Siegert 1999; Siegert & Dowdeswell 2002). On the Barents Sea margin, model predictions can also be tested against seismically-derived measurements of sediment volumes and dated cores where the rate of sedimentation has been established (e.g. Faleide *et al.* 1996; Solheim *et al.* 1998; Dowdeswell & Siegert 1999). Numerical-model results predict that ice extended to the shelf break, and that fast-flowing ice streams were active within bathymetric troughs, by around 25 000 years ago. For example, during full-glacial periods, the ice stream in the Bear Island Trough flowed at about 800 m a^{-1} and produced about 30 km^3 a^{-1} of icebergs. This was the largest single outlet of ice from the Eurasian Arctic Ice Sheet. Assuming that these ice streams had deformable beds (Alley *et al.* 1989), the modelled volumes of sediment predicted to accumulate at the continental margin of the Bear Island and Storfjorden troughs are similar to the volumes of Late Weichselian sediment measured over the respective fans using reflection-seismic methods (Dowdeswell & Siegert 1999). Thus, both observations of the marine geological and geophysical records, and numerical ice-sheet modelling, provide information on Late Weichselian ice-sheet dynamics and the nature and rate of glacier-influenced sedimentation in the Norwegian-Greenland Sea.

Marine geophysical investigations of the Norwegian–Greenland Sea have shown that this high-latitude margin is characterized by three distinctive styles of large-scale sedimentary architecture: submarine fans, slides and channel systems (Mienert *et al.* 1993; Dowdeswell *et al.* 1996; Vorren *et al.* 1998). The morphology and structure of each of these features is now outlined, and is discussed in terms of the sedimentary processes operating on these ice-influenced continental margins.

Submarine trough-mouth fans

Several large submarine fans are present on the continental margins of the Norwegian–Greenland Sea (Fig. 3). The area and volume of these fans have been investigated using a combination of geophysical tools, including reflection-seismic methods and side-scan sonar systems (e.g. Dowdeswell *et al.* 1996; Vorren *et al.* 1998). Each of the fans is made up of a prograding sediment wedge at the mouth of a cross-shelf trough or large fjord system. The largest is the Bear Island Fan, with an area of approximately 200 000 km², located at the mouth of the Bear Island Trough in the Barents Sea (Fig. 3).

A number of other fans are present on the Svalbard, Barents Sea and Norwegian margins (Fig. 3) (Vorren *et al.* 1998); Storfjorden Fan (35 000 km²), Bellsund Fan (6 000 km²), Isfjorden Fan (3 700 km²), Kongsfjorden Fan (2 700 km²) and, south of the Vøring Plateau, the North Sea Fan (142 000 km²) (King *et al.* 1996). On the East Greenland margin, the Scoresby Sund Fan, 19 000 km² in area, has also been investigated (Dowdeswell *et al.* 1997), and in Denmark Strait there is a prograding fan offshore of the Kangerlugssuaq Trough at 67 °N (Stein & Syvitski 1997). In addition, north of Eurasia, on the Arctic Ocean margin of the Quaternary Eurasian Ice Sheet, there is also evidence of large fans off the Franz-Victoria, St Anna and Voronin troughs, between the archipelagos of Svalbard, Franz Josef Land and Severnaya Zemlya (e.g. Vågnes 1996; Polyak *et al.* 1997; Kleiber *et al.* 2000).

These high-latitude fans appear to have a similar external geometry and internal architecture, according to geophysical investigations of

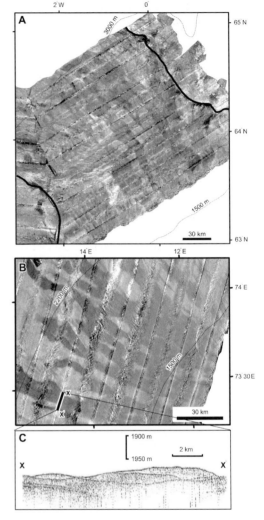

Fig. 4. GLORIA 6.5 kHz side-scan sonar image mosaics of (a) the North Sea Fan (thick lines indicate fan margins), and (b) the Bear Island Fan. Note the elongate downslope debris flows in each image. Lower backscatter is shown as darker tones. (c) 3.5 kHz sub-bottom profile through debris flows on the Bear Island Fan, showing the stacked nature of the flows and their convex upper surface.

several fans using a combination of side-scan sonar, 3.5 kHz and reflection-seismic methods. Their gross form is illustrated by images of the North Sea and Bear Island fans (Fig. 4a,b). The main element of high-latitude fan architecture is a series of stacked debris flows. The uppermost debris flows have been imaged clearly on long-range side-scan sonar records and high resolution 3.5 kHz profiles of several glacier-influenced fans (e.g. Vogt *et al.* 1993; Dowdeswell *et al.* 1996; Vorren *et al.* 1998).

The character of these debris flows, in plan and cross-section, is illustrated through 6.5 kHz side-scan sonar imagery and a 3.5 kHz profile from the Bear Island Fan (Fig. 4b,c). The acoustically transparent debris flows range between about 2–10 km in width, 10–50 m in thickness, and are 30–200 km in length (Dowdeswell *et al.* 1996; Taylor *et al.* 2002). In seismic lines over several fans, a number of acoustic packages are identified, each made up of a group of stacked debris flows and separated from the underlying packages by a major reflector (e.g. Vorren *et al.* 1989; Laberg & Vorren 1995; King *et al.* 1996). Each acoustic unit is interpreted to represent debris flows from a single glacial–interglacial, or stadial–interstadial, cycle (Laberg & Vorren 1995). The debris flows are assumed to be derived from the intermittent failure of relatively clay-rich, glacier-derived diamictic sediments, deposited relatively rapidly on the upper slope at the edge of large cross-shelf troughs during full-glacial conditions when the ice-sheet margin was at the shelf break (e.g. Elverhøi *et al.* 1995*b*; Laberg & Vorren 1995, 2000; Dowdeswell *et al.* 1996). Sometimes acoustically-stratified sediments are seen between debris flows within a given unit, probably indicating sedimentation from turbid glacial meltwater (Taylor *et al.* 2002).

On both the Bear Island and Scoresby Sund fans, it appears that debris-flow activity during the Late Weichselian was confined to parts of each fan system. Off Scoresby Sund, debris flows on the northern part of the fan are overlain on 3.5 kHz records by acoustically stratified sediments (Dowdeswell *et al.* 1997), and Late Weichselian conditions on the southern part of the Bear Island Fan are similar (Taylor *et al.* 2002). This suggests that the locations of palaeo-ice streams, and of rapid sediment delivery to the margin, has varied between glacial stages.

The debris-flow dominated architecture of high-latitude, glacier-influenced submarine fans is markedly different from the internal structure of the larger low-latitude, non-glacial fan systems (Dowdeswell *et al.* 1996). The latter are typically built up of channels, overbank deposits and sandy channel-mouth lobes, with predominantly fine-grained sediments delivered from fluvial catchments as density-controlled underflows (e.g. Pickering *et al.* 1995). The high-latitude fans described here are, however, derived from glacially-eroded debris of heterogeneous grain size which characteristically forms the debris flows that are the building blocks of these fans (Fig. 4).

Seismic stratigraphic studies and isopach mapping on the Bear Island and Storfjorden fans have estimated overall volumes of 340 000 km^3 and 116 000 km^3 of sediment, respectively, since the onset of extensive glaciation from 2.3 million years ago (Fiedler & Faleide 1996; Hjelstuen *et al.* 1996). Sedimentation rates of the order of 10–100 cm 1000 a^{-1} on these two large submarine fan systems contrast with the rate for the earlier, pre-glacial Cenozoic, from 55 to 2.3 million years ago, when average rates of fluvially-dominated sediment delivery were between about 2.2 and 3.2 cm 1000 a^{-1} for the Bear Island and Storfjorden margin (Faleide *et al.* 1996). This difference demonstrates clearly the important role of ice sheets in the Late Cenozoic evolution of the Norwegian–Greenland Sea margin. Furthermore, there is a marked difference in the size of drainage basin required to build high- and low-latitude fans of a given size. The Bear Island Fan has an area and volume, at 10^5 km^2 and 10^5 km^3, that is similar to the huge low-latitude fluvially-derived Amazon and Mississippi fans, but has a drainage basin area that is an order of magnitude smaller, indicating the effectiveness with which glaciers erode their substrate (Elverhøi *et al.* 1998*b*).

Large submarine slides

On the eastern continental margin of the Norwegian–Greenland Sea, three relatively large slide areas occur between the Vøring Plateau and Svalbard; the Traenadjupet, Andøya and Kveithola slide complexes (Fig. 3c) (Vorren *et al.* 1998). In addition, the 110 000 km^2 Storegga slide complex (Bugge *et al.* 1988) and smaller North Faeroes slide complex (van Weering *et al.* 1998; Taylor *et al.* 2000*a*) are located south of the Vøring Plateau and adjacent to the North Sea Fan. These large slope failures, first mapped by Damuth (1978) from 3.5 kHz records, occur in inter-fan areas on the continental margin. There is also a large slide on the Bear Island Fan itself (Laberg & Vorren 1993). By contrast, no large slide systems have been mapped on the East Greenland margin, which has been imaged extensively by 6.5 kHz long-range side-scan sonar and 12 kHz swath bathymetry systems between 69 and 79°N (Mienert *et al.* 1993; Dowdeswell *et al.* 1996, 1997).

The Traenadjupet and Andøya slides cover areas of 8–14 000 km^2 (Fig. 3), and appear as patterns of overall high backscatter on side-scan

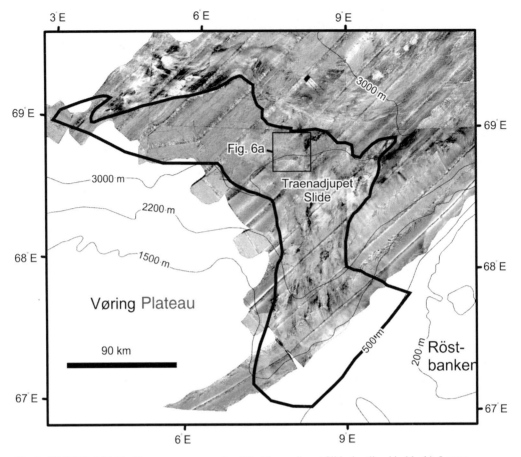

Fig. 5. GLORIA 6.5 kHz side-scan sonar mosaic of the Traenadjupet Slide (outlined in black). Lower backscatter is shown as darker tones.

sonar records. A 6.5 kHz side-scan sonar mosaic of the Traenadjupet Slide is shown in Figure 5. The slides are mainly erosional in their upper reaches and depositional down-slope (Laberg & Vorren 2000; Laberg *et al.* 2000). Each slide originates in multiple headwalls on the upper continental slope, associated with across-slope tensional fractures and longitudinal fractures (Kenyon 1987; Laberg & Vorren 2000). Laberg & Vorren (2000) used 30 kHz TOBI side-scan data to describe and interpret the upper area of sediment removal within the Traenadjupet Slide, and illustrated slide-scar escarpments, sediment ridges, streams and large blocks, together with tensional fractures. We have imaged the lower, depositional area of the slide using both 30 kHz side-scan and 12 kHz swath bathymetry systems. A part of the depositional zone, characterized by areas of large blocks,

separated by relatively smooth sea floor, is illustrated in Figure 6. This imagery indicates that the processes of slide failure and down-slope debris transport are complex, and have produced a rugged and highly-differentiated sea-floor topography.

The large slides on the Norwegian and Barents Sea margin are located in a variety of settings relative to the rates of past sediment delivery from Quaternary ice sheet margins. The Traenadjupet Slide is located immediately offshore of the pronounced cross-shelf trough which gives the slide its name. The presence of a Plio-Pleistocene sedimentary depocentre on the margin (Henriksen & Vorren 1996), and of streamlined sea-floor bedforms in the trough (Ottesen *et al.* 2002), suggests that this has been an area of relatively high sediment supply from ice streams draining the interior ice sheet to the shelf break.

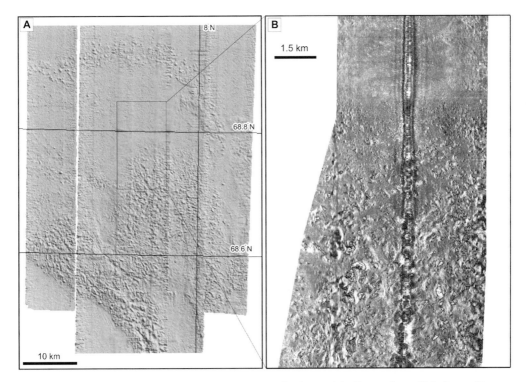

Fig. 6. Acoustic imagery of the blocky nature of the more distal part of the Traenadjupet Slide (located in Figure 5). (**a**) 12 kHz swath bathymetry of blocky areas within the distal part of the slide. (**b**) TOBI 30 kHz side-scan sonar image of block field on the slide (located in part **a**). Shadows are seen as black.

By contrast, the Andøya Slide is located offshore from an area of less rapid ice-sheet flow and, by implication, of lower sedimentation rates (Dowdeswell & Siegert 1999). In both these cases, however, diamictic glacial debris is likely to have been a significant component of overall sedimentation. Some large slide complexes are also sourced from silty contourite deposits on the upper continental slope, with the North Faeroes slide a clear example (van Weering *et al.* 1998; Taylor *et al.* 2000*a*).

Especially where sediment delivery rates to the Norwegian margin are relatively high, sediments may be underconsolidated and relatively weak, and methane build-up may also have increased pore pressure within the more organic-rich interstadial and interglacial muds (Laberg & Vorren 2000; Mienert *et al.* 2001). The immediate trigger for these large-scale slides is uncertain, but may be related to submarine earthquakes, which occur on parts of the eastern margin of the Norwegian–Greenland Sea (e.g. Bungum 1989), or sometimes to increased loading with continuing sediment build-up and sea-level rise associated with

ice-sheet retreat. Laberg & Vorren (2000) report a likely mid-Holocene age for the Traenadjupet Slide, and the three Storegga Slide events all occurred about 8200 years ago (e.g. Haflidason *et al.* 2001). Thus, it is the properties of the debris making up the slides, combined with external environmental triggers, that together cause the huge slides that characterize significant parts of the Norwegian and Barents Sea continental margin. The lack of major slides offshore of NE Greenland is probably related to two factors: first, very low rates of glacial-sediment delivery to the shelf break because, in the Weichselian at least, these ice-sheet margins did not reach far across the wide continental shelf (Funder *et al.* 1998; Evans *et al.* 2002); and, secondly, the downslope transport of sediment discussed below.

Submarine channel systems and sediment-wave fields

Few submarine channel systems have been identified on the eastern margin of the Norwegian–Greenland Sea (Vorren *et al.* 1998;

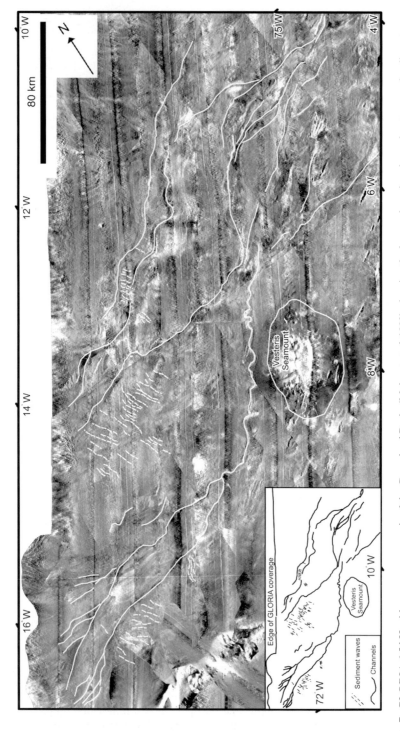

Fig. 7. GLORIA 6.5 kHz side-scan sonar mosaic of the Greenland Basin (Meinert *et al.* 1993), showing the large submarine channel systems and sediment-wave fields. The volcanic edifice of the Vesteris Seamount is also indicated. An interpretation of the GLORIA mosaic is inset.

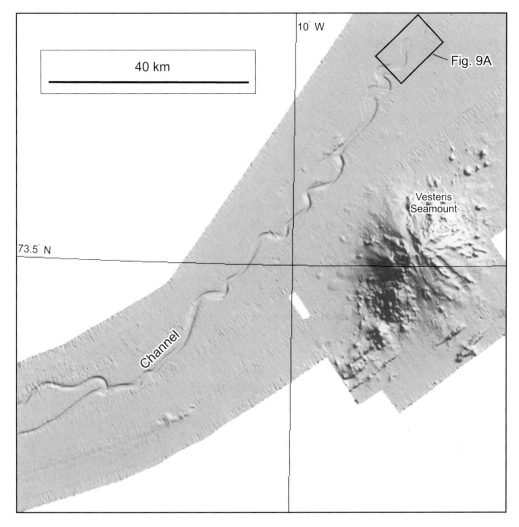

Fig. 8. EM120 12 kHz swath-bathymetric mosaic of part of the Greenland Basin. The confluence of two submarine channels, and meandering sections within the channel system, are shown. The volcanic Vesteris Seamount is clearly visible. The side-scan sonar image in Figure 9 is located within the mosaic.

Laberg *et al.* 2000). However, on the NE Greenland margin, the 250 000 km² Greenland Basin is dominated by well-developed networks of submarine channels that extend over 400 km from the upper continental slope to the deepest parts of the basin in water depths of up to 3500 m (Mienert *et al.* 1993). These large channel systems are clearly illustrated on 6.5 kHz long-range side-scan sonar imagery of the Greenland Basin (Fig. 7). Several channel systems are mapped (Fig. 8), orientated in a SE–NW direction that is consistent with flow from the Greenland continental slope to the deepest parts of the basin.

Extensive side-scan sonar, swath-bathymetric and acoustic-stratigraphic data from the Greenland Basin demonstrate that the submarine channels are up to about 70 m deep, several kilometres wide, up to several hundred kilometres long and have braided and sinuous reaches, probably related to changes in the gradient of the continental slope (Fig. 9a). The upper- to mid-slope consists of small rill-like channels, with channel width, depth and sinuosity increasing with distance into the abyssal basin. In this respect they differ from submarine channel systems at lower latitudes, which usually have relatively deep channelling in their upper

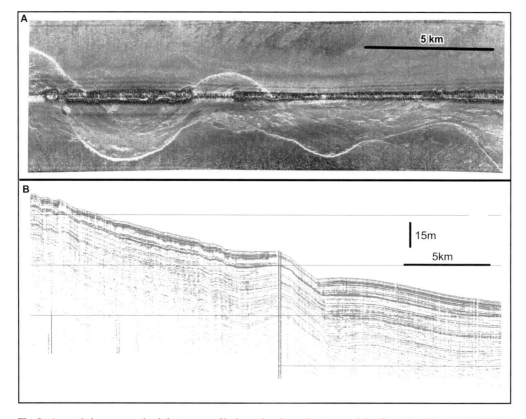

Fig. 9. Acoustic imagery and sub-bottom profile from the channel systems of the Greenland Basin. (**a**) TOBI 30 kHz image of channel with braided structures to the right (located in Fig. 8). (**b**) Channel overbank deposits on a TOPAS parametric sub-bottom profiler record.

reaches, often cutting back into the shelf edge. Immediately outside the channels, penetration echo-sounder records also show acoustically stratified overbank deposits (Fig. 9b).

In the lower reaches of the channel systems, approaching the very low gradient floor of the Greenland Basin, channel width increases and depth decreases, with the system becoming increasingly braided (Fig. 9a). In the deepest part of the basin, well-defined channels disappear and are replaced by channel-mouth lobes with both high and low backscattering areas (Fig. 7) which exhibit relatively poor acoustic penetration on parametric echo-sounder records. Sub-bottom profiler records and cores in these areas suggest that the lobes are sandy.

Between the channels, widespread fields of sediment waves are also present, especially at water depths of less than about 2600 m (Fig. 10). Sediment waves cover very large areas on the middle part of the slope between the channels

(Fig. 7). The wave crests are often orientated along the contours of the slope and basin, although some may also be related to along-slope currents. Those on the upper to middle slope are often asymmetric, with relatively low penetration on sub-bottom profiler records (Fig. 10b). Those on the lower slope are mainly located near channels and are less regular in form. The presence of these sediment-wave fields in the Greenland Basin suggests that down-slope flow and sediment-transfer processes are important not only in the channels themselves, but over much of the upper-mid slope in particular.

The large-scale channel systems and sediment-wave fields observed in the Greenland Basin are thought to be related to the intermittent downslope flow of dense water and assoicated turbidity currents. This is supported by the dominance of turbidite and bottom-current deposits in cores recovered from adjacent to the channel systems

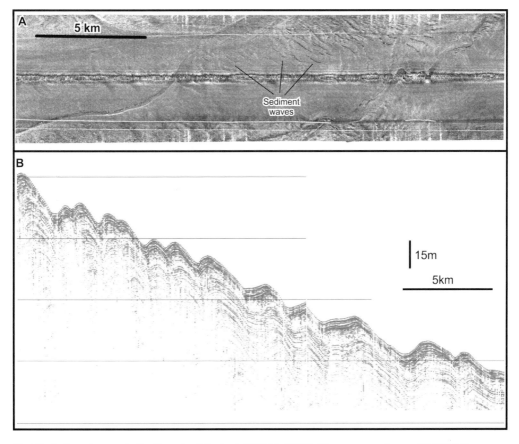

Fig. 10. Sediment waves in the Greenland Basin. (**a**) 30 kHz TOBI side-scan sonar image. (**b**) TOPAS parametric sub-bottom profiler record.

(Ó Cofaigh *et al.* 2002*a*). By contrast with the eastern margin of the Norwegian–Greenland Sea, the extensive NE Greenland shelf was not innundated by ice-sheet advance even during recent full-glacial conditions (Funder & Hansen 1996; Evans *et al.* 2002). This allowed sea-ice production and, by implication, dense-water production, to continue on the shelf and in the adjacent deep-sea basin, during both inter-glacial and full-glacial periods.

Dense, cold and saline water, produced along the 500 km-length of the East Greenland conti-nental shelf and adjacent slope by salt rejection during the formation of sea ice, is likely to cascade down the continental slope (e.g. Midttum 1985). It is probably responsible, together with the turbidity currents that such dense flows may trigger, for the channel systems and wave fields observed in the Greenland Basin (Fig. 7). Changes in the nature and rate of sedimentation within the Greenland Basin

should, therefore, provide important clues to the rate of deep-water production, which is of wider significance to the thermohaline circu-lation of the North Atlantic and the heat and momentum transfers associated with it.

Other features of the continental margin

A number of other erosional and depositional features are also present on the continental margins of the Norwegian–Greenland Sea, including canyons cut into the upper slope and contourite drifts (Fig. 11). In contrast to lower-latitude margins, canyons are relatively uncommon in this region and are confined mainly to a particularly steep area of the Nor-wegian margin at about 69°N (Taylor *et al.* 2000*b*).

Contourite drifts mark areas of the margin where along-slope ocean currents are important mechanisms of sediment transfer. Several drift

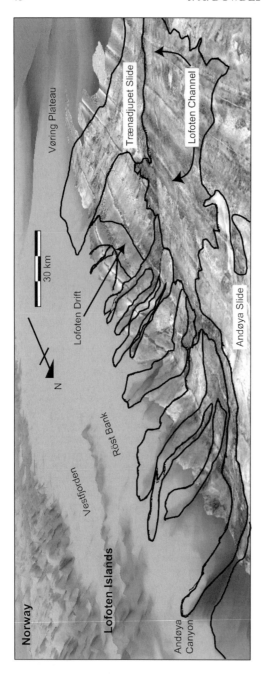

Fig. 11. Other features on the continental margins of the Norwegian–Greenland Sea include canyon systems and sediment drifts. A 6.5 kHz long-range side-scan sonar mosaic is superimposed upon a digital-elevation model of the north Norwegian margin. Canyons are outlined in black, and the Lofoten Drift and Traenadjupet Slide are also marked. (Adapted from Taylor *et al.* 2000*b*).

sheets have been mapped on the Norwegian margin, associated with the relatively strong northward flow of the Norwegian Current during interglacials in particular (Laberg *et al.* 1999, 2000). The Lofoten Drift is the clearest example (Fig. 11). Drift sheets have also been mapped on the margin north of the Faroes (van Weering *et al.* 1998) and in the Faroes–Shetland Channel (Masson 2000).

The relative scarcity of both canyons and drift sheets on the high-latitude margins of the Norwegian–Greenland Sea may be due to the prograding nature of the submarine fan systems and the associated high rates of sediment delivery, together with subsequent downslope transfer, where ice sheets intermittently reach the shelf break. There may also be a lack of meltwater and suspended sediment delivered to the ice-sheet margin. However, the slope off NE Greenland is the least well-investigated region of the Norwegian–Greenland Sea due to very difficult sea-ice conditions, and some drift sheets may remain undetected here.

In addition, a series of tectonic features of very high relief, associated with active and former spreading centres, are present in the Norwegian–Greenland Sea (Fig. 12). In many cases, these high-relief areas define the topographic boundaries between the large deep-sea basins making up the Norwegian–Greenland Sea (Fig. 3). They may be an important control on sediment reworking by turbidity currents at abyssal depths because they provide areas of steep topography adjacent to the otherwise low-energy setting on abyssal plains (Ó Cofaigh *et al.* 2002). However, such mid-ocean ridges, former spreading axes and seamounts are characteristic of all continental margins, and are not in any way restricted to high-latitude basins. They are not discussed further here.

Discussion

A simple model linking cryospheric processes and margin sedimentation

A simple conceptual model for glacier-influenced sedimentation on high-latitude continental margins summarizes the role of cryospheric processes, in the form of both ice sheets and sea ice, on the nature and rate of sediment build-up (Fig. 13). The model is based on almost 600 000 km^2 of long-range side-scan sonar, swath-bathymetry and seismic data from the Norwegian–Greenland Sea. During interglacial and interstadial periods, ice is far from the continental shelf edge, and sedimentation rates on

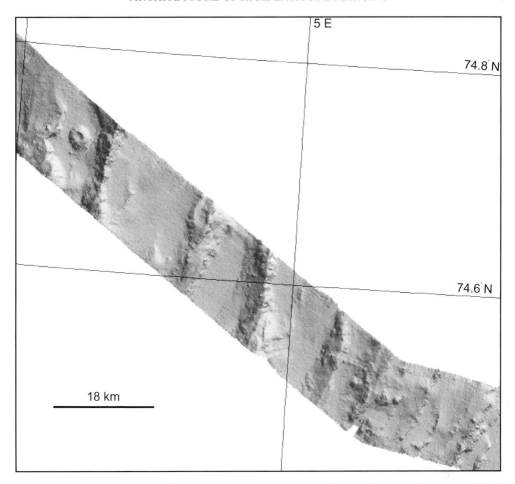

Fig. 12. EM120 swath bathymetric image from a traverse across the mid-ocean ridge bounding the east side of the Greenland Basin in the Norwegian–Greenland Sea. The water depth is between 2300 and 3500 m. There is a well-developed structural trend from north to south across the approximately 15 km-wide swath. Note the volcanic cone at 4.2°E.

the outer shelf and slope are low and predominantly hemipelagic in character (Andersen *et al.* 1996). However, in full-glacial and some stadial periods, ice advances onto and across the shelf to the shelf break (Fig. 13). Ice-stream activity is located preferentially within bathymetric troughs because, first, basal temperatures are more likely to reach the pressure melting point in these regions of relatively thick ice and, secondly, a reduction in basal drag may occur as a result of low buoyancy-induced effective basal pressures in such areas (Bentley 1987). Thus, the full-glacial ice sheet was characterized by a series of ice streams situated within troughs which delivered a large flux of ice and sediment to the shelf break (Fig. 13c).

The delivery of sediments from icebergs, meltwater and, in particular, from deforming subglacial sediment (Alley *et al.* 1989), is significantly higher at the margins of ice streams relative to slower moving ice. The development of fast-flowing ice streams in full-glacial ice sheets is, therefore, proposed to be fundamental to the growth of prograding submarine fan systems, and the debris flows which are their building blocks (Dowdeswell *et al.* 1996; King *et al.* 1996; Laberg & Vorren 1996).

Between ice streams, although ice may still reach the shelf break, the rate of ice-sheet flow is one to two orders of magnitude slower and sediment delivery to the continental slope is lower than at ice-stream margins (Fig. 13b)

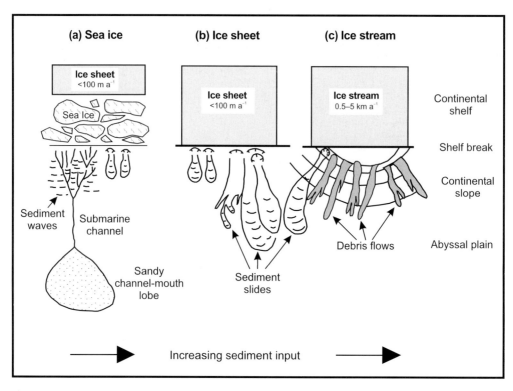

Fig. 13. Conceptual diagram of sedimentary architecture on ice-influenced continental margins in relation to: (**a**) sea-ice, (**b**) ice-sheet and (**c**) ice-stream processes. (Developed from Dowdeswell *et al.* 1996; Dowdeswell & Siegert 1999).

(Dowdeswell & Elverhøi 2002). In some areas between palaeo-ice streams on the eastern margin of the Norwegian–Greenland Sea, large slides and occasionally canyon systems have been observed (Figs 3, 11). There is increasing evidence that slides are not confined to these regions, and a few are located at the mouths of cross-shelf troughs which show evidence of the former presence of fast-flowing ice (e.g. Traenadjupet; Ottesen *et al.* 2002), or on submarine fans themselves (e.g. slide on the Bear Island Fan; Laberg & Vorren 1993). However, the slide on the Bear Island Fan is located in an area where debris flows were mainly absent during the last glacial maximum, probably due to ice-stream activity taking place further north. However, it appears that where the continental margin is relatively steep, sediment may be transported across much of the slope and extensive fans are less likely to form.

Finally, in some high-latitude areas, ice sheets fail to advance across the continental shelf to reach the shelf break even during full-glacials.

In these locations, sedimentation rates remain relatively low throughout glacial–interglacial cycles (Fig. 13a). These margins, and the deep-ocean basins beyond them, are dominated instead by large submarine channel systems and sediment-wave fields (Figs 7–10) (Mienert *et al.* 1993). This form of high-latitude sedimentary architecture is linked not to ice-sheet processes, but to the formation of dense and cold water as sea ice rejects salts during its formation at the surface of continental shelf waters. These dense waters flow down the continental slope and, together with associated turbidity currents, produce the channels and sediment waves observed in the Greenland Basin.

Comparison with other high-latitude ice-influenced margins

The geophysical data and cryospheric processes affecting sedimentation on high-latitude margins have been discussed so far mainly within the context of the Norwegian–Greenland

Sea. It is useful to consider how typical this continental margin might be of other passive margins at high latitudes in terms of both ice-sheet and sea-ice behaviour and of large-scale sedimentation. The cryospheric processes operating on and around the Late Cenozoic North American and Antarctic ice sheets were presumably similar to those taking place in the Greenland and Eurasian ice sheets surrounding the Norwegian–Greenland Sea. Ice-sheet and sea-ice edges waxed and waned over glacial–interglacial time (e.g. Dyke & Prest 1987; Anderson 1999). There is observational evidence for fast-flowing ice streams within the present Antarctic Ice Sheet (e.g. McIntyre 1985; Fahnstock & Bamber 2001), and marine geophysical evidence and numerical-model reconstructions indicate that fast ice flow also took place within cross-shelf troughs associated with these other palaeo-ice sheets (e.g. Marshall & Clarke 1996; Canals et al. 2000; Clark & Stokes in press; Ó Cofaigh et al. 2002b).

In terms of the sedimentary architecture of other high-latitude margins, prograding submarine fans or aprons are found at the edges of a number of cross-shelf troughs around Antarctica. Examples include sedimentary depocentres off the western Antarctic Peninsula (Larter & Vanneste 1995), and the Crary and Prydz Bay fans (e.g. Kuvaas & Kristoffersen 1991; Cooper et al. 1991; Hambrey et al. 1991). The Laurentian Fan off the Labrador Shelf provides an example from the North American Ice Sheet (e.g. Piper & Normark 1982), although both the Laurentian and Crary fans also include channel systems. Although geophysical data are sparse for the deep-sea basins around Antarctica, submarine channels and associated turbidites have been reported from some areas (e.g. Rebesco et al. 1996). These features may be associated with slope failure and dense-water production during sea-ice formation. Submarine slides and slope failures are also found on both other high-latitude and lower latitude margins (e.g. Bart et al. 1999; Weaver et al. 2000).

An inconsistency with the process-form relationship between fast glacier flow and submarine fans is, however, the lack of a major fan system offshore of the Hudson Strait in eastern North America. An ice stream which drained a 10^6 km^2 ice-sheet basin in Hudson Bay is inferred to have occupied the Hudson Strait, according to both numerical-model reconstructions (MacAyeal 1993; Marshall & Clarke 1997) and because of the very large numbers of icebergs produced intermittently from this source (e.g. Bond et al. 1992; Dowdeswell et al. 1995). However, extensive geophysical and geological investigations (e.g. Chough & Hesse 1976; Hesse et al. 1996, 1997) have shown that there is no well-defined fan offshore, but instead a continental slope comprising a mix of glacimarine sedimentation attributed to meltwater delivery of sediments, together with submarine turbidite and channel development feeding into the NW Atlantic Mid-Ocean Channel system (NAMOC). Neither have very marked ice-rafted debris layers, as thick and extensive as the predominantly Hudson Strait-derived North Atlantic 'Heinrich Layers' (Andrews & Tedesco 1992; Bond et al. 1992), been observed in other glacier-influenced ocean basins (e.g. Dowdeswell et al. 1999b). This could indicate differences in ice dynamics between Hudson Strait–Hudson Bay and other ice-sheet drainage basins, but there is as yet no agreed glaciological explanation to support such a view.

Conclusions

- The presence of ice during the Late Cenozoic distinguishes the nature and rates of operation of processes on high-latitude margins from those elsewhere. Glaciers and ice sheets deliver icebergs, meltwater and associated debris to high-latitude seas. During full-glacial periods, delivery is often focused on the continental shelf break and upper slope, whereas in interglacials many northern hemisphere glaciers terminate at fjord heads, and the shelf and slope therefore become more distal environments (Fig. 1). Sea ice also incorporates and transports sediments over long distances and, in addition, exerts a significant influence on ocean salinity structure and downslope water and sediment transfer. The operation of these cryospheric processes has led to the development of a sedimentary architecture on modern high-latitude continental margins that is, in many cases, distinctive from that observed at lower latitudes (e.g. Dowdeswell et al. 1996).

- Ice dynamics and the delivery of ice and sediments to the continental shelf edge during full-glacials were non-uniform along ice-sheet margins. Ice and sediment flux was focused mainly at fast-flowing ice stream termini, ranging from about 10 to 100 kilometres in width. In slower-flowing areas ice was frozen to the substrate. This pattern of flow-partitioning into fast- and slow-flowing areas appears to be typical of both Quaternary and modern ice sheets and ice caps (e.g. Bentley 1987; Dowdeswell & Siegert 1999; Stokes & Clark 2001). Where glacier ice failed to advance significantly across the continental shelf, sea ice and related oceanographic processes, together with contour-current sedimentation, provided the dominant control on margin sedimentary architecture.

- Large submarine fans are present beyond the mouths of cross-shelf troughs or major fjord systems on the continental margins of the Norwegian–Greenland Sea (Fig. 3b) (e.g. Vorren *et al.* 1998). Each is made up of a prograding sediment wedge, usually composed of stacked, relatively clay-rich debris flows (Fig. 4). Acoustically-stratified sediments are sometimes present between such debris flows, indicating sedimentation from turbid glacial meltwater. Debris flows were not active over all parts of the fans during the Late Weichselian, suggesting that the locations of palaeo-ice streams, and of rapid sediment delivery to the margin, has varied between glacial stages.

- Several large slides occur on the eastern continental margin of the Norwegian–Greenland Sea (Fig. 3c) (Vorren *et al.* 1998), originating in multiple headwalls on the upper continental slope (Kenyon 1987; Laberg & Vorren 2000). The slides are located in a variety of settings relative to the rates of past sediment delivery from Quaternary ice-sheet margins (Dowdeswell & Siegert 1999); both proximal to ice streams draining the interior ice sheet to the shelf break (e.g. Traenadjupet Slide; Figs 5, 6) and to areas of slower ice flow (e.g. Andøya, Storegga, North Faeroes slides). The lack of large slides offshore of NE Greenland is probably related to very low rates of glacial-sediment delivery to the shelf break combined with rapid down-slope transfer.

- Few submarine channel systems have been identified on the eastern margin of the Norwegian–Greenland Sea (Vorren *et al.* 1998). However, in the Greenland Basin a large system of submarine channels dominates basin morphology (Figs 7, 8). Between the channels, widespread fields of sediment waves are present, suggesting that downslope flow and sediment-transfer processes are important not only in the channels themselves, but over much of the upper-mid slope in particular (Figs 9, 10). The channels and sediment-wave fields are probably related to the intermittent downslope flow of dense water and associated turbidity currents.

- A number of other erosional and depositional features are also present on the continental margins of the Norwegian–Greenland Sea, including canyons cut into the upper slope and contourite drifts (Fig. 11). Contourite

drifts indicate areas where along-slope ocean currents are important mechanisms of sediment transfer. A series of tectonic features of very high relief, associated with active and former spreading centres (Fig. 12), are an important control on sediment reworking by turbidity currents at abyssal depths because they provide areas of steep topography adjacent to abyssal plains.

- A simple conceptual model for glacier-influenced sedimentation on high-latitude continental margins summarizes the role of cryospheric processes, in the form of both ice sheets and sea ice, on the nature and rate of sediment build-up (Fig. 13).

Geophysical data used in the production of figures for this paper were acquired on cruises JR-08 and JR-51 of the RRS *James Clark Ross*, funded by UK NERC grants GR3/8508 and GST/02/2198. Swath-bathymetric data acquisition was funded through NERC Grant GR3/JIF/02.

References

ALLEY, R. B., BLANKENSHIP, D. D., ROONEY, S. T. & BENTLEY, C. R. 1989. Sedimentation beneath ice shelves – the view from Ice Stream B. *Marine Geology*, **85**, 101–120.

ANDERSEN, E. S., DOKKEN, T. M., ELVERHØI, A., SOLHEIM, A. & FOSSEN, I. 1996. Late Quaternary sedimentation and glacial history of the western Svalbard continental margin. *Marine Geology*, **133**, 123–156.

ANDERSON, J. B. 1999. *Antarctic marine geology*. Cambridge University Press, Cambridge, 289 pp.

ANDREWS, J. T. & TEDESCO, K. 1992. Detrital carbonate-rich sediments, northwestern Labrador Sea: Implications for ice-sheet dynamics and iceberg rafting (Heinrich) events in the North Atlantic. *Geology*, **20**, 1087–1090.

ANDREWS, J. T., COOPER, T. A., JENNINGS, A. E., STEIN, A. & ERLENKEUSER, H. 1998. Late Quaternary iceberg-rafted detritus events on the Denmark Strait-Southeast Greenland continental slope (~65°N): related to North Atlantic Heinrich events? *Marine Geology*, **149**, 211–228.

BART, P. J., DE BATIST, M. & JOKAT, W. 1999. Interglacial collapse of the Crary Trough-Mouth Fan, Weddell Sea, Antarctica: implications for Antarctic glacial history. *Journal of Sedimentary Research*, **69**, 1276–1289.

BENTLEY, C. R. 1987. Antarctic ice streams: a review. *Journal of Geophysical Research*, **92**, 8843–8858.

BOND, G., HEINRICH, H., BROECKER, W., LABEYRIE, L. MCMANUS, J., ANDREWS, J., HUON, S., JANTSCHIK, R., CLASEN, S., SIMET, C., TEDESCO, K., KLAS, M., BONANI, G. & IVY, S. 1992. Evidence for massive discharges of icebergs into the North Atlantic ocean during the last glacial period. *Nature*, **360**, 245–249.

BOULTON, G. S. 1990. Sedimentary and sea level

changes during glacial cycles and their control on glacimarine facies architecture. *In:* DOWDESWELL, J. A. & SCOURSE, J. D. (eds) *Glacimarine environments: Processes and sediments*, Geological Society, London, Special Publications, **53**, 15–52.

BUGGE, T., BELDERSON, R. H. & KENYON, N. H. 1988. The Storegga Slide. *Philosophical Transactions of the Royal Society, London, Series A*, **325**, 357–388.

BUNGUM, H. 1989. Earthquake occurrence and seismotectonics in Norway and surrounding areas. *In:* GREGERSEN, S. & BASHAM, P. W. (eds) *Earthquakes at North Atlantic passive margins: Neotectonics and postglacial rebound.* Kluwer, Dordrecht, 501–509.

BUTT, F. A., ELVERHØI, A., SOLHEIM, A. & FORSBERG, C. F. 2000. Deciphering Late Cenozoic development of the western Svalbard margin from ODP Site 986 results. *Marine Geology*, **169**, 373–390.

CADMAN, V. M. 1996. Glacimarine sedimentation and environments during the Late Weichselian and Holocene in the Bellsund Trough and Van Keulenfjorden, Svalbard. Ph.D. Thesis, University of Cambridge, 250 pp.

CANALS, M., URGELES, R. & CALAFAT, A. M. 2000. Deep sea-floor evidence of past ice streams off the Antarctic Peninsula. *Geology*, **28**, 31–34.

CHOUGH, S. K. & HESSE, R. 1976. Submarine meandering thalweg and turbidity currents flowing for 4000 km in the Northwest Atlantic Mid-Ocean Channel. *Geology*, **4**, 529–533.

CLARK, C. D. & STOKES, C. R. in press. Extent and basal characteristics of the M'Clintock Channel Ice Stream. *Quaternary International*.

COOPER, A. K., STAGG, H. M. J. & GEIST, E. 1991. Stratigraphy and structure of Prydz Bay, Antarctica: implications from ODP Leg 119 drilling. *Proceedings of the Ocean Drilling Program*, **119**, 5–26.

DAMUTH, J. E. 1978. Echo character of the Norwegian-Greenland Sea: relationship to Quaternary sedimentation. *Marine Geology*, **18**, 1–36.

DAVIS, N. 2000. Arctic oceanography, sea ice and climate. *In:* NUTTALL, M. & CALLAGHAN, T. V. (eds) *The Arctic: Environment, people, policy.* Harwood, Amsterdam, 97–115.

DOKKEN, T. M. & JANSEN, E. 1999. Rapid changes in the mechanism of ocean convection during the last glacial period. *Nature*, **401**, 458–461.

DOWDESWELL, J. A. & COLLIN, R. L. 1990. Fast-flowing outlet glaciers on Svalbard ice caps. *Geology*, **18**, 778–781.

DOWDESWELL, J. A. & ELVERHØI, A. 2002. The timing of initiation of fast-flowing ice streams during a glacial cycle inferred from glacimarine sedimentation. *Marine Geology*, **188**, 3–14.

DOWDESWELL, J. A. & SIEGERT, M. J. 1999. Ice-sheet numerical modeling and marine geophysical measurements of glacier-derived sedimentation on the Eurasian Arctic continental margins. *Geological Society of America, Bulletin*, **111**, 1080–1097.

DOWDESWELL, J. A., VILLINGER, H., WHITTINGTON, R.

J. & MARIENFELD, P. 1993. Iceberg scouring in Scoresby Sund and on the East Greenland continental shelf. *Marine Geology*, **111**, 37–53.

DOWDESWELL, J. A., MASLIN, M. A., ANDREWS, J. T. & MCCAVE, I. N. 1995. Iceberg production, debris rafting, and the extent and thickness of Heinrich layers (H-1, H-2) in North Atlantic sediments. *Geology*, **23**, 301–304.

DOWDESWELL, J. A., KENYON, N. H., ELVERHØI, A., LABERG, J. S., HOLLENDER, F.-J., MIENERT, J. & SIEGERT, M. J. 1996. Large-scale sedimentation on the glacier-influenced Polar North Atlantic margins: long-range side-scan sonar evidence. *Geophysical Research Letters*, **23**, 3535–3538.

DOWDESWELL, J. A., KENYON, N. H. & LABERG, J. S. 1997. The glacier-influenced Scoresby Sund Fan, East Greenland continental margin: evidence from GLORIA and 3.5 kHz records. *Marine Geology*, **143**, 207–221.

DOWDESWELL, J. A., ELVERHØI, A. & SPIELHAGEN, R. 1998. Glacimarine sedimentary processes and facies on the Polar North Atlantic margins. *Quaternary Science Reviews*, **17**, 243–272.

DOWDESWELL, J. A., UNWIN, B., NUTTALL, A.-M. & WINGHAM, D. J. 1999a. Velocity structure, flow instability and mass flux on a large Arctic ice cap from satellite radar interferometry. *Earth and Planetary Science Letters*, **167**, 131–140.

DOWDESWELL, J. A., ELVERHØI, A., ANDREWS, J. T. & HEBBELN, D. 1999b. Asynchronous deposition of ice-rafted layers in the Nordic seas and North Atlantic Ocean. *Nature*, **400**, 348–351.

DREWRY, D. J., JORDAN, S. R. & JANKOWSKI, E. 1982. Measured properties of the Antarctic Ice Sheet: surface configuration, ice thickness, volume and bedrock characteristics. *Annals of Glaciology*, **3**, 83–91.

DYKE, A. S. & PREST, V. K. 1987. The late Wisconsin and Holocene history of the Laurentide ice sheet. *Géographie Physique et Quaternaire*, **41**, 237–263.

ELVERHØI, A., SVENDSEN, J. I., SOLHEIM, A., ANDERSEN, E. S., MILLIMAN, J., MANGERUD, J. & HOOKE, R.LeB. 1995a. Late Quaternary sediment yield from the High Arctic Svalbard area. *Journal of Geology*, **103**, 1–17.

ELVERHØI, A., ANDERSON, E. S., DOKKEN, T., HEBBELN, D., SPIELHAGEN, R., SVENDSEN, J. I., SØRFLATEN, M., RØRNES, A., HALD, M. & FORSBERG, C. F. 1995b. The growth and decay of the Late Weichselian ice sheet in western Svalbard and adjacent areas based on provenance studies of marine sediments. *Quaternary Research*, **44**, 303–316.

ELVERHØI, A., DOWDESWELL, J. A., FUNDER, S., MANGERUD, J. & STEIN, R. 1998a. Glacial and oceanic history of the Polar North Atlantic margins: an overview. *Quaternary Science Reviews*, **17**, 1–10.

ELVERHØI, A., HOOKE, R.LeB. & SOLHEIM, A. 1998b. Late Cenozoic erosion and sediment yield from the Svalbard-Barents Sea region: implications for the understanding of glacierized basins. *Quaternary Science Reviews*, **17**, 209–241.

EVANS, J., DOWDESWELL, J. A., GROBE, H., NIESSEN,

F., STEIN, R., HUBBERTEN, H.-W. & WHITTINGTON, R. J. 2002. Late Quaternary sedimentation in Kejser Franz Joseph Fjord and on the continental margins of East Greenland. *In:* DOWDESWELL, J. A. & Ó COFAIGH, C. (eds) *Glacier-Influenced Sedimentation on High-Latitude Continental Margins.* Geological Society, London, Special Publications, **203**, 149–179.

EYLES, N. 1993. Earth's glacial record and its tectonic setting. *Earth Science Reviews*, **35**, 1–248.

EYLES, C. H., EYLES, N. & MIALL, A. 1985. Models of glaciomarine sedimentation and their application to the interpretation of ancient glacial sequences. *Palaeogeography, Palaeoclimatology, Palaeoecology*, **51**, 15–84.

FAHNESTOCK, M. & BAMBER, J. 2001. Morphology and surface characteristics of the West Antarctic Ice Sheet. *In:* ALLEY, R. B. & BINDSCHADLER, R. A. (eds) *The West Antarctic Ice Sheet: Behavior and Environment.* American Geophysical Union, Washington D.C. *Antarctic Research Series*, **77**, 13–27.

FALEIDE, J. I., SOLHEIM, A., FIEDLER, A., HJELSTUEN, B. O., ANDERSEN, E. S. & VANNESTE, K. 1996. Late Cenozoic evolution of the western Barents Sea-Svalbard continental margin. *Global and Planetary Change*, **12**, 53–74.

FIEDLER, A. & FALEIDE, J. I. 1996. Cenozoic sedimentation along the southwestern Barents Sea margin in relation to uplift and erosion of the shelf. *Global and Planetary Change*, **12**, 75–93.

FUNDER, S. & HANSEN, L. 1996. The Greenland ice sheet – a model for its culmination and decay during and after the last glacial maximum. *Bulletin, Geological Society of Denmark*, **42**, 137–152.

FUNDER, S., HJORT, C., LANDVIK, J. Y., NAM, S.-L., REEH, N. & STEIN, R. 1998. History of a stable ice margin – East Greenland during the Middle and Upper Pleistocene. *Quaternary Science Reviews*, **17**, 77–123.

HAFLIDASON, H., SEJRUP, H. P., BRYN, P. & LIEN, R. 2001. The Storegga Slide: chronology and flow mechanism. *European Union of Geosciences XI, Conference Abstract*, **6**, 740.

HAMBREY, M. J., EHRMANN, W. U. & LARSEN, B. 1991. Cenozoic glacial record of the Prydz Bay continental shelf, East Antarctica. *Proceedings of the Ocean Drilling Program*, **119**, 77–132.

HEBBELN, D., DOKKEN, T., ANDERSEN, E. S., HALD, M. & ELVERHØI, A. 1994. Moisture supply for northern ice-sheet growth during the Last Glacial Maximum. *Nature*, **370**, 357–360.

HEBBELN, D., HENRICH, R. & BAUMANN, K.-H. 1998. Paleoceanography of the Last interglacial/glacial cycle in the Polar North Atlantic. *Quaternary Science Reviews*, **17**, 125–153.

HENRIKSEN, S. & VORREN, T. O. 1996. Late Cenozoic sedimentation and uplift history on the mid-Norwegian continental shelf. *Global and Planetary Change*, **12**, 171–199.

HESSE, R., KLAUCKE, I., RYAN, W. B. F., EDWARDS, M. B. & PIPER, D. J. W. 1996. Imaging Laurentide Ice Sheet drainage into the deep sea: impact on sediments and bottom water. *GSA Today*, **6**, 3–9.

HESSE, R., KLAUCKE, I., RYAN, W. B. F. & PIPER, D. J. W. 1997. Ice-sheet sourced juxtaposed turbidite systems in the Labrador Sea. *Geoscience Canada*, **24**, 3–14.

HJELSTUEN, B. O., ELVERHØI, A. & FALEIDE, J. I. 1996. Cenozoic erosion and sediment yield in the drainage area of the Storfjorden Fan. *Global and Planetary Change*, **12**, 95–117.

HONJO, S., MANGANINI, S. & WEFER, G. 1988. Annual particle flux and a winter outburst of sedimentation in the northern Norwegian Sea. *Deep-Sea Research*, **35**, 1223–1234.

JANSEN, E. & SJØHOLM, J. 1991. Reconstruction of glaciation over the past 6 Myr from ice-borne deposits in the Norwegian Sea. *Nature*, **349**, 600–603.

KENYON, N. H. 1987. Mass-wasting features on the continental slope of Northwest Europe. *Marine Geology*, **74**, 57–77.

KING, E. L., SEJRUP, H. P., HAFLIDASON, H., ELVERHØI, A. & AARSETH, I. 1996. Quaternary seismic stratigraphy of the North Sea Fan: glacially-fed gravity flow aprons, hemipelagic sediments, and large submarine slides. *Marine Geology*, **130**, 293–316.

KLEIBER, H. P., KNIES, J. & NIESSEN, F. 2000. The Late Weichselian glaciation of the Franz Victoria Trough, northern Barents Sea: ice sheet extent and timing. *Marine Geology*, **168**, 25–44.

KUVAAS, B. & KRISTOFFERSEN, Y. 1991. The Crary Fan: a trough-mouth fan on the Weddell Sea continental margin, Antarctica. *Marine Geology*, **97**, 345–362.

LABERG, J. S. & VORREN, T. O. 1993. A Late Pleistocene submarine slide on the Bear Island Trough Mouth Fan. *Geo-Marine Letters*, **13**, 227–234.

LABERG, J. S. & VORREN, T. O. 1995. Late Weichselian submarine debris flow deposits on the Bear Island Trough Mouth Fan. *Marine Geology*, **127**, 45–72.

LABERG, J. S. & VORREN, T. O. 1996. The Middle and Late Pleistocene evolution of the Bear Island Trough Mouth Fan. *Global and Planetary Change*, **12**, 309–330.

LABERG, J. S. & VORREN, T. O. 2000. The Traenadjupet Slide, offshore Norway – morphology, evacuation and triggering mechanisms. *Marine Geology*, **171**, 95–114.

LABERG, J. S., VORREN, T. O. & KNUTSEN, S.-M. 1999. The Lofoten contourite drift off Norway. *Marine Geology*, **159**, 1–6.

LABERG, J. S., VORREN, T. O. DOWDESWELL, J. A., KENYON, N. H. & TAYLOR, J. 2000. The Andøya Slide and the Andøya Canyon, north-eastern Norwegian-Greenland Sea. *Marine Geology*, **162**, 259–275.

LANDVIK, J. Y., BONDEVIK, S., ELVERHØI, A., FJELDSKAAR, W., MANGERUD, J., SIEGERT, M. J., SALVIGSEN, O., SVENDSEN, J. I. & VORREN, T. O. 1998. The last glacial maximum of Svalbard and the Barents Sea area: ice sheet extent and configuration. *Quaternary Science Reviews*, **17**, 43–75.

LARSEN, H. C., SAUNDERS, A. D., CLIFT, P. D., BEGET, J., WEI, W., SPEZZAFERRI, S. & ODP Leg 152 Scientific Party 1994. Seven million years of glaciation in Greenland. *Science*, **264**, 952–955.

LARTER, R. D. & VANNESTE, L. E. 1995. Relict subglacial deltas on the Antarctic Peninsula outer shelf. *Geology*, **23**, 33–36.

MACAYEAL, D. R. 1993. Binge/purge oscillations of the Laurentide Ice Sheet as a cause of the North Atlantic's Heinrich Events. *Paleoceanography*, **8**, 775–784.

MANGERUD, J. & SVENDSEN, J. I. 1992. The last inter-glacial/glacial period on Spitsbergen, Svalbard. *Quaternary Science Reviews*, **11**, 633–664.

MARSHALL, S. J. & CLARKE, G. K. C. 1996. Geologic and topographic controls on fast flow in the Laurentide and Cordilleran Ice Sheets. *Journal of Geophysical Research*, **101**, 17827–17839.

MARSHALL, S. J. & CLARKE, G. K. C. 1997. A continuum mixture model of ice stream thermo-mechanics in the Laurentide Ice Sheet. 2. Appli-cation to the Hudson Strait Ice Stream. *Journal of Geophysical Research*, **102**, 20615–20637.

MASSON, D. G. 2000. Sedimentary processes shaping the eastern slope of the Faeroe-Shetland Channel. *Continental Shelf Research*, **21**, 825–857.

MCINTYRE, N. F. 1985. The dynamics of ice-sheet outlets. *Journal of Glaciology*, **31**, 99–107.

MIDTTUM, L. 1985. Formation of dense bottom water in the Barents Sea. *Deep-Sea Research*, **32**, 1233–1241.

MIENERT, J., KENYON, N. H., THIEDE, J. & HOLLEN-DER, F.-J. 1993. Polar continental margins: studies off East Greenland. *EOS Transactions of the American Geophysical Union*, **74**, 225, 234, 236.

MEINERT, J., POSEWANG, J. & LUKAS, D. 2001. Changes in the hydrate stability zone on the Norwegian margin and their consequences for methane and carbon releases into the oceano-sphere. *In:* SCHAFER, P., RITZRAU, W., SCHLUTER, M. & THIEDE, J. (eds) *The northern North Atlantic: a changing environment*, 281–290.

NÜRNBERG, D., WOLLENBURG, I., DETHLEFF, D., EICKEN, H., KASSENS, H., LETZIG, T., REIMNITZ, E. & THIEDE, J. 1994. Sediments in Arctic sea ice: implications for entrainment, transport and release. *Marine Geology*, **119**, 185–214.

Ó COFAIGH, C., TAYLOR, J., DOWDESWELL, J. A., ROSELL-MELÉ, A., KENYON, N. H., EVANS, J. & MIENERT, J. 2002a. Sediment reworking on high-latitude continental margins and its implications for palaeoceanographic studies: insights from the Norwegian–Greenland Sea. *In:* DOWDESWELL, J. A. & Ó COFAIGH, C. (eds) *Glacier-Influenced Sedimentation on High-Latitude Continental Margins*, Geological Society, London, Special Publications, **203**, 325–348.

Ó COFAIGH, C., PUDSEY, C. J., DOWDESWELL, J. A. & MORRIS, P. 2002b. Evolution of subglacial bedforms along a paleo-ice stream, Antarctic Peninsula continental shelf. *Geophyscial Research Letters*, **29**, 10.1029/2001. GL014488.

OTTESEN, D., DOWDESWELL, J. A., RISE, L.,

ROKOENGEN, K. & HENRIKSEN, S. 2002. Large-scale morphological evidence for past ice-stream flow on the Norwegian continental margin. *In:* DOWDESWELL, J. A. & Ó COFAIGH, C. (eds) *Glacier-Influenced Sedimentation on High-Latitude Continental Margins*. Geological Society, London, Special Publications, **203**, 245–258.

PICKERING, K. T., HISCOTT, R. N., KENYON, N. H., RICCI LUCCI, F. & SMITH, R. D. A. 1995. *Atlas of deep water environments: Architectural style in turbidite systems*. Chapman & Hall, London.

PIPER, D. J. W. & NORMARK, W. R. 1982. Acoustic interpretation of Quaternary sedimentation and erosion on the channelled upper Laurentian Fan, Atlantic margin of Canada. *Canadian Journal of Earth Sciences*, **19**, 1974–1984.

POLYAK, L., FORMAN, S. L., HERLIHY, F. A., IVANOV, G. & KRINITSKY, P. 1997. Late Weichselian deglacial history of the Svyataya (Saint) Anna Trough, northern Kara Sea, Arctic Russia. *Marine Geology*, **143**, 169–188.

REBESCO, M., LARTER, R. D., CAMERLENGHI, A. & BARKER, P. F. 1996. Giant sediment drifts on the continental rise west of the Antarctic Peninsula. *Geo-Marine Letters*, **16**, 65–75.

SHIPP, S., ANDERSON, J. & DOMACK, E. 1999. Late Pleistocene-Holocene retreat of the West Antarctic Ice-Sheet system in the Ross Sea: Part 1 – Geophysical results. *Geological Society of America Bulletin*, **111**, 1486–1516.

SHIPP, S., WELLNER, J. S. & ANDERSON, J. B. 2002. Retreat signature of a polar ice stream: sub-glacial geomorphic features and sediments from the Ross Sea, Antarctica. *In:* DOWDESWELL, J. A. & Ó COFAIGH, C. (eds) *Glacier-Influenced Sedimentation on High-Latitude Continental Margins*. Geological Society, London, Special Publications, **203**, 277–304.

SIEGERT, M. J. & DOWDESWELL, J. A. 2002. Late Weichselian iceberg, meltwater and sediment production from the Eurasian Ice Sheet: results from numerical ice-sheet modelling. *Marine Geology*, **188**, 109–127.

SIEGERT, M. J., DOWDESWELL, J. A. & MELLES, M. 1999. Late Weichselian glaciation of the Russian High Arctic. *Quaternary Research*, **52**, 273–285.

SIEGERT, M. J., DOWDESWELL, J. A., HALD, M. & SVENDSEN, J. I. 2001. Modelling the Eurasian Ice Sheet through a full (Weichselian) glacial cycle. *Global and Planetary Change*, **31**, 367–385.

SOLHEIM, A., FALEIDE, J. I., ANDERSEN, E. S., ELVERHØI, A., FORSBERG, C. F., VANNESTE, K., UENZELMANN-NEBEN, G. & CHANNELL, J. E. T. 1998. Late Cenozoic seismic stratigraphy and glacial geological development of the East Greenland and Svalbard-Barents Sea continental margins. *Quaternary Science Reviews*, **17**, 155–184.

STEIN, A. B. & SYVITSKI, J. P. M. 1997. Glacially-influenced debris flow deposits: East Greenland slope. *In:* DAVIES, T. A., BELL, T., COOPER, A. K., JOSENHANS, H., POLYAK, L., SOLHEIM, A., STOKER, M. S. & STRAVERS, J. A. (eds) *Glaciated continental margins: An atlas of acoustic images.* Chapman & Hall, London, 134–135.

STOKER, M. S. 1990. Glacially-influenced sedimentation on the Hebridean slope, northwest United Kingdom continental margin. *In:* DOWDESWELL, J. A. & SCOURSE, J. D. (eds) *Glacimarine Environments: Processes and Sediments.* Geological Society, London, Special Publications, **53**, 349–362.

STOKES, C. R. & CLARK, C. D. 2001. Palaeo-ice streams. *Quaternary Science Reviews,* **20**, 147–1457.

SYVITSKI, J. P. M., STEIN, A. B., ANDREWS, J. T. & MILLIMAN, J. D. 2001. Icebergs and the sea floor of the East Greenland (Kangerlussuaq) continental margin. *Arctic, Antarctic and Alpine Research,* **33**, 52–61.

TAYLOR, J., DOWDESWELL, J. A., KENYON, N. H., WHITTINGTON, R. J., VAN WEERING, T. C. E. & MIENERT, J. 2000*a*. Morphology and Late Quaternary sedimentation on the North Faeroes slope and abyssal plain, North Atlantic. *Marine Geology,* **168**, 1–24.

TAYLOR, J., DOWDESWELL, J. A. & KENYON, N. H. 2000*b*. Canyons and Late Quaternary sedimentation on the North Norwegian margin. *Marine Geology,* **166**, 1–9.

TAYLOR, J., DOWDESWELL, J. A., KENYON, N. H. & Ó COFAIGH, C. 2002. Late Quaternary architecture of trough–mouth fans: debris flows and suspended sediments on the Norwegian margin. *In:* DOWDESWELL, J. A. & Ó COFAIGH, C. (eds) *Glacier-Influenced Sedimentation on High-Latitude Continental Margins.* Geological Society, London, Special Publications, **203**, 55–71.

VÅGNES, E. 1996. Cenozoic deposition in the Nansen Basin, a first-order estimate based on present-day bathymetry. *Global and Planetary Change,* **12**, 149–157.

VAN WEERING, T. C. E., NIELSEN, T., KENYON, N. H., AVENTIEVA, K. & KUIPERS, A. H. 1998. Sediments and sedimentation at the NE Faeroes continental margin: contourites and large-scale sliding. *Marine Geology,* **152**, 159–176.

VOGT, P. R., CRANE, K. & SUNDVOR, E. 1993. Glacigenic mudflows on the Bear Island submarine fan. *EOS Transactions of the American Geophysical Union,* **74**, 449, 452, 453.

VORREN, T. O., LEBESBYE, E., ANDREASSEN, K. & LARSEN, K.-B. 1989. Glacigenic sediments on a passive continental margin as exemplified by the Barents Sea. *Marine Geology,* **85**, 251–272.

VORREN, T. O., LABERG, J. S., BLAUME, F., DOWDESWELL, J. A., KENYON, N. H., MIENERT, J., RUMOHR, J. & WERNER, F. 1998. The Norwegian-Greenland Sea continental margins: morphology and Late Quaternary sedimentary processes and environment. *Quaternary Science Reviews,* **17**, 273–302.

WADHAMS, P. 2000. *Ice in the Ocean.* Gordon and Breach, Amsterdam, 351 pp.

WEAVER, P. P. E., WYNN, R. B., KENYON, N. H. & EVANS, J. 2000. Continental margin sedimentation, with special reference to the north-east Atlantic. *Sedimentology,* **47**, 239–256.

Late Quaternary architecture of trough–mouth fans: debris flows and suspended sediments on the Norwegian margin

J. TAYLOR[1], J. A. DOWDESWELL[2], N. H. KENYON[3], & C. Ó COFAIGH[2]

[1]Bristol Glaciology Centre, School of Geographical Sciences, University of Bristol, Bristol BS8 1SS, UK (e-mail: Justin.Taylor@bristol.ac.uk)

[2]Scott Polar Research Institute and Department of Geography, University of Cambridge, Lensfield Road, Cambridge CB2 1ER, UK

[3]Southampton Oceanography Centre, Empress Dock, Southampton SO14 3ZH, UK

Abstract: Trough-mouth fans are the main marine depocentres for glacier-derived sediments in the Polar North Atlantic, but their growth through the Late Quaternary is complex. Glacigenic debris flows (GDFs) are sourced from a common and homogeneous part of the upper fan and only develop as coherent individual flows after downslope transport. Their genesis and mode of deposition mean that GDFs are confined to particular areas of trough-mouth fans; accumulation of these subglacial sediments is controlled by a combination of margin glaciology and fan morphology. Although most of the fan sediment is deposited as GDFs, during glacials considerable areas of trough-mouth fans are dominated by sedimentation of suspension deposits, associated with extensive meltwater release from a warm-based ice sheet and probable contour current activity. The depositional sequence of these two sediment types may be important in generating the long run-out distances of GDFs, which are initiated and sustained over low gradients. Furthermore, emplacement of GDFs is interpreted to be a relatively low-frequency event, and temporally, at least, fans are not dominated by this mode of sediment emplacement whilst ice sheets are at the shelf break. Large-scale trough-mouth fan development is therefore asynchronous and non-uniform, a result of the interaction between glaciology, morphology, and oceanography.

Introduction

Trough-mouth fans (TMFs) have been one of the main depocentres of glacigenic sediments during the (late) Quaternary on many glaciated margins (Fig. 1; Vorren *et al.* 1989; Cooper *et al.* 1991; Kuvaas & Kristoffersen 1991; Laberg & Vorren 1995; Stoker 1995; Dowdeswell *et al.* 1996; King *et al.* 1996; Vorren & Laberg 1997; Dowdeswell *et al.* 1998; Vorren *et al.* 1998; Taylor *et al.* 2002). TMFs are thus an important morphological feature of the glacially influenced passive margins of the Polar North Atlantic. Late Weichselian input of sediment in the form of glacigenic debris flows (GDFs) has been estimated at approximately 1000 km[3] on each of the North Sea TMF and Bear Island TMF (henceforth the North Sea Fan and Bear Island Fan, respectively). This represents 15–20% of the Late Weichselian sediment delivered to the deep-sea basins in which each fan lies (Taylor *et al.* 2002). Fluxes of sediment delivered to the deep ocean at fan locations are high during glacial maxima because debris is emplaced in significant quantities only when ice sheets are located at or near the shelf break. The

duration of sediment input to the upper slope, and GDF emplacement, was short (a few thousand years) during the Late Weichselian (Sejrup *et al.* 1996; Vorren & Laberg 1996; Elverhøi *et al.* 1997; King *et al.* 1998; Dowdeswell & Siegert 1999; Dimakis *et al.* 2000).

GDF depositional processes are a focus of research because of their role as the 'building blocks' of TMFs (Laberg & Vorrren 1995; Dowdeswell *et al.* 1996; Elverhøi *et al.* 1997; Dimakis *et al.* 2000; Laberg & Vorren 2000). GDFs are a unique mid- to high-latitude form of debris flow, sharing characteristics with fluidized, liquefied, and classical debris flows, but with important differences. These differences relate mainly to their extremely long run-out distances (over 100 km) on very low slope gradients (<1°) and the approximately uniform thickness along their length (e.g. Laberg & Vorren 1995; Elverhøi *et al.* 1997; Vorren *et al.* 1998). Understanding the emplacement of these features has important implications for studies of other long-run-out distance processes, particularly the extensive deposits often associated with large-scale failure

From: DOWDESWELL, J. A. & Ó COFAIGH, C. (eds) 2002. *Glacier-Influenced Sedimentation on High-Latitude Continental Margins.* Geological Society, London, Special Publications, **203**, 55–71. 0305-8719/02/$15.00

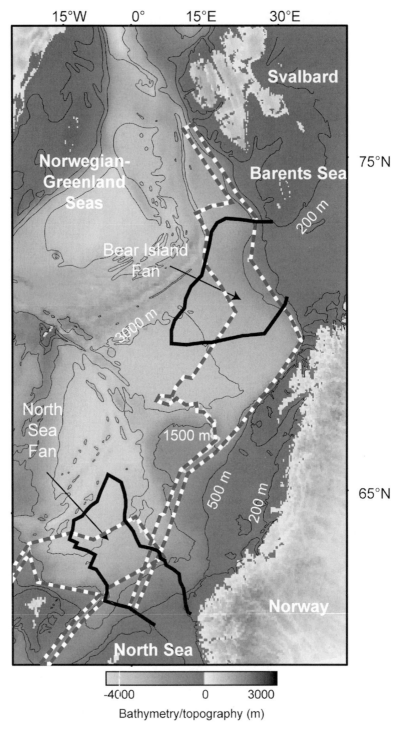

Fig. 1. The North Sea and Bear Island fans (black outlines) are important prograding features in the Norwegian–Greenland Sea, extending from the mouths of cross-shelf troughs in the North and Barents Seas to abyssal depths. Data coverage is indicated by the grey dashed outlines. Outer fan limits are based on the interpretations of Damuth (1978).

(e.g. Bugge *et al.* 1988). However, an important and hitherto relatively neglected part of this process is to constrain the broad-scale depositional environment of these glacier-influenced fans more clearly.

The objective of this paper is to provide a contextual view of glacigenic debris flows from detailed geophysical mapping of the two largest trough-mouth fans in the Polar North Atlantic, the Bear Island and North Sea fans, where the high degree of analogy implies that conclusions drawn from either are equally applicable to the other (Fig. 1). The study is based on an extensive geophysical dataset of long-range side-scan sonar and shallow seismic lines, suited to the investigation of late Quaternary architecture and deposition on these fans.

Geological setting and database

The Bear Island and North Sea fans are significant prograding features within the Norwegian–Greenland Seas. The fans are located in front of large cross-shelf troughs developed in the epicontinental Barents and North seas, respectively (Fig. 1). These troughs served as important drainage paths for the Fennoscandian and Barents ice sheets, and, therefore, as foci for sediment output from the Eurasian Ice Sheet (Vorren *et al.* 1989; Dowdeswell *et al.* 1996; Sejrup *et al.* 1996; Dowdeswell & Siegert 1999). Although fan build-up began more than 2 million years ago (Vorren *et al.* 1991; Andersen *et al.* 1996; Faleide *et al.* 1996; King *et al.* 1996; Laberg & Vorren 1996), they did not attain their present morphology until relatively recently, around the mid-Pleistocene (discussion in Vorren *et al.* 1998). Geomorphically, the most recently (latest Quaternary) active areas of the Bear Island Fan cover an area of 125 000 km², with a width of about 350 km, a run-out distance of 490 km, and upper, middle and lower slope gradients of 0.8°, 0.5°, and 0.2°, respectively. The North Sea Fan is of a similar size and morphology, covering 108 000 km², being 250 km wide and 490 km in downslope extent. The upper, middle and lower fan have gradients of 0.6°, 0.3°, and 0.2°.

The seismic stratigraphy of these fans is relatively well constrained. Deeper packages of sediments on both fans are interpreted as a combination of hemipelagic and disturbed slide sequences (Faleide *et al.* 1996; King *et al.* 1996, 1998; Kuvaas & Kristoffersen 1996). The uppermost sequences in both middle and lower fans consist of stacked, mounded lens-shaped features in cross-section, interpreted as GDFs. They are separated by reflectors of medium to high amplitude in the case of the Bear Island Fan and by thick, only occasionally seismically structured, sequences in the case of the North Sea Fan. These North Sea Fan sequences are interpreted by King *et al.* (1996) to be hemipelagic units. GDFs often nest in the depressions between previous flows and are generally of positive relief (e.g. Vorren *et al.* 1998).

We use an extensive dataset of 120 000 km² of 6.5 kHz GLORIA side-scan sonar imagery with a swath width of about 20 km and over 45 000 trackline kilometres of contemporaneously collected 3.5 kHz sub-bottom profiler records to characterize the sedimentation processes which have occurred across the Bear Island and North Sea fans during the recent geological past (Fig. 1; Dowdeswell 1996; Dowdeswell *et al.* 1996). These data form part of a much larger survey database. The 3.5 kHz data typically have a penetration of less than 60 m through the sediment column, using an assumed velocity of approximately 1500 m s⁻¹.

Flow location and fan morphology

Distribution of glacigenic debris flows

Mapping of the geometry and extent of GDFs, using a combination of the 3.5 kHz sub-bottom profiler cross-sections connected by surface mapping from the 6.5 kHz GLORIA side-scan sonar mosaic, produces a clear picture of the distribution of the uppermost (and therefore most recent) flows on each of the main fans investigated (Figs 2 & 3). On the North Sea Fan 24 flows are identified, five for significant lengths, and 19 only partially. The distribution of flows is relatively even across the fan, showing a slightly radial pattern of deposition (Fig. 2). However, large-scale flow development is limited to the central fan area, and the weakly radial pattern of deposition suggests that the fan did not have a significantly more lobate appearance prior to failure of the Storegga Slide on its northeastern flank. Thus, although the path and morphology of a few flows are constrained strongly by the presence of the Storegga Slide, this is principally an erosive contact, truncating several flows. It does not appear to be a significant influence on fan development (Fig. 2; Evans *et al.* 1996; King *et al.* 1996). Indeed, the location and form of the North Sea Fan has been an important factor in constraining margin instability between this depocentre and the tectonically related Vøring Plateau to the north (Evans *et al.* 1996).

On the Bear Island Fan 30 flows are identified, all for significant lengths (over

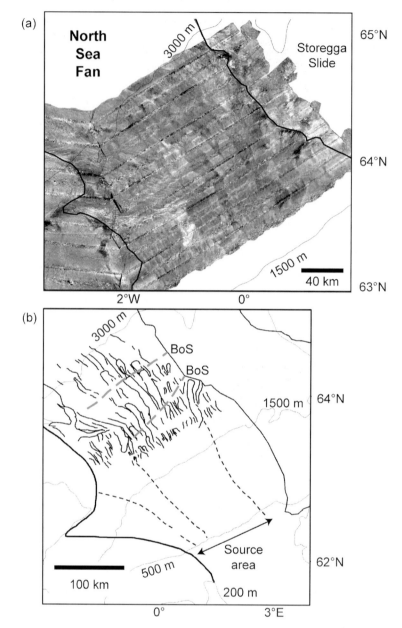

Fig. 2. (**a**) The distribution of GDFs on the North Sea Fan mapped from GLORIA and 3.5 kHz profiler data superimposed on the GLORIA mosaic. (**b**) Interpretation of the North Sea Fan. Flows display a generally radial pattern but are not evenly distributed across the fan. Breaks of slope (BoS) on the fan surface are indicated by the grey dashed lines.

50 km). These flows include several GDFs that have been studied previously (Vogt *et al.* 1993; Crane *et al.* 1997; Fig. 3). The distribution of flows across the Bear Island Fan is not as even as that of the North Sea Fan. The entire fan area south of approximately 72°N, including the region of the Bjørnøyrenna Slide (Vorren *et al.* 1989; Laberg & Vorren 1993), is devoid of flows in the uppermost part of the sediment column. Recent flows are confined to the middle and

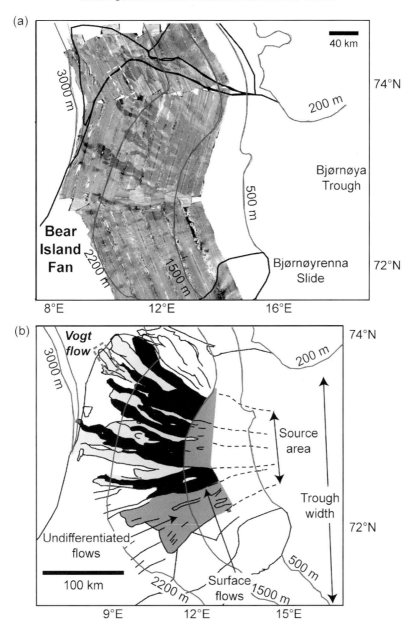

Fig. 3. (**a**) GLORIA mosaic of GDFs on the Bear Island Fan. (**b**) Flows display a generally radial pattern but are not evenly distributed across the fan. Surface (black), buried (light grey), and undifferentiated (grey) flows are distinguished. Location of the GDF imaged by SeaMARC II by Vogt *et al.* (1993) is indicated by grey dashed boxes. The data are inconclusive in mapping detail of the northernmost flows of the fan.

northern sections of the fan (Laberg & Vorren 1996). Many extend the full length of the fan from upslope of our data coverage (approximately 1500 m water depth) to the abyssal plain

and spreading axis in this region, at approximately 3000 m water depth (Fig. 3a).

GDFs display a strongly radial pattern on the Bear Island Fan, and tracing the source of flows

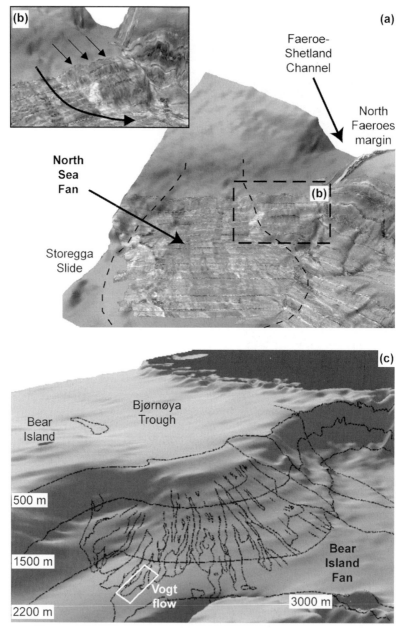

Fig. 4. The relationship between fan morphology and GDF location is shown on both the (**a**) North Sea and (**c**) Bear Island fans. Even minor changes in gradient have a profound effect on flow pathways and character (**b**). Bathymetry from the North Sea Fan from a regional compilation in Taylor *et al.* (2000).

through upslope extrapolation to the shelf break indicates that the likely source areas for GDFs is a relatively confined section of the Bjørnøya Trough only some 100 km in width (Fig. 3b). Furthermore, sets of flows appear to trace back to four or five zones within this source area. Limited areal input of sediment to the top of the fan therefore appears to translate into limited subsequent fan development on the middle and lower slopes.

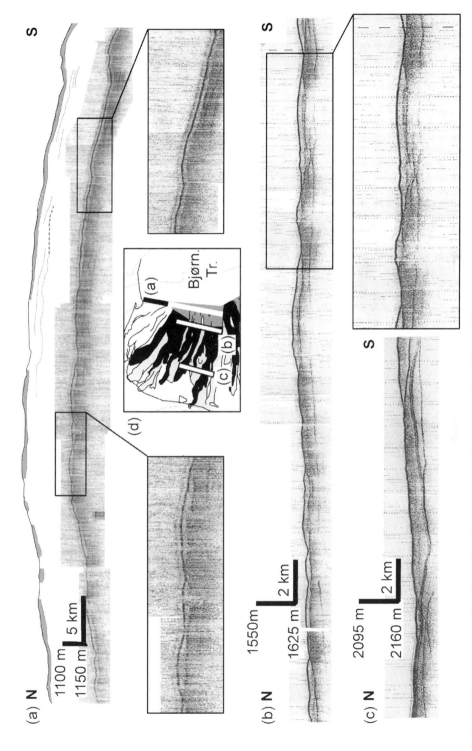

Fig. 5. 3.5 kHz sections from Bear Island Fan, showing the transition from (**a**) sheet-like, undifferentiated flows above 1500 m to (**b**) lenticular cross-sectioned individual lobes at greater water depths. Note the definition of GDF packages by buried high amplitude and parallel reflectors (**c**). It may be possible to identify individual flows on the upper slope on the basis of morphology ((**a**); left panel) (cf. Laberg & Vorren 2000), but that this is confined to the steepest, northernmost section of the fan (**d**).

Relation to fan morphology

The uneven distribution of flows across the face of both the North Sea and Bear Island fans, when viewed three-dimensionally, can be seen to be related strongly to fan topography (Fig. 4). Although the surface gradients of TMFs are extremely low (less than 1°), and therefore variations in this slope are even lower, even subtle changes in surface morphology appear to influence flow behaviour. For example, GDFs on the North Sea Fan appear to originate from a quite confined, steeper area in the central fan, defined by very small variations in slope on either side (Fig. 4a). Whereas the main source area has a gradient of 0.6°, the fan to either side has a slope of less than 0.5°. The influence of small changes in topography on GDF pathways and fan development is illustrated further by a plateau on the mid-fan, which diverts major GDFs to its eastern side and is covered by thin and much smaller flows running over its surface and cascading down its steep downslope face (Fig. 4b). The top of this plateau has a gradient of about 0.5°, whereas the face of the plateau has a gradient of 0.8°, compared with the gradient of the main fan depositional pathway immediately to the east of only 0.6°. Minor changes in topography translate into significant variations in depositional path.

This pattern is also seen on the Bear Island Fan, where the slightly steeper overall gradient of the northern fan margin is associated with strongest flow development and aggradation (Fig. 4c). Flows studied as part of the broader investigation into GDF behaviour (e.g. Vogt *et al.* 1993; Laberg & Vorren 1995; Elverhøi *et al.* 1997; Laberg & Vorren 2000) come from this steepest part of the fan.

Sea-floor gradient also appears to play a significant role in the termination of flows. There is a strong correlation between the terminus depth of GDFs on both the North Sea and Bear Island fans and main breaks of slope in fan morphology. On the North Sea Fan, where 13 GDF termini are imaged, there is a generally good correspondence between terminus depth and two breaks of slope identified at about 2000 m and about 2700 m water depth (Fig. 2b). The relationship is more strongly developed on the Bear Island Fan, where more than 85% of the termini imaged on the GLORIA side-scan mosaic terminate within the range 2000–2400 m water depth, 2200 m dividing the middle and lower fan (Laberg & Vorren 1995). However, not all flows cease at these points, and several flows on both fans continue beyond the extent of the survey data into deeper water (Figs 2 & 3).

Geophysical character of flows

The seismic character of GDFs on our geophysical datasets, and the mapping derived from this, is an important parameter in interpreting the processes by which they are deposited. However, in this investigation we attempt only to determine the overall style of deposition and we do not deal in detail with particular flow processes.

The first main observation is that flows are indistinct on the upper fan (approximately above 1500 m water depth), but become well-defined further downslope. This is apparent from both GLORIA (Figs 2 & 3) and 3.5 kHz records (Fig. 5) from the Bear Island Fan, where the data are clearest. Flows have no lateral definition on the middle slope on either GLORIA imagery, where they appear as a continuous and uniformly low backscatter surface, or on 3.5 kHz records, where they appear as extensive and thin (~15 m) 'sheets' (Fig. 5a). At approximately 1500 m water depth, there is a sharp transition, whereby separate flows become identifiable and elongate in plan on GLORIA records and lenticular in cross-section on 3.5 kHz profiles (Figs 3b & 5b). This depth is notable because it is also the morphological divide between upper and middle fan (Laberg & Vorren 1995).

Although most flows are uniformly low in backscatter throughout their width and length on the GLORIA mosaic (Fig. 3a), a 'marbled' textured surface, with high and low backscatter halos around its outer edges, and an overall higher level of backscatter, is observed on one flow towards the northern edge of the Bear Island Fan (Vogt *et al.* 1993; Crane *et al.* 1997). Laberg & Vorren (2000) indicate that these arcuate lineations probably correspond to locally high relief. This flow is marked in Figures 3 and 4 and, perhaps because it was the first of these flows to be described in detail, has been used subsequently to form hypotheses on flow processes and development. However, the description above contrasts strongly with the appearance of other flows on the Bear Island Fan from the less steep, and thus arguably also more representative morphologically, areas immediately to the south (Fig. 3). This implies that such an appearance (and therefore deposit?) is atypical of debris flows emplaced on trough mouth fans. Similarly, no such marbled or 'halo' pattern is discernible on flows from any section of the North Sea Fan side-scan mosaic (Fig. 2a).

Non-GDF sedimentation on glacier-influenced fans

GDFs are a spatially restricted process on and close to the surface of the Bear Island Fan. Large parts of the fan have not been subject to debris flow processes during the recent geological past (Fig. 3b), and different processes may have been operating in these areas. The location of the Bjørnøyrenna Slide on the margin is indicative that (predominantly glacial) sediment supply in this area is not as rapid as that associated with flow development further north, but higher, perhaps, than hemipelagic sedimentary rates elsewhere in the Polar North Atlantic during the Late Weichselian. This is because very rapidly delivered sediment is moved away quickly to abyssal depths through GDFs processes, whereas low sedimentation rates do not 'prime' the sediment column for large-scale reworking (Taylor *et al.* 2002).

Geophysical evidence

The side-scan sonar records indicate that the entire southern part of the Bear Island Fan outside of the slide is a featureless, medium backscatter area, with no indications of large-scale downslope movement or sediment reworking (Fig. 3a). The 3.5 kHz profiles show a sediment column characterized by frequent acoustic stratification, traceable across considerable distances (Fig. 6). This stratified sequence is up to at least 20 m thick and is cut by the erosive margins of the Bjørnøyrenna Slide. A low energy depositional environment is indicated.

Critically, GDF lenses are visible beneath these stratified reflectors in the central part of the fan (e.g. Fig. 6a and c), indicating that GDF activity has not always been absent from this part of the margin, but that it no longer occurs here (also cf. Laberg & Vorren 1996). It is also possible to trace several of the reflectors from within this stratified unit northwards into the uppermost package of stacked GDFs on the central and northern Bear Island Fan (Fig. 6a). Such reflectors have been interpreted previously from reflection seismic records as relating to relatively low sedimentation rates/erosion during interstadials or interglacials (Vorren *et al.* 1989).

The extensive presence of an acoustically stratified unit across this southern end of the fan is interpreted to indicate that a low energy depositional environment has persisted here throughout at least the last glacial/interglacial cycle. Laminated or stratified sediments are not uncommon in glacially influenced environments and are generally indicative of the presence of meltwater and sediment-laden plumes, whether from turbidity or contouritic currents, or some cyclicity in sedimentation involving ice-rafted debris (e.g. Powell 1990; Hesse 1992; Stoker *et al.* 1998; Hesse *et al.* 1999; Shipp *et al.* 1999; Dowdeswell *et al.* 2000b; Knutz *et al.* 2001; Ó Cofaigh & Dowdeswell 2001). Importantly, the presence of cyclic deposition is evidence for relatively high glacial depositional rates, as bioturbation occurs at low sedimentation rates (cf. Ashley & Smith 2000).

Sedimentary evidence of non-GDF deposition

Sediment cores provide evidence for a geologically based interpretation of the southern fan area. Although the region is poorly sampled, three cores are available to constrain the depositional environment between the area of GDF deposition and the Bjørnøyrenna Slide (Fig. 6d). Cores MD95-2012 and M23260 are described by Dreger (1999) and show a consistent, interpretable stable isotope and foraminiferal record of oceanographic and environmental change. Core JR51 GC07 was collected during the summer of 2000 aboard the R.R.S. *James Clark Ross* (Dowdeswell *et al.* 2000a) and allows a sedimentological constraint to be put on the proxy evidence provided by the other two core records. None of these cores exhibit any obvious signs of hiatuses or major reworking.

Core MD95-2012 is approximately 16 m long and covers the last 40 000 years, suggesting that large-scale deposition of GDFs has not occurred on the southern Bear Island Fan for at least this period. This is probably true back to at least 50 000 years ago, given that the 3.5 kHz profiles show the stratified sediment sampled by the core to be at least 20 m thick. The core contains a strong ice-rafted debris (IRD) signal, a significant component of reworked carbonate in the form of ^{14}C-free foraminiferal tests from the Barents Sea, and evidence of highly variable sea-surface temperatures, $\delta^{18}O$, and $\delta^{13}C$ proxies (Fig. 7). These all reflect a strong glacial influence and record significant inputs of glacial meltwater and associated sediment onto the slope from the Barents Sea ice sheet (Dreger 1999).

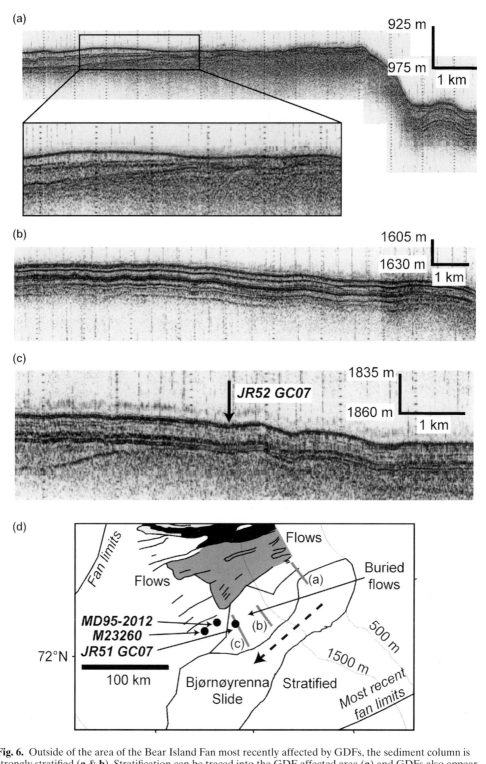

Fig. 6. Outside of the area of the Bear Island Fan most recently affected by GDFs, the sediment column is strongly stratified (**a** & **b**). Stratification can be traced into the GDF affected area (**a**) and GDFs also appear buried beneath these sediments (**c**). Profiles and cores located in (**d**).

However, these palaeoceanographic records provide no useful sedimentological information in determining what processes are responsible for sediment delivery to the Bear Island Fan in those areas outside the influence of GDFs. For this we must use core JR51 GC07, which, although as yet unsupported by ^{14}C or δ^{18}O dating, may be interpreted to show the inclusion of a single glacial sequence, probably indicative of Late Weichselian ice-sheet behaviour (Fig. 7), through tentative correlation with these other cores in the following manner.

Identification of the termination of the last glacial maximum (LGM) in core JR51 GC07 may be done with some confidence by direct correlation with other cores nearby because of the strongly coherent regional sedimentation rate (Taylor et al. 2002; Ó Cofaigh et al. 2002). In core M23260, approximately 40 km to the west, the transition from LGM to Holocene occurs at approximately 1 m, although at a considerably deeper depth in core MD95-2012, probably relating to stretching of the latter during its acquisition (Dreger 1999). In core JR51 GC07, an approximately 1m thick sequence of diamict, thin clayey or silty mud, topped by an olive grey sandy mud with clasts, is overlain by 13 cm of massive brown silty mud containing abundant dispersed foraminifera. This capping unit is similar to Holocene hemipelagic sediments described elsewhere in this part of the Polar North Atlantic (e.g. Laberg & Vorren 1995) and its massive structure and presence of abundant foraminifera supports a postglacial hemipelagic origin (Stow & Piper 1984). The diamict and mud sequence beneath is thus interpreted to reflect large-scale glacimarine sediment input associated with deglaciation of the Barents Sea at approximately 15–17 cal. ka BP. Diamicts some 30 cm below this probably therefore record sediment input as minor sediment gravity flows (based on their sharp contacts and poorly sorted texture) associated with the onset of the maximum extent of the Barents Sea ice sheet in the Bjørnøya Trough. This suggests emplacement during the period 21–19 cal. ka BP (Fig. 7). Dates are derived from Mangerud et al. (1998). The period in between is characterized by 25 cm of grey clayey mud, again interpreted to be glacimarine in origin.

Prior to this, a thick (160 cm), massive grey clayey mud sequence containing scattered sand grains is interpreted as an ice-rafted debris-rich glacimarine unit. There is some indication of cyclicty in the deposition of these sediments, with minor, weak, and diffuse laminae of various darker colours interspersed throughout the section. Occasional zones of sandier mud (2–3 cm thick) that are gradational with bounding sediments, and bioturbation-induced mottling, also occur in this unit. These characteristics suggest that the clayey mud formed through a combination of meltwater-related sedimentation (explaining the weak cyclicity) (Ó Cofaigh & Dowdeswell 2001), and the action of contour-parallel sediment transfer (explaining the diffuse zones of sandier mud and bioturbation-induced mottling) (Stoker et al. 1998; Stow et al. 1998). The latter process is also consistent with regional evidence for alongslope transport in this region (Kenyon 1986; Laberg et al. 1999; Knutz et al. 2001).

The visual absence of sand-grains in the clayey mud towards the base of the core (Fig. 7) suggests that ice-rafted debris is not a significant component of this part of the core, and that the Barents Sea ice sheet was not present. Hence, the basal clayey mud may pre-date approximately 27 cal. ka BP, when the ice-sheet had only just begun expanding onto the shelf (Fig. 7). This is a preliminary observation, however, and awaits confirmation by X-radiograph-based logging of the core.

Therefore, despite the presence of minor reworked sediment, it is the thick unit of weakly laminated mud with ice-rafted debris that is representative of the depositional environment of 'quiescent' areas of the Bear Island Fan. Such sediments may have been deposited across the entire fan during those periods when the ice sheet was at the shelf break but no GDFs were being emplaced. There are also striking resemblances between the deposits described here and those characterizing the upper Labrador Slope, where such deposits are laterally continuous on acoustic records. Although they are sedimentologically somewhat variable, these sediments are interpreted as having been deposited by turbid surface-plumes, developed from sediment-rich meltwater from the Laurentide Ice Sheet (Hesse et al. 1997, 1999).

Discussion: Implications for TMF environments and growth

Large-scale patterns

Trough-mouth fans are a complex depositional environment, characterized by two main depositional processes: glacigenic debris flows, and glacimarine suspension and turbidity current sediments ('plumites': Hesse et al. 1997). GDFs are not distributed evenly across the face of fans, they are highly sensitive to fan morphology, and are sourced from relatively narrow

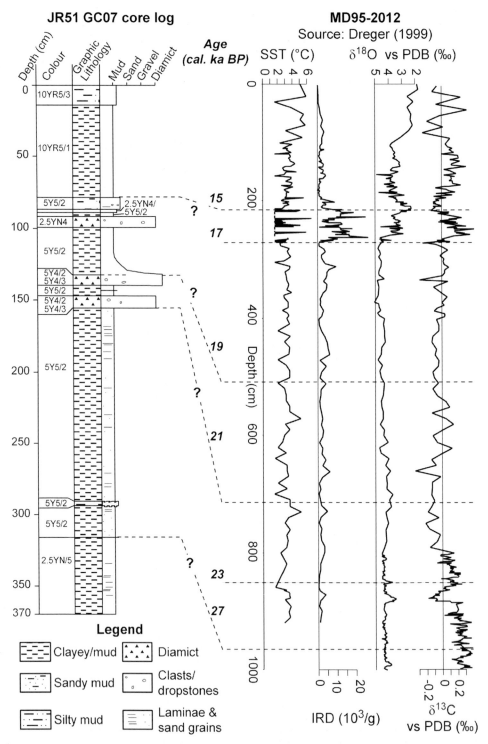

Fig. 7. Lithostratigraphy of JR51 GC07 (located in Fig. 6b) interpreted in the context of proxies derived from isotope and foraminiferal data from core MD95-2012 (Dreger 1999). Diamict and clasts correspond to the LGM and deglaciation, whilst the presence of sand grains is interpreted to indicate ice in the Barents Sea.

regions (a few tens of kilometres wide) on the upper slope (Figs 2, 3 & 4). Outside areas where GDFs predominate, glacier-influenced fans are dominated by extensive lower energy turbiditic (and contouritic?) deposits, that are rich in ice-rafted debris and, on the basis of isotopic characterization, are associated with meltwater input. Sedimentation rates are increased during glacials (Dreger 1999).

Sediments deposited rapidly by turbid plumes have a tendency to be rheologically weak, partly because of the increased rates of deposition (tens of centimetres to metres per 1000 years e.g. Elverhøi et al. 1983; Powell 1990; Haflidason et al. 1998; Fig. 7) and partly because they have no inherent structure or cohesiveness. The entire southern region of the Bear Island Fan has thus been built up during the later part of the Quaternary (at least the last 50 ka) through accumulation of these types of sediments. From direct observation of the continuation of strati-fied sediments into the GDF package on the northern Bear Island Fan, and from process similarities and extensive glacimarine deposits recorded on the upper North Sea Fan (King et al. 1996, 1998), it seems that flow areas defined from the geophysical mapping presented here are influenced by these processes during periods when GDFs are not being emplaced.

Linking GDFs and suspension deposits

The presence of such a weak, extensive layer of suspension-transport sediment across the entire fan is potentially important in understanding GDF transport mechanisms. If there is a long time gap between episodes of GDF emplace-ment, there is scope for a significant thickness of such sediments to be built up across the transport paths of GDFs. Deposition of such sediments would provide a potential low shear-strength substrate for travel over extremely low gradient slopes. This observation is made in view of the changing character of GDF deposits with distance downslope on the fan, and how this may be interpreted in terms of process. Such a process interpretation is in contrast with but does not exclude the more widely cited model for GDF transport, whereby individual lobes from the upper slope are transported long distances over shallow gradients by the estab-lishment of an overrun and thus lubricating 'film' of water at the debris flow base (hydroplaning; Elverhøi et al. 1997). However, flows defined by individual lobes on the middle and lower fan, below 1500 m, are linked to a single sediment body above. There is, therefore, a strong likelihood that all GDFs at a given level

on the Bear Island Fan in particular, but presumably also on the North Sea Fan where a similar seismic stratigraphy is found, have been deposited in a single episode.

Several implications arise. Individual flows would not be sourced from the shelf break and would therefore not be expected to leave small erosional scars, explaining previously puzzling observations of the lack of extensive, morpho-logically defined source areas for flows (Laberg & Vorren 1995; Vorren et al. 1998). Also, surface flows mapped on the Bear Island Fan were presumably the latest unit in the Late Weich-selian package to be laid down (Fig. 3). Seismic profiles show that, on both the North Sea (King et al. 1996, 1998) and Bear Island fans (Fig. 4), GDFs are stacked three or four high over regionally extensive, interstadial or interglacial stratified hemipelagic sediments, defining the complete Late Weichselian sequence of flows. Given that the Eurasian Ice Sheet is thought to have been present at the shelf break at both the Norwegian and Bjørnøya Troughs for approximately 5–6 ka (e.g. Elverhøi et al. 1995; Dowdeswell & Siegert 1999), GDF activity is suggested to have occurred approximately every 2000 years (assuming three upslope sheet-like GDF-initiation events over 6 ka). This is a radically different interpretation of the frequency of GDF emplacement from that held previously, whereby n flows deposited over the same time period were thought to have resulted in the development of a GDF once every 35–75 years (Laberg & Vorren 1995, 2000). Import-antly, our calculated frequency is compatible with estimated periods of gravitational failure of sediment numerically modelled to have accumulated at the shelf break, having been delivered by the Eurasian Ice Sheet during the Late Weichselian (cf. Dowdeswell & Siegert 1999; Taylor et al. 2002).

Assuming a period of 2000 years between episodes of GDF activity, we can estimate how much glacimarine material could potentially be deposited across the fan to act as a lubricating surface for flow transport, excluding any poten-tial lateral export of sediment associated with contour currents. Dreger (1999) reported a long-term sedimentation rate of 64 cm/ka for core MD95-2012, reaching 143 cm/ka in the lower 10 m of the core, although these values may have been overestimated due to extension of the core section. Core M23260 records a linear sedimentation rate of approximately 35 cm/ka through the Late Weichselian, and a similar value is obtained from core JC51 GC07, if our sedimentologically interpreted timescale is correct (Fig. 7). We estimate that 0.5 to 2 m of

weak glacimarine sediments may accumulate
across the face of the Bear Island Fan in
between periods of GDF formation, equivalent
to 2–10% of the average thickness of GDFs
themselves (Taylor 2000). Such a thickness
would presumably be sufficient to prevent GDF
deposition processes from influencing or being
influenced by underlying flows, even assuming
some basal erosion.

Fan development and ice-sheet dynamics

The link between GDF formation and ice-sheet
activity therefore extends beyond the fact that
flows consist of material derived from the
subglacial environment (Laberg & Vorren 1995;
King *et al.* 1996, 1998; Elverhøi *et al.* 1997;
Vorren *et al.* 1998). There may be a link between
direct ice-sheet subglacial sedimentation and
GDF initiation, but also a link between GDF
transport processes and meltwater-related
deposits found across the entire fan. Furthermore, the inferred behaviour of the ice sheet
reinforces the strong control that sediment
delivery at the shelf break has on TMF
development through the location of GDF
processes (Fig. 8). Numerically modelled behaviour supports the notion of ice-stream development in cross-shelf troughs during the Late
Weichselian (Dowdeswell & Siegert 1999).
These ice streams are strongly controlled topographically, to the extent that modelling of the
width of the Bjørnøya Trough ice stream indicates that it does not fill the full width of the
trough. This is supported fully by the relatively
limited lateral source of GDFs identified above
from mapping on the fan itself (Fig. 8a). The
shelf upslope of the 'quiescent' southern half of
the Bear Island Fan is suggested to be characterized by relatively fast ice flow, indicative of
warm basal conditions and ample meltwater
availability, but not to the extent that very rapid
sediment fluxes are reworked as GDFs. Indeed,
the erosive head of the Bjørnøyrenna Slide may
have played a crucial role here, re-routing
subglacially derived sediments into the upper
scar, which cuts back substantially into the shelf
break. Comparison of GDF sources and
modelled ice-stream behaviour in the Norwegian Trough indicate a similar scenario,
whereby GDF transport paths are extremely
sensitive to the exact ice-stream location
(Fig. 8b).

Finally, the long-term influence of these
interactions between morphology, ice-sheet
dynamics and sediment deposition should also
be addressed. GDF development across fans is
not even. This is a function of both morphology,

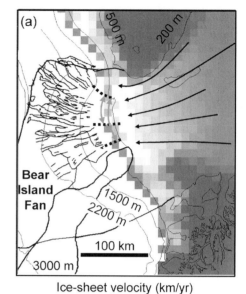

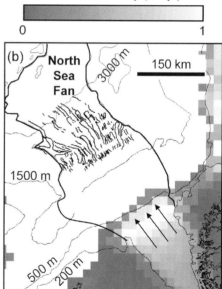

Fig. 8. The relationship between numerically
modelled ice-sheet velocity and GDF deposition is
clear and provides support for a directly glacial
source for these features. Confined ice-stream
development leads to focused GDF emplacement
and uneven fan progradation on (**a**) the Bear Island
and (**b**) North Sea fans.

and its control on gravity-driven processes, and
the limited source area of flows. This would
suggest that growth of TMFs is not likely to be
even in all directions. This is especially evident
on the Bear Island Fan, where no recent GDF

emplacement has occurred in the south, and Late Weichselian emplacement appears to be trending towards to the steeper, northern margin. The lateral extent of the Bear Island Fan during the Quaternary therefore appears to incremental with time, with associated implications for changing foci of sea-floor stability and instability. A further consequence of this uneven build-up is that any horizon within the glacigenic sequence of the fan may well be asynchronous with neighbouring sediment bodies. This in turn makes dating of fan-building episodes difficult and has implications for analogue models in hydrocarbon studies.

Conclusions

Using a combination of geophysical datasets, we have developed a new model for the late Quaternary depositional environment across the two main trough-mouth fans in the Polar North Atlantic. Although the Bear Island and North Sea fans are quantitatively dominated by glacigenic debris flow processes (Taylor *et al.* 2002), these processes do not act uniformly across the fans in any given glacial period, nor do they occur in isolation from depositional processes which are otherwise associated with warm basal ice-sheet conditions. Approximately 40% of the area of the most recently active Bear Island Fan has not been characterized by GDF activity for at least the past 50 ka, for example.

Furthermore, the nature of debris-flow deposits leads us to conclude that GDF emplacement is probably a relatively low-frequency event on fans, occurring approximately only once every 2000 years, whilst ice sheets are at the shelf break. In temporal terms at least, although not in volume, trough–mouth fans are typically dominated by deposition of fine-grained sediments from turbid plumes of subglacial meltwater released into the marine environment, with probably some laterally derived contouritic component. Hesse *et al.* (1999) indicate that such an environment must be associated with at least seasonally ice-free oceanographic conditions at the ice-sheet margin, and this is consistent with the developing palaeoceanographic picture in the Polar North Atlantic (e.g. Hebbeln *et al.* 1994; Bauch *et al.* 2001). Fan architecture therefore consists of an interacting mixture of debris flow and suspension-derived glacimarine sediments.

Finally, because of the spatially limited nature of progradation associated with GDF build-up, fan expansion is unlikely to be equal in all directions at all times. This leads to the conclusion that, at the large scale, any one level of such a glacially influenced fan is likely to have developed diachronously.

Data were collected under UK NERC grants GR3/8508 and GST/02/2198 (R.R.S. James Clark Ross cruises 08 and 51) and EU MAST III, ENAM II grant MAS3-CT95-0003 (R.V. Siren). We thank both the captains and crews concerned, and Dr T. P. Le Bas, SOC. We also thank M. Stoker and J. S. Laberg for their careful and thoughtful reviews.

References

ANDERSEN, E. S., DOKKEN, T. M., ELVERHØI, A., SOLHEIM, A. & FOSSEN, I. 1996. Late Quaternary sedimentation and glacial history of the western Svalbard continental margin. *Marine Geology*, **133**, 123–156.

ASHLEY, G. M. & SMITH, N. D. 2000. Marine sedimentation at a calving margin. *Bulletin of the Geological Society of America*, **112**, 657–667.

BAUCH, H. A., ERLENKEUSER, H., SPIELHAGEN, R. F., STRUCK, U., MATTHIESSEN, J., THIEDE, J. & HEINEMEIER, J. 2001. A multiproxy reconstruction of the evolution of deep and surface waters in the subarctic Nordic seas over the last 30,000 yr. *Quaternary Science Reviews*, **20**, 659–678.

BUGGE, T., BELDERSON, R. H. & KENYON, N. H. 1988. The Storegga Slide. *Philosophical Transactions of the Royal Society, London, A*, **325**, 357–388.

COOPER, A. K., STAGG, H. M. J. & GEIST, E. 1991. Stratigraphy and structure of Prydz Bay, Antarctica: Implications from ODP Leg 119 drilling. *In:* BARRON, J., LARSEN, B., & SHIPBOARD SCIENTIFIC PARTY. *Ocean Drilling Program, Scientific Results*. Ocean Drilling Program, College Station, Texas, **119**, 5–26.

CRANE, K., VOGT, P. R. & SUNDVOR, E. 1997. Glacigenic mudflows on the Bear Island trough mouth fan. *In:* DAVIES, T. A., BELL, T., COOPER, A. K., JOSENHANS, H., POLYAK, L., SOLHEIM, A., STOKER, M. S. & STRAVERS, J. A. (eds) *Glaciated continental margins. An atlas of acoustic images.* Chapman and Hall, London, 120–121.

DAMUTH, J. E. 1978. Echo character of the Norwegian-Greenland Sea: relationship to Quaternary sedimentation. *Marine Geology*, **28**, 1–36.

DIMAKIS, P., ELVERHØI, A., HØEG, K., SOLHEIM, A., HARBITZ, C., LABERG, J. S., VORREN, T. O. & MARR, J. 2000. Submarine slope stability on high-latitude glaciated Svalbard-Barents Sea margin. *Marine Geology*, **162**, 303–316.

DOWDESWELL, J. A. 1996. *Long-range side-scan sonar investigations of the North Sea Fan and North Faroes Margin, North Atlantic.* Cruise report, Centre for Glaciology, University of Wales, Aberystwyth. 36 pp.

DOWDESWELL, J. A. & SIEGERT, M. J. 1999. Ice-sheet numerical modeling and marine geophysical measurements of glacier-derived sedimentation on the Eurasian Arctic continental margins. *Bulletin of the Geological Society of America*, **111**, 1080–1097.

DOWDESWELL, J. A., KENYON, N. H., ELVERHØI, A.,

LABERG, J. S., HOLLENDER, F.-J., MIENERT, J. & SIEGERT, M. J. 1996. Large-scale sedimentation on the glacier-influenced Polar North Atlantic margins: Long-range side-scan sonar evidence. *Geophysical Research Letters*, **23**, 3535–3538.

DOWDESWELL, J. A., ELVERHØI, A. & SPIELHAGEN, R. 1998. Glacimarine sedimentary processes and facies on the Polar North Atlantic margins. *Quaternary Science Reviews*, **17**, 243–272.

DOWDESWELL, J. A., Ó COFAIGH, C., KENYON, N. H. & ROSELL-MELE, A. 2000*a*. *Geophysical and geological investigations of sedimentation and ice-ocean variability on Arctic continental margins.* R. R. S. James Clark Ross Cruise 51 Cruise Report. University of Bristol. 62pp.

DOWDESWELL, J. A., WHITTINGTON, R. J., JENNINGS, A. E., ANDREWS, J. T., MACKENSEN, A. & MARIEN-FELD, P. 2000*b*. An origin for laminated glaci-marine sediments through sea-ice build-up and suppressed iceberg rafting. *Sedimentology*, **47**, 557–576.

DREGER, D. 1999. *Decadal-to-centennial-scale sediment records of ice advance on the Barents Shelf and meltwater discharge into the northeastern Norwegian Sea over the last 40 kyr.* Berichte-Reports Institut für Geowissenschaften, Üniversität Kiel, Nr. 3.

ELVERHØI, A., LØNNE, Ø. & REINERT, S. 1983. Glaciomarine sedimentation in a modern fjord environment, Spitsbergen. *Polar Research*, **1**, 127–149.

ELVERHØI, A., ANDERSEN, E. S., DOKKEN, T., HEBBELN, D., SPIELHAGEN, R., SVENDSEN, J. I., SØRFLATEN, M., RØRNES, A., HALD, M. & FORSBERG, C. F. 1995. The growth and decay of the Late Weich-selian ice sheet in western Svalbard and adjacent areas based on provenance studies of marine sediments. *Quaternary Research*, **44**, 303–316.

ELVERHØI, A., NOREM, H., ANDERSEN, E. S., DOWDESWELL, J. A., FOSSEN, I., HAFLIDASON, H., KENYON, N. H., LABERG, J. S., KING, E. L., SEJRUP, H. P., SOLHEIM, A. & VORREN, T. 1997. On the origin and flow behaviour of submarine slides on deep-sea fans along the Norwegian-Barents Sea continental margin. *Geo-Marine Letters*, **17**, 119–125.

EVANS, D., KING, E. L., KENYON, N. H., BRETT, C. & WALLIS, D. 1996. Evidence for long-term instability in the Storegga Slide region off western Norway. *Marine Geology*, **130**, 281–292.

FALEIDE, J. I., SOLHEIM, A., FIEDLER, A., HJELSTEUN, B. O., ANDERSEN, E. S. & VANNESTE, K. 1996. Late Cenozoic evolution of the western Barents Sea-Svalbard continental margin. *Global and Planetary Change*, **12**, 53–74.

HAFLIDASON, H., KING, E. L. & SEJRUP, H. P. 1998. Late Weichselian and Holocene sediment fluxes of the northern North Sea Margin. *Marine Geology*, **152**, 189–215.

HEBBELN, D., DOKKEN, T., ANDERSEN, E. S., HALD, M. & ELVERHØI, A. 1994. Moisture supply for northern ice-sheet growth during the Last Glacial Maximum. *Nature*, **370**, 357–370.

HESSE, R. 1992. Continental slope sedimentation adjacent to an ice margin. I. Seismic facies of Labrador Slope. *Geo-Marine Letters*, **12**, 189–199.

HESSE, R., KHODABAKHSH, S., KLAUCKE, I. & RYAN, W. B. F. 1997. Asymmetrical turbid surface-plume deposition near ice-outlets of the Pleistocene Laurentide ice sheet in the Labrador Sea. *Geo-Marine Letters*, **17**, 179–187.

HESSE, R., KLAUCK, I., KHODABAKHSH, S. & PIPER, D. 1999. Continental slope sedimentation adjacent to an ice margin. III. The upper Labrador Slope. *Marine Geology*, **155**, 249–276.

KENYON, N. H. 1986. Evidence from bedforms for a strong polewards current along the upper continental slope of Northwest Europe. *Marine Geology*, **72**, 187–198.

KING, E. L., SEJRUP, H. P., HAFLIDASON, H., ELVERHØI, A. & AARSETH, I. 1996. Quaternary seismic stratigraphy of the North Sea Fan: glacially-fed gravity flow aprons, hemipelagic sediments, and large submarine slides. *Marine Geology*, **130**, 293–315.

KING, E. L., HAFLIDASON, H., SEJRUP, H. P. & LØVLIE, R. 1998. Glacigenic debris flows on the North Sea Trough Mouth fan during ice stream maxima. *Marine Geology*, **152**, 217–246.

KNUTZ, P. C., AUSTIN, W. E. N. & JONES, E. J. W. 2001. Millenial-scale depositional cycles related to British Ice Sheet variability and North Atlantic paleocirculation since 45 kyr B.P., Barra Fan, U.K. margin. *Paleoceanography*, **16**, 53–64.

KUVAAS, B. & KRISTOFFERSEN, Y. 1991. The Crary Fan: A trough-mouth fan on the Weddell Sea continental margin, Antarctica. *Marine Geology*, **97**, 345–362.

KUVAAS, B. & KRISTOFFERSEN, Y. 1996. Mass movements in glaciomarine sediments on the Barents Sea continental slope. *Global and Planetary Change*, **12**, 287–307.

LABERG, J. S. & VORREN, T. O. 1993. A Late Pleistocene submarine slide on the Bear Island Trough Mouth Fan. *Geo-Marine Letters*, **13**, 227–234.

LABERG, J. S. & VORREN, T. O. 1995. Late Weichselian submarine debris flow deposits on the Bear Island Trough Mouth Fan. *Marine Geology*, **127**, 45–72.

LABERG, J. S. & VORREN, T. O. 1996. The Middle and Late Pleistocene evolution of the Bear Island Trough Mouth Fan. *Global and Planetary Change*, **12**, 309–330.

LABERG, J. S. & VORREN, T. O. 2000. Flow behaviour of the submarine glacigenic debris flows on the Bear Island Trough Mouth Fan, western Barents Sea. *Sedimentology*, **47**, 1105–1117.

LABERG, J. S., VORREN, T. O. & KNUTSEN, S.-M. 1999. The Lofoten contourite drift off Norway. *Marine Geology*, **159**, 1–6.

MANGERUD, J., DOKKEV, T., HEBBELN, D., *ET AL.* 1998. Fluctuations of the Svalbard-Barents Sea Ice Sheet during the last 150,000 years. *Quaternary Science Reviews*, **17**, 11–42.

Ó COFAIGH, C. & DOWDESWELL, J. A. 2001. Laminated sediments in glacimarine environments: diagnostic criteria for their interpretation. *Quaternary Science Reviews*, **20**, 1411–1436.

Ó COFAIGH, C., TAYLOR, J, DOWDESWELL, J. A., KENYON, N. H., ROSELL-MELE, A. & MIENERT, J. 2002. Sediment reworking on high latitude continental margins and its implications for palaeoceanographic studies: insights from the Norwegian–Greenland Sea. In: DOWDESWELL, J. A. & Ó COFAIGH, C. (eds) Glacier-Influenced Sedimentation in High-Latitude Continental Margins. Geological Society, London, Special Publications, 203, 325–348.

POWELL, R. D. 1990. Glacimarine processes at grounding-line fans and their growth to ice-contact deltas. In: DOWDESWELL, J. A. & SCOURSE, J. D. (eds) Glacimarine Environments: Processes and Sediments. Geological Society, London, Special Publications, 53, 53–73.

SEJRUP, H. P., KING, E. L., AARSETH, I., HAFLIDASON, H. & ELVERHØI, A. 1996. Quaternary erosion and depositional processes: western Norwegian fjords, Norwegian Channel and North Sea Fan. In: DE BATIST, M. & JACOBS, P. (eds) Geology of Siliciclastic Shelf Seas. Geological Society, London, Special Publications, 117, 187–202.

SHIPP, S., ANDERSON, J. & DOMACK, E. 1999. Late Pleistocene–Holocene retreat of the West Antarctic Ice-Sheet system in the Ross Sea: Part 1 – Geophysical results. Bulletin of the Geological Society of America, 111, 1486–1516.

STOKER, M. S. 1995. The influence of glacigenic sedimentation on slope-apron development on the continental margin off Northwest Britain. In: SCRUTTON, R. A., STOKER, M. S., SHIMMIELD, G. B. & TUDHOPE, A. W. (eds) The Tectonics, Sedimentation, and Palaeoceanography of the North Atlantic Region. Geological Society, London, Special Publications, 90, 159–177.

STOKER, M. S., AKHURST, M. C., HOWE, J. A., & STOW, D. A. V. 1998. Sediment drifts and contourites on the continental margin off northwest Britain. Sedimentary Geology, 115, 33–51.

STOW, D. A. V. & PIPER, D. J. W. 1984. Deep-water fine-grained sediments: facies models. In: STOW, D. A. V. & PIPER, D. J. W. (eds) Fine-Grained Sediments: Deep-Water Processes and Facies. Geological Society, London, Special Publications, 15, 611–646.

STOW, D. A. V., FAUGERES, J. C., VIANA, A., GONTHIER, E. 1998. Fossil contourites: a critical review. Sedimentary Geology, 115, 3–31.

TAYLOR, J. 2000. Large-scale sedimentation and ice sheet dynamics in the Polar North Atlantic. PhD thesis, University of Bristol.

TAYLOR, J., DOWDESWELL, J. A., KENYON, N. H., WHITTINGTON, R. J. & VAN WEERING, TJ. C. E. 2000. Morphology and Late Quaternary sedimentation on the North Faeroes slope and abyssal plain, North Atlantic. Marine Geology, 168, 1–24.

TAYLOR, J., DOWDESWELL, J. A. & SIEGERT, M. J. 2002. Late Weichselian depositional processes, fluxes, and sediment volumes on the margins of the Norwegian Sea (62–75°N). Marine Geology, 188, 61–78.

VOGT, P. R., CRANE, K. C. & SUNDVOR, E. 1993. Mudflows on the Bear Island Submarine Fan. EOS, Transactions of the AGU, 74, 449, 452, 453.

VORREN, T. O. & LABERG, J. S. 1996. Late glacial air temperature, oceanographic and ice sheet interactions in the southern Barents Sea region. In: ANDREWS, J. T., AUSTIN, W. E. N., BERGSTEN, H. & JENNINGS, A. E. (eds) Late Quaternary Paleoceanography of the North Atlantic Margins. Geological Society, London, Special Publications, 111, 303–321.

VORREN, T. O. & LABERG, J. S. 1997. Trough mouth fans – Palaeoclimate and ice-sheet monitors. Quaternary Science Reviews, 16, 865–881.

VORREN, T. O., LEBESBYE, E., ANDREASSEN, K. & LARSEN, K.-B. 1989. Glacigenic sediments on a passive continental margin as exemplified by the Barents Sea. Marine Geology, 85, 251–272.

VORREN, T. O., LABERG, J. S., BLAUMME, F., DOWDESWELL, J. A., KENYON, N. H., MIENERT, J., RUMOHR, J. & WERNER, F. 1998. The Norwegian-Greenland Sea continental margins: morphology and Late Quaternary sedimentary processes and environment. Quaternary Science Reviews, 17, 273–302.

VORREN, T. O., RICHARDSEN, G., KNUTSEN, S.-M. & HENRIKSEN, E. 1991. Cenozoic erosion and sedimentation in the western Barents Sea. Marine and Petroleum Geology, 8, 317–340.

Submarine mass-wasting on glacially-influenced continental slopes: processes and dynamics

ANDERS ELVERHØI[1], FABIO V. DE BLASIO[1,2], FAISAL A. BUTT[1,6], DIETER ISSLER[3], CARL HARBITZ[2], LARS ENGVIK[4], ANDERS SOLHEIM[2], & JEFFREY MARR[5]

[1]*Department of Geology, University of Oslo, P.O. Box 1047, Blindern, N-0316 Oslo, Norway (e-mail: anders.elverhoi@geologi.uio.no)*
[2]*Norwegian Geotechnical Institute, P.O. Box 3930, Ullevål Stadion, N-0806 Oslo, Norway*
[3]*NaDesCoR, Promenade 153, CH-7260 Davos Dorf, Switzerland*
[4]*Sør-Trøndelag University College, N-7004 Trondheim, Norway*
[5]*St Anthony Falls Laboratory, University of Minnesota, Mississippi River at 3rd Avenue, Minneapolis, MN 55414, USA*
[6]*Present address: Dept. of Geography and Geology, University of the West Indies, Mona Campus, Kingston 7, Jamaica*

Abstract: Submarine slides and debris flows are common and effective mechanisms of sediment transfer from continental shelves to deeper parts of ocean basins. They are particularly common along glaciated margins that have experienced high sediment flux to the shelf break during and after glacial maxima. During one single event, typically lasting for a few hours or less, enormous sediment volumes can be transported over distances of hundreds of kilometres, even on very gentle slopes. In order to understand the physics of these mass flows, the process is divided into a release phase, followed by break-up, flow and final deposition. Little is presently known regarding release and break-up, although some plausible explanations can be inferred from basic mechanics of granular materials. Once initiated, the flow of clay-rich or muddy sediments may be assumed to behave as a (non-Newtonian) Herschel-Bulkley fluid. Fluid dynamic concepts can then be applied to describe the flow provided the rheological properties of the material are known. Numerical modelling supports our assertion that the long runout distances observed for large volumes of sediments moving down gentle slopes can be explained by partial hydroplaning of the flowing mass. Hydroplaning might also explain the sharp decrease of the friction coefficient for submarine mass flows as a function of the released volume. The paper emphasizes the need for a better understanding of the physics of mass wasting in the submarine environment.

Introduction

Sediment mass-wasting in the form of slides, debris flows and turbidity currents on continental slopes represents an important process for transporting large volumes of sediments in the submarine environment. We use the term 'mass-wasting' in a broad sense to describe failure of a mass at the shelf break or on the upper slope, which subsequently starts to move downslope under the influence of gravity. Along the flow path, the mass may disintegrate and, depending on its rheology and external conditions, liquefy into a debris flow and also produce turbidity currents. Numerous subsea images of mass flows reveal a gradual transition from a typically blocky composition in the upper part to a more remoulded form in the middle and lower part of the mass flow (e.g. Booth *et al.* 1993).

Typically, large sediment volumes are involved in submarine mass-wasting, even on gentle slopes. A well-known example is the Storegga Slide, involving about 6000 km[3] of sediments, that took place on the Norwegian margin on a slope of about 1° or even less (Bugge *et al.* 1988). Of particular interest in the case of glacial margins are the often well defined mud-rich debris flows (up to 70% clay and silt) commonly associated with outlets of former large ice streams along the Eastern Canadian, East Greenland, Norwegian and Svalbard–Barents Sea margins (Aksu & Hiscott 1989, 1992; King *et al.* 1998; Dowdeswell *et al.* 1997; Solheim *et al.* 1998; Vorren *et al.* 1998).

From: DOWDESWELL, J. A. & Ó COFAIGH, C. (eds) 2002. *Glacier-Influenced Sedimentation on High-Latitude Continental Margins.* Geological Society, London, Special Publications, **203**, 73–87. 0305-8719/02/$15.00
© The Geological Society of London 2002.

The individual debris flows along the Norwegian and the Svalbard–Barents Sea margins involve volumes ranging from less than 1 to about 50 km³ of sediments and cover areas of up to 2000 km² (Vorren *et al.* 1998).

A characteristic of submarine mass wasting is the long runout distance that, at times, can reach hundreds of kilometres. Information combined from glacially-influenced as well as non-glacial margins shows that the long runout distances (more than 100 km) generally occur on slopes of less than 2° (Booth *et al.* 1993; Vorren *et al.* 1998) and that the largest volumes of displaced sediments are associated with long runout distances (Fig. 1). Based on information from cable breaks, flow velocities in the range from 20 to 100 km hr⁻¹ are calculated, even on slopes less than 1° (Heezen & Ewing 1952; Bjerrum 1971). The same data also show a tendency for the flow velocities to increase with the mobilized mass. Although the measured velocities may be those of the turbidity current in most cases, laboratory experiments (Mohrig *et al.* 1999) suggest that the dense debris flow may be even faster than the accompanying turbidity current, at least in the steeper parts of the flow path. Comparison with subaerial gravity mass flows (Simpson 1987) is difficult, given the steeper flow paths and smaller release masses typically encountered on land. Comparing the runout lengths of subaqueous

and subaerial debris-flow events of similar release volume and drop height (Table 1), it appears that subaqueous debris flows are potentially more mobile than their subaerial counterparts, despite the reduced gravitational force and higher viscous drag in water. Understanding the dynamics of submarine mass flows thus represents a considerable challenge.

In this paper, we examine recent progress in modelling the muddy flows typical of high-latitude margins. We focus on the control exerted by the rheological properties of the material on processes that facilitate the long runout distances observed in debris flows on high-latitude margins, which may be applicable to submarine mass flows in general.

High-latitude debris flows: release mechanism and initial break up

Sediment release

Earthquakes are believed to be the most common cause of submarine landslides. Some directly documented landslides have taken place after major earthquakes. Examples include the Newfoundland slope failure following the 1929 Grand Banks earthquake (Heezen & Ewing 1952; Piper *et al.* 1999) and the catastrophic slides in Seward and Valdez, Alaska, associated with the major 1964 earthquake

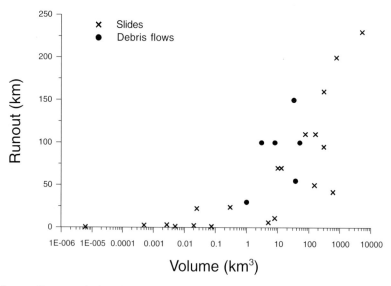

Fig. 1. Runout distances of submarine mass movements plotted against mobilized sediment volumes. Note the significant increase in run out as a function of volume when dealing with mass flows larger than 1 km³. See Table 1 for data.

Table 1. *Characteristics of selected submarine slides including some data on subaerial slides*

Location	Slope angle (degrees)	Volume (km^3)	Runout L (km)	Height H (m)	H/L ratio	Reference
Submarine slides						
Grand Banks	3.5	7.60x10^1	110	365	0.00332	Prior & Coleman 1982
Hawaii	6.0		160	2000	0.0125	Prior & Coleman 1982
Kidnappers	2.5	0.08x10^{-2}	11	50	0.00455	Prior & Coleman 1982
Bay of Biscay			21	250	0.0119	Prior & Coleman 1982
Rockall	2.0	3.00x10^2	160	330	0.00206	Prior & Coleman 1982
Bassein	6.0		37	360	0.00973	Prior & Coleman 1982
Agulhas			106	375	0.00354	Prior & Coleman 1982
Copper River Delta	1.0		18	115	0.00639	Prior & Coleman 1982
Albatross Bank	7.0		5.3	300	0.0566	Prior & Coleman 1982
Portlock Bank	4.0		6.5	200	0.03077	Prior & Coleman 1982
Kayak Trough	1.0		15	115	0.00767	Prior & Coleman 1982
Magdalena	2.0	0.30x10^{-2}	24	1400	0.05833	Edgers & Karlsrud 1982
Valdez	6.0	7.50x10^{-2}	1.3	168	0.13125	Edgers & Karlsrud 1982
Mississippi River Delta	0.5	4.00x10^{-2}		20		Edgers & Karlsrud 1982
Suva	3.0	1.50x10^{-1}		100		Edgers & Karlsrud 1982
Sagami Wan	11.0	0.70x10^2				Edgers & Karlsrud 1982
Scripps Canyon	7.0	5.00x10^{-5}		6		Edgers & Karlsrud 1982
Orkadalsfjord		2.50x10^{-2}	22.5	500	0.02222	Edgers & Karlsrud 1982
Sandnesjoen		5.00x10^{-3}	1.2	180	0.15	Edgers & Karlsrud 1982
Sokkelvik		5.00x10^{-4}	2.5	120	0.048	Edgers & Karlsrud 1982
Helsinki		6.00x10^{-6}	0.4	11	0.0275	Edgers & Karlsrud 1982
Storegga 1		5.50x10^3	400	1700	0.00425	Bugge *et al.* 1988
Storegga 2		5.50x10^3	850	1700	0.002	Bugge *et al.* 1988
Trænadjupet	1.25	7.60x10^2	200	250	0.00125	Laberg & Vorren 2000
Cape Fear	4.2		30	700	0.02333	Poponoe *et al.* 1993
Blake Escarpment	8.6	6.00x10^2	42	3600	0.08571	Dillon *et al.* 1993
East Break East	1.5	0.13x10^2	70	1150	0.01643	McGregor *et al.* 1993
East Break West	1.5	1.60x10^2	110	1100	0.01	McGregor *et al.* 1993
Navarin Canyon	3.0	0.05x10^2	6	175	0.02917	Carlson *et al.* 1993
Seward		2.70x10^{-3}	3	200	0.06667	Hampton *et al.* 1993
Alsek	1.3		2	20	0.01	Schwab & Lee 1993
Sur	0.5	0.10x10^2	70	750	0.01071	Gutmacher & Normark 1993
Santa Barbara	4.8	2.00x10^{-2}	2.3	120	0.05217	Edwards *et al.* 1993
Alika 2		3.00x10^2	95	4800	0.05053	Normark *et al.* 1993
Nuuanu		5.00x10^3	230	5000	0.02174	Normark *et al.* 1993
Tristan de Cunha		1.50x10^2	50	3750	0.075	Holocomb & Searle 1991
Debris flows						
Isfjorden	3.5	0.01x10^{-2}	30	1830	0.061	Fossen 1996
Storfjorden	1.5	0.08x10^{-2}	100	2000	0.02	Laberg & Vorren 1995
Bear Island	0.6	0.32x10^2	150	1600	0.01067	Vorren *et al.* 1998
North Sea	0.7	0.50x10^2	100	750	0.0075	King *et al.* 1996
Newfoundland	0.75	0.03x10^{-2}	100	1700	0.017	Aksu & Hiscott 1992
Baffin Bay	1.8	0.037x10^{-2}	55	1000	0.01818	Aksu & Hiscott 1989
Subaerial slides						
Mount Rainier		1.00x10^0	120	4800	0.04	Vallance & Scott 1997
Nevados Huascaran		1.00x10^{-1}	120	6000	0.05	Plafker & Ericksen 1978
Nevado del Ruiz		1.00x10^{-2}	103	5190	0.05039	Pierson *et al.* 1990
Mount St. Helens		1.00X10^{-2}	44	2350	0.05341	Fairchild & Wigmosta 1983
Mount St. Helens 2		1.00X10^{-2}	31	2150	0.06935	Pierson 1985
Wrightwood		1.00X10^{-3}	24	1524	0.0635	Sharp & Nobles 1953
Mount Thomas		1.00x10^{-4}	3.5	600	0.17143	Pierson 1980
Wrightwood 2		1.00x10^{-4}	2.7	680	0.25185	Morton & Campbell 1974
Santa Cruz		1.00x10^{-4}	0.6	200	0.33333	Wieckzorek *et al.* 1988

(Coulter & Migliaccio 1966; Lemke 1967). Cyclic loading by storm-wave action and expansion of gas hydrates are the other commonly cited causes (Kvenvolden 1994; Kvalstad *et al.* 2001). In all these cases, failure of sediments is caused by instability induced by an external mechanism or event and a failure plane can normally be recognized.

However, in the case of glaciated continental margins, an external trigger may not be required. High sedimentation rates have been recognized as a potential mechanism for failure (e.g. Vorren *et al.* 1998; Dimakis *et al.* 2000). On glacial margins, where high sedimentation rates are the norm rather than the exception, the depositional environment itself could be an inherently unstable system: the sediments fail when the rate of sedimentation exceeds some threshold value, depending on factors such as the rheology of the sediments, the mechanism of sediment transport, and the bed slope. This may be true particularly in cases where fast-flowing ice streams deliver sediments with a high clay content to the upper slope in the form of deformation till, as opposed to delta-fed deep-sea fans where the sediments are first suspended in water upon delivery before settling out. Sediment supply in the form of deformation till, as is interpreted to be the case for northern North Atlantic margins, favours under-consolidation and rapid build-up of excess pore pressure, eventually leading to failure.

For the deposits off Bear Island and Storfjorden, average sedimentation rates during glacial periods are estimated to be in the range of tens of centimetres per year (Laberg & Vorren 1995). For the Bear Island Fan, Dimakis *et al.* (2000) showed that the build-up of excess pore pressure due to rapid sedimentation rates under glacial advance could lead to failures on the upper continental slope. By using an infinite slope stability analysis together with excess pore pressure, Dimakis *et al.* (2000) argued that the observed uniformly thick, layered slope deposits seen on the Bear Island Fan can be explained through a regenerative process. In this process, glacially-derived sediments accumulate over a source area until they become unstable through build-up of excess pore pressure. A minimum sedimentation rate of 20 cm a^{-1} is found to be required for excess pore pressure to build up in this case. Failure takes place along a plane within the accumulated sediment mass such that the sediments above the plane move downslope without generating any slide scars, while the sediments below it remain and become the new surface for sediment deposition (Dimakis *et al.* 2000). The cycle is repeated whenever the sediment mass becomes thick enough to exceed the failure criteria. Failures are calculated to occur after 95–170 years of high-rate sedimentation and to remove 25–30% of the sediments deposited during the period. The analysis also shows that an increase in slope will cause more frequent but smaller failures. This may also explain why larger volumes of sediments are mobilized on gentler slopes.

Although the model of Dimakis *et al.* (2000) was presented for a glaciated margin during periods of glacial expansion, it may also be valid in typical delta environments where the combination of sedimentation rate and sediment composition (high clay content) leads to periodical build-up of sufficient excess pore pressure so as to cause instability and failure.

Initial break-up

The debris flows recorded on high-latitude margins do not show the presence of any slide scars but rather a well-defined flow mass all the way along the slope (e.g. Vorren *et al.* 1998). The flow often forms characteristic elongated bodies running all the way from the upper to the lower slope. This suggests liquefaction immediately following failure. Thus, although soil-mechanical parameters define the state of sediments on the upper slope, the behaviour of the moving mass in the mid- and lower-slope regions is better explained by principles of fluid flow regardless of initial processes (Fig. 2). Failure and the initial break-up are only discussed briefly here. It is the mid- and lower-slope processes that we attempt to describe in detail.

After the release of a debris flow, the flow dynamics will depend on evolving properties of the flowing sediment such as granulometry, yield strength, viscosity, and pore fluid pressure. Little is known about the progression of the break-up and the early phase of flow. Theoretical considerations suggest that a wide variety of scenarios may occur in different slide events depending on grain size, degree of consolidation and strength/intensity of the initiating mechanism. For example, in strongly underconsolidated, homogeneous soils subject to a powerful earthquake, liquefaction may occur almost instantaneously throughout most of the released mass because the jolt from the earthquake propagates through the soil and is strong enough to break its weak texture. If the soil has a high content of fines, a debris flow with only relatively small clasts is expected to result, whose properties approach those of a mudflow. Conversely, significant small-scale variations of

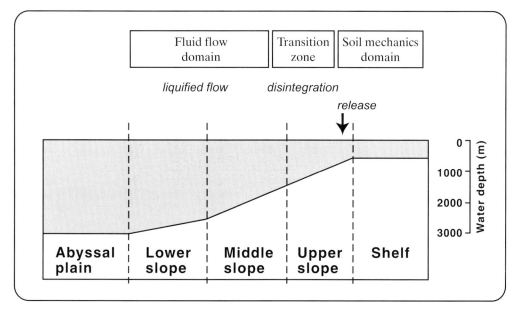

Fig. 2. The various stages of sediment mass behaviour along the flow avenue, starting as a block (soil mechanics domain) and gradually being transformed in a liquefied flow characterized by fluid dynamics.

soil properties will lead to a form of debris flow with a large fraction of clasts up to boulder size.

If distinct layers exist in slightly underconsolidated or normally-consolidated sediment, a situation may occur where the seismic waves are too weak to break the inter-particle bonds in the bulk of the soil, but shear fractures are generated along the weakest layers. Where the weak layer collapses, it can no longer support the shear stresses generated by the overlying slab. At the circumference of the collapsed area, the shear stresses are significantly increased and drive fracture propagation along the weak layer. The slab is released when the downslope gravitational force on the slab above the collapsed part of the weak layer exceeds the tensile, shear and compressive forces that the surrounding area can exert on the slab to hold it in place. The gravitational force grows with the collapsed area, but the resistive forces increase only with its circumference. Note also that the maximum size of the slab grows with its strength and diminishes with increasing slope angle. This scenario indicates that huge slabs of intact soil might be released instantaneously under certain conditions; it is closely analogous to the release mechanism of snow-slab avalanches (e.g. McClung & Schaerer 1993, for an introduction). Two notable differences should be pointed out, however. In subaqueous landslides, the

triggering earthquake is effective in a large area and excess pore pressure in the weak layer plays a crucial role, whereas it is of little importance in snow avalanche release.

The work required for breaking up a large slab ultimately comes from gravitational energy released when the slab moves down the incline. The central question is whether this work is a large fraction of the total initial potential energy. In the case of brittle tensile fracture, the work required for breaking a slab of volume V and tensile strength σ_t into two halves is $W_h = V\sigma_t^2 / (2E)$, assuming the elastic energy to be completely dissipated after fracture. Here, E is Young's modulus, and we consider a block of length l, width w and height h. As the slab is stretched to failure along l, at stress σ_t, its length increases by $\Delta l = l \times \sigma_t / E$. The mechanical work is $\sigma_t \times \Delta l \times w \times h / 2 = V\sigma_t^2 / (2E)$, as indicated above. The same result is obtained for fracture in the two other directions.

Dividing a one-dimensional object of length l by cutting n times in the middle of each newly-formed piece, one obtains a total of 2^n pieces of length d, namely, $n = \log_2(l / d)$. Extending the reasoning to three dimensions, if the break-up mechanisms were to break each block repeatedly into two parts of equal size, a total of $n = \log_2(l / d) + \log_2(w / d) + \log_2(h / d) = \log_2(V / V_p)$ fracture sequences involving all

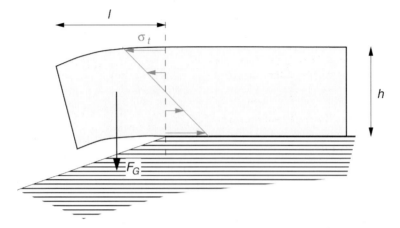

Fig. 3. Simplified model for the effect of sea-floor undulations found, for example, below the Storegga Slide headwall. Tensile and compressional stresses in a large flowing slab encountering an abrupt slope change.

particles and total work $W_{\text{total}} \approx W_h \times \log_2 (V / V_p)$ would be required to reduce a slab of volume V to particles of volume $V_p = d^3$. Alternatively, if mostly single particles were chipped off the slab by collisions, the total work would be $W_{\text{total}} \approx W_1 \times V / V_p$, with W_1 the average binding energy of a surface particle. Lacking a detailed model for calculating W_1 at present, we expect it to be proportional to the volume affected by the collision-induced deformation, which should be somewhat larger than V_p, and the square of the (shear) strength $\sigma_s > \sigma_t$. The total break-up work is then similar in order of magnitude to the first case. For Young's modulus $E \geq 10^7$ Pa and tensile strength $\sigma_t \approx 10^5$ Pa, typical of submarine sediments, a drop of only 3 m releases gravitational energy equal or larger than the work needed to break a slab of 1 km^3 into particles of 1 mm^3. Despite the large uncertainties in such estimates, we tentatively conclude that the break-up rate is not so much limited by the energy supply as by the efficiency of the various processes that contribute to break-up.

One such break-up process may be due to ocean-bed undulations which are found, for example, in bathymetric data from slide areas such as Storegga (P. Bryn pers. comm.), with amplitudes of several tens of metres at length scales of 0.1 to 1 km. The curvature radius is then not much larger than the slab height. Large tensile and compressive stresses are induced in a sediment slab sliding over such terrain. For a crude estimate, we equate the gravitational torque from the part of a slab protruding over the crest of an undulation to the torque created

by the tensile and compressive stresses in a cross-section of the slab as in elementary beam theory (Fig. 3). Assuming a tensile strength of about 100 kPa and a slab height of 100 m, we find that pieces about 10–30 m long should break off at the front of the slab. Where such bed undulations are present, this appears to be a rather effective first-stage break-up mechanism. It is accompanied by strong agitation of the fragments, facilitating remoulding and incorporation of water.

At a later stage, collisions between blocks may be the driving break-up mechanism. The processes involved are quite complex and presumably depend strongly on a number of parameters, among them the soil properties, the fines content that strongly influences the viscosity of the pore fluid, and the velocity, size and shape distributions of the fragments. Of importance here, and as yet poorly constrained, is the role of interstitial fluid (Iverson 1997). Its presence on the one hand dampens velocity fluctuations and on the other hand increases the mobility of the flowing material and the velocity gradients which, in turn, drive the generation of disordered motion of particles (granular temperature). More detailed investigation of this subtle interplay between particles and interstitial fluid will be important not only for describing the break-up process, but also for modelling velocities, runout distances and deposits.

To summarize, depending on the soil properties and the external triggering mechanism, the initial slab may liquefy quasi-instantaneously or undergo a more gradual transition from large blocks to a more fine-grained composition. The duration of the transition may vary from a small fraction to all of the flow duration. Present knowledge of the possible scenarios, the conditions of their occurrence, and their dynamics is still very rudimentary.

High-latitude debris flows: flow dynamics and flow simulations

Flow models

Presently, two classes of model are widely used to simulate subaqueous mass flows, namely visco-plastic models (e.g. Johnson 1970; Huang & Garcia 1998, 1999) and granular models (Savage & Hutter 1989) incorporating the effects of viscous pore fluid (Iverson 1997) as well as Bagnold's dispersive pressure caused by grain–grain and grain–fluid interaction (Norem et al. 1987). Both types of model have been discussed widely (Norem et al. 1990; Hampton et al. 1996; Locat & Lee 2001; Iverson 1997; Iverson & Vallance 2001), the latter mostly with a view towards terrestrial debris flows characterized by a very low content of fines (less than 5% silt and clay). The visco-plastic models, on the other hand, have been developed for clay-rich or muddy materials with cohesion and a very low content of coarse particles capable of particle–particle interactions, a situation outside the range of applicability of a true granular flow model.

Grain size distribution in the debris flow deposits along the North Atlantic margins is influenced strongly by the input of fine-grained glacial erosion products. Typical particle distribution for the debris flows and other mass displacements is 30–40% clay, 30–40% silt and 20–30% sand (Vorren et al. 1998). Gravel is almost missing. Due to this almost complete lack of particles large enough for particle–particle interaction and the high clay content, we are inclined to favour the visco-plastic model for explaining debris-flow dynamics along these margins. Recent modelling of muddy debris flows using the visco-plastic approach shows a high degree of correspondence between observed and modelled runout distances for small debris flows (Huang & Garcia 1999). The sediment composition in modelling by Mohrig et al. (1998, 1999) was selected to reflect field data, supporting our choice for muddy, almost clast-free debris flows.

In its simplest form, visco-plastic behaviour can be described by the so-called Bingham rheology where the stress τ and strain $\partial U/\partial y$ (where U is the velocity component parallel to the bed and y is the co-ordinate perpendicular to it) are linearly related as follows:

$$|\tau| = \tau_y + \mu_B \left| \frac{\partial U}{\partial y} \right|^n \text{ for } |\tau| \geq \tau_y, n = 1$$

where τ_y is the yield stress and μ_B the dynamical Bingham viscosity. The stress–strain relation for a Bingham fluid in laminar flow implies that no deformation takes place until a specified yield stress is applied to the material, after which the deformation is driven by the excess stress beyond this yield stress. The visco-plastic rheological relation dictates the division of the flow into a plug layer on top of a shear layer.

More general relations such as a Herschel-Bulkley rheology, where the stress depends non-linearly on the strain rate (exponent $n > 0$ instead of $n = 1$ in the equation above), and a bi-viscous rheology (which reduces to a Newtonian flow for small shear rates and a Bingham fluid for high shear rates: Locat 1997) have also been proposed. These constitutive relations provide a more general rheological behaviour. For $0 < n < 1$, the Herschel-Bulkley rheology describes shear thinning, namely the observed tendency of yielded mud to become less viscous with increasing shear rate. Both the Herschel-Bulkley and the bi-viscous rheologies have been implemented in particular in the BING code (Imran et al. 2001). However, experiments on materials with high clay content (about 40% clay, 40% silt and 20% sand) reveal an exponent n very close to unity (Huang & Garcia 1999). Since this is the kind of material we consider in our study, we make use of a simple $n = 1$ Bingham fluid. Considering the uncertainties in the rheology of submarine mass flows, many of which have variable compositions, a Bingham rheology is suitable as a first-order approximation of the flow. In addition, it will be shown that hydroplaning (see below) is not strongly dependent on the rheology of the sediment, provided the characteristic critical velocity is attainable.

A related problem concerns the variation of pore water pressure during the flow. In what follows we do not explicitly consider excess pore pressure, but more appropriate models, especially for sandy debris flow, should incorporate the generation and diffusion of excess pore pressure as a dynamic process that influences the rheology of the material. We assume that the very low permeability of clay-rich sediment hinders efficient diffusion of excess pressure through the moving mass, at least during the relevant time scale of a few hours. The chosen yield stress and Bingham viscosity are to be interpreted as mean values over the flow episode.

Flow models: hydroplaning

Recent studies have shown that the classical visco-plastic concept, too, falls short of fully simulating the long runout distances for subaqueous debris flows, at laboratory scale as well as for the debris flows observed on the Svalbard–Barents Sea margin (Huang & Garcia 1999; Marr *et al.* 2002). In recent years, hydroplaning has been suggested as a possible mechanism for debris flows covering long distances on low-angle slopes (Mohrig *et al.* 1998, 1999; Harbitz *et al.* 2001). Hydroplaning is shown to occur in a cohesive mass once that mass exceeds a critical velocity such that the flow cannot displace the ambient fluid fast enough. As a result, the head of the flow is lifted and a water layer intrudes beneath the moving mass. The intruding water layer acts as a lubricant, reducing basal friction and increasing head velocity. Although some mud may be mixed into the thin water layer, increasing its viscosity, the resulting slurry will still have a significantly lower viscosity than the mud.

The concept of hydroplaning is of particular interest since it produces large runout distances even at relatively high yield stresses. As the water film greatly reduces the basal shear stress, rapid flow without or with negligible erosion of the underlying strata can be explained quite naturally. Although hydroplaning has been observed directly only in the laboratory so far, the same effect must occur in nature as well. At the heart of the phenomenon is the balance between the pressure in the basal water layer, characterized by the stagnation pressure $\rho_w U^2 / 2$, and the overburden load of the mud layer, $(\rho_d - \rho_w) g H$, where ρ_d and ρ_w are the mud and water densities, respectively. The dimensionless ratio of these pressures is, up to a factor of 2, the square of the densimetric Froude number. In going from the laboratory to the field scale one can use distorting modelling relations between the physical and geometrical quantities. The appropriate scaling factors can be obtained, in principle, by keeping the same value for the relevant dimensionless numbers, like the Froude and the Reynolds numbers. It has been shown that present experiments are well scaled except for the dimensions of the microstructure: that is, the grain size (G. Parker pers. comm). Focusing on the rheological properties of the sediment (viscosity and yield strength) rather than the grain size circumvents

the problem of scaling the microstructure. Through scaling down the values of viscosity and yield strength and scaling up the slope of the experimental facility, it is possible to model field scale debris flows in the laboratory. Thus, although the scaling problem is still under investigation, we believe that a direct application of the experimental results to natural debris flows is appropriate.

Other explanations for high mobility of compacted sediments have been put forward recently by Gee *et al.* (1999). Making specific reference to the Saharian debris flow, which is composed of two different layers, a volcaniclastic layer overlain by pelagic mud, these authors proposed that long runouts are attained due to an increase of pore water in the volcaniclastic sediments by overloading from the muddy layer. Although the model of Gee *et al.* (1999) might be applied to this specific case, in view of the relative commonness of long run-out distances of subaqueous debris flows, we believe that a more general process is likely to be at work.

For a closer look at the effects of hydroplaning, the BING program, describing the movement of non-hydroplaning flow (Imran *et al.* 2001), was modified to include a water layer at the bottom of the moving sediment mass (Water-BING). Hydroplaning starts when the velocity at the front of the flow is sufficiently high for the water to lift the moving mass. Subsequent evolution of the water layer is calculated numerically from the vertically-integrated Navier-Stokes and continuity equations for the water layer and appropriate boundary conditions. Hydroplaning is invoked once the water thickness becomes larger than a minimal thickness representing the size of the bed and sediment irregularities.

When the mud or debris is hydroplaning, the stress at the boundary with the basal water layer is approximately $\mu_w U / D_w$, where U is the velocity of the hydroplaning slab, D_w is the thickness of the water layer and μ_w is the water viscosity. Using values of the order of $U \approx 10$ m s^{-1}, $\mu_w \approx 0.005$–0.01 kg m^{-1} s^{-1} and $D_w \approx 1$ cm, one finds a shear stress between 5 and 10 Pa, a value which is much smaller than the yield stress of the sediment, $\tau_y \approx 8$–15 kPa. The hydroplaning mass thus moves rigidly on top of the water layer, essentially without shearing.

To understand the forces at work as the sediment is moving, we can imagine dividing the flowing material into vertical elements, the positions of which change in time. If the material is hydroplaning, the bed-parallel forces acting on a slice of sediment are: 1) the component of the gravity force parallel to the bed; 2) the earth pressure gradient; a force arising from variations in the height of the material and directed from thick toward thin sediment; 3) the internal resistance force of the material, determined by both the yield stress and viscosity; 4) the drag force due to the inter-action of the moving mass with embedding water. The equation of motion for a hydro-planing gravity flow can be written simply from Newton's law as

$$\frac{dU}{dt} = -\frac{\rho_d - \rho_w}{\rho_d} \frac{\partial}{\partial x}(D + D_w)g_y$$
$$-\frac{\tau_w}{D\rho_d} + \frac{\rho_d - \rho_w}{\rho_d}g_x - f_{drag}$$

where the co-ordinate x denotes the distance along the flow, U is the velocity of an element of flowing sediment, D is the local sediment thick-ness, ρ_d and ρ_w are the sediment and water densities, respectively, g_x and g_y are the components of the gravity acceleration parallel and perpendicular to the bed, respectively, f_{drag} is the specific drag force due to the resistance of the water on top of the flowing mass and τ_w is the frictional stress between the sediment and the water below. The specific drag force can be written as

$$f_{drag} = \frac{1}{2}U^2 \frac{\rho_w}{\rho_d D}\left[C_F + \left|\frac{\partial D}{\partial x}\right|C_P\right]$$

where C_F and C_P are the frictional and pressure drag coefficients, respectively. The value of these constants can be estimated from flow experiments against bodies of given shape at fixed Reynolds number. At Reynolds numbers of the order 10^7–10^9, we estimate from standard tables (Newman 1977) values of about $C_F = 0.003$ and $C_P = 0.01$ or smaller. It is possible that higher pressure drag coefficients should be used for well-developed sediment heads. However, due to the pronounced flatness of the sediment, an arbitrary increase in the value of the pressure drag coefficient C_P does not produce significant changes. The roughness of the solid material at the interface with the liquid might change the effective value of drag coefficients (Schlichting 1968). The space derivative results from the calculation of the total sediment surface directed perpendicular to the ambient water velocity. One should keep in mind that our approximation for the drag force is rather crude. We are not aware of any complete model for the drag force in this problem. Furthermore, the interaction of the sediment with ambient water

generates a region of the size of the boundary layer where water is in a turbulent regime. The strong mixing with mud, which might later produce a turbidity current, changes locally and unpredictably the properties of both the water and the flowing sediment. In addition, a shear region exists in the sediment due to the shearing effect induced by the drag whereas, in calculat-ing the equations of motion, we considered the sediment as a rigid body.

In the case of non-hydroplaning, the rigid 'plug' layer is coupled with a shear layer below, the flow being determined by the rheological properties of the material. As a result, the resist-ance is in general substantially increased. The complete equations for the shear flow can be found in Huang & Garcia (1998) and Imran et al. (2001). When hydroplaning takes place, the frictional stress between the sediment and the bottom is sensitive to the properties of the water layer rather than to the rheology of the sediment. In order to calculate the stress between the mud and the water, one needs to solve the mass and momentum equations in the water layer as a function of time.

Modelling results

In the following, we briefly present results from numerical simulations of a mass flow obtained with the BING code (without hydroplaning) and the modified version Water-BING that includes hydroplaning. The initial configuration consists of a 20 km wide and 300 m thick deposit lying on a less than 1° slope, from the position $x = -20$ km to $x = 0$. Starting from rest, the sediment accelerates and within ten minutes reaches velocities of 36 m s^{-1} and 60 m s^{-1} without and with hydroplaning, respectively. Note that a substantial contribution to the acceleration at the beginning of the flow comes from the earth pressure gradient. The material properties used in the simulation are taken from Elverhøi et al. (1997) and Marr et al. (2002): density $\rho_d = 1800$ kg m^{-3}, viscosity $\mu_B = 300$ Pa s and yield stress $\tau_y = 15$ kPa.

To estimate the yield stress of a debris flow at rest one can use the concept that a viscoplastic fluid on a slope φ will stay at rest if the shear stress, $\Delta\rho\, g\, D \sin \varphi$, where $\Delta\rho$ is the density difference between the sediment and water and D is the thickness, is equal or smaller than the yield stress, τ_y (Johnson 1970). This leads to a direct relation between sediment thickness and yield stress, $D = \tau_y /(\Delta\rho\, g \sin \varphi)$. Without hydroplaning, the mass covers about 25 km before coming to rest, whereas the hydro-planing mass continues to flow (Fig. 4a). The

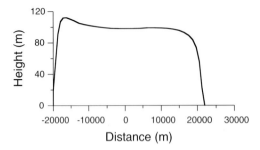

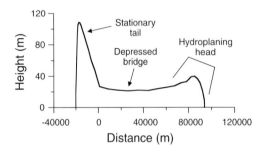

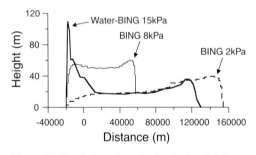

Fig. 4. (**a**) Simulation of non-hydroplaning debris flow using BING. Final geometry of the deposit after 25 minutes of run-time is shown. Sediment properties: density, $\rho_d = 1800$ kg m^{-3}; viscosity, $\mu = 300$ Pa s; and yield stress, $\tau_y = 15$ kPa. (Data from Elverhøi *et al.* 1997; Marr *et al.* 2002). (**b**) Simulation of hydroplaning debris flows using Water-BING. Sediment properties as in (a). Run-time 32 minutes. (**c**) Comparison of final deposits after the completion of hydroplaning and non-hydroplaning simulation runs. For the non-hydroplaning case, the yield strength has to be reduced by one order of magnitude to achieve the same runout distance as the hydroplaning case.

non-hydroplaning mass stops essentially because the sediment thickness has become too small to maintain a shear layer. On the other hand, the hydroplaning sediments continue flowing as, in this case, the flow is determined by the physical conditions of the water layer rather than the rheological properties of the sediments. The hydroplaning flow comes to rest at 130 km (Fig. 4c), which is closer to the runout length observed for the debris flows on the western Barents Sea continental margin (Elverhøi *et al.* 1997).

The numerical simulations also reveal that sediments undergo non-uniform acceleration during the flow. Water intrudes from the head of the mass and then migrates under the main body of the sediment. The front of the flow begins to hydroplane earlier and moves forward; the main body accelerates moderately while the tail remains stationary. The head of the deposit tends to be detached from the rest of the body. This concurs with the results of small-scale laboratory experiments (Mohrig *et al.* 1999).

The numerical simulations also demonstrate the effects of variations in yield strength and viscosity on the flow (Fig. 4c). In the absence of hydroplaning, yield stress has to be reduced by one order of magnitude (to 2 kPa) in order to reproduce the distances reached by hydroplaning sediments. Such low values are unrealistic for such clay-rich sediments and would not allow deposits of the thickness commonly observed along the Norwegian continental slope (Marr *et al.* 2002).

Runout distances with and without hydroplaning are shown as functions of yield strength (Fig. 5). The reason for the dependence of hydroplaning flow on the yield strength, as seen from the figure, is that the critical velocity is reached early with small yield stresses, causing hydroplaning to start early. As shown by experimental studies (Mohrig *et al.* 1999) hydroplaning sediment can be treated as a rigid block flowing in a low-viscosity medium exerting drag forces. Once initiated, hydroplaning flow is controlled by the behaviour of the ambient water and is independent of the sediment rheology. Thus, simulations confirm that hydroplaning flows cover much longer runout distances than non-hydroplaning flows, as has been observed earlier in laboratory experiments (Mohrig *et al.* 1999).

The value of viscosity is more uncertain, because it is not constrained strongly by the height of the final deposit. However, in the presence of high yield stress, the flow is insensitive to viscosity. In the simulated runs, a decrease of the viscosity by one order of magnitude increases the total runout distance by only 8%. Even decreasing the viscosity unrealistically by four orders of magnitude does not change this conclusion. The results concur with Locat & Lee (2001), who concluded that the contribution of yield strength to flow resistance is three orders of magnitude larger than

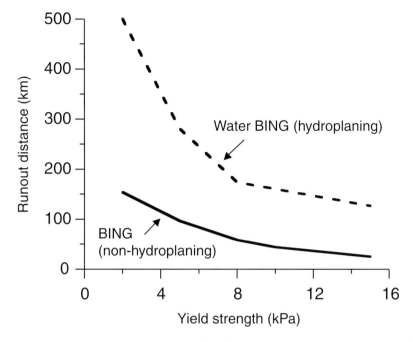

Fig. 5. Plot showing the run-out distances with and without hydroplaning as a function of sediment yield stress.

viscosity. Therefore, a small value of the viscosity does not seem to be the cause of the large runout distance of debris flows on low-angle slopes.

The results of our simulations depend on a set of parameters and initial conditions, some of which are not well-known in nature. When the water penetrates below the mass flow, it mixes with the sediments to form a slurry with different rheological properties. The Water-BING model accounts for this by attributing a viscosity value to the water that is ten to twenty times larger than the viscosity of pure liquid (Mohrig *et al.* 1998). In addition, water residing under the sediment for a long time will become completely mixed with it, a process made more effective by the turbulence in the water layer. This effect, which is not incorporated in our model, may eventually bring hydroplaning to an end. Water penetration through the debris is possible, but very limited if the material is mostly composed of clay, with permeability probably as low as 10^{-6} cm s^{-1}. This quantity might be substantially higher in debris flows with more sandy composition. This is one of the reasons why sandy debris flows do not hydroplane in the laboratory and are dominated by the effect of pore water pressure diffusion through the bulk sediment.

Our model presents some uncertainties associated with the initial water profile, the front and surface drag force and especially the initial size and shape of the sediment slab. The larger the initial height, the higher are the velocities and the runout distances. The dynamical problem is thus connected closely to the still problematic question of the triggering mechanism discussed earlier.

Mobility and runout of submarine mass wasting: discussion and concluding remarks

One of the main challenges in studies of submarine mass-wasting is to explain the high mobility of debris flows. The fundamental process with regard to the mobility of submarine sediments is the transformation of potential energy of the sediment mass into work required to, firstly, dislocate (break-up) and, subsequently, to move the displaced block downslope while overcoming the various resistances (e.g. friction, drag). The displacement of the centre of mass of a debris flow is characterized by the horizontal distance L from source to deposit and the vertical elevation of the debris flow source above the deposit, H.

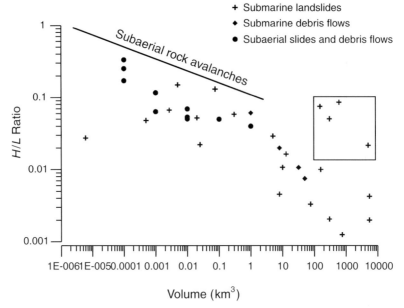

Fig. 6. Mobility of submarine mass flows, expressed as *H/L* ratio plotted against sediment volume (for data see Table 1). Data for selected subaerial slides/debris flows are also included (source: Iverson, 1997). Data points in the rectangle represent submarine volcanic debris and carbonate rock falls, for details see text.

Defining the effective friction coefficient, f, by the relation $f = H/L \tan \beta$ where β is the average slope angle from the release zone to the deposit (Scheidegger 1973), one readily obtains $f = \tan \phi$ if only Coulombian (dry) friction is present — the effective friction coefficient is equal to the tangent of the bed friction angle ϕ for granular material. Since ϕ is an intrinsic property of a granular material and the bed, this relationship implies that the net efficiency (or mobility) of a debris flow is independent of the mass involved. Observations, however, show a clear increase of the net efficiency (or decrease of the effective friction) with increasing mobilized mass. Iverson (1997) argued that, although not yet clearly understood, the increase in mobility with increasing volume may be attributed largely to changes in sediment composition and rheology. Edgers & Karlsrud (1981) developed a simple flow analysis to estimate the runout velocities of submarine slides. They used data from five submarine slides to back-calculate the equivalent sediment viscosities and compared these values with estimates of equivalent viscosity in subaerial quick clay slides back-calculated in a similar way. Good correspondence between the two viscosity estimates led them to conclude that viscous flow was an important mechanism in submarine slides just as in the case of quick clay slides. However, Edgers & Karlsrud's

model does not incorporate yield strength. As shown by this work, yield strength is a much more significant parameter than viscosity in determining flow behaviour of debris flows. The Edgers & Karlsrud model, therefore, is not adequate for analysing the mobility of submarine mass-wasting.

The ratio of *H/L* and volume for different mass flows is shown in Figure 6. It includes slides, debris flows and rock avalanches from different parts of the world, both in submarine and in subaerial environments. Submarine slides smaller than about 1 km³ all have ratios larger than 0.02. The same value represents instead an *upper* limit when the volume is larger than 1 km³. The exceptions, marked in the figure by a rectangle, are primarily debris *avalanches* containing both sediments and rock fragments that could be as large as 1 km in the horizontal direction (Normark *at al.* 1993). The other outlier represents limestone blocks, up to 10 km across and bounded by faults, that fall off when carbonate cliffs become oversteepened due to erosion (Dillon *et al.* 1993). These outliers, thus, do not represent the muddy mass flows considered in the present study. Interestingly, they fall on the continuation of the line for subaerial rock avalanches (Scheidegger 1973), indicating a similarity of processes between subaerial and submarine rock avalanches. For slides and debris flows, the critical volume of $1 km^3$ marks a change in the dynamic behaviour of the flow, as can be seen from the steepening of the curve. For volumes above1 km^3, a fit to the available data indicates H/L to be proportional to $V^{-0.34}$ and to decay both much more rapidly than below the $1 km^3$ limit, and also more rapidly than for rock avalanches. We suggest that for volumes smaller than $1 km^3$, the flow is controlled primarily by the rheology of the moving mass. When the volume exceeds this value, the flow is associated with an increase in the velocity beyond the critical value required to initiate hydroplaning. Once the mass begins to hydroplane, its mobility becomes independent of the rheology. Note that this conclusion is based on the analysis of a limited number of data points. It will be interesting to see if this view can be maintained as more data become available.

We have examined some of the problems regarding the initiation phase and dynamic behaviour of submarine mass flows, particularly in relation to glacier-influenced continental margins. For clay-rich sediments which do not reach the critical velocity for hydroplaning, the visco-plastic model appears to be suitable for describing the flow in terms of fluid dynamics of a non-Newtonian fluid. This study indicates that large mass flows (larger than $1 km^3$ in volume) might reach a sufficient velocity to initiate hydroplaning. This would explain the long runout distances reached on gentle slopes, without assigning unreasonably small values for yield strength, and the steep decrease of the H/L ratio as a function of the volume. On the other hand, for small slides and debris flows, covering short distances, a visco-plastic model without hydroplaning is well suited. In this case the flow behaviour is determined by the sediment rheology.

The work presented in this paper was carried out as part of projects funded by the EU (COSTA project no. EVK3-CT-1999-0006), VISTA (project 6241) and the Norwegian Research Council (NFR project 1333975/431). The authors wish to thank R. Hiscott and P. Talling for their careful and extensive reviews of the manuscript, which helped to improve the paper considerably.

References

AKSU, A. E. & HISCOTT, R. N. 1989. Sides and debris flows on the high-latitude continental slopes of Baffin Bay. *Geology*, **17**, 885–888.

AKSU, A. E. & HISCOTT, R. N. 1992. Shingled Quaternary debris flow lenses on the north-east Newfoundland Slope. *Sedimentology*, **39**, 193–206.

BJERRUM, L. 1971. *Subaqueous slope failures in Norwegian fjords*. Norwegian Geotechnical Institute, **88**, 1–8.

BOOTH, J. S., O'LEARY, D. W., POPENOE, P. & DANFORTH, W. W. 1993. U.S. Atlantic continental slope landslides: Their distribution, general attributes, and implictions. *In:* SCHWAB, W. C., LEE, H. J. & TWICHELL, D. C. (eds) *Submarine Landslides: Selected Studies in the US Exclusive Economic Zone.* USGS Bulletin 2002, 14–22.

BUGGE, T., BELDERSON, R. H. & KENYON, N. H. 1988. The Storegga slide. *Transactions of the Royal Society of London*, **325**, 357–388.

CARLSON, P. R., KARL, H. A., EDWARDS, B. D. & GARDNER, J. V. 1993. Mass movement related to large submarine canyons along the Beringian margin, Alaska. *In:* SCHWAB, W. C., LEE, H. J. & TWICHELL, D. C. (eds) *Submarine Landslides: Selected Studies in the US Exclusive Economic Zone.* USGS Bulletin 2002, 104–116.

COULTER, H. W. & MIGLIACCIO, R. R. 1966. *Effects of the earthquake of March 27, 1964, at Valdez, Alaska.* US Geological Survey Professional Paper **542–C**, 36.

DILLON, W. P., RISCH, J. S., SCANLON, K. M., VALENTINE, P. C. & HUGGETT, Q. J. 1993. Ancient crustal fractures control the location and size of collapsed blocks at the Blake Escarpment, east of Florida. *In:* SCHWAB, W. C., LEE, H. J. & TWICHELL, D. C. (eds) *Submarine Landslides: Selected Studies in the US Exclusive Economic Zone.* US Geological Survey Bulletin 2002, 54–59.

DIMAKIS, P., ELVERHØI, A., HØEG, K., SOLHEIM, A., HARBITZ, C., LABERG, J. S., VORREN, T. O. & MARR, J. 2000. Submarine slope stability on high-latitude glaciated Svalbard-Barents Sea margin. *Marine Geology*, **162**, 303–316.

DOWDESWELL, J. A., KENYON, N. H. & LABERG, J. S. 1997. The glacier influenced Scoresby Sund Fan, East Greenland continental margin: evidence from GLORIA and 3.5 kHz records. *Marine Geology*, **143**, 207–221.

EDGERS, L. & KARLSRUD, K. 1981. *Stability Evaluations for Submarine Slides: Viscous Analysis of Soil Flows Generated by Submarine Slides and Quick-Clay Slides.* Norwegian Geotechnical Institute Report 52207–8, 14.

EDGERS, L. & KARLSRUD, K. 1982. Soil flows generated by submarine slides – case studies and consequences. *In:* CHRYSSOSTOMIDIS, C. & CONNOR, J. J. (eds) *Proceedings of the Third International Conference on the Behaviour of Offshore Structures.* Hemisphere, Bristol, 425–437.

EDWARDS, B. D., LEE, H. J. & FIELD, M. F. 1993. Seismically induced mudflow in Santa Barbara Basin, California. *In:* SCHWAB, W. C., LEE, H. J. & TWICHELL, D. C. (eds) *Submarine Landslides: Selected Studies in the US Exclusive Economic Zone.* USGS Bulletin 2002, 167–175.

ELVERHØI, A., NOREM, H., ANDERSEN, E. S., DOWDESWELL, J. A., FOSSEN, I., HAFLIDASON, H., KENYON, N. H., LABERG, J. S., KING, E. L., SEJRUP, H. P., SOLHEIM, A. & VORREN, T. 1997. On the origin and flow behaviour of submarine slides on deep-sea fans along the Norwegian-Barents Sea continental margin. *Geo-Marine Letters*, **17**, 119–125.

FAIRCHILD, L. H. & WIGMOSTA, M. 1983. Dynamic and volumetric characteristics of the 18 May 1980 lahars on the Toutle River, Washington. *In: Proceedings of the Symposium on Erosion Control in Volcanic Areas*. Technical Memoir 108, Japanese Public Works Research Institute, Tokyo, pp. 131–153.

FOSSEN, I. 1996. *Stability and Geotechnical Properties of Glacial Sediments on the Western Svalbard Continental Slope*. Masters' thesis, University of Oslo (in Norwegian).

GEE, M. J. R., MASSON, D. G., WATTS, A. B. & ALLEN, P. A. 1999. The Saharan debris flow: an insight into the mechanics of long runout submarine debris flows. *Sedimentology*, **46**, 317–355.

GUTMACHER, C. E. & NORMARK, W. R. 1993. Sur Submarine Slide, a Deep-Water Sediment Slope Failure. *In:* SCHWAB, W. C., LEE, H. J. & TWICHELL, D. C. (eds) *Submarine Landslides: Selected Studies in the US Exclusive Economic Zone*. USGS Bulletin, 2002, 158–166.

HAMPTON, M. A., LEMKE, R. W. & COULTER, H. W. 1993. Submarine landslides that had a significant impact on man and his activities: Seward and Valdez, Alaska. *In:* SCHWAB, W. C., LEE, H. J., & TWICHELL, D. C. (eds) *Submarine Landslides: Selected Studies in the US Exclusive Economic Zone*. USGS Bulletin, 2002, 123–134.

HAMPTON, M. A., LEE, H. J. & LOCAT, J. 1996. Submarine slides. *Reviews of Geophysics*, **34**, 33–59.

HARBITZ, C. A., PARKER, G., ELVERHØI, A., MARR, J. M., MOHRIG, D. & HARFF, P. in press. Hydroplaning of subaqueous debris flows and glide blocks. *Journal of Geophysical Research – Oceans*.

HEEZEN, B. C. & EWING, M. 1952. Turbidity currents and submarine slumps and the 1929 Grand Banks earthquake. *American Journal of Science*, **250**, 849–873.

HOLOCOMB, R. T. & SEARLE, R. C. 1991. Large landslides from oceanic volcanoes. *Marine Geotechnology*, **10**, 19–32.

HUANG, X. & GARCIA, M. H. 1998. A Herschel-Bulkley model for mud flow down a slope. *Journal of Fluid Mechanics*, **374**, 305–333.

HUANG, X. & GARCIA, M. H. 1999. Modeling of non-hydroplaning mud flows on continental slopes. *Marine Geology*, **154**, 132–142.

IMRAN, J., HARFF, P. & PARKER, G. 2001. A numerical model of submarine debris flows with graphical user interface. *Computers and Geosciences*, **27**, 717–729.

IVERSON, R. M. 1997. The physics of debris flows. *Reviews of Geophysics*, **35**, 245–296.

IVERSON, R. M. & VALLANCE, J. W. 2001. New views of granular mass flows. *Geology*, **29**, 115–118.

JOHNSON, A. M. 1970. *Physical Processes in Geology*. Freeman, San Francisco.

KING, E. L., SEJRUP, H. P., HAFLIDASON, H., ELVERHØI, A. & AARSETH, I. 1996. Quaternary seismic stratigraphy of the North Sea Fan: glacially-fed gravity flow aprons, hemipelagic sediments, and large submarine slides. *Marine Geology*, **130**, 293–315.

KING, E. L., HAFLIDASON, H., SEJRUP, H. P. & LØVLIE, R. 1998. Glacigenic debris flows on the North Sea Trough Mouth Fan during ice stream maxima. *Marine Geology*, **152**, 217–216.

KVALSTAD, T. J., NADIM, F. & HARBITZ, C. B. 2001. *Deepwater Geohazards: Geotechnical Concerns and Solutions*. 2001 Offshore Technology Conference, Texas.

KVENVOLDEN, K. A. 1994. Natural gas hydrate occurrence and issues. *International Conference On Natural Gas Hydrates*, New York.

LABERG, J. S. & VORREN, T. O. 1995. Late Weichselian submarine debris flow deposits on the Bear Island Trough Mouth Fan. *Marine Geology*, **127**, 45–72.

LABERG, J. S. & VORREN, T. O. 2000. The Traenadjupet Slide, offshore Norway; morphology, evacuation and triggering mechanisms. *Marine Geology*, **171**, 95–114.

LEMKE, R. W. 1967. *Effects of the earthquake of March 27, 1964, at Seward, Alaska*. USGS Professional Paper, **542–E**.

LOCAT, J. 1997. Rheological behavior of fine muds and their flow properties in a pseudo-plastic regime. *In: Proceedings of the First International Conference on Debris-Flow Hazard Mitigation: Mechanics, Prediction, and Assessment*. Water Resources Division, ASCE, 260–269.

LOCAT, J. & LEE, H. J. 2002. Submarine landslides: Advances and challenges. *Canadian Geotecnical Journal*, **39**, 193–212.

MARR, J., ELVERHØI, A., HARFF, P., IMRAN, J., PARKER, G. & HARBITZ, C. B. 2002. Numerical simulation of mud-rich subaqueous debris flows on the glacially active margins of the Svalbard-Barents Sea. *Marine Geology*, **159**, 351–364.

McCLUNG, D. & SCHAERER, P. 1993. *The Avalanche Handbook*. The Mountaineers, Seattle.

McGREGOR, B. A., ROTHWELL, R. G., KENYON, N. H. & TWICHELL, D. C. 1993. Salt tectonics and slope failure in an area of salt domes in the northwestern Gulf of Mexico. *In:* SCHWAB, W. C., LEE, H. J. & TWICHELL, D. C. (eds) *Submarine Landslides: Selected Studies in the US Exclusive Economic Zone*. USGS Bulletin, 2002, 92–96.

MOHRIG, D., WHIPPLE, K. X., HONDZO, M., ELLIS, C. & PARKER, G. 1998. Hydroplaning of subaqueous debris flows. *Bulletin of the Geological Society of America*, **110**, 387–394.

MOHRIG, D., ELVERHØI, A. & PARKER, G. 1999. Experiments on the relative mobility of muddy subaqueous and subaerial debris flows and their capacity to remobilize antecedent deposits. *Marine Geology*, **154**, 117–129.

MORTON, D. M. & CAMPBELL, R. H. 1974. Spring mudflows at Wrightwood, southern California. *Quarterly Journal of Engineering Geology*, **7**, 377–384.

NEWMAN, J. N. 1977. *Marine Hydrodynamics*. The Massachusetts Institute of Technology, Cambridge.

NORMARK, W. R., MOORE, J. G. & TORRESAN, M. E. 1993. Giant volcano-related landslides and the development of the Hawaiian Islands. In: SCHWAB, W. C., LEE, H. J. & TWICHELL, D. C. (eds) Submarine Landslides: Selected Studies in the US Exclusive Economic Zone. USGS Bulletin 2002, 184–196.

NOREM, H., IRGENS, F. & SCHIELDROP, B. 1987. A continuum model for calculating snow avalanche velocities. In: Proceedings of Avalanche Formation, Movements and Effects. Davos, IAHS Publications, 162, 363–379.

NOREM, H., LOCAT, J. & SCHIELDROP, B. 1990. An approach to the physics and the modeling of submarine flowslides. Marine Geotechnology, 9, 93–111.

PIERSON, T. C. 1980. Erosion and deposition by debris flows at Mt Thomas, North Canterbury, New Zealand. Earth Surface Processes and Landforms, 5, 227–247.

PIERSON, T. C. 1985. Initiation and flow behaviour of the 1980 Pine Creek and Muddy river lahars, Mount St Helens, Washington. Bulletin of the Geological Society of America, 96, 1056–1069.

PIERSON, T. C., JANDA, R. J., THOURET, J. C. & BORRERO, C. A. 1990. Perturbation and melting of snow and ice by the 13 November 1985 eruption of Nevado del Ruiz, Colombia, and consequent mobilization, flow and deposition of lahars. Journal of Volcanological Geothermal Resources, 41, 17–66.

PIPER, D. J. W., COCHONAT, P. & MORRISON, M. L. 1999. The sequence of events around the epicentre of the 1929 Grand Banks earthquake: initiation of debris flows and turbidity current inferred from sidescan sonar. Sedimentology, 46, 79–97.

PLAFKER, G. & ERICKSEN, G. E. 1978. Nevado Huascaran avalanches, Peru. In: VOIGT, B. (ed.) Rockslides and Avalanches. Vol. 1, Natural Phenomena. Elsevier, Amsterdam, 277–314.

POPENOE, P., SCHMUCK, E. A. & DILLON, W. P. 1993. The Cape Fear landslide: Slope failure associated with salt diapirism and gas hydrate decomposition. In: SCHWAB, W. C., LEE, H. J. & TWICHELL, D. C. (eds) Submarine Landslides: Selected Studies in the US Exclusive Economic Zone. US Geological Survey Bulletin 2002, pp.40–53.

PRIOR, D. B. & COLEMAN, J. M. 1982. Submarine landslides – geometry and nomenclature. Zeitschrift für Geomorphology N.F., 23, 415–426.

SAVAGE, S. B. & HUTTER, K. 1989. The motion of a finite mass of granular material down a rough incline. Journal of Fluid Mechanics, 199, 177–215.

SCHEIDEGGER, A. E. 1973. On the prediction of the reach and velocity of catastrophic landslides. Rock Mechanics, 5, 231–236.

SCHLICHTING, H. 1968. Boundary Layer Theory. McGraw-Hill, New York.

SCHWAB, W. C. & LEE, H. J. 1993. Processes controlling the style of mass movement in glaciomarine sediment: Northeastern Gulf of Alaska. In: SCHWAB, W. C., LEE, H. J. & TWICHELL, D. C. (eds) Submarine Landslides: Selected Studies in the US Exclusive Economic Zone. USGS Bulletin 2002, 135–142.

SHARP, R. P. & NOBLES, L. H. 1953. Mudflow of 1941 at Wrightwood, southern California. Bulletin of the Geological Society of America, 64, 547–560.

SIMPSON, J. E. 1987. Gravity Currents: In the Environment and the Laboratory. Ellis Horwood Ltd., Chichester, England.

SOLHEIM, A., FALEIDE, J. I., ANDERSEN, E. S., ELVERHØI, A., FORSBERG, C. F., VANNESTE, K., UENZELMANN-NEBEN, G. & CHANNELL, J. E. T. 1998. Late Cenozoic seismic stratigraphy and glacial geological development of the East Greenland and Svalbard-Barents Sea continental margins. Quaternary Science Reviews, 17, 155–184.

VALLANCE, J. W. & SCOTT, K. M. 1997. The Osceola mudflow from Mount Rainier: Sedimentology and hazard implications of a huge clay-rich debris flow. Bulletin of the Geological Society of America, 109, 143–163.

VORREN, T. O., BLAUME, F., DOWDESWELL, J. A., LABERG, J. S., MIENERT, J., RUMOHR, J. & WERNER, F. 1998. The Norwegian-Greenland Sea continental margins: morphology and late Quaternary sedimentary processes and environments. Quaternary Science Reviews, 17, 273–302.

WIECZOREK, G. F., HARP, E. L., MARK, R. K. & BHATTACHARYA, A. K. 1988. Debris flows and other landslides in San Mateo, Santa Cruz, Contra Costa, Alameda, Napa, Solano, Sonoma, Lake and Yolo counties and factors influencing debris-flow distribution. In: ELLEN, S. D. & WIECZOREK, G. F. (eds) Landslides, Floods and Marine Effects of the Storm of January 3–5, 1982, in the San Francisco Bay Region, California. US Geological Survey Professional Paper, 1434, 133–162.

Experimental constraints on shear mixing rates and processes: implications for the dilution of submarine debris flows

P. J. TALLING[1], J. PEAKALL[2], R. S. J. SPARKS[1], C. Ó COFAIGH[3],
J. A. DOWDESWELL[3], M. FELIX[2], R. B. WYNN[5], J. H. BAAS[2], A. J. HOGG[4],
D. G. MASSON[5], J. TAYLOR[6] & P. P. E. WEAVER[5]

[1]Department of Earth Sciences, University of Bristol, Queens Road, Bristol, BS8 1RJ, UK
(e-mail: Peter.Talling@bris.ac.uk)
[2]School of Earth Sciences, University of Leeds, Leeds, West Yorkshire LS2 9JT, UK
[3]Scott Polar Research Institute and Department of Geography, University of Cambridge,
Cambridge CB2 1ER, UK
[4]Department of Mathematics, University of Bristol, Bristol BS8 1TW, UK
[5]Challenger Division, Southampton Oceanography Centre, European Way, Southampton,
Hampshire SO14 3ZH, UK
[6]Centre for Glaciology, Institute of Geography and Earth Sciences, University of Wales,
Aberystwyth SY23 3DB, Wales, UK

Abstract: Submarine debris flows show highly variable mixing behaviour. Glacigenic debris flows travel hundreds of kilometres along the sea floor without undergoing significant dilution. However, in other locations, submarine slope failures may transform into turbidity currents before exiting the continental slope. Rates and processes of mixing have not been measured directly in submarine flow events. Our present understanding of these rates and processes is based on experimental and theoretical constraints. Significant experimental and theoretical work has been completed in recent years to constrain rates of shear mixing between static layers of sediment and overlying turbulent flows of water. This work was driven by a need to predict transport of fluid mud and the erosion of cohesive mud beds in shallow water settings such as estuaries, docks and shipping channels. These experimental measurements show that the critical shear stress necessary to initiate shear mixing (around 0.1 to 2 Pa) is typically several orders of magnitude lower than the yield strength of the debris. Shear mixing should initiate at relatively low velocities (about 10–200 cm s^{-1}) on the upper surface of a submarine debris flow, at even lower velocities at its head (about 1–10 cm s^{-1}), and play an important role in mixing over-ridden water into the debris flow. Addition of small amounts of mud (approximately 3% kaolin) to a sand bed dramatically reduces the rate of mixing at its boundary, and changes the processes by which sediment is removed. Estimates are presented for rates of shear mixing at a given flow velocity, and for the critical velocity necessary for hydroplaning or a transition from laminar to turbulent flow. Although these estimates are crude, and highlight the need for further experimental work, they illustrate the potential for highly variable mixing behaviour in submarine flow events.

Observations from the modern sea-floor show that, in certain situations, submarine debris flows can travel several hundred kilometres along the sea floor without entraining significant amounts of sea water and diluting to form turbidity currents (Gee *et al.* 1999; Masson *et al.* 1997; Schwab *et al.* 1996). Glacigenic debris flows are particularly good examples of minimally-diluting debris flows (Fig. 1). Their deposits extend for 100–200 km without significant textural variation (e.g. Laberg & Vorren 1995; Dowdeswell *et al.* 1996; King *et al.* 1998;

Taylor *et al.* 2002; Fig. 1). However, in other locations, subaqueous debris flows have been diluted rapidly to form turbidity currents. Evidence from the modern sea floor shows that complete transformation of slope failures into turbidity currents can occur before the flows exit canyons on the continental slope (Piper *et al.* 1999; Mulder *et al.* 1997; Wynn *et al.* 2000, 2002). Understanding the reasons for this variable mixing behaviour is important, both for petroleum exploration and prediction of natural hazards. Clean turbiditic sandstones can form

From: DOWDESWELL, J. A. & Ó COFAIGH, C. (eds) 2002. *Glacier-Influenced Sedimentation on High-Latitude Continental Margins.* Geological Society, London, Special Publications, **203**, 89–103. 0305-8719/02/$15.00
© The Geological Society of London 2002.

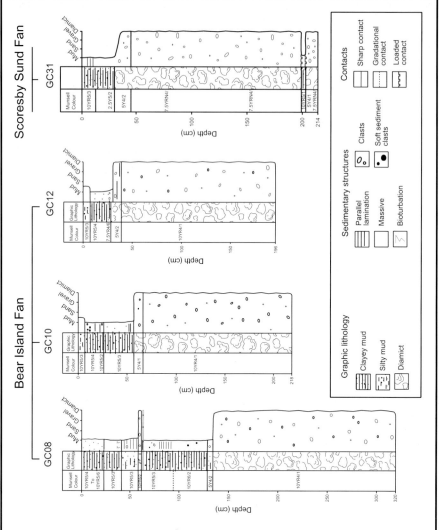

Fig. 1. Gravity cores from the Bear Island and Scoresby Sund trough mouth fans, Norwegian–Greenland Sea. The cores are dominated by a dark grey, massive diamict facies that was deposited by downslope remobilization of glacigenic debris (Glacigenic Debris Flow – GDFs) during periods when ice sheets expanded to the continental shelf break. Note the lack of turbidites between glacigenic debrites. The uppermost GDFs in each core are sharply overlain by thin units of olive grey and dark grey massive diamict or poorly-sorted silty mud with clasts, which occasionally exhibit stratification (e.g. core GC12). An abrupt change in sedimentation is indicated by an abrupt colour change, poor sorting, sharp lower contacts, occasional weak stratification and, in cores from the Bear Island Fan, increasing water content up-core compared to the underlying dark grey GDFs. Laberg & Vorren (1995) interpret this type of sediment and the overlying grey muds as glacimarine deposits, presumably, although not explicitly stated as such, due to suspension settling and iceberg rafting. However, the sharp lower contacts, poor sorting and occasional stratification may also reflect subsequent downslope resedimentation by debris flow processes. Cores from the fans are capped by massive, yellowish-brown muds with abundant foraminifera, which record the most recent transition to hemipelagic sedimentation. In core GC31 from the Scoresby Sund Fan, a lowermost GDF unit at the base of the core is separated from a second dark grey GDF at 45–199 cm depth by a thin (4 cm) unit of lighter grey, silty mud. The contact between the silty mud and overlying GDF is abrupt, whilst the contact between the mud and underlying GDF is gradational. Thus, the upper contacts of GDFs are not ubiquitously sharp.

excellent reservoir units, with high porosity and permeability. Debris flow or slurry beds are much less attractive reservoirs, due to their higher mud content and much lower permeability and porosity, and may represent significant baffles to flow within a reservoir. Mixing behaviour will determine the down-flow geometry of reservoir units comprising debris flow deposits (debrites) and turbidites. Dilution of debris flows into turbidity currents will favour deposition of thick turbiditic sandstones in distal locations. Transitions from debris flow to turbidity current dramatically alter flow velocity, density, impact pressures and run-out distances, and also have important implications for assessing the hazards such events pose to sea-floor communications equipment and structures.

The processes by which debris flows transform into turbidity currents are likely to be complex. For instance, progressive dilution may cause transitions from laminar to turbulent flow and strongly non-linear changes in debris rheology, which in turn feed back to influence mixing rates. Submarine debris flows have not been observed directly, due to their infrequent occurrence, destructive nature and inaccessible location. Our understanding of these flow events is based primarily on indirect observations (e.g. cable breakage and deposit geometries) and laboratory experiments allied to theoretical analysis.

A significant amount of experimental and theoretical work has been completed in recent years in order to constrain rates of shear mixing between static layers of sediment and overlying turbulent flows of water. This work was driven by a need to predict transport of fluid mud and the erosion of cohesive mud beds in shallow water settings such as estuaries, docks and shipping channels. Our contribution explores, for the first time, how this body of work can better constrain rates and processes of shear mixing along the margins of submarine debris flows. Experiments have shown that there are three primary mechanisms by which debris flows mix with water; shear mixing at the head, shear mixing along the upper surface of the flow, and over-riding water below the head. Qualitative observations of these three mechanisms are reviewed briefly, and the factors that control their rates are outlined. A synthesis of experimental constraints on rates of shear mixing is then presented for turbulent flow of water above static sediment beds with variable (but finite) yield strength, volumetric sediment concentration and grain-size. The applicability of such data for predicting the shear mixing rates experienced by submarine debris flows is discussed critically. Shear mixing rates at variable flow velocity are then estimated crudely and compared to the critical velocity necessary for (i) hydroplaning and (ii) the transition from laminar to turbulent flow. This comparison is used to investigate the mixing behaviour of debris flows with a 'strong' or 'weak' rheology. The 'strong' rheology approximates to that of glacigenic debris flows on high-latitude continental margins. This quantitative analysis illustrates why submarine debris flows may display highly variable mixing behaviour.

Mixing processes for submarine debris flows

Laboratory experiments indicate that subaqueous debris flows mix with ambient water through three distinct processes (Fig. 2; Allen 1971; Hampton 1972; Mohrig et al. 1998; Marr et al. 2001). Shearing occurs at the outer surface of the debris flow due to velocity differences between the debris flow and surrounding water. Shear mixing at the head of the debris flow may be distinguished from shear mixing along the upper surface of the flow. The head of the debris flow has a blunt overhanging nose due to drag at the bed and at the contact with the overlying water. Water flowing under this overhanging nose is over-ridden by the debris flow (Allen 1971). Hydroplaning may occur if the over-ridden water does not penetrate or mix into the overlying debris flow (Mohrig et al. 1998).

Shear mixing on the upper surface

Shear occurs at the boundary between ambient water and the upper surface of the debris flow. Experimental observations illustrate that shear mixing is typically much less vigorous along the upper surface of the debris flow than at the head (Hampton 1972; van Kessel & Kranenburg 1996; Mohrig et al. 1998). Indeed, in some experiments, a distinct interface with no visible mixing was reported along the upper surface of the flow (van Kessel & Kranenburg 1996).

Shear stress on the boundary of the debris flow (τ) may be estimated using,

$$\tau = C_D(0.5\rho U^2), \qquad (1)$$

where C_D is the drag coefficient, ρ is the density of the overlying water, and U is the speed of the debris flow. Schlichting (1980, p. 653–654) provides an equation based on empirical results for roughened pipe-flow for estimating the average value of the drag coefficient for an

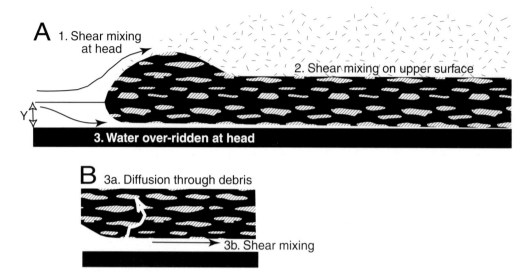

Fig. 2. (**A**) Schematic summary of the three main mixing processes for submarine debris flows; shear mixing at the head, shear mixing on the upper surface of the body, and incorporation of water over-ridden below the head. (**B**) Water over-ridden below the head may mix into the body of the debris flow either by diffusive percolation (as discussed by Mohrig *et al.* 1998) or by shear mixing. Hydroplaning may occur if the over-ridden water, of thickness *Y*, is not mixed into the body (Mohrig *et al.* 1998).

entirely turbulent boundary layer above a rough plate,

$$C_D = (1.89 + 1.62 \log (L/k)^{-2.5}), \qquad (2)$$

where *L* is the length of the flow or 'plate', and *k* is a roughness length that will range from 0.01 to 1.0 m (Norem *et al.* 1990; Harbitz 1992). This indicates an average value of 0.0014–0.0019 for C_D along the upper surface of the debris flow, a significant distance from the head (Norem *et al.* 1990; Harbitz 1992). Drag coefficients ranging between 0.0015 and 0.0070 are quoted by Young (1989) for turbulent flows above smooth flat plates, with values increased by less than three fold for rough flat plates (Young 1989, p. 178). This work indicates that drag coefficients ranging between about 0.001 and 0.02 are a reasonable characterization of the upper surface of submarine debris flows, using the maximum and minimum estimates from this previous work. Such estimates of C_D are subsequently used to calculate shear stress and shear mixing rates for submarine debris flows.

Shear mixing at the head

Shear stresses experienced by the head are much greater than those on the upper boundary of the flow. The head resembles the upper half

of a sphere being towed through water. The drag coefficient for a non-deforming sphere towed through water is of the order 0.5 (e.g. Middleton & Wilcock 1994; Young 1989). The drag coefficient of the head will approximate this value, although the lateral extent of a debris flow may well be greater than that of a half-sphere. From equation 1, using drag coefficients of 0.5 and 0.002, the shear stress at the head will be approximately 250 times that on the flow's upper surface.

Hampton (1972) described in detail the vigorous mixing observed at the head of subaqueous debris flows, and suggested that it was the primary process that caused their dilution. He noted that small changes in the volume concentration of kaolin, in flows comprising mixtures of kaolin and water, profoundly influenced rates of mixing at the head. Changes in volume concentration also influenced flow density and hence velocity. Flows with 20–25% by weight of kaolin had a head that was thoroughly mixed in its entirety. When kaolin concentrations were increased to 25–30% by weight, a distinct thin layer deformed by shear was observed on the front of the head. As the material within this layer was swept back along the head, it was thrown into a dilute cloud or wake that overlay the debris flow. The abrupt transition from a thin layer of reverse shear to an expanded dilute wake was attributed to variations in the dynamic normal stress imposed by the flow on the boundary of the head. The presence of these normal stresses at the head may be an important difference between shear mixing at the head and on the upper surface of the flow. Flows with higher kaolin concentrations (30–35% by weight) were observed to experience very little shear mixing at the head, although chunks of material were occasionally ripped off the head and flung into the wake (Hampton 1972).

Marr *et al.* (2001) recently extended Hampton's (1972) experimental work to investigate mixing processes in debris flows comprising sand, kaolin and water. Again, relatively small variations in both clay and water content were found to change the 'coherency' of the debris mixture and significantly alter the character of mixing processes. Coherent debris flows had a laminar head and body, little shear mixing occurred at the head, and the flows were likely to hydroplane (Marr *et al.* 2001). Shear mixing and deformation of the head increased for moderately coherent flows, which generated a substantial turbidity current due to shear mixing at the head. Both moderately and weakly coherent flows had a turbulent head and a laminar body. The head of weakly coherent flows was strongly turbulent and the location of very vigorous mixing.

Water over-ridden at the head

A pronounced overhang is observed at the front of experimental subaqueous debris flows and turbidity currents, and in other natural gravity currents such as atmospheric fronts and powder snow avalanches, due to viscous deformation of the head (Allen 1971; Simpson 1972, 1997). Ambient fluid reaching the gravity current below the nose of this overhang will flow under the current and form lobes and clefts, assuming it cannot be deflected around the sides of the flow. This produces a buoyantly unstable situation, with low-density fluid underlying higher density fluid. This over-ridden water is either mixed into and dilutes the head or forms a layer at the base of the debris flow (Fig. 2b; Hallworth *et al.* 1996; Mohrig *et al.* 1998; Stix 2000). Experiments have shown that in low density turbulent gravity currents this ingested ambient fluid is mixed vigorously within the head, and that it plays a crucial role in producing transverse irregularities in the flow front (e.g. Simpson 1972, especially his figure 6; Britter & Simpson

1978; Hallworth *et al.* 1996; Stix 2000). Lobes and clefts have also been observed when ambient fluid is over-run by highly viscous non-particulate gravity currents (Snyder & Tait 1998).

Experiments using mixtures of mud, silt and clay that are similar to those comprising submarine debris flows have illustrated the potential for hydroplaning (Mohrig *et al.* 1998, 1999; Marr *et al.* 2001). Hydroplaning occurs when the dynamic pressure gradient at the nose of a flow, that is driving ambient water under the flow, exceeds the weight of the overlying flow (Mohrig *et al.* 1998). A thin layer of water is driven under the head, drastically reducing drag at the base of the head and increasing the head's velocity. Hydroplaning can occur when a critical Froude number (F_c) of 0.3 to 0.4 is exceeded (Mohrig *et al.* 1998).

$$F_c = U/[(\rho_d/\rho_w)gD\cos\theta/\rho_d]^{0.5} \qquad (3)$$

where ρ_d and ρ_w are the densities of the debris flow and water, D is the flow thickness and θ is the sea-floor gradient. In certain cases, this can lead to the faster-moving head breaking off from and out-running the body of the flow (Mohrig *et al.* 1998; Marr *et al.* 2001). For hydroplaning to occur, the ingested water at the base of the flow must also be unable to mix into the main body of the head, at least over time scales similar to the duration of the flow (Mohrig *et al.* 1998). This ensures that hydroplaning tended to occur in the experiments for high-concentration flows with high viscosities, yield strengths, and low permeabilities (Mohrig *et al.* 1998; Marr *et al.* 2001). Although the experiments have clearly demonstrated the potential for hydroplaning to occur, inevitable issues of scaling between the experiments and natural flows should be noted (e.g. Marr *et al.* 2001).

Mohrig *et al.* (1998) provide a thorough discussion of diffusive mixing of over-ridden water into the base of the flow. However, it seems probable that shear mixing, and not diffusion, is the dominant mixing process at the base of the flow. As analysed quantitatively in a later section, shear mixing is favoured by the high velocity gradients and shear stresses at the base of the flow. Shear mixing of ambient water may lead to a relatively dilute layer, with high pore fluid pressures, underlying the impermeable body of the debris flow. This is a similar vertical flow-structure to that proposed by Gee *et al.* (1999), although they inferred that it was sea-floor material with high water contents that was sheared, rather than material comprising the debris flow itself. Moreover, in laterally

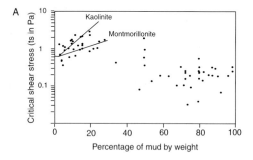

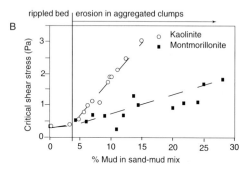

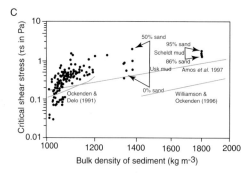

Fig. 3. Experimental data showing controls on critical shear stress (τ_s) for surface erosion of a cohesive debris mixture. (**A**) Critical shear stress for surface erosion plotted against sand content for artificially produced mixed sediment. (**B**) Critical shear stress for erosion plotted against percentage of silt and clay in a mixture of sand, silt and clay from the Scheldt Estuary, Belgium. (**C**) Critical shear stress for surface erosion plotted against wet bulk density for artificially produced sediment beds. From Mitchener & Torfs (1996). Also shown are the trends for similar data plotted by Amos *et al.* (1997), Williamson & Ockenden (1996) and Ockenden & Delo (1991).

extensive debris flows, the over-run fluid may also well-up along discrete zones (Snyder & Tait 1998; Marr *et al.* 2001).

Synthesis of experimental studies of shear-mixing

Recent experimental work has improved our understanding of the factors controlling the initiation and rate of shear mixing between debris and water significantly. The implications of this work have not been summarized previously in the context of submarine debris flow. The following section synthesizes relevant work on shear mixing. This synthesis forms the basis for discussion of rates and processes of shear mixing and their implications for the evolution of submarine debris flows. Experiments involving both cohesive beds (with a significant yield strength) and fluid mud (with a much smaller yield strength) are discussed. Submarine debris flows, especially glacigenic debris flows, will typically comprise material that is comparable to that making up cohesive beds rather than fluid mud. However, the flow velocities and shear stresses generated in flume experiments are typically much less than in submarine flow events. Fluid mud experiments may provide insight into mixing processes when the shear stress on the bed greatly exceeds the critical shear stress for shear mixing, or indeed the yield strength of the sediment mixture. These scaling issues are discussed more fully in a later section (also see Marr *et al.* (2001) and Coussot (1997) for a fuller discussion of the scaling issues involved in debris flow experiments).

Cohesive beds (shear mixing across a large density interface)

Rates at which cohesive sediment beds are eroded have been determined experimentally and in shallow marine field locations using annular (circular- or racetrack-shaped) flumes (e.g. Parchure & Mehta 1985; Mitchener & Torfs 1996; Amos *et al.* 1996, 1997; Houwing 1999; Fig. 3). In these experiments, water is recirculated above the bed around such flumes at pre-determined rates, using an upper rotating shear plate, propellers, or pumps. Measurements of progressively increasing suspended sediment concentrations above the bed are used to constrain erosion rates, and the depth of erosion. The shear stress applied to the bed (τ) is varied and calculated from measured velocity profiles. These studies show that a critical shear stress (τ_s) must be exerted on the cohesive bed to initiate erosion. For consolidated beds, this critical shear stress increases significantly with depth below the sediment surface. The critical shear stress typically lies between 0.1 and 2 Pa (Fig. 3; Mitchener & Torfs 1996; Houwing 1999).

Such critical shear stresses are one or two orders of magnitude lower than the yield strength of the same debris mixture, as measured by vane shear, falling cone or other rheometrical tests (e.g. Amos *et al.* 1996; Zreik *et al.* 1998). Although Marr *et al.* (2001) compared the 'coherency' of a debris mixture to its yield strength, the ability of debris to resist mixing may be better quantified by its critical shear stress. The critical shear strength is dependent on material properties and increases systematically with increasing density of the sediment mixture (Mitchener & Torfs 1996; Houwing 1999).

Two types of erosion behaviour are commonly observed (Amos *et al.* 1996, 1997; Mitchener & Torfs 1996). The erosion rate gradually decreases during type I erosion from a peak value to a very low but non-zero value, termed the floc erosion rate (E_f). Type I erosion is sometime termed 'benign' erosion (Mitchener & Torfs 1996). This decrease is attributed to erosion into progressively lower parts of the consolidated bed that have progressively higher critical shear stresses (Parchure & Mehta 1985). The very low but constant floc erosion rate occurs when the applied shear stress reaches the critical shear stress ($\tau = \tau_s$). Observations of the bed during type I erosion indicate that it remains smooth and that individual particles or flocs were detached (Amos *et al.* 1996). In contrast, type II ('chronic') erosion occurs at higher shear stresses (e.g. > 0.35 Pa, Houwing 1999) and is observed to occur through excavation and undermining of defects on a rough bed. Rip-up clasts and large aggregates may be released during type II erosion. The erosion rate is seen to be highly variable over short periods of a few seconds, but averaging over longer periods shows it attains a constant value that does not change as deeper levels of the consolidated bed are reached (Amos *et al.* 1996, 1997). Thus, in type II erosion, increasing critical shear stresses at deeper levels within a consolidated bed do not influence erosion rates significantly.

A number of different types of equation have been proposed to characterize cohesive bed erosion rates. Parchure & Mehta (1985) proposed that erosion rates increased in a nonlinear fashion with increased excess shear stress, such that the erosion rate (E in kg m^{-2} s^{-1}) may be calculated as,

$$E/E_f = e^{\,\alpha(\tau-\tau_s)^{0.5}}. \qquad (4)$$

The floc erosion rate (E_f in kg m^{-2} s^{-1}) is reached when there is no excess shear stress ($\tau-\tau_s = 0$) and is typically between 2×10^{-4} to 3.3×10^{-6}. The empirical constant α (with units of Pa$^{0.5}$) has values between about 4 and 28 (Parchure & Mehta 1985; Kuiper *et al.* 1989; Houwing 1999). A simpler type of equation has been proposed, such that

$$E = M[(\tau-\tau_s) / \tau_s]. \qquad (5)$$

The empirical constant M (in kg m^{-2}s^{-1}) is found to have values between 0.0005 and 0.0015 (Amos *et al.* 1996). This equation has been shown to provide a slightly better fit to field data for both type I and II erosion (Teisson *et al.* 1993; Amos *et al.* 1997). Due to this slightly better fit and its simplicity, Equation 5 is employed subsequently to estimate erosion rates in submarine debris flows. Both Equations 4 and 5 imply a non-linear increase in erosion rates, with increasing excess shear stress, for low excess shear stresses (i.e. type I erosion). However, only Equation 5 is compatible with the constant erosion rates observed at higher excess shear stresses (i.e. $\tau \gg \tau_s$) during type II erosion. As discussed subsequently, it seems likely that type II erosion occurs for rapidly moving submarine debris flows. However, excess shear stresses of more than about 10 Pa have not been generated in these laboratory and field experiments (Mitchener & Torfs 1996; Amos *et al.* 1997). Erosion rates and processes at such high shear rates have not been observed and must be inferred from the lower shear rate data. Further work is also needed to investigate how the empirical constant (M) in Equation 5 varies with sediment concentration in the bed.

Addition of variable amounts of sand to a muddy cohesive bed influences rates and processes of erosion in a complex fashion (Fig. 3). The addition of mud to a sand bed produced a significant increase in the critical shear stress at which shear mixing initiates (τ_s). The addition of only a few percent of kaolin can increase τ_s tenfold. However, a slight decrease in critical shear stress was observed as mud percentages increased from about 50 to 100%, such that maximum critical shear stresses occurred for 50–70% mud by weight (Fig. 3; Mitchener & Torfs 1996, figs 6 and 7). Addition of a small percentage of mud (e.g. 3% of kaolin) to a sand bed can reduce the erosion rate constant (M in Eqn. 5) by a factor of between two and ten (Mitchener & Torfs 1996). The way in which erosion and sediment transport occurred was also observed to undergo a fundamental change after only 3% of kaolin was added to a sand bed (Mitchener & Torfs 1996). For beds comprising almost entirely sand, the small percentage of mud rapidly washed out and dispersed into suspension leaving a rippled bed along which sand was moved as bedload. Once

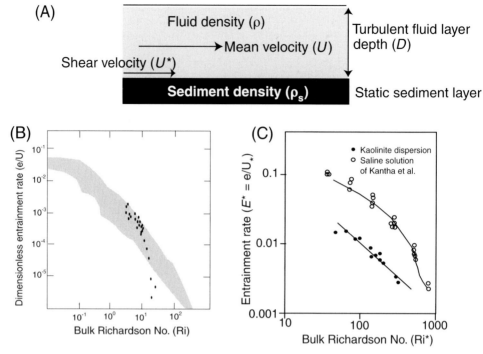

Fig. 4. (**A**) Experimental set-up used to investigate shear mixing in annular flumes, and parameters used to define the bulk Richardson number (*Ri* or *Ri*∗), as defined in the text. (**B**) Dimensionless entrainment rate (*E* = e/*U*) plotted against bulk Richardson number (*Ri*) for fluid mud experiments (dots). Also shown in grey shading is the region occupied by data from non-particulate mixing experiments reported by Fernando (1991). In this plot, the bulk Richardson number and dimensionless entrainment rate are calculated using the mean flow velocity (*U*). (**C**) Dimensionless entrainment rate (*E* = e/*U*∗) plotted against bulk Richardson number (*Ri*∗). In this case the bulk Richardson number and dimensionless entrainment rate are calculated using the shear velocity (*U*∗). From Kranenburg & Winterwerp (1996) and Kantha *et al.* (1977).

the percentage of kaolin exceeded about 3% by weight, type II cohesive behaviour was observed with aggregated clumps of material removed from the bed and transported as bedload. A similar change in erosion style was observed for montmorillonite at slightly higher percentages (7–13% weight of montmoril-lonite).

Fluid mud beds (shear mixing across low density interface)

Fluid mud comprises relatively dilute suspensions (with density < c. 1300 kg m^{-3}; Mehta & Srinivas 1993) of cohesive mud particles. The occurrence of fluid mud has been documented in a range of shallow marine settings, where it can be a significant problem for navigation channels and harbours (Mehta & Srinivas 1993; Winterwerp & van Kranenburg 1997). A number of experimental and theoretical

analyses have addressed the entrainment of stationary fluid mud into an overlying turbulent layer of water. The elevation of the upper free surface of the turbulent layer is fixed, and entrained mud is well mixed within this layer. There is an abrupt change in density (ΔB) and velocity (ΔU) across the mixing interface (Fig. 4a). These analyses define entrainment rates in terms of a bulk Richardson number, rather than excess shear stress. A bulk Richardson number (*Ri* ∗ or *Ri*) can be defined using a shear velocity (U_*) or the flow velocity (U). If the shear velocity is used to calculate *Ri*∗,

$$Ri_* = gT(\Delta B/\rho) / (U_*)^2 \qquad (6)$$

where T is the thickness of the upper turbulent well mixed layer, ρ is the density of the upper turbulent layer, and U_* is the shear velocity. The shear velocity is related to the shear stress (τ),

$$\tau = \rho \, U_*^2. \qquad (7)$$

Experiments in annular flumes (Fig. 4; Winterwerp & Kranenburg 1997; Mehta & Srinivas 1993), similar to those described previously for erosion of cohesive beds, have been conducted. These show a strong negative correlation between non-dimensionalized entrainment rates (E_s / U^* where E_s is the vertical velocity of the sharp mixing interface) and Richardson number, such that,

$$E_s \, / \, U^* = c \, Ri^{*-n}. \tag{8}$$

Winterwerp & Kranenburg (1997) found n to be 0.5 (Fig. 4). A similar relationship between Richardson number and entrainment rate has been documented for mixing between non-particulate fluids of different density, with a stationary lower fluid overlain by a turbulent well-mixed fluid (Fernando 1991; Strang & Fernando 2001). For such fluids, the value of n increases with increasing Richardson number (Fig. 4). The data of Mehta & Srinivas (1993) illustrate how entrainment rates for fluid mud may deviate from those of Newtonian fluids lacking a yield strength (Fig. 4). At high Richardson numbers, entrainment rates for fluid mud are significantly lower than those for non-particulate layers (Fig. 4).

Application of such Richardson number-based entrainment functions to submarine sediment gravity currents is attractive, due to the large amount of experimental work that may then be used to constrain mixing rates. However, applying a Richardson-number based entrainment function to submarine debris flows is problematic. Without direct measurements, it is difficult to estimate the thickness of the upper turbulent well-mixed layer (T) for a given debris flow thickness or velocity, particularly as mixing with oceanic water may cause T to increase progressively as the flow progresses. In the fluid mud experiments, a free surface fixes the upper boundary of this layer and this is not the case for submerged density flows. The length scale T is related to turbulence properties, such as eddy size and momentum. However, its role in determining mixing rates may be relatively small in comparison to the shear stress velocity or density contrast. It is difficult to assess quantitatively the importance of T for mixing rates, as only values of the Richardson number and not values of T are typically reported (Fernando 1991; Strang & Fernando 2001 and references therein). In any case, the size of experimental apparatus has resulted in a rather narrow range of T between about 0.05 and 0.40 m. A wider range of T would be needed to determine its role in controlling mixing rates for submarine debris flows. If it is assumed that variations in T

have a minimal influence on mixing rates, Equation 8 could then be used for estimating mixing rates for submarine debris flows. However, due to the uncertainty in making this assumption, such estimates are not undertaken here.

The Richardson number-based approach to quantifying entrainment rates can be rewritten in a form comparable to the entrainment rate function proposed (Equation 5) for cohesive bed erosion, by combining Equations 6, 7 and 8 and setting n equal to 0.5 (as found by Winterwerp & Kranenburg 1997) and τ_s equal to zero.

$$E_s = C(\tau - \tau_s) \tag{9}$$

where $C = 1/T\Delta B$ for fluid mud and $C = M/\tau_s$ for the cohesive mud beds. This type of equation may provide a holistic approach to quantifying mixing rates for debris with, and without, a yield strength where C is defined as a function of critical shear stress (τ_s), excess density (ΔB), depth of the upper turbulent layer (T), or roughness of the interface. Such an approach is promising but would need to recognize the importance of changes in the detailed processes by which mixing occurred (q.v. Amos 1996, 1997; Strang & Fernando 2001).

Fluid mud gravity currents

The experimental data summarized in the previous two sections is for mixing between a turbulent layer that overlies a stationary layer of variable sediment concentration. A number of studies have quantified entrainment and mixing rates in subaqueous gravity currents themselves (Fig. 5). Most of this work addresses mixing between turbulent non-particulate gravity currents and surrounding water (e.g. Britter & Linden 1980; Hallworth et al. 1993, 1996). Van Kessel & Kranenburg (1996) provide the only presently available study of entrainment rates into particle-laden density flows. Their study quantified entrainment rates into laminar and turbulent gravity currents comprising fluid mud (densities), using sequential measurements of density and velocity profiles. They found that entrainment rates were proportional to the inverse of the bulk Richardson number (Fig. 5). Although their currents could dilute through mixing both at the head and upper surface of the flow, this result is broadly consistent with the form of relationships for shear mixing of static fluid mud beds (Fig. 4). They showed that the Richardson number is much larger for laminar flows with high mud concentrations, and that this results in particularly low entrainment rates.

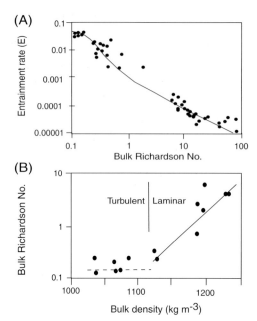

Fig. 5. Experimental data from subaqueous density flows (from Van Kessel & Kranenburg 1996). (**A**) Plot of dimensionless entrainment rate E (e/U) against bulk Richardson number (Ri_*) in flows comprising both non-particulate solutions and kaolin dispersions. (**B**) Plot of bulk Richardson number (Ri_*) against kaolin volume concentration.

Discussion

Applicability of shear mixing experiments to submarine debris flow

Differences in flow geometry between laboratory shear mixing experiments and submarine flow events need to be stated clearly and considered carefully. The experimental shear mixing data come from a geometry in which the static sediment bed is not deforming internally (i.e. flowing). Any deformation of the sediment bed is due to shear from the overlying turbulent flow of water. This situation may be reproduced in the upper part of a plug flow in a submarine debris flow, and it is this type of plug flow that is best represented by the experimental geometry. Laminar, and especially turbulent, flow of the debris at the margins of the debris flow may result in different shear mixing rates. Internal deformation of the debris may be particularly common at the head of the debris flow. Experiments on shear mixing rates are needed in which the lower sediment layer is flowing.

The sediment entrained from the upper surface of a submarine debris flow forms a dilute turbidity current. This turbidity current will experience a gravitational force in the downslope direction due to its excess density. The dilute upper layer in the horizontal shear mixing experiments will not experience such a downslope force. As the velocity of a submarine turbidity current approaches that of the debris flow, the shear stress experienced by the upper surface of the debris flow will decrease. If the velocity of the debris flow and turbidity current are similar, there will be minimal shear on the upper surface of the debris flow. This protection of the upper surface of the debris flow by a co-genetic turbidity current is an important issue that needs to be addressed by further experiments or numerical modelling. The following analysis may overestimate mixing rates on the upper surface of the debris flow.

The preceding analysis has implicitly treated mixtures of sediment and water as a single Bingham-Plastic type fluid with spatially and temporally invariant yield strength and viscosity. However, observations of subaerial debris flows in the field and large-scale flumes indicate that this is a significant simplification (Iverson 1997; Iverson & Vallance 2001). In subaerial debris flows, size segregation often occurs so that larger clasts migrate to the front and sides of the flow (Iverson 1997; Parsons *et al.* 2001). This may produce a relatively fluid central portion of the flow confined within a coarse granular dam (Iverson 1997; Parsons *et al.* 2001). If such size segregation occurs in a submarine flow, such a granular dam may protect the flow from vigorous mixing at its margins.

Scaling issues are important for all laboratory experiments. The shear stress imparted on the sediment bed in laboratory and field experiments ($< \sim 10$ Pa; Mitchener & Torfs 1996; Amos *et al.* 1997) is much less than that in most submarine debris flow events. For instance, using flow velocities of 10 m s^{-1} and drag coefficients of 0.5 and 0.001–0.2 in Equation 1, the head would experience a shear stress of about 25 000 Pa and the upper surface of the body a shear stress between 50 and 1000 Pa. Such large shear stresses will approach or exceed the yield strength of the material, and may favour mixing by 'ripping off' distinct aggregates of material (as sometimes seen in type II erosion of cohesive beds; Amos *et al.* 1996, 1997). The presence of variable-sized clasts on the surface of the debris flow may also generate localized erosion, such that aggregates are undermined and detached. Marr *et al.* (2001) and Coussot (1997) discuss the general scaling of laboratory debris flow experiments, based on the simplifying assumptions that the debris can be treated as a Bingham plastic. Considering the ratio of material shear strength (τ_c) and the shear stress experienced by the debris flow material, this scaling analysis suggests that weakly coherent laboratory flows with low yield strengths ($\tau_c \sim$ 10 Pa) reproduce the behaviour of thicker and higher velocity submarine flows with much higher yield strengths ($\tau_c \sim$7500 Pa). A similar analysis may be undertaken, comparing the ratio of the critical shear stress necessary for shear mixing (τ_s) and the shear stress imparted on the surface of the debris flow. Such an analysis indicated that laboratory experiments involving material with low τ_s may reproduce the mixing behaviour of submarine flows with a higher τ_s.

The following analysis of mixing behaviour uses a number of empirically derived constants. In particular, the constant M in Equation 5 relates excess shear stress on the bed to mixing rate. The value of M is likely to vary with the characteristics of the debris, such as mud content and sediment volume concentration. The single value of M used in the following analysis is a very crude approximation, and further experimental work is needed to assess how M varies with a range of factors. As previously discussed, the estimates of the drag coefficients for the head and upper body of the flow are based on turbulent flow over a rough rigid plate. Further work is needed to determine more accurately the drag coefficients for turbulent flow over deformable debris flow material.

Reduction of shear mixing by small percentages of mud

It has been suggested, for both lahars and submarine debris flows, that the presence or absence of a small amount of cohesive mud determines whether the debris flow does or does not undergo significant dilution (Hampton 1972; Vallance & Scott 1997). The observation that glacigenic debris flows undergo minimal dilution is consistent with this argument, as they comprise up to 70% mud. The addition of a small amount of mud does significantly increase the critical shear stress needed to initiate shear mixing (Fig. 3; Mitchener & Torfs 1996). An order of magnitude reduction in shear mixing rates, and a change to erosion of large clasts, also occurs when a small amount of mud (~3%) was added to the bed (Mitchener & Torfs 1996; their fig. 3). Thus, a small amount of mud has a dramatic effect on shear mixing rates and processes. A small increase in the concentration

of a mud suspension will also produce significant increases in yield strength and viscosity that will increase the critical velocity at which the debris flow becomes turbulent and inhibit mixing of water over-ridden by the head. These arguments for emphasizing the importance of small percentages of mud differ somewhat to that of Hampton (1972, 1975) and Shanmugam & Miola (1995). These authors emphasized how small amounts of mud can produce a yield strength sufficient to suspend dispersed sand grains, rather than their ability to inhibit mixing processes in debris that may have relatively high concentrations of sand.

Comparative analysis of mixing behaviour

The mixing behaviour of submarine debris flows is illustrated in Figure 7. The analysis was carried out for flows comprising 'strong' (ρ = 1800 kg m^{-3}, τ_s = 2 Pa, τ_c = 5000 Pa, υ = 200 Pa s) or 'weak' (ρ = 1100 kg m^{-3}, τ_s = 0.1 Pa, τ_c = 10 Pa, υ = 2 Pa s) material. Estimates of the drag coefficient (C_D = 0.5 for the head and 0.001–0.02 for the upper surface) allow the shear stress on the flow boundary to be estimated for a given velocity (Eqn. 1). If this shear stress exceeds that needed to initiate shear mixing, the excess shear stress is used to calculate shear mixing rates using Eqn. 5. The critical velocity at which hydroplaning may occur, if over-ridden water is not mixed into the head, can be estimated using Eqn. 3. The transition from laminar to turbulent flow within a submarine debris flow is likely to have important implications for its mixing behaviour (Kuenen & Migliorini 1950; Morgenstern 1967). Mixing can occur much more rapidly at the margins of a turbulent flow, due to the diffusion of momentum and fluid across the flow's boundary. Turbulent flow began in clay-water suspension slurry flows within pipes when the Hampton number (H) reached a critical value of about 1000 (Hampton 1972; Enos 1977). The Hampton number is defined as,

$$H = \rho U^2 / \tau_c \qquad (10)$$

where U is the mean flow velocity, ρ is the fluid density, and τ_c is the yield strength of the material.

An alternative criterion for the onset of turbulence in non-Newtonian flows has been provided by Liu & Mei (1990). An effective Reynolds number (R^*) is defined empirically, such that

$$1/R^* = 1/R^\upsilon + 1/R^\tau \qquad (11)$$

and

$$R^\upsilon = 4\rho U D / \upsilon \text{ and } R^\tau = 8\rho U^2 / \tau_c \qquad (12)$$

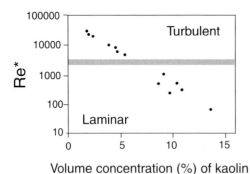

Volume concentration (%) of kaolin

Fig. 6. Experimental data from subaqueous density flows (from Van Kessel & Kranenburg 1996). Plot of effective Reynolds number (Re_* defined in Eqn. 11) against volume concentration of kaolin in dispersion. Flows that were visually observed to be laminar or turbulent are indicated.

where D is the flow thickness and v is viscosity. At high sediment concentrations, viscosities and yield strengths, R^T determines the effective Reynolds number. (q.v. Enos 1977, his fig. 4). A critical value of the effective Reynolds number (R^*) of about 2000–3000 has been observed to separate laminar and turbulent flow of fluid mud in subaqueous gravity currents and sub-aerial mudflows (Fig. 6; Liu & Mei 1990; van Kessel & Kranenburg 1996). This latter approach may be favoured as it has been able to predict the laminar-turbulent flow in sub-aqueous fluid mud flows (Fig. 6; Van Kessel & Kranenburg 1996).

Mixing of over-ridden water into the head is not quantified in this analysis (Fig. 7), although its effects are discussed qualitatively. The analysis is very crude. Primarily, it shows the need for further experimental work, for instance to determine how the value of M in Equation 5 varies with material properties. However, some robust conclusions can be drawn, and the potentially variable nature of mixing processes is well illustrated. Glacigenic debris flows have a relatively 'strong' rheology that will be approximated by the 'strong material' in this analysis.

Initiation and rates of shear mixing

Shear mixing initiates at relatively modest velocities, first when velocities reach 1.5 to 8.5 cm s^{-1} at the head and then between 7.5 to 190 cm s^{-1} for the upper surface. This indicates that submarine debris flows have to travel at very slow velocities to avoid at least a minimal degree of mixing. For a given flow velocity, rates of shear mixing per unit area at the head are 25 to 500 times those along the upper surface, assuming uniform material properties (i.e. uniform τ_s). The rate of shear mixing at a head comprising strong material is similar to that on the upper surface of a debris flow comprising weak material. Thus, changes in material properties can counterbalance differences in drag coefficients. The total amount of material eroded from a debris flow by mixing at the head and along the upper surface is also dependent on the area of the flow boundary involved in each process.

The length of a glacigenic debris flow greatly exceeds the thickness of the flow at the head, typically by a factor of well over 500 (Taylor *et al.* 2002). Thus, it might be expected that most material is eroded from the upper boundary of such a flow, despite much lower erosion rates per unit area. It is apparent that the depth of erosion due to shear mixing on the upper surface, experienced by a submarine debris flow that travels 100 km, is highly variable. The analysis predicts that a 50 m thick debris flow comprising strong material will not be completely eroded unless it has a velocity greater than 100 m s^{-1}. (However, at such high velocities the flow would be turbulent, and mixing rates would exceed those predicted by Eqation 5). A weak debris flow of the same thickness may be completely diluted if its velocity exceeds 2 m s^{-1}. This variability may partly explain our observation that some debris flows transform to turbidity currents whereas others do not. It may also show why glacigenic debris flows with a 'strong' rheology fail to dilute significantly over run-out distances of 100–200 km. For the weak material in a 10 m thick flow, a transition from laminar to turbulent flow can occur at velocities of about 2–3 m s^{-1}. Hydroplaning also theoretically occurs for such weak flows at these velocities (Fig. 7); however, hydroplaning is unlikely to occur because over-ridden water will mix with the head. Hydroplaning is more likely to occur for the 'strong' 10 m thick flow at velocities of about 7 m s^{-1} (Fig. 7). Somewhat greater flow velocities are needed (~20 m s^{-1}) before this 'strong' material undergoes a transition from laminar to turbulent flow. If it is assumed that a glacigenic (or other) submarine debris flow remains in a laminar state, a constraint can be placed on the maximum shear-mixing rate it experienced. This maximum shear rate is about 0.5 m min^{-1} for the body and 10 m min^{-1} for the head.

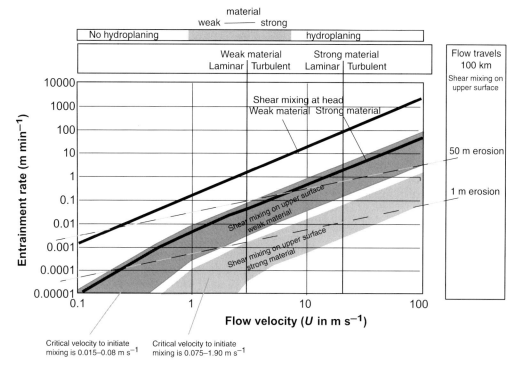

Fig. 7. Plot showing entrainment rate (in m min⁻¹) for a debris flow travelling at velocity U (m s⁻¹) for shear mixing at the head of the flow and on the upper surface of the flow. Drag coefficients of between 0.001 and 0.02 are used for the upper surface of the flow, and a value of 0.5 for the drag coefficient at the front of the flow. Entrainment rates are shown for debris flows comprising 'strong' ($\rho = 1800$ kg m⁻³, $\tau_s = 2$ Pa, $\tau_c = 5000$ Pa, $\upsilon = 200$ Pa s) or 'weak' ($\rho = 1100$ kg m⁻³, $\tau_s = 0.1$ Pa, $\tau_c = 10$ Pa, $\upsilon = 2$ Pa s) material. The velocity at which there is a transition from laminar to turbulent flow (Eqn. 11 and 12) and at which hydroplaning may occur for a 10 m thick debris flow on a 1° slope (Eqn. 3; Mohrig *et al.* 1998) are marked for strong and weak material. The velocity and entrainment rates corresponding to 1 and 50 m of erosion on the upper surface of the flow are indicated for a flow that travelled 100 km.

Shear mixing of water over-ridden by the head

Water over-ridden by a submarine debris flow can either mix into and dilute the head of the flow or remain at base of the flow. If over-ridden water cannot diffuse through the body of a submarine debris flow, over the timescale of the flow, it will remain near the base of the flow. Hydroplaning may then occur due to the presence of a thin layer of water or dilute sediment between the debris flow and the sea floor (Mohrig *et al.* 1998). The empirical data synthesized in this study suggest strongly that water over-ridden by a debris flow into a basal layer is likely to undergo shear mixing. Assuming drag coefficients between 0.01 to 0.20 and densities between 1800 and 1100 kg m⁻³, Equations 1 and 5 indicate that shear mixing will

be initiated at velocities between 0.075 – 1.9 m s⁻¹. Thus, it is predicted that shear mixing will frequently produce a relatively dilute sediment suspension at the base of the flow (*q.v.* Stix 2000), which will have high pore fluid pressures if water cannot escape through the body of the debris flow. A dilute layer at the base of the flow may also originate from shearing of soft pre-existing sea-floor sediment that has a high water-content (Gee *et al.* 1999).

Conclusions

Recent experimental data allow rates of shear mixing to be predicted for beds of cohesive sediment (or fluid mud) overlain by a turbulent dilute flow. These data show that shear mixing can initiate at relatively low velocities and that the addition of small amounts of mud (about

3%) can significantly decrease shear mixing rates. Shear mixing rates at variable flow velocity can be estimated crudely and compared to the critical velocity necessary for hydroplaning and the transition from laminar to turbulent flow. This analysis, carried out for 'strong' or 'weak' debris, illustrates why submarine debris flows undergo highly variable degrees of mixing. The analysis is crude and also emphasizes the need for further experimental and numerical analysis. In particular, the degree to which co-genetic turbidity currents can protect the upper surface of a submarine debris flow needs to be investigated. Better constraints on the drag coefficients for the surface of the debris flow, and the way in which debris properties influence the constant M in Equation 5, are also needed to produce a more accurate quantitative analysis of shear mixing behaviour.

This collaborative work formed part of NERC grant GR8/4894 awarded to Talling and NERC Ocean Margins LINK programme grant NER/T/S/2000/01403 and 01402 awarded to Bristol University, Leeds University, and Southampton Oceanography Centre. The LINK award was co-funded by CONOCO. The cores described in Figure 1 were collected under NERC grant GST/02/2198. We thank Anders Elverhøi for his extremely helpful comments on a draft of this paper.

References

ALLEN, J. R. L. 1971. Mixing at turbidity current heads, and its geological implications. *Journal of Sedimentary Petrology*, **41**, 97–113.

AMOS, C. L., SUTHERLAND, T. F. & ZEVENHUIZEN, J. 1996. The stability of sublittoral, fine grained sediments in a subarctic estuary. *Sedimentology*, **43**, 1–19.

AMOS, C. L., FEENEY, T., SUTHERLAND, T. F. & LUTERNAUER, J. L. 1997. The stability of fine-grained sediments from the Fraser River Delta. *Estuarine, Coastal and Shelf Science*, **45**, 507–524.

BRITTER, R. E. & LINDEN, P. F. 1980. The motion of the front of a gravity current travelling down an incline. *Journal of Fluid Mechanics*, **99**, 531–543.

BRITTER, R. E. & SIMPSON, J. E. 1978. Experiments on the dynamics of a gravity current head. *Journal of Fluid Mechanics*, **88**, 223–240.

COUSSOT, P. 1997. *Mudflow rheology and dynamics*. IAHR Monograph, A.A. Balkema, Rotterdam.

DOWDESWELL, J. A., KENYON, N. H., ELVERHØI, A., LABERG, J. S., HOLLENDER, F.-J., MIENERT, J. & SIEGERT, M. J. 1996. Large-scale sedimentation on the glacier-influenced Polar North Atlantic margins: long-range side-scan sonar evidence. *Geophysical Research Letters*, **23**, 3535–3538.

ENOS, P. 1977. Flow regimes in debris flow. *Sedimentology*, **24**, 133–142.

FERNANDO, H. J. S. 1991. Turbulent mixing in stratified fluids. *Annual Reviews of Fluid Mechanics*, **23**, 455–493.

GEE, M. J. R., MASSON, D. G., WATTS, A. B. & ALLEN, P. A. 1999. The Saharan debris flow: an insight into the mechanics of long runout submarine debris flows. *Sedimentology*, **46**, 317–355.

HALLWORTH, M. A., PHILLIPS, J. C., HUPPERT, H. E. & SPARKS, R. S. J. 1993. Entrainment in turbulent gravity currents. *Nature*, **362**, 829–831.

HALLWORTH, M. A., HUPPERT, H. E., PHILLIPS, J. C. & SPARKS, R. S. J. 1996. Entrainment into two-dimensional and axisymmetric turbulent gravity currents. *Journal of Fluid Mechanics*, **308**, 289–311.

HAMPTON, M. A. 1972. The role of subaqueous debris flow in generating turbidity currents. *Journal of Sedimentary Petrology*, **42**, 775–793.

HAMPTON, M. A. 1975. Competance of fine-grained debris flows. *Journal of Sedimentary Petrology*, **45**, 834–844.

HARBITZ, C. B. 1992. Model simulations of tsunamis generated by the Storegga slides. *Marine Geology*, **105**, 1–21.

HOUWING, E-J. 1999. Determination of the critical erosion threshold of cohesive sediments on inter-tidal mudflats along the Dutch Wadden Sea coast. *Estuarine, Coastal and Shelf Science*, **49**, 545–555.

IVERSON, R. M. 1997. The physics of debris flows. *Reviews of Geophysics*, **35**, 245–296.

IVERSON, R. M. & VALLANCE, J. W. 2001. New views of granular mass flow. *Geology*, **29**, 115–118.

KANTHA, L. H., PHILLIPS, O. M. & AZAD, R. S. 1977. On turbulent entrainment at a stable density interface. *Journal of Fluid Mechanics*, **79**, 753–768.

KING, E. L., HAFLIDASON, H., SEJRUP, H. P. & LØVLIE, R. 1998. Glacigenic debris flows on the North Sea Trough Mouth Fan during ice stream maxima. *Marine Geology*, **152**, 217–246.

KUENEN, P. H. & MIGLIORINI, C. I. 1950. Turbidity currents as a cause of graded bedding. *Journal of Geology*, **58**, 91–127.

KUIPER, C., CORNELISSE, J. M. & WINTERWERP, J. C. 1989. Research on erosive properties of cohesive sediments. *Journal of Geophysical Research*, **94**, 14341–14350.

LABERG, J. S. & VORREN, T. O. 1995. Late Weichsalian submarine debris flow deposits on the Bear Island Trough Mouth Fan. *Marine Geology*, **127**, 45–72.

LIU, K. F. & MEI, C. C. 1990. Approximate equations for the slow spreading of a thin sheet of Bingham plastic fluid. *Physics of Fluids*, **A2**, 30–36.

MARR, J. G., HARFF, P. A., SHANMUGAM, G. & PARKER, G. 2001. Experiments on subaqueous gravity flows: the role of clay and water content in flow dynamics and depositional structures. *Geological Society of America Bulletin*, **113**, 1377–1386.

MASSON, D. G., VAN NIEL, B. & WEAVER, P. P. E. 1997. Flow processes and sediment deformation in the Canary Debris Flow on the NW African Continental Rise. *Sedimentary Geology*, **110**, 163–179.

MEHTA, A. J. & SRINIVAS, R. 1993. Observations on the entrainment of fluid mud by shear flow. *In*: MEHTA, A. J. (ed.) *Coastal and estuarine studies*,

American Geophysical Union, Washington D.C., **42**, 224–246.

MIDDLETON, G. V. & WILCOCK, P. R. 1994. *Mechanics in the earth and environmental sciences*, Cambridge University Press, Cambridge, 459 pp.

MITCHENER, H. & TORFS, H. 1996. Erosion of mud/sand mixtures. *Coastal Engineering*, **29**, 1–25.

MOHRIG, D., WHIPPLE, K. X., HONDZO, M., ELLIS, C. & PARKER, G. 1998. Hydroplaning of subaqueous debris flows. *Geological Society of America Bulletin*, **110**, 387–394.

MOHRIG, D., ELVERHØI, A. & PARKER, G. 1999. Experiments on the relative mobility of muddy sediments and subaerial debris flows, and their capacity to mobilise antecedent deposits. *Marine Geology*, **154**, 117–129.

MORGENSTERN, N. R. 1967. Submarine slumping and the initiation of turbidity currents. *In:* RICHARDS, A. F. (ed.) *Marine Geotechnique*. University of Illinois Press, Urbana, 189–210.

MULDER, T., SAVOYE, B. & SYVITSKI, J. P. M. 1997. Numerical modelling of a mid-sized gravity flow: the Nice 1979 turbidity current (dynamics, processes, sediment budget, and sea-floor impact). *Sedimentology*, **44**, 305–326.

NOREM, H., LOCAT, J. & SCHIELDROP, B. 1990. An approach to the physics and the modelling of submarine flow slides. *Marine Geotechnology*, **9**, 93–111.

OCKENDEN, M. C. & DELO, E. A. 1991. Laboratory testing of muds. *Geo-Marine Letters*, **11**, 138–142.

PARCHURE, T. M. & MEHTA, A. J. 1985. Erosion of soft cohesive sediment deposits. *Journal of Hydraulic Engineering ASCE*, **111**, 1308–1326.

PARSONS, J. D., WHIPPLE, K. X. & SIMIONI, A. 2001. Experimental study of the grian-flow, fluid-mud transition in debris flow. *Journal of Geology*, **109**, 427–447.

PIPER, D. J. W., COCHONAT, P. & MORRISON, M. L. 1999. The sequence of events around the epicentre of the 1929 Grand Banks earthquake: initiation of the debris flows and turbidity current inferred from side scan sonar. *Sedimentology*, **46**, 79–97.

SCHLICHTING, H. 1980. *Boundary layer theory*. McGraw-Hill, New York. 817 pp.

SCHWAB, W. C., LEE, H. J., TWICHELL, D. C., LOCAT, J., NELSON, H. C., MCARTHUR, W. G. & KENYON, N. H. 1996. Sediment mass-flow processes on a depositional lobe, outer Mississippi Fan. *Journal of Sedimentary Petrology*, **66**, 916–927.

SHANMUGAM, G. & MIOLA, R. J. 1995. Reinterpretation of depositional processes in a classic flysch sequence (Pennsylvanian Jackfork Group) Ouachita Mountains, Arkansas and Oklahoma. *American Association of Petroleum Geologists Bulletin*, **79**, 672–695.

SIMPSON, J. E. 1972. Effects of the lower boundary on the head of a gravity current. *Journal of Fluid Mechanics*, **53**, 759–768.

SIMPSON, J. E. 1997. *Gravity currents: in the environment and the laboratory*. Cambridge University Press, Cambridge, 244 pp.

SNYDER, D. & TAIT, S. 1998. A flow front instability in viscous gravity currents. *Journal of Fluid Mechanics*, **369**, 1–21.

STIX, J. 2000. Flow evolution of experimental gravity currents: implications for pyroclastic flows at volcanoes. *Journal of Geology*, **109**, 381–398.

STRANG, E. J. & FERNANDO, H. J. S. 2001. Entrainment and mixing in stratified shear flows. *Journal of Fluid Mechanics*, **428**, 349–387.

TAYLOR, J., DOWDESWELL, J. A., KENYON, N. H. & Ó COFAIGH, C. 2002. Late Quaternary architecture of trough–mouth fans: debris flows and suspended sediments on the Norwegian margin. *In:* DOWDESWELL, J. A. & Ó COFAIGH, C. (eds) *Glacier Influenced Sedimentation on High Latitude Continental Margins*, Geological Society, London, Special Publications, **203**, 55–71.

TEISSON, C., OCKENDEN, M., HIR, P. L. E., KRANENBURG, C. & HAMM, L. 1993. Cohesive sediment transport processes. *Coastal Engineering*, **21**, 129–162.

VALLANCE, J. W. & SCOTT, K. M. 1997. The Osceola Mudflow from Mt Rainier: Sedimentology and hazard implications of a huge clay-rich debris flow. *Geological Society of America Bulletin*, **109**, 143–163.

VAN KESSEL, T. & KRANENBURG, C. 1996. Gravity current of fluid mud on sloping bed. *Journal of Hydraulic Engineering ASCE*, **122**, 710–717.

WINTERWERP, J. C. & KRANENBURG, C. 1997. Erosion of fluid mud layers. II: Experiments and model validation. *Journal of Hydraulic Engineering ASCE*, **123**, 512–519.

WYNN, R. B., MASSON, D. G., STOW, D. A. V. & WEAVER, P. P. E. 2000. The Northwest African slope apron: a modern analogue for deep-water systems with complex seafloor topography. *Marine and Petroleum Geology*, **17**, 253–265.

WYNN, R. B., WEAVER, P. P. E., STOW, D. A. V. & MASSON, D. G. 2002. Turbidite depositional architecture across three interconnected deep-water basins on the Northwest African margin. *Sedimentology*, in press.

YOUNG, A. D. 1989. *Boundary layers*. BSP Professional Books, Oxford, 269 pp.

ZREIK, D. A., KRISHNAPPAN, B. G., GERMAINE, J. T., MADSEN, O. S. & LADD, C. C. 1998. Erosional and mechanical strengths of deposited cohesive sediments. *Journal of Hydraulic Engineering ASCE*, **124**, 1076–1085.

Late Oligocene and early Miocene glacimarine sedimentation in the SW Ross Sea, Antarctica: the record from offshore drilling

MICHAEL J. HAMBREY[1], PETER J. BARRETT[2] & ROSS D. POWELL[3]

[1]Centre for Glaciology, Institute of Geography & Earth Sciences, University of Wales, Aberystwyth, Ceredigion SY23 3DB, Wales (e-mail: mjh@aber.ac.uk)

[2]Antarctic Research Centre, Victoria University of Wellington, PO Box 600, Wellington, New Zealand

[3]Department of Geology & Environmental Geosciences, Northern Illinois University, DeKalb, Illinois 60115, USA

Abstract: Recent offshore drilling in the SW Ross Sea has recovered three cores with a cumulative thickness of 1500 m of Oligocene–early Miocene strata. Together with data from earlier drilling, notably from the CIROS-1 drillhole 70 km to the south (702 m of core), the cores record shallow marine glacigenic sedimentation at the margin of a rift basin bordering a mountain range that was breached by outlet glaciers from an ice sheet. This paper focuses on the late Oligocene–early Miocene record, which preserves sedimentary facies that represent warmer glacier ice and climatic conditions than are evident in the Antarctic today. Strata were cored near the South Victoria Land coast, and sedimentary facies include diamictite, conglomerate, breccia, sandstone, siltstone, mudstone and rhythmite. Facies associations, combined with seismic stratigraphic data, indicate an alternating proximal and distal marine record of glacigenic sedimentation, including phases of glacier grounding and variable degrees of iceberg rafting. Reworking by gravity-flow processes and near-shore submarine currents is also evident. These facies, together with evidence for periglacial vegetation, provide evidence of a late Oligocene–early Miocene climatic regime resembling that in the high-Arctic today. Thus, the climate was transitional between cool-temperate conditions of early Oligocene and cold-polar conditions of Quaternary time.

The last three decades have revolutionized our understanding of the pre-Quaternary history of Antarctica, largely as a result of offshore drilling by the Ocean Drilling Program, and by New Zealand- and American-led programmes, with onshore investigations of deposits of controversial age. The history of glaciation is now known to extend back to at least earliest Oligocene time, but the older record is quite different in terms of glacier thermal regime from that which typifies the ice sheet today. Although lithological logging has already defined the main depositional environments in the context of proximity to the glaciers and to sea-level change (Hambrey *et al.* 1989; Fielding *et al.* 1998, 2000; Powell *et al.* 1998, 2000), there is a need to identify more precisely the facies associations with the wide variety of possible sub-environments.

The aims of this paper are twofold:

(i) To evaluate the overall context for Cenozoic strata in the western Ross Sea, based on drilling data, and review the key evidence that bears on our understanding of climatic evolution; and

(ii) To summarize characteristic facies associations, and interpret them in the light of equivalent modern environments, by focusing on the late Oligocene–early Miocene interval, which offers the most varied range of glacimarine environments in the Cenozoic sequence. Two main locations are compared, both about 12 km offshore, but in different geomorphological settings with respect to the termini of major outlet glaciers from the East Antarctic Ice Sheet.

Rationale for drilling in the Ross Sea

For establishing a Cenozoic palaeoenvironmental record, the Ross Sea continental shelf has been the most intensively investigated part of Antarctica. Although the central themes have been the inception and evolution of the Antarctic ice sheet and their relationship with climate, tectonic history has also been seen as an important and related focus. Most work has concentrated on the southwestern part of the Ross Sea, in McMurdo Sound, where there is a considerable logistical infrastructure associated

From: DOWDESWELL, J. A. & Ó COFAIGH, C. (eds) 2002. *Glacier-Influenced Sedimentation on High-Latitude Continental Margins.* Geological Society, London, Special Publications, **203**, 105–128. 0305-8719/02/$15.00
© The Geological Society of London 2002.

Table 1. *Drill holes of the Ross continental shelf and coastal Victoria Land from which Cenozoic strata have been recovered.*

Hole	Year	Latitude	Longitude	Elevation (m)	Penetration (m)	Percent recovered	Oldest Cenozoic strata
DSDP270	1973	77°26.48'S	178°30.19'W	−634	422.5*	62.4	late Oligocene
DSDP271	1973	76°43.27'S	175°02.86'W	−554	265.0	5.8	Pliocene
DSDP272	1973	77°07.62'S	176°45.61'W	−629	443.0	36.6	early Miocene
DSDP273 & 273A	1973	74°32.29'S	174°37.57'E	−495	346.5	24.1	early Miocene
DVDP10	1974	77°34.72'S	163°30.70'E	+2.8	185.9	83.4	early Pliocene
DVDP11	1974	77°35.40'S	163°24.67'E	+80	328.0	94.1	late Miocene
DVDP12	1974	77°38.37'S	162°51.22'E	+75	166*		Pliocene
DVDP15	1975	77°26.65'S	164°22.82'E	−122	65	52.0	?Pliocene–Pleistocene
MSSTS−1	1979	77°33.43'S	163°23.21'E	−195	229.6	56.1	late Oligocene
CIROS−2	1984	77°41'S	163°32'E	−211	168.1	67.0	early Pliocene
CIROS−1	1986	77°04.91'S	164°29.93'E	−197	702.1	98.0	late Eocene
CRP−1	1997	77.008°S	163.755°E	−153.5	147.69	86	early Miocene
CRP−2/2A	1998	77.006°S	163.719°E	−177.94	624.15	95 (below −45.97m)	early Oligocene
CRP−3 (at 823 mbsf)	1999	77.006°S	163.719°E	−295	939	97	earliest Oligocene

* excludes basement

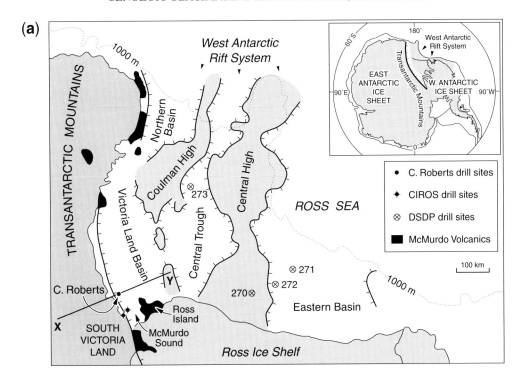

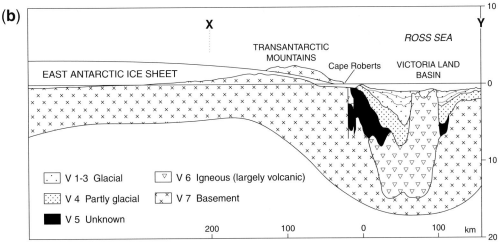

Fig. 1. Tectonic setting of drill sites in the western Ross Sea region. (**a**) Generalized map of rifts and bathymetric highs, with locations of drill sites; inset of Antarctica shows location of the West Antarctic rift system. (**b**) Cross-section through the Transantarctic Mountains and adjacent Victoria Land basin in the vicinity of the Cape Roberts drill sites, illustrating nature of basement, sedimentary basin fill and volcanic rocks (after Cape Roberts Science Team 1999).

with America's McMurdo Station and New Zealand's Scott Base.

Our knowledge of the Cenozoic history of the region stems from drilling at 14 sites since the early 1970s from the western side of McMurdo Sound and the more open waters of the central Ross Sea (Table 1). Four phases of drilling have now taken place in the region:

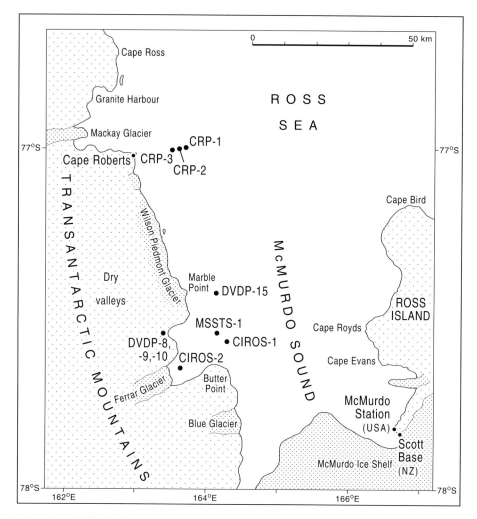

Fig. 2. Location of drill sites in the McMurdo Sound region.

(i)   The USA-led Deep Sea Drilling Project of
      1973;
(ii) The USA/New Zealand/Japan Dry Valleys
      Drilling Project of 1974;
(iii) New Zealand-led drilling in McMurdo
      Sound (1979–1986) and
(iv) The multinational Cape Roberts Project in
      NW McMurdo Sound in 1997–1999 (Table
      1, Figs 1, 2).

History of drilling through glacimarine strata in the western Ross Sea

Several comprehensive reviews of drilling in the
Ross Sea region have been published, embracing all operations except the most recent, Cape

Roberts, phase of drilling (McKelvey 1991;
Hambrey & Barrett 1993; Barrett, 1996). Thus
only brief summaries of the earlier drilling
results are given here. The stratigraphic ranges
for each of the key drill sites, plotted against the
Berggren *et al.* (1995) time scale (Fig. 3),
indicate predominantly proximal to distal
glacimarine sedimentation through the earliest
Oligocene to Quaternary interval, although
numerous hiatuses occur in the record.

Deep Sea Drilling Project of 1973

Four cores were recovered during Leg 28 of the
Deep Sea Drilling Project (DSDP) in 1973 on
the continental shelf of the central Ross Sea

(Hayes & Frakes 1975) (Fig. 1). The sites, DSDP 270, 271, 272 and 273, lie on a transect running from the SW near the edge of the Ross Ice Shelf to the NE, sampling progressively younger strata that were truncated during the main glacial advances of Plio-Pleistocene time. Core recovery was best in the oldest strata at Site 270 (62%), where the sediments consist of diamictite and mudstone with iceberg-rafted debris, resting on a thin greensand unit. Diatoms and foraminifera, together with magnetostratigraphy provide a chronology that indicates that the oldest glacimarine strata are of late Oligocene age. Younger sediments with variable proportions of iceberg-rafted debris occur at the other sites. The most striking finding of DSDP Leg 28 was the recognition that the history of Antarctic glaciation was an order of magnitude longer than that of northern hemisphere ice sheets, contrary to what most geologists believed at the time.

Dry Valleys Drilling Project of 1974

Onshore drilling adjacent to McMurdo Sound was undertaken by the Dry Valley Drilling Project (DVDP) in 1974 (McGinnis 1981) (Fig. 2). The resulting cores illustrate primarily a fjordal sequence. The longest, DVDP 11 from Taylor Valley, spans parts of the interval from late Miocene to Holocene. Facies are dominated by diamictite, conglomerate and sandstone (Powell 1981). The diamictites were believed to have formed close to the grounding line of a tidewater glacier ice front, whereas the conglomerate and sandstone suggest the influence of abundant meltwater. Following the onshore drilling, a 65 m hole was obtained from the sea ice offshore at DVDP 15. Although of only limited success, this hole indicated the potential of drilling from sea ice.

New Zealand-led drilling of 1979–1986

The next phase of drilling in McMurdo Sound recovered cores named MSSTS-1 in 1979 (Barrett 1986) and CIROS-1 in 1986 (Barrett 1989a) to the east of the major fault that separates the Transantarctic Mountains from the Victoria Land Basin. CIROS-1 was drilled offshore from the central flow-line of Ferrar Glacier (Fig. 2), and the 702m-long core remained the deepest and most complete recovered from Antarctica until the drilling of Cape Roberts hole CRP-3 in 1999. With 98% recovery, the CIROS-1 core permitted detailed integration of palaeoenvironmental, biostratigraphic and palaeomagnetic data. The CIROS-1

core is dominated by diamictite, sandstone and mudstone, and there are lesser amounts of conglomerate and intraformational breccia. Based on siliceous and calcareous microfossils, this core initially established a glacial record extending back into the early Oligocene Epoch, but subsequent work has demonstrated that the lower part of the core is probably of late Eocene age (Hannah 1997; Wilson et al. 1998). The discovery of a well-preserved beech leaf in the upper Oligocene part of the core indicated that shrubby vegetation was growing as glaciers reached the sea, thus recording a much warmer climatic regime than that of today (probably similar to the modern maritime high-Arctic).

To complement the CIROS-1 site, CIROS-2 (Fig. 2) was drilled close to the terminus of Ferrar Glacier, and yielded a late Miocene–Pleistocene record of glacial fluctuations on the Transantarctic Mountains side of the large rift bounding the Victoria Land Basin (Barrett & Hambrey 1992). Facies in this core suggest fjordal and lake settings depending on whether glaciers entered Ferrar Fjord from the interior, or blocked the entrance from the east, respectively. Facies further indicate a cold climate.

Cape Roberts Project of 1997–1999

In the early 1990s a multinational group was established to drill 70 km north of the earlier McMurdo Sound sites, near the western margin of the Victoria Land Basin off Cape Roberts (Barrett & Davey 1992). Financed largely by Italy, New Zealand and the USA, with additional contributions from Australia, Germany, The Netherlands and the UK, drilling of three holes, CRP-1, -2 and -3 took place in 1997, 1998 and 1999 respectively (Table 1, Fig. 2). The sites were located seawards of the terminus of Mackay Glacier, which like the Ferrar is another outlet from the East Antarctic Ice Sheet. The three holes recovered a cumulative thickness of some 1500 m of slightly overlapping strata, and obtaining a record of early Oligocene to early Miocene (34 to 17 Ma) sedimentation and climate, as well as a thin Pliocene–Quaternary cover. Early results are reported in the Initial Reports of CRP-1, CRP-2/2A and CRP-3 (Cape Roberts Science Team 1998, 1999, 2001), and in companion Scientific Results volumes (Hambrey et al. 1998; Barrett et al. 2000, in press).

At site CRP-1, a Quaternary sequence, 43 m thick, was cored that included glacigenic sediments similar to those of today, as well as a 2m-thick interglacial fossiliferous carbonate

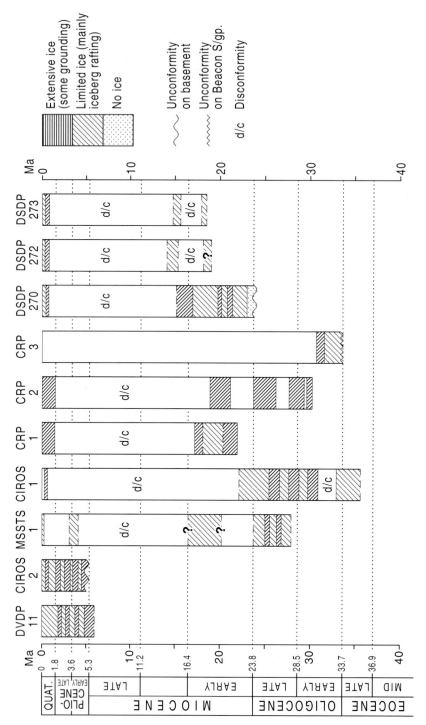

Fig. 3. Summary of core data, plotted against time, for drill sites in the western Ross Sea, illustrating relative extent of ice and disconformities. Time scale follows Berggren *et al.* (1995).

sequence at 32 mbsf around 1 Ma old. The lower part of the core, which is early Miocene in age, comprises diamictite, sandstone and siltstone reflecting eight cyclic fluctuations of glaciers and perhaps sea level, under conditions warmer than those of today (Fielding *et al.* 1998). Palynological studies indicate a herbmoss tundra in the coastal foothills at this time (Raine 1998). These fluctuations have been determined using the twin approaches of sequence stratigraphy and process-based sedimentology, supported by palaeoecological evidence. Unfortunately, a severe storm weakened the sea ice platform so much that drilling was abandoned prematurely (Cape Roberts Science Team 1998).

CRP-2/2A cored lower Oligocene to lower Miocene strata, unexpectedly recovering parts of the same interval as CIROS-1 70 km to the south, (Cape Roberts Science Team 1999; Barrett *et al.* 2000). The combination of volcanic ash, biostratigraphic, isotopic and palaeomagnetic data is providing a stratigraphic record with temporal resolution in part comparable with that from mid and low latitudes, and in places to within ±0.05 Ma. Through sedimentological investigations, glacimarine depositional environments, representing the open coast and continental shelf, have been recognized. Also sedimentary features indicate that the climatic setting becomes progressively warmer downcore, so in late Oligocene to early Miocene times the glaciers were probably polythermal (comparable to those in the high-Arctic today) with considerable meltwater discharge, whereas older strata indicate temperate glacial conditions. Sedimentation rates calculated for the interval 39–56 mbsf from magnetic data, that could be equally representative of other parts of the core, are 2 m Ka^{-1}, which is of the same order as modern Arctic systems (Elverhøi *et al.* 1983; Laberg & Vorren 1996). Palynological studies suggest that on-land vegetation was a herb moss tundra from 19 to 24 Ma with increasing elements of a low woodland *Nothofagus* downcore in strata from 29 to 31 Ma, implying a cool temperate climate (Askin & Raine 2000). Once again, long cycles of sedimentation can be recognized (24 in this case), which are interpreted as recording glacial and probably some sea-level fluctuations. Detailed sedimentological study and ash-based magneto-biostratigraphy of three particularly thick cycles (9, 10, 11 from 130 to 307 mbsf) indicate that the cycles were deposited in 100 Ka, 40 Ka and 40 Ka respectively, from 23.7 to 24.1 Ma ago, the first demonstration of orbital forcing of the Antarctic Ice Sheet in the distant past (Naish *et al.* 2001). CRP-2/2A also helps to elucidate the early tectonic history of the Victoria Land Basin and the adjacent Transantarctic Mountains, with progressive erosion of the mountains being recorded in clasts and minerals in the core. Clast lithologies (Talarico *et al.* 2000) and sandstone mineralogy (Smellie 2000) show that Beacon Supergroup and Ferrar dolerite dominate the lower Oligocene part of the core. In contrast, the upper part of the core contains mainly basement-derived products, possibly indicative of mid-Cenozoic erosion of an already established mountain range. The shallow-water character of the early Miocene Cape Roberts strata indicates that there has been virtually no offshore (marine) subsidence (Cape Roberts Science Team 2000, Chapter 7).

The target of CRP-3 was recovery of strata transitional from those influenced by polythermal glaciers, already demonstrated in CRP-2/2A, down through temperate glacial facies, and into warm marine or terrestrial facies (Cape Roberts Science Team 2000, Ch. 7). Instead, the 939 m-long core represented only 3 million years of Cenozoic time (early Oligocene, 31–34 Ma), an unconformity representing over 300 Ma, and over 100 m of Devonian strata beneath.

The Cenozoic succession consists of diamictite, clast-supported conglomerate, a basal breccia, sandstone and mudstone. Numerous sequence boundaries have been recognized, interpreted in the context of ice fluctuations (including grounding) across an open marine shelf or contributing to the development of a proglacial delta within the lower strata. Based on the abundance of meltwater-influenced facies, combined with palaeoecological evidence, a cool-temperate climate, with glaciers debouching into the sea most of the time, is envisaged.

Geographical and geological setting of late Oligocene to Miocene strata

Geographical setting

Drill sites CIROS-1, CRP-1 and CRP-2/2A (Table 1), which yielded the relevant interval of upper Oligocene to lower Miocene strata, lie immediately seawards of the Transantarctic Mountains. Through this range, numerous outlet glaciers flow from the East Antarctic Ice Sheet (Fig. 1b). The mountains here represent a rift shoulder that rises to over 4000 m, and they extend across most of the continent after running along the coast of Victoria Land. To the east lies a series of north–south-trending basins that dissect the Ross Sea continental shelf. The southernmost drill site, CIROS-1 lies on the axis of the central flow-line of an outlet glacier, the Ferrar Glacier, whereas the Cape Roberts sites like 70 km to the north, offset from the central flow-line of another outlet glacier, the Mackay Glacier. The cores were taken from what is interpreted as a near-shore sedimentary prism that extends along the coast, although much of the prism has been eroded at the Cape Roberts sites. Sediments from both these sites reveal a long history of glacial and deltaic deposition, as well as reworking by waves and by sediment gravity flows.

Geological setting of western Ross Sea

The Ross Sea continental shelf consists of a series of four north–south-trending rift basins with intervening highs (e.g. Cooper *et al.* 1994). DSDP sites 271 and 272 were drilled in the Eastern Basin, and site 273 on the Central High to the west. The other offshore drill sites, including those that are the main focus of this paper, were drilled near the western margin of the Victoria Land Basin, just offshore from the rising Transantarctic Mountain front in McMurdo Sound (Figs 1a, 2). This basin is itself a complex structure, comprising a large basin some 14 km deep in the west, separated from a sub-basin to the east by a volcanic zone that apparently continues into the volcanic Ross Island (Fig. 1b) (Cape Roberts Science Team 1998; Hamilton *et al.* 2001). The age of these basins is not well constrained, but two main extensional events have been recognized:

(i)   An early non-magmatic rifting event over the whole of the Ross Sea associated with late Mesozoic break-up of Gondwana.
(ii) A later event previously presumed to be associated with volcanic activity in the Victoria Land Basin in Eocene and later time (Cooper *et al.* 1987). Now it is known from Cape Roberts drilling to be un-related to the initiation of volcanism (mid-Oligocene), and to date only from earliest Oligocene times (Cape Roberts Science Team 1999, 2000).

The Cenozoic strata in McMurdo Sound dip gently eastwards from the western shelf, and continue under Ross Island. Seven seismic stratigraphic units have been recognized (V1–V7, top to bottom: Fig. 1b), and their extent defined by multichannel seismic investigations (Cooper *et al.* 1987; Brancolini *et al.* 1994; Bartek *et al.* 1996), and constrained by drilling. The Cenozoic/Beacon Supergroup

boundary cannot be identified because Oligocene and Devonian strata beneath Roberts ridge both have similar seismic velocities and similar attitudes.

In the Transantarctic Mountains, the dominant source of sediment for the Victoria Land Basin, the rock succession comprises granitic basement, separated from mainly flat-lying Devonian–Triassic terrestrial sedimentary rocks of the Beacon Supergroup, and basic igneous rocks of the Jurassic Ferrar Supergroup. Exposures of pre-Quaternary diamictites, known as the Sirius Group are scattered throughout the area, but their age (variously interpreted as Miocene or Pliocene) has been the subject of considerable debate (Barrett 1996), and remains unresolved.

For more focused investigations of glacial sedimentary environments, we compare two Cape Roberts cores (CRP-1 and CRP-2/2A) with CIROS-1 (Fig. 4). These cores are largely coeval and, with near-complete recovery, allow comparisons of glaciological settings.

Facies analysis

The three cores described here were all logged lithologically on-site. Routine observations included: texture, bedding characteristics, sedimentary structures, clast shape and fabric characteristics and structural analysis. These observations were supported by laboratory-based studies, focusing on particle-size analysis and microstructural (or 'micromorphological') analysis. In addition, a wide range of bio-stratigraphical, palaeoecological, provenance, diagenetic and chronological studies were undertaken, some of which amplify our understanding of depositional conditions. All these data are presented in a series of '*Initial Reports*' and '*Scientific Results*' volumes (Barrett 1989; Cape Roberts Science Team 1998, 1999, 2001; Barrett *et al.* 2000; Hambrey *et al.* 1998).

Lithofacies and their interpretation

Eleven lithofacies are defined for the upper Oligocene/lower Miocene interval in the three cores. Representative facies are illustrated in Figure 5. The basic descriptions and generalized interpretations are given in Table 2; detailed accounts are given elsewhere: CIROS-1 (Hambrey *et al.* 1989; Fielding *et al.* 1997), CRP-1 (Fielding *et al.* 1998; Howe *et al.* 1998; Powell *et al.* 1998), and CRP-2/2A (Fielding *et al.* 2000; Powell *et al.* 2000). The eleven facies represent the entire spectrum of facies from conglomerates and breccias, through diamictites, to

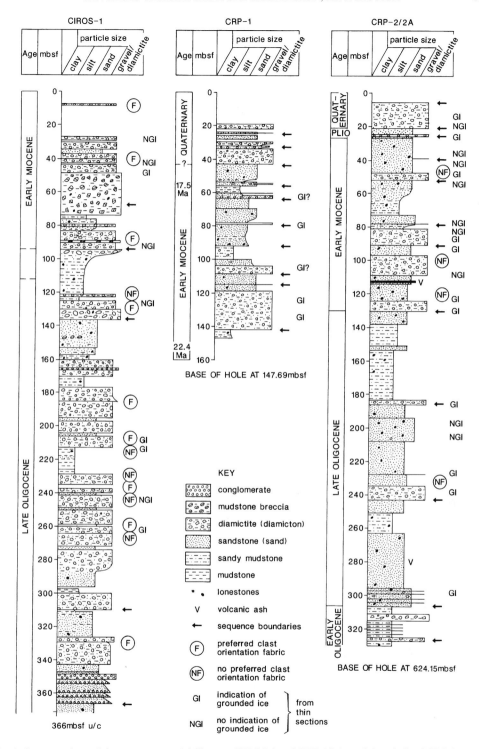

Fig. 4. Summary logs of the upper parts of drill-cores CIROS-1 and CRP-2/2A, and the whole of CRP-1. Sequence boundaries, clast fabric information and indications of glacier-grounding from thin sections are also shown.

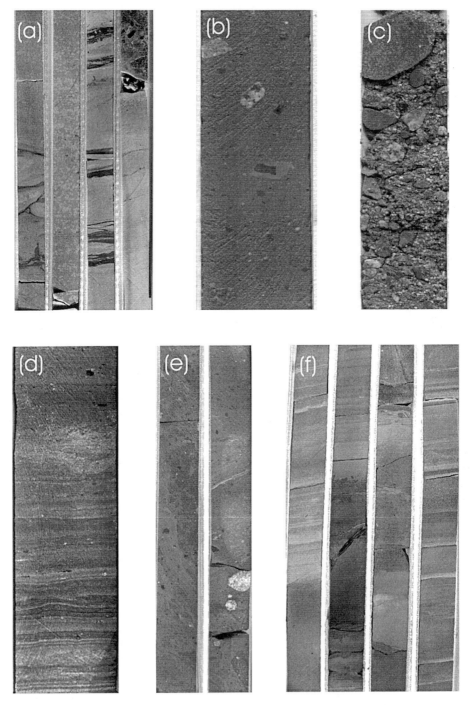

Fig. 5. Representative lithofacies from CRP-1 and CRP-2/2A drill cores: (a) well-sorted sandstone, (b) massive diamictite, (c) clast-supported conglomerate, (d) rhythmites of interlaminated fine-grained sandstone and mudstone, with lonestones up to pebble size, (e) stratified muddy diamictite showing deformation by soft-sediment folding, (f) inter-stratified, moderately sorted sandstone (light colour) and mudstone (dark); note local bioturbation in mudstone. Core is 61mm wide in all cases. Typical facies in CIROS-1 are illustrated in Hambrey *et al.* (1989).

Table 2. *Upper Oligocene and Lower Miocene lithofacies and their interpretation in CIROS-1 and Cape Roberts Project cores. Percentages of lithofacies as a proportion of each core for these stratigraphic intervals are also given. Data for CRP-2/2A are from Powell et al. (2000).*

Lithofacies	Depositional process	Lithofacies percentages CIROS-1		CRP-1	CRP-2/2A	
		Lower Miocene	Upper Oligocene	Lower Miocene	Lower Miocene	Upper Oligocene
Mudstone; minor lonestones	Hemipelagic suspension settling, with minor iceberg rafting	4.3	17.4	26.3	33	11
Interstratified sandstone and mudstone	Sediment gravity-flow deposition; combined wave and current action	1.5	4.0	18.8	4	2
Poorly sorted (muddy) very fine to coarse sandstone	Sediment gravity-flow (turbidite) deposition; some mixing by bioturbation, loading	15.9	15.6	14.6	18	24
Moderately to well sorted, stratified or massive, fine to coarse sandstone	Marine currents/wave influence	5.5	12.4	9.7	21	47
Stratified diamictite	Subglacial or debris-flow deposition, or rain-out from icebergs	32.6	32.5	3.4	1	3
Massive diamictite	Subglacial or debris-flow deposition, or rain-out from icebergs	10.2	10.5	26.5	21	12
Rhythmite of sandstone and siltstone	Suspension settling from turbid plumes	0	0.9	0	0	1
Clast-supported conglomerate	Settling from submarine discharge of subglacial streams; fluvial or shallow marine deposition; redeposition of conglomerate by mass-flow	1.4	0.6	0	Tr	0
Matrix-supported conglomerate	High-density mass flow deposit; hyperconcentrated flow from subglacial stream; suspension settling and rainout	0	3.1	0	Tr	0
Mudstone breccia	Mass-flow deposition	28.6	3.1	0.6	Tr	0
Non-welded lapillistone	Airfall of volcanic ash through water; reworked by marine currents and sediment gravity flows	0	0	0	2	0
Total core recovered		**58 m**	**268 m**	**104 m**	**159 m**	**121 m**

sandstones and mudstones. Poorly sorted facies dominate. Bedding characteristics range from massive to well-stratified, and in places there is clear evidence of bioturbation and deformation.

The interpretation of these lithofacies has been made on the basis of our understanding of modern glacimarine environments, such as Alaska (e.g. Powell & Molnia 1989; Cai *et al.* 1997; Powell & Alley 1997), East Greenland (Dowdeswell *et al.* 1994; Syvitski *et al.* 1996; Ó Cofaigh *et al.* 2001) and Svalbard (Elverhøi *et al.* 1983; Bennett *et al.* 1999). Almost all the lithofacies can be explained in the context of a proximal to distal glacimarine environment in which large amounts of sediment are emplaced on the sea floor by direct deposition from a glacier, *via* iceberg rafting, from sediment plumes emanating from the glacier, and as a result of remobilization as subaqueous gravity flows. In addition, textural studies of the cores show progressive changes in sediment texture (most obviously recorded as mud content) that are closely comparable with fining offshore trends on sediment-fed wave-dominated coasts (Barrett 1989*b*; Dunbar *et al.* 1997; Dunbar & Barrett 2001).

Critical to determining the depositional environment is the proximity and style of glaciation, especially whether the sea floor was overridden by a glacier and the substrate subject to deformation, which would have influenced the stability of the overlying glacier (Boulton & Hindmarsh 1987; Alley *et al.* 1997; Murray 1997). This question may be resolved by undertaking two types of laboratory investigation:

Analysis of soft-sediment microstructures (Passchier *et al.* 1998; Passchier 2000). Subglacial deformation results in the development of shear zones, whilst hydrofaulting subglacially results in brecciation. Brecciation takes two forms: hairline fractures in sand and silt, and clastic dykes in diamictite. These structures have been described from the Cape Roberts cores, and their presence supports interpretations of glacier-grounding derived from lithofacies analysis, and suggests additional horizons where this has taken place.

Analysis of 'micromorphology' using thin sections. This approach focuses on the structures (rotational, shear and 'comet') and plasmic fabric (birefringence characteristics of clay-sized sediment) evident under cross-polarized light. The trigger for this investigation was the desire to test two contrasting hypotheses concerning the glacial signature in the upper part of the CIROS-1 core (Hiemstra 1999). The

initial interpretation (Hambrey 1989) was made at the drill-site at a time when little was known about processes in modern glacimarine environments, especially the roles of glacier grounding and sediment gravity flow. A general glacimarine setting was inferred, but with indications of lodgement till deposition at various levels on the basis of grain-fabric studies and texture (Fig. 4). This early interpretation of glacier-grounding was refuted by Fielding *et al.* (1997) using a sequence stratigraphic approach. Through thin-section studies of 12 diamictite samples, Hiemstra (1999) was able to recognize evidence for subglacial deformation, some of which coincided with Hambrey and colleagues' (1989) lodgement tills (Fig. 4). However, all the samples analysed showed a glacimarine aspect, thus indicating a deformable bed rather than direct deposition from a basal debris layer. Other samples did not show subglacial deformation; instead they were interpreted as sediment gravity flows. Thus, elements of both hypotheses proved to be correct. The same technique was also applied to the Cape Roberts cores. Five samples were analysed in CRP-1 (van der Meer & Hiemstra 1998), three of which showed clear evidence of glacier-grounding, whereas the other two showed only limited evidence (Fig. 4). In CRP-2/2A, twenty-six samples were analysed throughout the upper Oligocene/lower Miocene section (van der Meer 2000), half of which showed convincing evidence of grounding. It is relevant to note that these glacier-grounding indicators commonly coincide with sequence boundaries (Fig. 4) that have been interpreted as 'glacial surfaces of erosion' (Fielding *et al.* 1998, 2000), as discussed below.

These studies demonstrate the importance of laboratory investigations in complementing visual core descriptions for interpreting the origin of diamictites and other facies.

Lithofacies proportions

The relative proportions of the different lithofacies have been calculated for each site and for each interval; upper Oligocene and lower Miocene (Table 2). There do not appear to be any systematic differences, but a few general points can be made. Diamictite (stratified and massive) accounts for over 40% of the core in CIROS-1, which is located in line with the central flow-line of Ferrar Glacier. This figure contrasts with 15–30% for CRP-1 and CRP-2/2A, which are less centrally placed with reference to the Mackay Glacier. In contrast, the finer grained lithofacies are more important in

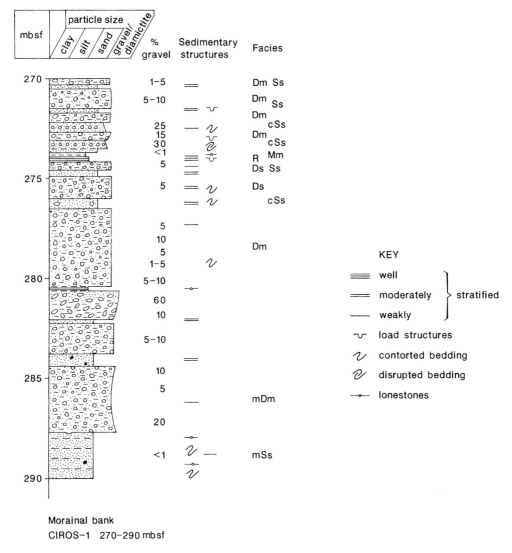

Fig. 6. Lithological log through an inferred morainal bank succession in the CIROS-1 core (270–290 mbsf), illustrating particle size, percent of gravel, and sedimentary facies.

the Cape Roberts cores, reflecting perhaps the greater input of subglacial meltwater-derived sediments, both as direct discharge and as suspended sediment. Mudstone breccias form a significant part (29%) of the lower Miocene section in CIROS-1, and suggest slope failure of previously deposited sediments. Passchier *et al.* (1998) concluded that the shallower breccias (<85 mbsf) in CRP-1 were most likely related to subglacial shearing, whereas those at greater depth were caused by slope failure.

Lithofacies were separated according to

stratigraphic interval to determine whether they could be linked to changing glaciological, oceanographic or climatic regime. In CIROS-1, the balance between ice-contact/ice-proximal and other lithofacies does not appear to change significantly between upper Oligocene and lower Miocene strata. There is, however, a strong contrast between the two parts of CRP-2/2A: a substantially higher proportion of ice-contact and a lesser proportion of well-sorted, marine-dominated sediments is found in the lower Miocene section than in the upper

Oligocene. These differences may reflect a cooling glaciological thermal regime through time that is clearly evident in the changing flora. Less easy to assess are other factors that might influence facies changes, such as changing oceanographic conditions and water depths.

These data show that sedimentation during late Oligocene and early Miocene times was quite unlike that of today (cf. Anderson 1999), where over-compacted diamictons and diatom-rich sediments have dominated deposition on the continental shelf since the Pleistocene Epoch. Rather, similar facies associations are evident in high-Arctic regions where climate is characterized by short summers with above zero temperatures, and the generation of large volumes of glacial meltwater, but nevertheless with substantial deposition and survival of both basal till and ice-proximal glacimarine sediment (cf. Dowdeswell *et al.* 1994; Syvitski *et al.* 1996; Bennett *et al.* 1999; Ó Cofaigh *et al.* 2001).

Facies associations

The lithofacies outlined above are commonly grouped into vertical associations throughout the cores. These associations have been recorded separately using both a process-based approach (Powell *et al.* 1998, 2000), and a sequence-stratigraphic approach (Fielding *et al.* 1998, 2000). The former is emphasized here, as it links with our understanding of depositional processes in a variety of contemporary glacial settings. These associations are interpreted as representing sedimentary subenvironments embracing a complementary suite of processes. Although a standard vertical succession may be inferred for a particular subenvironment, in practice parts are commonly missing because of glacial erosion, resedimentation or sudden switches in sedimentation regime, processes which are common in glacimarine settings. Detailed descriptions of all facies associations have been provided for CRP-1 (Powell *et al.* 1998) and CRP-2/2A (Powell *et al.* 2000), so only key representative subenvironments are described here.

Moralnal bank association. This association, exemplified by 20 m of upper Oligocene core from CIROS-1 (Fig. 6), represents the clearest indications of direct deposition from a glacier in the glacimarine environment. The moraine bank builds up below water-level where basal glacial sediment in the form of massive diamicton is delivered to the grounding-line (Powell & Alley 1997). In its lithified state in the core, massive diamictite is interbedded with a variety of minor lithofacies, including stratified diamictite representing iceberg rain-out sedimentation, sandstone derived from subglacial conduits, mudstone derived from suspension and rhythmite which represents tidally-controlled, rhythmically alternating laminae of sand (cyclopsams) and mud (cyclopels) (Mackiewicz *et al.* 1984; Cowan *et al.* 1998, 1999). The presence of soft-sediment deformation structures indicates that slumping was commonplace. Similar facies associations in the Cape Roberts cores also reveal evidence of loading, glacial pushing, thrusting and possible injection. The association was built out from an ice cliff resting on the sea bottom, and, by analogy with modern environments, may have accumulated in just a few years. As such, the association forms a sharp contact with distal glacimarine sediment, typically sandy mudstone or muddy sandstone.

Grounding-line fan association. Grounding-line fans, built up from subglacial conduits (Powell 1990), are represented several times in each of the three cores (Fig. 7). Diamictite is much less abundant, and lithofacies are dominated by subglacially-derived meltwater products, including sandstone, mudstone and rhythmites. Slumping and bioturbation structures are also present. Dropstones indicate background iceberg rafting. As with moraine banks, the grounding-line fan association commonly builds rapidly and forms a sharp contact with distal marine lithofacies beneath.

Distal glacimarine association. Distal marine sediments are characterized by sandy mudstone or muddy sandstone derived from suspended sediment plumes and bottom currents (Fig. 7). Added to these is a scattering of clasts (typically of pebble size), rafted by icebergs to the site of deposition. Slope failure may have reworked these facies to give intraformational breccias, typically mudstone clasts in a muddy sandstone matrix. Emplacement of these breccias may have resulted in loading of sediments beneath. Although some distal glacimarine associations are largely barren of fossils, others contain abundant diatoms and various other micro- and macrofossils. Bioturbation is evident in places, commonly destroying any stratification.

Gravity-flow associations. Using grain-size data, detailed core logging, X-radiography and thin-section analysis, coupled with statistical grouping of grain-size data, three main styles of fine-grained gravity-flow sedimentation were distinguished by Howe *et al.* (1998) in CRP-1:

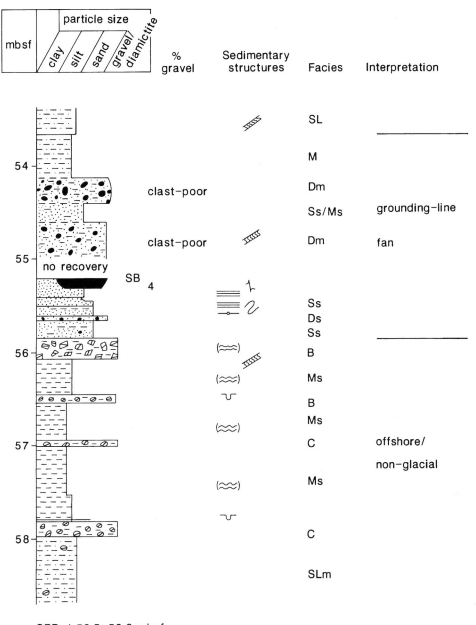

CRP-1 53.5–53.6 mbsf

Grounding-line fan above offshore (glacier recessional association)

Fig. 7. Lithological log through an inferred grounding-line fan succession, underlain by offshore, non-glacial sediments; CRP-1 (53.5–58.6 mbsf), illustrating particle size, percent of gravel and sedimentary facies.

(i) faintly laminated muddy sandstones with granules towards the base, with contacts ranging from sharp to gradational, interpreted as debris-flow deposits; (ii) rhythmically stacked sequences of pebbly coarse sandstones, representing successive thin debris flows on an unstable slope in a shallow-marine ice-proximal setting; and (iii) normally graded,

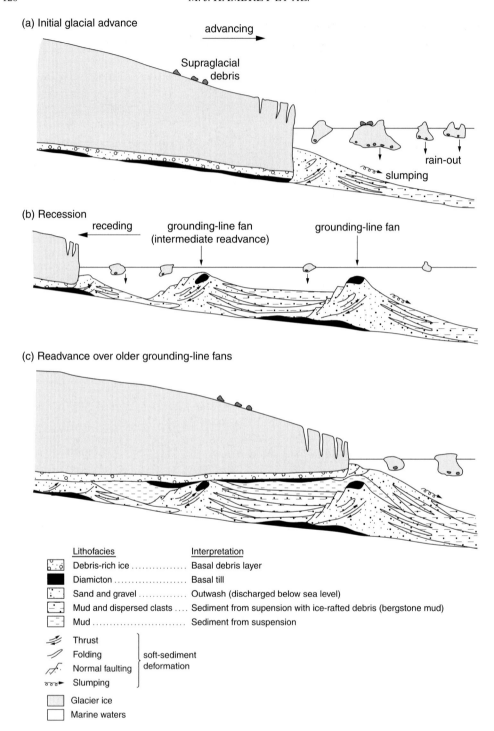

Fig. 8. Facies model for the development of typical glacimarine successions in the western Ross Sea, with emphasis on grounding-line fan development. The model shows hypothetical glacier advance, recession and readvance sequences. This represents a simplified version of the model presented by Powell *et al.* (2000) for the CRP-2/2A core, but is equally applicable to the other sites.

well-laminated siltstones with occasional lone-stones, representing turbidites deposited in a more distal shallow-marine environment. The glaciological context of these associations needs to be explored further. In addition, there are diamictite units with some indications of coarse-tail grading, which are interpreted as glacier-proximal gravity flows, such as would be expected on the slopes of the morainal banks and grounding-line fans.

Facies model

The upper Oligocene–lower Miocene interval, as represented in the three cores, is character-ized by facies associations indicative of a grounded ice-cliff at the terminus of glaciers emanating from the Transantarctic Mountains (Fig. 8). The model developed in detail by Powell *et al.* (1998, 2000) is broadly applicable to all three cores, and is simplified here. Large amounts of poorly sorted, gravelly sediment, carried in basal glacier ice, and *via* englacial and subglacial streams, were delivered to a rela-tively shallow marine environment, building up unstable aprons of sediment. Reworking of sediment as a result of slope instability was common, as was variable input of iceberg-rafted sediment (Fig. 8a). During periods of recession, mud deposition was dominant, although rafting of sediment by icebergs continued in the background (Fig. 8b). A stillstand or minor readvance generated a new grounding-line fan or morainal bank inshore of an older feature (Fig. 8b). Readvance over the whole site trun-cated the various fans and banks, resulting in sediment deformation and over-compaction, and producing a disconformity (Fig. 8c).

Cenozoic stratigraphy

Lithostratigraphy

Ten cores with good recovery define the Cenozoic stratigraphy of the Ross Sea region (Fig. 3). Each core demonstrates long time gaps in the record, and none extends further back in time than latest Eocene. However, a wide variety of fossils of mid- to late Eocene age, preserved in erratics within coastal moraines in McMurdo Sound, indicates that, prior to the onset of glaciation, a cool-temperate climate prevailed (Harwood & Levy 2000). The Oligocene to early Miocene record is well represented in several cores, but everywhere there appears to be a significant hiatus repre-senting middle and late Miocene time. The offshore record from this period has been lost to

Pliocene–Quaternary erosion, but a few data for late Miocene and younger times come from inner fjord settings (CIROS-2, DVDP-11).

Focusing on upper Oligocene and lower Miocene lithostratigraphy, CIROS-1 and CRP-2/2A provide the most detailed records, with CRP-1 in addition providing a lower Miocene succession that overlaps with that of CRP-2/2A (Fig. 4). Lithologies are predominantly con-glomerate, intraformational breccia, diamictite, sandstone and mudstone (for details of the lithostratigraphy see Hambrey *et al.* 1989 for CIROS-1, Cape Roberts Science Team 1998 for CRP-1, and Cape Roberts Science Team 1999 for CRP-2/2A).

Sequence stratigraphy

Fielding *et al.* (1997, 1998, 2000) have provided a sequence-stratigraphic framework for each core. Such an approach, rarely undertaken for cored glacigenic sequences previously, is based on 'the recognition of vertically-stacked facies successions, bounded by sharp erosion surfaces that mark prominent lithological dislocations' (Fielding *et al.* 2000: 329). Sequences can be divided into four components in ascending order, as follows:

(1) a sharp-based, poorly sorted coarse-grained unit comprising diamictite, pebbly sandstone or conglomerate (or in combi-nation);
(2) a fining-upwards interval of muddy sand-stone, passing up into interbedded sand-stone and mudstone;
(3) a mudstone, sometimes with a condensed fossiliferous interval at the base, passing upwards into a more sandstone-rich section;
(4) a sharp-based, well-sorted, medium to coarse sandstone, with interbedded or interlaminated sandstone and mudstone in places.

The best example of this is 'Sequence 11' in CRP-2A, one of three 60 m thick sequences deposited within a 450 000-year period around the Oligocene–Miocene boundary (Naish *et al.* 2001). Here, a convincing case has been made that these sequences were a response to orbital forcing leading to variations in both ice extent and sea level. Fielding and colleagues (*op cit.*) interpret the other cycles in each of the three cores in a similar way, and their sequence boundaries are plotted in Figure 4 (indicated by arrows). Alternatively, some sequences may have been the result of a gradual, monotonic increase in relative sea level rise, arising from continuous and gradual tectonic subsidence to create accommodation space. Sedimentary cycles may then have been created by super-imposed erosion and sedimentation during glacial advances and recessions. Such glacier fluctuations also result in local changes in water depth and thus coarsening and fining of sequences (Powell *et al.* 2000; Powell & Cooper 2002).

Discussion

Glaciation styles and thermal regime

The facies associations and facies model developed above probably apply to both late Oligocene and early Miocene times in the vicinity of the drill sites, as there are no system-atic differences in the proportion of facies. We envisage that the style of glaciation is one domi-nated by outlet glaciers, breaching the Transantarctic Mountains from the ice sheet, and expanding out onto a gently sloping ramp offshore. The nature of the sediment reflects progressive down-cutting of the bedrock through Beacon Supergroup strata, with its sheets of Ferrar Dolerite, into granitic basement. These components were incorpor-ated into the basal ice zone before release. There are few indications of a substantial supraglacial debris load at any time, unlike in the early Oligocene core in CRP-3. Contrasts between the CIROS-1 and Cape Roberts sites can be attributed to their position relative to the central flow-lines of the respective ancestral glaciers, Ferrar and Mackay respectively.

Cores comprise ten main facies, with diamic-tite accounting for a substantial proportion of each interval and each core. In contrast, older cores (early Oligocene) have relatively little diamictite, whereas younger cores (Pliocene to Quaternary) are dominated by diamictite. Previous models predict that the colder the climate, the greater the proportion of poorly sorted sediment (including diamicton) and the lesser the proportion of well-sorted sediments derived from meltwater (Dowdeswell *et al.* 1996; Powell & Alley 1997). By this reasoning, the climatic and thermal regimes during late Oligocene to early Miocene time, recovered in CIROS-1, CRP-1 and CRP-2 2A, are inter-mediate between a temperate setting such as Alaska (Powell & Molnia 1989) and a cold polar setting exemplified by Antarctica today (Powell *et al.* 1996). Modern high-Arctic environments, with fjordal, polythermal tidewater glaciers and tundra vegetation on the surrounding land best

explain the facies associations described here. Contemporary examples include East Green-land (Dowdeswell *et al.* 1994; Syvitski *et al.* 1996; Ó Cofaigh *et al.* 2001) and Svalbard (Elverhøi *et al.* 1983; Bennett *et al.* 1999). Such inferences concerning climate are supported by the indi-cations of herb moss tundra with low-growing shrubby plants, including *Nothofagus* and podocarp conifers (Raine 1998; Askin & Raine 2000).

Depositional model for the western Ross Sea

Summing evidence derived from the drill-site reports, we envisage a series of outlet glaciers emanating from an ice sheet in the hinterland to the west and breaching the rising Transantarctic Mountains, before entering the western Ross Sea (Fig. 9). The sediment carried by these glaciers was mainly basal, but their flanks carried minor amounts of supraglacial debris. The glaciers, grounded on the sea floor, probably spread out as piedmont lobes, deliver-ing icebergs with debris across a wide front and producing grounding-line fans and morainal banks (see Fig. 3 of Powell *et al.* 1998). Sediment plumes emerged from subglacial conduits below water level, and the resulting sand and mud was largely reworked by gravity-flow and bottom-current processes, much of it above wave base. It is probable that not all valleys contained glaciers that reached the sea, so some sediment may have been delivered *via* local braided rivers. Such rivers created small deltas and, when glacier-fed, also produced sediment plumes. For the few late Oligocene–early Miocene intervals that bear no evidence of ice activity, the coast was dominated by waves and related currents of sufficient strength to grade the sediments from near-shore sand to offshore mud.

Shore-normal seismic lines close to the Cape Roberts drill sites show lateral continuity of strata on a scale of kilometres, and a broad parallelism on a scale of tens of metres in thick-ness. There is no obvious or deep (>50 m) chan-nelling within the cored strata from rivers or glaciers during the period of deposition. This observation, along with the palaeontological evidence of shallow marine conditions and evidence of occasional grounded ice above cycle-bounding unconformities, suggests a picture of sediment accumulating as a prism with sediment supply keeping pace roughly with basin margin subsidence. To this we add facies being modulated by advance and recession of late Oligocene–early Miocene polythermal

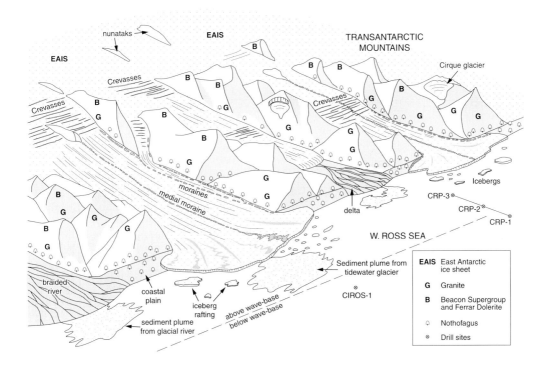

Fig. 9. Generalized three-dimensional perspective of palaeoenvironmental setting for sedimentation along the flanks of the Transantarctic Mountains in late Oligocene time.

glaciers (or early Oligocene temperate glaciers) forming piedmont lobes seaward of the mountains, and at times extended beyond the coast.

Incompleteness of the stratigraphic record

The drill-site data, plotted on a linear time scale (Fig. 3), indicate that there are still many gaps in the Cenozoic record. This is to be expected where the shelf has been affected by a succession of glacier-grounding events that would have been accompanied by considerable erosion. The most prominent gap spans the middle and late Miocene Epoch, and this record may only be obtainable beyond the continental shelf break or in deep shelf basins. There are many other shorter time gaps in the record at individual sites, again related to glacier-grounding events. Indeed, given the rapidity of sedimentation in glacial environments, the time represented by sediment preserved in the cores may only be a small fraction of Cenozoic time. The presence of glacigenic sediments indicates

the proximity of glaciers, whereas hiatuses represent times of glacial erosion and more advanced glacial conditions. There are no indications of full interglacial conditions in the late Oligocene to early Miocene record, although it is conceivable that any such evidence could have been destroyed by subsequent glacial advances.

Preservation potential of sequence and ancient analogues

The sequences described above have a high potential for preservation in the long-term geological record. Sediments were delivered to a deep, subsiding extensional basin (the Victoria Land Basin; Fig. 1), bordering a mountain range that provided a ready supply of debris, allowing several hundred metres of glacially-influenced sediment to accumulate (Cape Roberts Science Team 1998). Although there are large erosional hiatuses, resulting from glacial advances across the inclined shelf, leading to the loss of entire glacial cycles, there is generally a net gain of sediment through each advance/recession cycle. Tectonic subsidence is a key factor in preservation of the rapidly deposited glacial and marine sediments by providing accommodation space at a sufficiently rapid rate. This also enhances the likelihood that the sequence is not completely eroded by a succeeding glacial advance. In fact, significant erosional gaps in the stratigraphic record in these basins may be a good indicator that tectonic subsidence ceased or slowed dramatically to allow significant glacial erosion.

Among the best pre-Cenozoic examples of a similar setting to the Ross Sea, are those inferred for Neoproterozoic (Vendian) glaciation around the North Atlantic (Eyles 1993). Detailed facies analyses have been undertaken for Scottish (Spencer 1971; Eyles 1988), East Greenland (Hambrey & Spencer 1987; Moncrieff & Hambrey 1990), Svalbard (Fairchild & Hambrey 1984, 1995; Harland *et al.* 1993) and Norwegian (Edwards 1975) sequences. The common theme here is that up to several hundred metres of glacimarine (and some glaciterrestrial) sediment were deposited in an ensialic basin in close proximity to an elevated, possibly mountainous landmass. Similar facies associations to those in the western Ross Sea occur, and the inferences concerning glaciological and climatic regimes are similar, although closer comparisons are now warranted.

Future drilling prospects

Despite the successes of drilling to date, important unresolved questions remain concerning Cenozoic tectonic and palaeoenvironmental evolution of the Antarctic region. Chief amongst these is the timing of the onset of glaciation. This has been an elusive target for both CIROS and Cape Roberts drilling. Indeed, in CRP-3, Oligocene strata passed *via* an unconformity into the pre-glacial Devonian–Triassic Beacon Supergroup rocks. The search continues for sites that will recover an Eocene and earlier Cenozoic, or even a Cretaceous record. This issue, along with others, will be addressed in a new multinational programme named ANDRILL (Wilson *et al.* 2001).

Conclusions

The various drill cores from the western Ross Sea have now been subjected to a comprehensive suite of investigations, involving geophysics and physical properties, sedimentary environments, biostratigraphy and palaeoecology,

provenance and climate from petrology, chronology and chronostratigraphy. In total, well over 2 km of Cenozoic strata, dating back to late Eocene have now been drilled in the McMurdo Sound region, with an average recovery rate of 97%. This paper has focused only on sedimentary environments for the late Oligocene and early Miocene interval, but has drawn relevant palaeoenvironmental information from other fields as appropriate. The main conclusions concerning sedimentary environments are as follows.

(1) Lithofacies in the upper Oligocene – lower Miocene strata (28–17 Ma) are dominated by well-sorted and poorly sorted varieties of mudstone and sandstone, commonly with dispersed lonestones, stratified and massive diamictite, and mudstone breccia.

(2) Seismically determined geometry of these strata along strike, along with textural and palaeoecological evidence, point to the long term accumulation as a nearshore sediment prism, with facies patterns influenced by the delivery of sediment by glaciers and rivers from the mountains to the west, and by a strong influence from waves, related currents and some resedimentation processes.

(3) The facies are interpreted as having formed in proximal to distal glacimarine settings, with substantial evidence of glacier grounding and resedimentation. Ice-contact aprons of sediment (morainal banks and grounding-line fans) are significant components in the build-up of the Antarctic margin.

(4) The proportions of lithofacies for late Oligocene to early Miocene times suggest a climate significantly warmer than that of today, with the production of abundant meltwater. East Greenland or Svalbard, where the glaciers are commonly polythermal, offer suitable modern analogues. This climate is transitional from the rather more temperate climate with glaciers of the early Oligocene, to the frigid climate we have today, where many glaciers are frozen to the bed. No marked changes in climate are apparent within this interval, however.

(5) These facies are associated with herb moss tundra, typical of periglacial areas. The tundra succeeded a low, woodland vegetation that prevailed in early Oligocene time. These and other palaeoecological data, support the sedimentological inference for a regional climate that cooled, perhaps in stepwise fashion, through time.

(6) There are many gaps in the stratigraphic record resulting from numerous lowstands of sea level and glacial erosional events. Where well-constrained chronology exists, sedimentation rates are shown to have been rapid, and are in line with values in modern high-Arctic glacial environments.

(7) The accumulation of the glacigenic sediments in a tectonically subsiding rift basin has produced a sequence several hundred metres thick. This type of tectonic setting is particularly favourable for preservation of glacial sequences in the older geological record.

The success of drilling from sea ice in the western Ross Sea is attributable to the impressive teamwork of many individuals. In particular we would like to acknowledge drilling managers Kevin Jenkins (for CIROS-1) and Pat Cooper (for the Cape Roberts Project) and their teams; Alex Pyne, Science Support Manager for both the CIROS and Cape Roberts Projects; Jim Cowie, Cape Roberts Project Manager; the staff at McMurdo Station and Scott Base; our 60 or so scientific colleagues who have contributed to the various reports. Hambrey thanks the Victoria University of Wellington, Antarctica New Zealand, the Leverhulme Trust and the Transantarctic Association, all of which supported his participation in the CIROS-1 and CRP-1 drilling projects, and the UK Natural Environment Research Council (Grant GR9/04/194) for the writing-up phase. Powell was funded by US National Science Foundation Grant OPP-9527481. We thank referees P. Barker and J. Howe for helpful comments on the manuscript.

References

ALLEY, R. B., CUFFEY, K. M., EVENSON, E. B., STRASSER, J. C., LAWSON, D. E. & LARSON, G. J. 1997. How glaciers entrain and transport basal sediment: physical constraints. *Quaternary Science Reviews*, **16**, 1017–1038.

ANDERSON, J. B. 1999. *Antarctic marine geology*. Cambridge: Cambridge University Press.

ASKIN, R. A. & RAINE, J. I. 2000. Oligocene and early Miocene terrestrial palynology of the Cape Roberts drillhole CRP-2/2A, Victoria Land Basin, Antarctica. *In:* BARRETT, P. J., RICCI, C. A., DAVEY, F. J., *ET AL.* (eds) *Studies from the Cape Roberts Project, Ross Sea, Antarctica: Scientific Report of CRP-2/2A*. Terra Antartica, **7**(4), 493–502.

BARRETT, P. J. (ed.) 1986. *Antarctic Cenozoic history from the MSSTS-1 drillhole, McMurdo Sound*. DSIR Bulletin, New Zealand, **237**.

BARRETT, P. J. (ed.) 1989a. *Antarctic Cenozoic history from the CIROS-1 drillhole, McMurdo Sound*. DSIR Bulletin, New Zealand, **245**.

BARRETT, P. J. 1989b. Sediment texture. *In:* BARRETT, P. J. (ed.) *Antarctic Cenozoic history from the CIROS-1 drillhole, McMurdo Sound*. DSIR Bulletin, New Zealand, **245**, 49–58.

BARRETT, P. J. 1996. Antarctic palaeoenvironment through Cenozoic times – a review. *Terra Antartica*, **3**, 103–119.

BARRETT, P. J. & DAVEY, F. J. 1992. *Cape Roberts Project Workshop Report*. Royal Society of New Zealand, Miscellaneous Series, **23**.

BARRETT, P. J. & HAMBREY, M. J. 1992. Plio-Pleistocene sedimentation in Ferrar Fiord, Antarctica. *Sedimentology*, **39**, 109–123.

BARRETT, P. J., RICCI, C. A., DAVEY, F. J., EHRMANN, W. U., HAMBREY, M. J., JARRARD, R., VAN DER MEER, J. J. M., RAINE, J., ROBERTS, A. P., TALARICO, F. & WATKINS, D. H. (eds) 2000. *Studies from the Cape Roberts Project, Ross Sea, Antarctica: Scientific Report of CRP-2/2A*. Terra Antartica, **7** (3&4), 213–412 & 413–665.

BARRETT, P. J., RICCI, C. A., BUCKER, C., *ET AL.* (eds) in press. *Studies from the Cape Roberts Project, Ross Sea, Antarctica: Scientific Report of CRP-2/2A*. Terra Antartica, in press.

BARTEK, L. R., HENRYS, S. A., ANDERSON, J. B. & BARRETT, P. J. 1996. Seismic stratigraphy in McMurdo Sound: implications for glacially influenced early Cenozoic eustatic change. *Marine Geology*, **130**, 79–98.

BENNETT, M. R., HAMBREY, M. J., HUDDART, D., GLASSER, N. F. & CRAWFORD, K. 1999. The landform and sediment assemblage produced by a tidewater glacier surge in Kongsfjorden, Svalbard. *Quaternary Science Reviews*, **18**, 1213–1246.

BERGGREN, W. A., KENT, D. V., SWISHER, C. C. III & AUBREY, M. P. 1995. A revised Cenozoic geochronology and biostratigraphy. *In:* BERGGREN, W. A., KENT, D. V., AUBREY, M. P. & HARDENBOL, J. (eds) *Geochronology, time scales, and stratigraphic correlation framework for an historical geology*. Society for Economic Paleontologists and Mineralogists, Special Publications, **54**, 129–212.

BOULTON, G. S. & HINDMARSH, R. C. A. 1987. Sediment deformation beneath glaciers: rheology and geological consequences. *Journal of Geophysical Research*, **92**(B9), 9059–9082.

BRANCOLINI, G., DE SANTIS, L. & BUSETTI, M. 1994. Structural evolution across a section south of the Drygalski Ice Tongue (Victoria Land Basin). *In:* VAN DER WATEREN, F. M., VERBERS, A. L. L. M. VERBERS, F. TESSENSOHN (eds) *Landscape evolution of the Ross Sea region, Antarctica*. Rijks Geologische Dienst, The Netherlands, 69–75.

CAI, J., POWELL, R. D., COWAN, E. A. & CARLSON, P. 1997. Lithofacies and seismic-reflection interpretation of temperate glacimarine sedimentation in Tarr Inlet, Glacier Bay, Alaska. *Marine Geology*, **143**, 5–37.

CAPE ROBERTS SCIENCE TEAM 1998. *Initial Report on CRP-1, Cape Roberts Project, Antarctica*. (eds P. J. BARRETT, C. R. FIELDING & S. W. WISE). Terra Antartica, **5**(1), 1–187.

CAPE ROBERTS SCIENCE TEAM 1999. *Studies from*

Cape Roberts Project. *In:* FIELDING, C. R. & THOMSON, M. R. A. (eds) *Initial Report on CRP2/2A, Ross Sea Antarctica.* Terra Antarctica, **6**(1/2), 1–173.

CAPE ROBERTS SCIENCE TEAM 2001. *Studies from Cape Roberts Project. In:* BARRETT, P. J., SARTI, M. & WISE, S. (eds) *Initial Report on CRP-3, Ross Sea Antarctica.* Terra Antartica, **7**(1/2), 1–209.

COOPER, A. K., DAVEY, F. J. & BEHRENDT, J. C. 1987. Seismic stratigraphy and structure of the Victoria Land basin, western Ross Sea, Antarctica. *In:* COOPER, A. K. & DAVEY, F. J. (eds) *The Antarctic Continental Margin: Geology and Geophysics of the Western Ross Sea.* Circum-Pacific Council for Energy & Mineral Resources, Earth Science Series, **5B**, Houston, 27–65.

COOPER, A. K., DAVEY, F. J. & HINZ, K.1994. Crustal extension and origin of sedimentary basins beneath the Ross Sea and Ross Ice Shelf, Antarctica. *In:* THOMSON, M. R. A., CRAME, J. A. & THOMSON, J. W. (eds) *Geological Evolution of Antarctica.* Cambridge: Cambridge University Press, 285–291.

COWAN, E. A., CAI, J., POWELL, R. D., SERAMUR, K. C. & SPURGEON, V. L. 1998. Modern tidal rhythmites deposited in deep water. *Geo-marine Letters,* **18**, 40–48.

COWAN, E. A., SERAMUR, K. C., CAI, J. & POWELL, R. D. 1999. Cyclic sedimentation produced by fluctuations in meltwater discharge, tides and marine productivity in an Alaskan fjord. *Sedimentology,* **46**, 1109–1126.

DOWDESWELL, J. A., WHITTINGTON, R. J. & MARIENFELD, P. 1994. The origin of massive diamicton facies by iceberg rafting and scouring, Scoresby Sund, East Greenland. *Sedimentology,* **41**, 21–35.

DOWDESWELL, J. A., KENYON, N. H., ELVERHØI, A., LABERG, J. S., HOLLENDER, F.-J., MEINERT, J. & SIEGERT, M. J. 1996. Large-scale sedimentation on the glacier-influenced polar North Atlantic margins: long-range side-scan sonar evidence. *Geophysical Research Letters,* **23**, 3535–3538.

DUNBAR, G. & BARRETT, P. J. 2001. Sediment Texture and Palaeobathymetry Changes off Cape Roberts during the Late Oligocene. *In:* FLORINDO, F. & COOPER, A. K. (eds) *The Geological Record of the Antarctic Ice Sheet from Drilling Coring and Seismic Studies.* Quaderni di Geofisica, **16**, 55–58.

DUNBAR, G., BARRETT, P. J., GOFF, J., HARPER, M. A. & IRWIN, S. 1997. Estimating vertical tectonic movement using sediment texture. *Holocene,* **7**, 213–221.

EDWARDS, M. B. 1975. Glacial retreat sedimentation of the Smalford Formation, Late Precambrian, North Norway. *Sedimentology,* **22**, 75–94.

ELVERHØI, A., LONNE, O. & SELAND, R. 1983. Glaciomarine sedimentation in a modern fjord environment, Spitsbergen. *Polar Research,* **1**, 127–149.

EYLES, C. H. 1988. Glacially and tidally influenced shallow marine sedimentation of the Late Precambrian Port Askaig Formation, Scotland.

Palaeogeography, Palaeoecology and Palaeoclimatology, **68**, 1–25.

EYLES, N. 1993. Earth's glacial record and its tectonic setting. *Earth-Science Reviews,* **35**, 1–248.

FAIRCHILD, I. J. & HAMBREY, M. J. 1984. The Vendian succession of north-eastern Spitsbergen: petrogenesis of a dolomite-tillite association. *Precambrian Research,* **26**, 111–167.

FAIRCHILD, I. J. & HAMBREY, M. J. 1995. Vendian basin evolution in East Greenland and NE Svalbard. *Precambrian Research,* **73**, 217–233.

FIELDING, C. R., WOOLFE, K. J., PURDON, R. G., LAVELLE, M. A. & HOWE, J. A. 1997. Sedimentological and stratigraphical re-evaluation of the CIROS-1 core, McMurdo Sound, Antarctica. *Terra Antartica,* **4**, 149–160.

FIELDING, C. R., WOOLFE, K. J., HOWE, J. A. & LAVELLE, M. 1998. Sequence stratigraphic analysis of CRP-1, Cape Roberts Project, McMurdo Sound, Antarctica. *Terra Antartica,* **5**, 353–361.

FIELDING, C. R., NAISH, T. R., WOOLFE, K. J. & LAVELLE, M. A. 2000. Facies analysis and sequence stratigraphy of CRP-2/2A, Victoria Land Basin, Antarctica. *Terra Antartica,* **7**, 323–338.

HAMBREY, M. J. 1989. Grain fabric studies on the CIROS-1 core. *In:* BARRETT, P. J. (ed.) *Antarctic Cenozoic History from the CIROS-1 Drillhole, McMurdo Sound, Antarctica.* DSIR Bulletin, **245**, 59–62.

HAMBREY, M. J. & BARRETT, P. J. 1993. Cenozoic sedimentary and climatic record, Ross Sea region, Antarctica. *In:* KENNETT, J. P. & WARNKE, D. A. (eds) *The Antarctic Paleoenvironment: A Perspective on Global Change, 2.* Antarctic Research Series, **60**, 91–124. American Geophysical Union, Washington.

HAMBREY, M. J. & SPENCER, A. M. 1987. *Late Precambrian glaciation in central East Greenland.* Meddelelser om Grönland. Geoscience series no. **19**.

HAMBREY, M. J., BARRETT, P. J., HALL, K. J. & ROBINSON, P. H. 1989. Stratigraphy. *In:* BARRETT, P. J. (ed.) *Antarctic Cenozoic History from the CIROS-1 Drillhole, McMurdo Sound, Antarctica.* DSIR Wellington, New Zealand, Bulletin, **245**, 23–48.

HAMBREY, M. J., WISE, S. W., BARRETT, P. J., DAVEY, F. J., EHRMANN, W. U., SMELLIE, J. L., VILLA, G. & WOOLFE, K. J. 1998. Studies from the Cape Roberts Project, Ross Sea, Antarctica: Scientific Report of CRP-1. *Terra Antartica,* **5**(3), 255–713.

HAMILTON R. J., LUYENDYK, B. P. & SORLEIN, C. C., 2001. Cenozoic tectonics of the Cape Roberts rift basin and Transantarctic Mountains front, southwest Ross Sea, Antarctica. *Tectonics,* **20**, 325–342.

HANNAH, M. 1997. Climate controlled dinoflagellate distribution in late Eocene-earliest Oligocene strata from CIROS-1 drillhole, McMurdo Sound, Antarctica. *Terra Antartica,* **4**, 73–78.

HARLAND, W. B., HAMBREY, M. J. & WADDAMS, P. 1993. *Vendian Geology of Svalbard.* Norsk Polarinstitutt Skrifter, **193**.

HARWOOD, D. M. & LEVY, R. H. 2000. The McMurdo
erratics: introduction and overview. *In:* STILWELL,
J. D. & FELDMANN, R. M. (eds) *Paleobiology and
Paleoenvironments of Eocene Rocks, McMurdo
Sound, East Antarctica.* AGU, Antarctic
Research Series, **76**, 1–18.

HAYES, D. E. & FRAKES, L. A. 1975. General synthesis, Deep Sea Drilling Project, Leg 28. *Initial
Report Deep Sea Drilling Project*, **28**, 919–942.

HIEMSTRA, J. F. 1999. Microscopic evidence of
grounded ice in the sediments of the CIROS-1
core, McMurdo Sound, Antarctica. *Terra Antartica*, **6**, 365–376.

HOWE, J. A., WOOLFE, K. J. & FIELDING, C. R. 1998.
Lower Miocene glacimarine gravity flows, Cape
Roberts Drillhole-1, Ross Sea, Antarctica. Terra
Antartica, **5**, 393–399.

LABERG, J. S. & VORREN, T. O., 1996. The middle and
late Pleistocene evolution of the Bear Island
trough mouth fan. *Global and Planetary Change*,
12, 309–330.

MACKIEWICZ, N. E., POWELL, R. D., CARLSON, P. R. &
MOLNIA, B. F. 1984. Interlaminated ice-proximal
glacimarine sediments in Muir Inlet, Alaska.
Marine Geology, **57**, 113–147.

McKELVEY, B. C. 1991. The Cainozoic glacial record in
South Victoria Land: a geological evaluation of
the McMurdo Sound drilling projects. *In:* TINGEY,
R. J. (ed.) *The Geology of Antarctica.* Oxford
University Press, Oxford, 434–454.

McGINNIS, L. D. (ed.) 1981. *Dry Valley Drilling
Project.* Antarctic Research Series, **33**, AGU,
Washington DC.

MONCRIEFF, A. C. M. & HAMBREY, M. J. 1990.
Marginal-marine glacial sedimentation in the
Late Proterozoic succession of East Greenland.
In: DOWDESWELL, J. A. & SCOURSE, J. D.
(eds) *Glacimarine Environments: Processes and
Sediments.* Geological Society, London, Special
Publications, **53**, 387–410.

MURRAY, T. 1997. Assessing the paradigm shift:
deformable glacier beds. *Quaternary Science
Reviews*, **16**, 995–1016.

NAISH, T., WOOLFE, K. J. *ET AL.*, 2001. Orbitally
induced oscillations in the East Antarctic Ice
Sheet at the Oligocene-Miocene boundary.
Nature, **413**, 719–723.

Ó COFAIGH, C., DOWDESWELL, J. A. & GROBE, H.
2001. Holocene glacimarine sedimentation, inner
Scoresby Sund, East Greenland: the influence of
fast-flowing ice-sheet outlet glaciers. *Marine
Geology*, **175**, 103–129.

PASSCHIER, S. 2000. Soft-sediment deformation
features in core from CRP-2/2A, Victoria Land
Basin, Antarctica. *Terra Antartica*, **7**, 401–412.

PASSCHIER, S., WILSON, T. J. & PAULSEN, T. S. 1998.
Origin of breccias in the CRP-1 core. *Terra
Antartica*, **5**, 401–409.

POWELL, R. D. 1981. Sedimentation conditions in
Taylor Valley, Antarctica, inferred from textural
analysis of DVDP cores. *In:* McGINNIS, L. D.
(ed.) *Dry Valley Drilling Project.* American
Geophysical Union, Washington D.C., 63–94.

POWELL, R. D. 1990. Glacimarine processes at

grounding-line fans and their growth to ice-
contact deltas. *In:* DOWDESWELL, J. A. &
SCOURSE, J. D. (eds) *Glacimarine Environments:
Processes and Sediments.* Geological Society,
London, Special Publication, **53**, 53–73.

POWELL, R. D. & ALLEY, R. B. 1997. Grounding-line
systems: process, glaciological inferences, and the
stratigraphic record. *In:* BARKER, P. F. & COOPER,
A. C. (eds) *Geology and seismic stratigraphy of
the Antarctic margin, 2.* Antarctic Research
Series, American Geophysical Union, Washington DC, 169–187.

POWELL, R. D. & COOPER, J. M. 2002. A glacial
sequence stratigraphic model for temperate
glaciated continental shelves. *In:* DOWDESWELL, J.
A. & Ó COFAIGH, C. (eds) *Glacier-Influenced
Sedimentation on High-Latitude Continental
Margins.* Geological Society, London, Special
Publications, **203**, 215–244.

POWELL, R. D. & MOLNIA, B. F. 1989. Glacimarine
sedimentation processes, facies and morphology
of the south-east Alaska shelf and fjords. *Marine
Geology*, **85**, 359–390.

POWELL, R. D., DAWBER, M., MCINNES, J. N. & PYNE,
A. R. 1996. Observations of the grounding-line
area at a floating glacier terminus. *Annals of
Glaciology*, **22**, 217–223.

POWELL, R. D., HAMBREY, M. J. & KRISSEK, L. A. 1998.
Quaternary and Miocene glacial and climatic
history of the Cape Roberts drillsite region,
Antarctica. *In:* HAMBREY, M. J., WISE, S. W.,
BARRETT, P. J., DAVEY, F. J., EHRMANN, W. U.,
SMELLIE, J. L., VILLA, G. & WOOLFE, K. J. (eds)
*Studies from the Cape Roberts Project, Ross Sea,
Antarctica, Scientific Report of CRP-1.* Terra
Antartica, **5**, 341–351.

POWELL, R. D., KRISSEK, L. A. & VAN DER MEER,
J. J. M. 2000. Preliminary depositional environmental analysis of CRP-2/2A, Victoria Land
Basin, Antarctica: Palaeoglaciological and
palaeoclimatic inferences. *Terra Antartica*, **7**,
313–322.

RAINE, J. I. 1998. Terrestrial palynomorphs from Cape
Roberts Project Drillhole CRP-1, Ross Sea,
Antarctica. *In:* HAMBREY, M. J., WISE, S. W.,
BARRETT, P. J., DAVEY, F. J., EHRMANN, W. U.,
SMELLIE, J. L., VILLA, G. & WOOLFE, K. J. (eds)
*Studies from the Cape Roberts Project, Ross Sea,
Antarctica: Scientific Report of CRP-1.* Terra
Antartica, **5**(3), 539–548.

SMELLIE, J. 2000. Erosional history of the Transantarc-
tic Mountains deduced from sand grain detrital
modes in CRP-2/2A, Victoria Land Basin,
Antarctica. *In:* BARRETT, P. J., RICCI, C. A.,
DAVEY, F. J., EHRMANN, W. U., HAMBREY, M. J.,
JARRARD, R., VAN DER MEER, J. J. M., RAINE, J.,
ROBERTS, A. P., TALARICO, F. & WATKINS, D. H.
(eds) *Studies from the Cape Roberts Project, Ross
Sea, Antarctica: Scientific Report of CRP-2/2A.*
Terra Antartica **7**(4), 545–552.

SPENCER, A. M. 1971. *Late Precambrian glaciation in
Scotland.* Geological Society of London,
Memoir, **6**.

SYVITSKI, J. P. M., ANDREWS, J. T. & DOWDESWELL,

J. A. 1996. Sediment deposition in an iceberg-dominated glacimarine environment, East Greenland: basin fill implications. *Global and Planetary Change*, **12**, 251–270.

TALARICO, F., SANDRONI, S., FIELDING, C., ATKINS, C. 2000. Variability, petrography and provenance of basement clasts in core from CRP-2/2A, Victoria Land Basin, Antarctica. *In:* BARRETT, P. J., RICCI, C. A., DAVEY, F. J., EHRMANN, W. U., HAMBREY, M. J., JARRARD, R., VAN DER MEER, J. J. M., RAINE, J., ROBERTS, A. P., TALARICO, F. & WATKINS, D. H. (eds) *Studies from the Cape Roberts Project, Ross Sea, Antarctica: Scientific Report of CRP-2/2A.* Terra Antartica, **7**(4), 529–544.

VAN DER MEER, J. J. M. 2000. Microscopic observations on the first 300 metres of CRP-2/2A, Victoria Land Basin, Antarctica. *Terra Antartica*, **7**, 339–348.

VAN DER MEER, J. J. M. & HIEMSTRA, J. F. 1998. Micromorphology of Miocene diamicts: indications for grounded ice. *Terra Antartica*, **5**, 363–366.

WILSON, G. S., ROBERTS, A. P., VEROSUB, K. L., FLORINDO, F. & SAGNOTTI, L. 1998. Magnetobiostratigraphic chronology of the Eocene-Oligocene transition in the CIROS-1 core, Victoria Land margin, Antarctica: Implications for Antarctic glacial history. *Geological Society of America Bulletin*, **110**, 35–47.

WILSON, G. S., FLORINDO, F., HARWOOD, D. H., NAISH, T. R. & POWELL, R. D. 2001. Future Antarctic margin drilling – the ANDRILL initiative and McMurdo Sound portfolio. *ANTOSTRAT Workshop*, Erice, Italy, Abstracts, 201.

Late Pleistocene glacially-influenced deep-marine sedimentation off NW Britain: implications for the rock record

STEPHEN DAVISON[1,2] & MARTYN S. STOKER[2]

[1]*Department of Geology & Geophysics, University of Edinburgh, West Mains Road,*
Edinburgh EH9 3JW, UK
[2]*British Geological Survey, Murchison House, West Mains Road, Edinburgh EH9 3LA,*
UK (e-mail: SDAV@bgs.ac.uk)

Abstract: A new deep-water borehole at the foot of the West Shetland Slope revealed a sequence of Late Pleistocene glacimarine sediments. This section comprises an inter-bedded sequence of muddy diamictons, with subordinate sandy muds with dropstones and matrix-poor sands and gravels. Integration of core data, seabed imagery and high-resolution seismic-reflection records indicate the succession is dominated by stacked glacigenic debris flows.

Individual debris flows are composed of massive, clast-poor muddy diamicton, and are seperated by sandy muds and the sands and gravels. The mud units are interpreted as a combination of hemipelagic-glacimarine and distal ice-rafted deposits. The sands and gravels are interpreted as the result of bottom-current reworking of glacial and pre-glacial sediments. The new data support previous studies which suggest that Pleistocene ice sheets reached the edge of the West Shetland Shelf on several occasions, delivering large volumes of glacigenic sediment directly to the shelf edge and upper slope. These deposits became unstable, forming debris flows which transported the glacigenic sediments to the deep-water environment. Seabed imagery shows that the area around the borehole formed a focus for this style of deposition.

A depositional model constructed from the new data is enhanced by the application of data from an ancient deep-water, glacially-influenced, succession from the Neoproterozoic Macduff Formation NE Scotland.

Previous studies of the deep marine environment of glaciated continental margins have been largely based upon seismic reflection profiles and short gravity cores. This has allowed the characterization of the sedimentary architecture and identification of shallow lithologies. However, the sediments below the depth of gravity core penetration are rarely sampled and interpretation is somewhat conjectural. This scarcity of lithological data has restricted the development of more detailed models of the deep-water setting.

A new deep-water borehole core from the Faroe–Shetland Channel off NW Britain, combined with high resolution seismic reflection profiles and sea-bed imagery, provides a new insight into the sedimentary processes and deposits of this deep-marine glacially-influenced environment. This combination of data has not been available previously and has only arisen as a result of intensive commercial exploration in the Faroe–Shetland area.

The purpose of this paper is to integrate the new, and existing datasets as a case study for the glacially-influenced lower slope and basin floor

setting, using seismic and sea-bed image data calibrated lithologically and chronologically by a long borehole. Such well-constrained datasets are essential to the accurate interpretation and reconstruction of ancient glacial successions in the rock record. The potential of the dataset for analogue modelling is shown by applying the results to a Late Proterozoic glacially-influenced deep-water sequence of sediments from the Dalradian Macduff Slate Formation, NE Scotland. In addition to aiding the interpretation of successions from the rock record, the model can also be improved by reference to the ancient sequences, allowing features seen in outcrop but not in recent cores or seismic surveys, to be incorporated into the overall model and enhance current understanding of the deep-water glacimarine environment.

Existing models of glaciated margins are largely the result of studies from high-latitude areas such as the Antarctic, Svalbard and Greenland. The mid-latitude margins of western Norway and the NW UK also provide good examples of non-polar margins which are no longer subject to a glacial influence.

From: DOWDESWELL, J. A. & Ó COFAIGH, C. (eds) 2002. *Glacier-Influenced Sedimentation on High-Latitude Continental Margins.* Geological Society, London, Special Publications, **203**, 129–147. 0305-8719/02/$15.00
© The Geological Society of London 2002.

Although these areas share many common features, they also have many diverse characteristics that contribute to the overall view of glaciated margins.

Seismic investigations have shown the large scale architecture of most glaciated margins comprises a prograding wedge of glacigenic sediment extending from the pre-glacial shelf edge (e.g. Cooper *et al.* 1991; Hambrey 1994; Kristofferson *et al.* 2000). The specific characteristics of the glacigenic wedge vary between different margins, but generally consist of a combination of acoustically transparent lensoid bodies, interpreted as debris flows, and parallel to sub-parallel acoustically stratified units interpreted as hemipelagic–glacimarine sediments. On a shelf-to-basin transect, the main seismic packages are a roughly sigmoidal shape in vertical section, being thin on the shelf, thickening markedly at the shelf edge and thinning towards the slope base into the deep water environment. This architecture is particularly well developed on the SE Greenland margin (Clausen 1998) and the West Shetland margin of the UK (Stoker 1999).

On the Svalbard and NW Norwegian margins the wedge shape progradation is modified by the presence of large glacially-fed trough-mouth fan systems which dominate the depositional architecture. These fans extend seawards from the shelf edge at the outlets of former ice streams and are composed mainly of stacked debris flows (King *et al.* 1996; Dowdeswell *et al.* 1998; Vorren *et al.* 1998).

Generally, lithological information from the deep water section of glacigenic prograding wedges is very limited. Cores from Ocean Drilling Program (ODP) sites on the lower slopes of the East Greenland and Svalbard margins have shown the presence of muddy glacimarine sediments, silty debris flows and diamictons (Solheim *et al.* 1998). Of particular significance are the results from ODP site 987 which recovered sediments dominated by dropstone-bearing silty muds, with an upward increase of turbidites and two large (60–80 m thick) debris flows (Jansen *et al.* 1996). Other recent ODP sites from the Antarctic margin have added to the lithological characterization of the glacimarine environments on the lower slope and basin (O'Brien *et al.* 2001; Barker *et al.* 1999). Data collected from ODP sites 1097 and 1103 indicate the importance of debris flow and turbidite processes in the progradation of the Antarctic continental margin (Eyles *et al.* 2001).

The ODP cores have added detail to the lower slope and basin settings. However, the ODP drill sites, particularly those on the Antarctic margin, are in regions which have not been completely free of a glacimarine influence even during the late Quaternary interglacials. In addition, the sites on the East Greenland margin are located on, or at the base of large trough-mouth fan systems and are not representative of more passive settings.

Construction of glaciated margin models, particularly from the Northern Hemisphere, has relied upon seismic facies architecture to infer the sedimentary processes involved in the deposition of the prograding wedges. This is particularly true of the deep water environments. Where deep cores have been available they have allowed more detailed analysis of processes but because of the wide spacing of seismic survey lines and the large scale of the margins involved, it is difficult to extrapolate the interpretations.

Geographical and geological setting

The geographical area of this study lies at the base of the West Shetland slope, on the SE side of the Faeroe–Shetland Channel (Fig. 1). Water depths in this area vary from 1000 m in the southwestern end of the channel to over 2000 m at the northeastern end where it joins the foot of the North Sea Fan. The new borehole (BGS Borehole 99/3) was drilled by the British Geological Survey and the UK Rockall Consortium in 1999 at a water depth of 986 m in an area of large debris flows described briefly by Stoker (1990). Results from the borehole proved a 56 m thickness of Quaternary sediments showing a glacial influence throughout.

The Quaternary deposits on the West Shetland margin form a thick prograding and slightly aggrading wedge resting unconformably on eroded Neogene rocks. The glacial component of this section is termed the Upper Nordland Unit of the Nordland Group (Stoker 1999). On the inner shelf sediment cover is very thin or absent but thickens rapidly on the outer shelf and upper slope where it locally reaches 180m thick (Fig. 2). On the lower slope and basin floor the pre-glacial Quaternary is absent. Previous studies of the Quaternary glacial section have suggested the margin has been influenced by repeated glacial events during the last 440 Ka (Stoker *et al*, 1993) and that ice sheets probably reached the contemporary shelf edge during glaciations (Stoker & Holmes 1991; Stoker *et al.* 1993; Holmes 1997). As a result of poor biostratigraphic control, however, it is not currently possible to distinguish with any certainty the different phases of glacial activity.

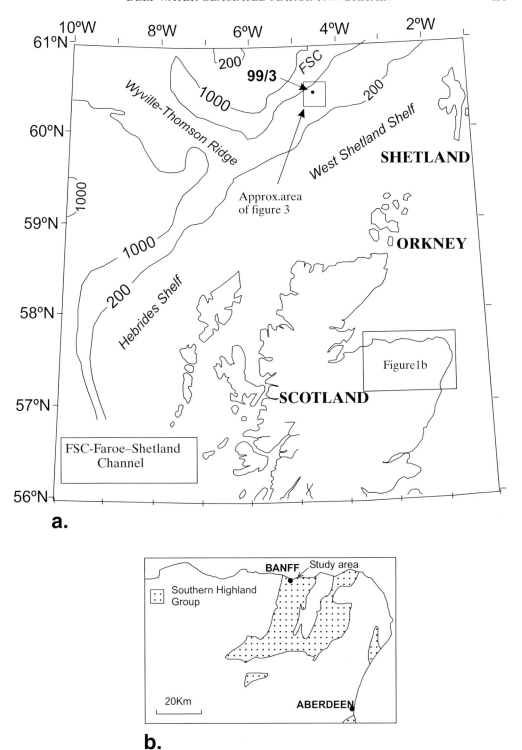

Fig. 1. (**a**) Location of borehole 99/3. (**b**) Location of the studied section of the Macduff Slate Formation.

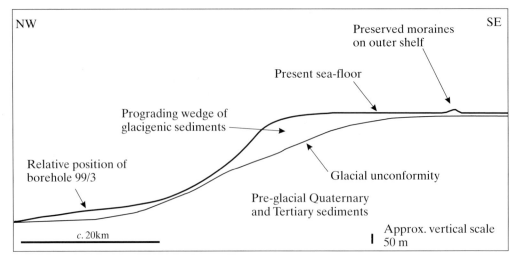

Fig. 2. Schematic section across the West Shetland Margin.

Ice-streaming on the continental shelf, up-slope of the current study area has been proposed as a means of sediment supply for the debris flows (Stoker 1995).

Survey and borehole data

Sea-bed image

The sea-bed image (Fig. 3) is part of a much larger shaded relief image of the West Shetland margin developed by the British Geological Survey and is derived from commercial 3-D seismic data using the sea-bed reflector. The image shows the existence of an extensive slope apron composed of numerous coalescent fan-like lobes which have been interpreted as debris flows (Bulat & Long 1998). The presence of the debris flows has been suggested from previous studies based upon seismic surveys (Stoker *et al.* 1991). The morphology of individual flows shows elongate lobes extending up to 16 km from the base of the continental slope, branching towards the distal end forming smaller lobes with steep terminal slopes. The interfingering and overlapping nature of these lobes demonstrates that repeated debris flow events took place.

Seismic reflection data

The two seismic reflection profiles presented in this study form part of a small grid from a sparker survey covering an area approximately 5 km × 5 km with the borehole located near the SW edge (Fig. 3). The profiles shown in Figure 4 are typical of the survey profiles in the remainder of the survey grid.

The base of the Quaternary sequence lies at a depth of about 56 m below the seabed at the borehole site. This depth is based upon micropalaeontological results from sediments at a depth of 54 m, which show abundant late Quaternary taxa, and sediments from 56.95 m which are dominated by Eocene taxa. Quaternary taxa are absent from sediments recovered from below 54 m (Hitchen 1999). The seismic data for the borehole location show a prominent reflector at 75 ms below the seafloor. This is interpreted as the unconformity at the base of the Quaternary sequence. Using an average velocity of 1.5 k ms⁻¹, the Quaternary unconformity lies at 56.25 m below the seabed, which is within the window 54.0–56.95 m indicated by the core data. A slower velocity is not thought to be credible, as this would be below that of seawater at approximately 1.495 k ms⁻¹. This interpretation agrees with the regional interpretation of Stoker (1990; 1999), the Quaternary stack resting upon eroded Eocene strata.

Broadly, the Quaternary sequence can be divided into two seismic facies: thin parallel or sub-parallel, acoustically stratified units; and acoustically transparent or chaotic units. Both the illustrated profiles show an alternation of these two basic types. The transparent and chaotic units have a lensoid shape in the strike sections and wedge shape in the dip sections. In some of the lower units there is also limited

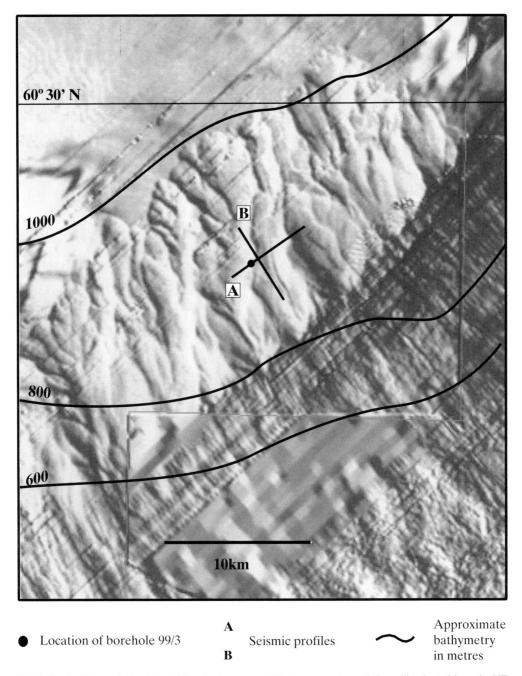

● Location of borehole 99/3   A
                                   B   Seismic profiles   〜   Approximate
                                                                      bathymetry
                                                                      in metres

Fig. 3. Sea-bed image derived from 3-D seismic surveys of the lower continental slope, illuminated from the NE.

evidence of discontinuous hyperbolic reflectors. Thickness of the transparent units varies greatly, reaching 17 m at the borehole site and over 30 m at fixed point 11/9 on profile B (Fig. 4). This style of architecture and the acoustic response is typical of debris flows and has been described from deep marine sequences and many glacigenic successions (e.g. Piper 1988; Stein & Syvitski 1997; Stravers 1997), including other parts of the West Shetland margin (e.g.

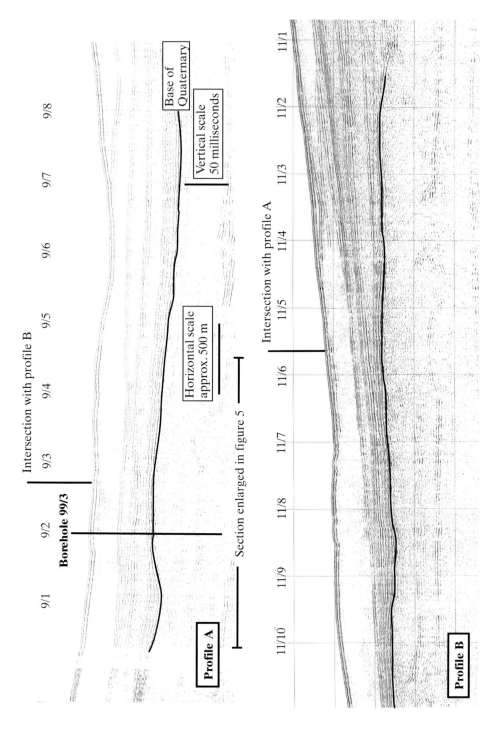

Fig. 4. Seismic (sparker) reflection profiles from the area of borehole 99/3 showing the stacked nature of the parallel reflections and transparent reflections. Profile A is strike-parallel, profile B is dip-parallel.

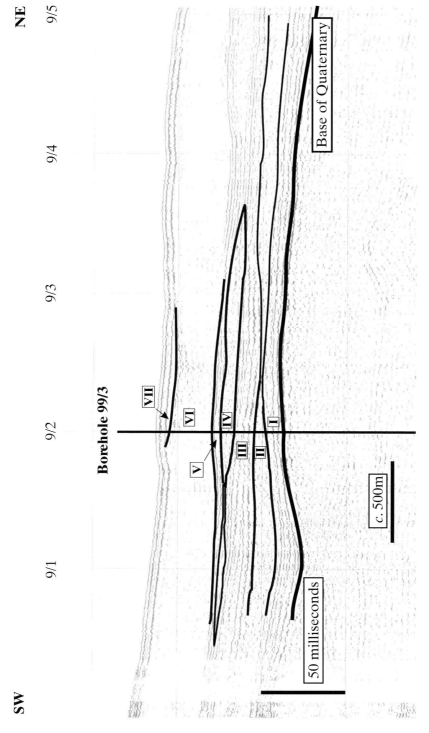

Fig. 5. Enlarged section of seismic reflection profile A, showing the location of borehole 99/3 and the sub-division into seven seismic packages.

Table 1. *Summary of interpreted seismic units and corresponding lithofacies.*

Seismic unit (depth below sea-floor in metres)	Seismic characteristics	Lithofacies correlation
VII 0 – 7.34 m	Thin transparent, possibly with thin parallel reflections at the base. Lens-shaped in both strike and dip section. Top of unit obscured by sea-bed reverberations.	Massive diamicton facies, overlain by sandy mud and sand and gravel facies. The upper two lithofacies are not visible on the seismic profiles.
VI 7.34 – 24.5 m	Thick, virtually reflection-free package showing lateral persistence over a distance of more than 4 km. Along slope to the NE this unit thins and is overlain by another similar package. Up-slope, the structureless nature changes abruptly to a series of thin acoustically stratified and transparent units, before returning to a single transparent unit.	Massive diamicton facies, although core recovery is very poor, especially from the lower part of this unit.
V 24.5 – 29.20 m	Thin parallel and sub-parallel, laterally continuous reflections, forming a package similar in character to I and III.	Sandy mud facies.
IV 29.20 – 34.0 m	Package of weak chaotic reflections forming a thin lens-shaped body in strike-parallel section and a thin, elongate wedge-shaped body in dip-parallel section.	Core recovery indicates sandy mud facies, although the seismic characteristics are that of the massive diamicton facies.
III 34.0 – 43.50 m	Acoustically stratified package of laterally persistent parallel and sub-parallel reflections. Reflections become more closely spaced towards the top. To the NE, the base of this unit becomes poorly defined where it overlies a chaotic lensoid package.	Predominantly sandy mud facies with a minor sand and gravel unit. Core recovery also shows the presence of a thin massive diamicton unit which does not show on the seismic profile, and is presumably below the seismic resolution.
II 43.50 – 48.5 m	Lens-shaped body in strike section, with a wedge-shaped profile which thins down-slope in dip section. The internal structure shows weak, chaotic reflections.	Massive diamicton facies.
I 48.5 – 56.25 m	Thin, laterally continuous parallel and sub-parallel reflections. Unit thickens along slope to the SW and NE over topographic depressions, but is relatively constant upslope to the NW and down slope to the SE.	Sand and gravel facies, overlain by sandy mud facies. The two lithofacies are indistinguishable on the seismic profiles.

Stoker *et al.* 1991). This interpretation is consistent with the coalescent fan-like lobes seen at the base of the continental slope on the sea-bed image. The profiles also indicate that the debris flows have been accumulating throughout the deposition of the preserved Quaternary sequence.

The parallel and sub-parallel reflectors form packages which separate the debris flows and represent the general background deposition between the debris flow events. It is not possible to determine the nature of these sediments from the seismic data alone. However, these packages and to a lesser extent individual reflectors, are laterally persistent along strike and tend to drape the underlying surface. The thickness of the individual packages varies between 2 m and 12 m, thickening slightly over

Table 2. *Summary of lithological characteristics and interpretation.*

Lithofacies	Sedimentary characteristics	Interpretation
Sand and gravel	Unconsolidated sands and gravels with a low mud content. Compositional ratios vary between 60:15:25 and 3:95:2 gravel: sand: mud. Sand fraction has quartz: lithic grain ratio of between 70:30 and 50:50. Up to 40% of quartz grains have Fe oxide coating. Grain morphology A–WR. Pebbles up to 110 mm, generally SR–R. Lithologies of pebbles and lithic grains include red and green sandstones, limestones, gneiss, amphibolite and basalt. No preserved sedimentary structures. Mixed microfossil assemblage from both shallow and deep water environments, cold and temperate water temperatures and ages from Cretaceous to the late Pleistocene.	Current-deposited or winnowed sediments. Maybe the product of contour or downslope currents, or possibly both. The diversity of lithologies represented and the wide age range suggests a different source area from that of the sediments of other facies.
Sandy mud	Sandy, structureless grey–green mud with rare pebbles. 80–95% mud with up to 20% sand. Pebbles reach up to 5 mm, forming < 1% of the total volume. Exceptional cobbles of gneiss and sandstone reaching 110 mm. Sand grain size varies between very fine and granular, but is dominated by the finer grain sizes. Composition of the sand fraction is typically 80: 20 quartz: lithics with a SA–SR grain morphology. Iron oxide coatings are present on < 20% of the sand grains. Microfossil assemblage includes frequent to abundant cold water foraminifera. Rare bivalve mollusc fragments and complete valves.	Hemipelagic-glacimarine sediments deposited from suspension. The fine-grained nature of this facies indi cates there was little bottom current activity at the time of deposition. The coarse component of this facies is interpreted as ice-rafted debris. The low proportion of sand and scarcity of pebbles suggests this material was being deposited in an ice-distal setting.
Massive diamicton	Red-brown, clast-poor muddy to intermediate diamicton. Comprises mud 60–75%, sand 25–35%, pebbles between 1 and 5%. Sand grain size varies from very fine to granular, with a modal grain size of fine to medium. Lithic grains comprise 20–35% of the sand fraction. Quartz grain morphology is A–R, with larger grains being more rounded. Pebbles reach an average diameter of 15–20 mm and a maximum of 80 mm. Lithologies include gneiss, amphibolite, red sandstones and siltstones, grey quartzite and dolomitic limestone. No evidence of internal structure. Contains arctic water temperature foraminifera dominated by shelf water depth species.	Debris flow(s). Position at the base of the slope precludes the possibility of deposition as a subglacial till. The presence of a cold water shelf micro-fauna indicates the sediments of this facies are not in their original depo sitional setting. This also indicates glacial conditions were established at the time of original accumulation, prior to being transported and redeposited by debris flow processes.

Abbreviations for Powers Roundness categories: A, angular; SA, sub-angular; SR, sub-rounded; R, rounded; WR, well-rounded.

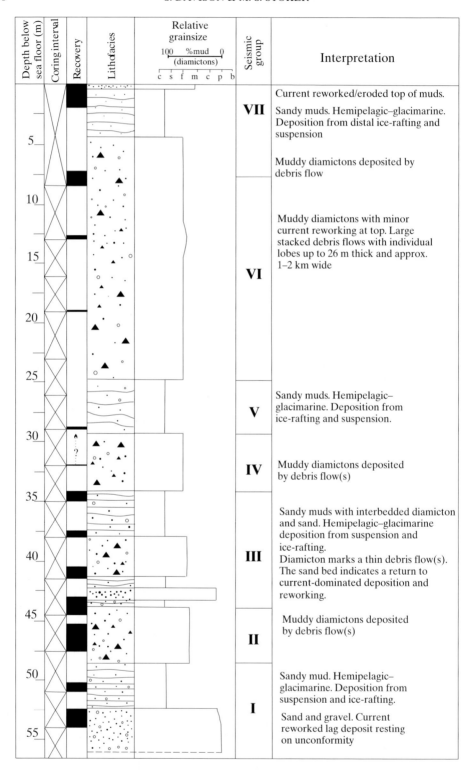

Fig. 6. Summary log of the Quaternary section of borehole 99/3.

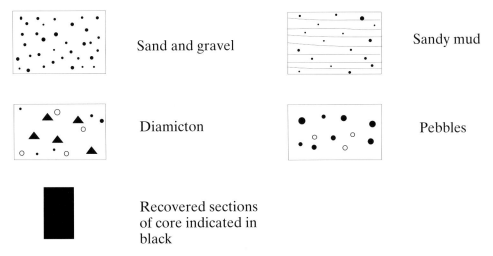

Fig. 6. Key to figure.

topographic lows and thinning over topographic highs. An enlarged interpreted section of profile A, covering the locality of borehole 99/3 is presented in Figure 5. The characteristics of the individual seismic units related to the borehole are summarized in Table 1.

Borehole data and lithofacies

The core from BGS borehole 99/3 is summarized in Figure 6. Core recovery was generally not good, with retrieval from above 28 m being poor. However, correlation with the seismic stratigraphy would suggest that all of the main seismic units are represented in the core, although contacts between different lithologies were rarely recovered. Within intervals of no recovery, the boundary separating lithologies is taken as the mid-point between the recovered sections, except where a boundary can be fixed using the seismic data. The core was drilled in separate runs of 3 m or 3.5 m. All core recovered was measured upwards from the base of each core run.

Broadly, the recovered sediments can be divided into 3 lithofacies: sand and gravel; sandy mud; and massive diamicton. These lithofacies are described in detail in Table 2. For the purposes of this paper, the term mud is taken to include all grain sizes from clay to silt, in common with accepted practice for the description of glacial sediments. The classification of the diamictons uses the system of Moncrieff (1989), as modified by Hambrey (1994).

Sand and gravel facies: The sand and gravel facies was recorded from three intervals, the uppermost 20 cm, 43.02–43.31 m and from approximately 51.50–56.25 m. The deepest recovered sediments from 54.0–53.50 m show a fining upwards trend from coarse gravel and cobbles to pebbly sand, and are interpreted here as a basal lag deposit which rests upon the unconformity surface. The remainder of the basal sand unit, and those higher in the sequence, are interpreted as lag deposits formed by current winnowing of glacimarine sandy muds. This is supported by the occurrence of sandy muds of glacimarine origin underlying both of the upper sand intervals and by the presence of Arctic water temperature foraminifera in the sands. Analysis of microfossils from the sand and gravel facies indicates substantial variation in the abundance and diversity of foraminifera, including forms from both deep and shallow water environments, and from cold and more temperate water temperatures. The basal sand is dominated by the shallow, cold water species *Cassidulina teteris*, together with the planktonic species *Neogloboquadrina pachyderma* (sinistral) which is abundant (Wilkinson *et al.* 2000). The diversity of environments represented by the microfossils, and the wide variety of clast lithologies and mineral grains present, suggests the sands may be derived from more than one source. A combination of *in-situ* winnowing of glacimarine sandy muds by contour/basin floor currents, and input of sediment from a more distant origin by the same currents, seems the most likely explanation. This is compatible with reported contour current activity from other localities in the Faroe–Shetland Channel (Stoker *et al.* 1993; Ackhurst 1991).

We interpret the thin sand and gravel facies from the top of the borehole as having developed since the end of glacial influence, representing the normal bottom conditions for the deep water environment of the present-day Faroe–Shetland Channel. This is important in considering the thickness of the sand and gravel facies between the debris flows at 43.02–43.31 m, and the much thicker basal unit below 51.50 m. The implications are that the same conditions existed on at least two previous occasions within the late Pleistocene, and that these conditions were established for a much longer period.

Sandy mud facies: The sandy mud facies was recovered from five different intervals, mostly below a depth of 29 m. The characteristics shown by this facies, of mud-dominated sediments with fine sand and out-sized pebble clasts, are typical of hemipelagic deposits of glacimarine origin. The sand component and the clasts are interpreted as ice-rafted debris, with the mud fraction derived from ice rafting and meltwater plume activity. The interpretation of this facies as hemipelagic–glacimarine is supported by a foraminiferal assemblage dominated by the frequent to abundant occurrence of the planktonic foraminifera *Neogloboquadrina pachyderma* (sinistral), which is a recognized indicator of cold (Arctic) water temperatures (Ericson 1959).

The absence of lamination or trace fossils within the sandy mud facies is unexpected. Whether internal structure was never present, or was destroyed by post depositional processes is not known. The destruction of any original structure by bioturbation is possible, but this seems unlikely given the scarcity of macrofossils.

Massive diamicton facies: The compositional and textural features of these diamictons are characteristic of subglacial tills and debris flows. Based upon lithology alone, it would be difficult to determine which of these depositional settings the massive diamicton represents. However, its position at the base of the slope in almost 1000 m of water precludes deposition as a subglacial till. In addition, the indication of large debris flow lobes on the sea-bed image and seismic profiles supports our interpretation of the massive diamicton facies as having been deposited by debris flows.

The arctic water foraminifera assemblage recovered from the massive diamicton includes the species *Elphidium excavatum calvatum*, *Cassidulina reniforme*, and *Haynesina orbiculare*. These species are all benthonic shallow water species generally found on the inner shelf (Wilkinson *et al.* 2000). The presence of these species in a deep water slope-base setting strongly supports both a redeposited origin for the massive diamicton facies, and also indicates that the debris flow sediments were originally deposited under glacial conditions, prior to their remobilization down the slope.

Integration of borehole and seismic data

The seismic data from the borehole site (Fig. 5) can be divided into seven units, consisting of stratified, chaotic or transparent reflectors. These broadly correspond to the lithological units defined from the core.

Unit I: 56.25–48.50 m

The lowermost seismic unit is a combination of the sand and gravel facies, overlain by the sandy mud facies. On seismic profile A (Fig. 5), the two different lithofacies are indistinguishable, and form a single group of parallel and sub-parallel reflectors. The asymmetric thickening of the unit over the topographic low at fixed point 9/1 (Fig. 5) suggests strike-parallel transport of sediments, indicating contour currents rather than downslope currents were more likely responsible for formation of the sand and gravel facies of Unit I.

The inclusion of cold water Pleistocene foraminifera within the sands of Unit I indicates that glacial conditions were already established in the source region of these sediments and that the pre-glacial Quaternary succession was either removed by erosion, or was never deposited. The well-preserved state of the foraminifera tests suggests that they have not been transported a great distance, and indicates a probable source, and hence glacial conditions, within the Faroe–Shetland Channel. The timing of the deposition of the sand and gravel component of Unit I is not known. Persistent erosion and winnowing of the channel floor has probably occurred throughout the Neogene–Quaternary interval.

The overlying sediments of the sandy-mud facies represent the onset of glacial conditions, with the delivery of suspended sediment and ice-rafted debris to the area. Deposition of the fine-grained sandy-mud facies also indicates the slowing down or halting of the current regime responsible for forming the underlying sand and gravel facies. Whether this is a local or regional effect is unknown.

Unit II: 48.50–43.50 m

Unit II is interpreted as a debris flow, derived from glacimarine sediments originally deposited higher up the continental slope. The interpretation as a debris flow is based upon the seismic architecture and lithological character. The position on the lower slope at a water depth of almost 1000 m excludes the possibility that Unit II is a subglacial till. The presence of cold water shelf microfossils in the unit indicates that glacial conditions prevailed at the time the original sediment was deposited, prior to redeposition by debris flow activity. This suggests a glacimarine origin for these sediments on the shelf, although this interpretation is not conclusive.

Unit III: 43.50–34.00 m

Unit III contains sediments of all three lithofacies and represents changing depositional, and possibly environmental conditions. The acoustically stratified reflectors hide the lithological changes seen in the core. Generally we interpret Unit III as the product of suspension sedimentation and iceberg rafting. However, the thin sand unit from 43.03 m–43.31 m represents a return to current reworking. This may be similar to the current regime which formed the sands at the base of Unit I. Whether this change represents a climatic warming and change to non-glacial conditions, or is a local current effect is not known.

Within the core recovered from Unit III there is also a thin bed of massive diamicton which is interpreted as a thin debris flow or the distal end of a larger flow. This interpretation is based solely upon the lithological similarity to Unit II as it does not appear on the seismic profiles and is therefore assumed to be below the resolution of the seismic survey. The inclusion of a debris flow within Unit III has important implications for the interpretation of acoustically stratified units, showing that without the control provided by the core, the presence of debris flows would not have been detected.

Unit IV: 34.00–29.20 m

Acoustically, Unit IV has the same characteristics as Unit II and is interpreted as a debris flow. However, sediments recovered from this interval are from the sandy-mud facies which conflicts with the interpretation. This apparent conflict seems to be the result of poor core recovery and the way in which the core was measured from the base of each coring run. If the core from 31.78–32.00 m is re-assigned to the top of the coring run, the conflict is resolved, suggesting that no sediments were in fact recovered from Unit IV. This solution is supported by the presence of a large (9 cm) cobble at the base of the recovered section which may have prevented further core recovery.

Unit V: 29.20–24.50 m

Unit V is interpreted as a series of ice-distal glacimarine muds deposited by suspension settling and iceberg rafting. Along strike to the NE on Profile A (Fig. 5) there is some evidence of slight thickening over topographic low points, as seen in Unit I. This suggests there may have been small scale current activity during the deposition of Unit V, which resulted in minor redistribution of sediment and localized thickening.

Unit VI: 24.50–7.34 m

Unit VI is of a different scale to the preceding units, comprising 30% of the entire Quaternary sequence. The seismic transparency, architecture and massive diamicton lithology indicate that this unit was deposited by debris flow. When the unit is traced laterally and upslope on the seismic profiles, it is evident that it comprises a series of large, stacked debris flows. The seismic profiles also illustrate the considerable surface topography (approximately 18 m) created by the individual flow lobes. This topography appears to be largely responsible for the present day surface morphology of the coalesced debris flows which form the slope apron observed on the seabed image (Fig. 3).

The thickness of these debris flows indicates that there must have been a considerable build up of sediment higher up the slope or on the shelf edge to provide sufficient source material. There are two possible causes for the magnitude of this build-up; (i) sediment accumulation occurred over an extended period as distal glacimarine deposits with a relatively stable slope; and (ii) the ice front was located near to, or reached the shelf edge (Stoker *et al.* 1993) and delivered large volumes of sediment directly to the slope in the form of meltwater plumes, till deltas and ice-rafting. The second of these possibilities seems more likely as it would have produced more rapid sediment accumulation which, in turn, is more likely to lead to slope instability.

Unit VII: 7.34 m–seabed

Unit VII may be a thin debris flow of the same structureless lensoid shape as Units II and IV, but of a smaller scale. Alternatively it may be the distal end of a larger flow, which is perhaps more likely given the position at the top of a series of large stacked debris flows. The thin concentration of shell debris from the base of this unit probably marks a brief hiatus in deposition, during which macrofauna colonization occurred prior to emplacement of the debris flow. Alternatively it is possible that the shell horizon is a lag deposit, formed during a period of increased current activity. However, the low abundance of macrofauna in the sandy muds generally, suggests that this is less likely. The uppermost section of Unit VII is lost in the seabed reflections, but the core evidence indicates a return to distal glacimarine deposition from suspension followed by the re-establishment of a current re-working regime which continues to the present (Stoker 1999).

Provenance of the sediments from borehole 99/3

The three lithofacies recorded in the core contain a wide variety of grains and pebbles of different lithologies. Although no detailed work has been undertaken on the provenance of the lithic grains and clasts, some aspects of their origin are worthy of comment. The dominant lithologies in the pebble and lithic grain suites are gneiss, red sandstone and dolomitic limestone. A cursory examination indicates that these bear a strong resemblance to the rocks of the mainland of NW Scotland. The dolomitic limestone clasts are most probably derived from the Cambro-Ordovician Durness Group from the north coast of Scotland, as this is the nearest potential source of such material. The gneisses and red sandstones are more problematic as there are more potential sources of these rocks both onshore and as seabed outcrop. The iron-stained quartz grains observed in all the lithofacies from borehole 99/3 are likely to be derived from the Devonian or Proterozoic red sandstones in the northern Scotland and Orkney region. Occurrences of iron stained grains within deep-marine sediments have been used to infer a continental origin by ice rafting (Bond & Lotti 1995).

Discussion

The preceding sections have summarized the sea-bed, seismic and lithological characteristics of the late Pleistocene deep-water sediments on the West Shetland margin. The data presented has enabled the development of a depositional model for glacially-influenced deep-water settings and to test its potential applicability to the pre-Pleistocene rock record

Glacially-influenced deep-water environment

The general characteristics of the West Shetland glacigenic deep-water margin demonstrated by borehole 99/3 can be summarized in three main facies associations based upon the sedimentary processes which formed them: glacigenic–hemipelagic facies (sandy muds); current re-worked facies (sand and gravel); and debris flow facies (massive diamicton).

Glacigenic–hemipelagic facies: This facies is composed of acoustically layered, laterally extensive silty and sandy muds with dropstones. These sediments are deposited by suspension settling from meltwater plumes, and by iceberg rafting. This style of deposition is thought to represent the pervasive background depositional scenario in deep-water settings during glacial events. A similar interpretation has been proposed for the glaciated margin off eastern Canada to explain accumulations of glacimarine muds on the Scotian slope (Piper 1988). The sedimentation rate in the basin environment from this style of deposition would be expected to increase as the ice advanced across the shelf, and the ice front became more proximal to the shelf edge. Conversely, the sedimentation rate in the basin would decrease during a retreat phase as the ice receded from the shelf edge.

Current re-worked facies: Acoustically these units form thin, continuous layered reflectors which are indistinguishable from the silty mud facies. Lithologically they are composed of mud-poor sands and gravels which apparently lack internal structure. Formed by bottom current reworking of pre-existing lithologies, these sediments represent the normal background conditions in the deep-water areas of the West Shetland margin during interglacials (Stoker *et al.* 1993). Potentially, the erosive and re-working processes of this facies could greatly reduce or completely remove lithological units or entire successions.

Debris flow facies: In the areas where debris flows are present, they dominate the glacigenic succession. Acoustically they are transparent or

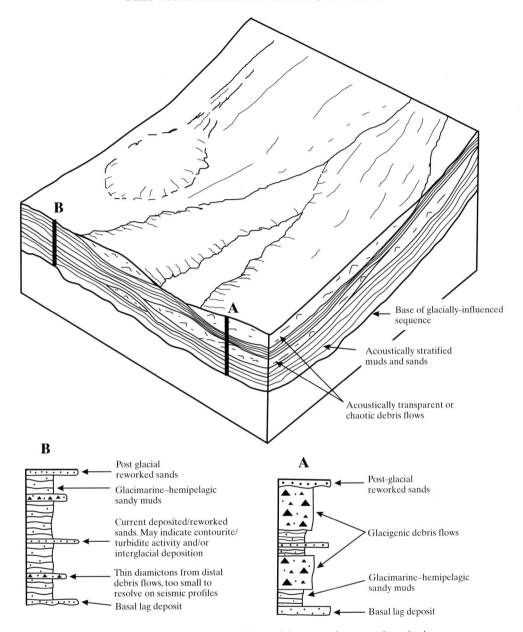

Fig. 7. Hypothetical model for a glacially-influenced base of slope area, incorporating seismic response, sea-bed morphology and sediment lithology. Log A represents a scenario similar to that seen in borehole 99/3. Log B represents the setting envisaged for the glacially-influenced Macduff sequence shown in Figure 8. The model also shows the relative positions of these sections on a glacially-influenced slope.

chaotic, laterally discontinuous and frequently form stacked successions. Lithologically they are structureless and are composed of muddy diamictons with very variable sand and pebble content and a mixed microfauna. The formation

of debris flows seems most likely to occur when the ice is at the shelf edge and consequently the rate of deposition is at its highest. The presence of debris flows at the slope base appears to be a strong indicator of the location of ice stream outflows at or near the shelf break, providing the volume of sediment to generate the mass flows (Stoker 1995). This possibility is supported by initial findings from the West Shetland Shelf, which show a buried trough structure on the shelf lying directly above the debris flows. The trough is up to 12 km wide and is traceable for over 30 km southeastwards from the shelf edge. Core and seismic data suggest it is filled with glacial sediments reaching up to 180 m in thickness at the shelf edge. This thickness of sediment appears to be the product of more than one phase of glacial activity.

From the data provided by borehole 99/3, the deep-water glacigenic sequence of the West Shetland margin can be characterized using a combination of the three facies associations. These characteristics can be used to build a hypothetical analogue model shown in Figure 7, which could be applied to other localities along the West Shetland margin, and potentially to other glaciated margins. This model consists of a sequence of silty muds with dropstones deposited by suspension settling and iceberg rafting, and massive diamictons deposited by debris flows. In the case of the West Shetland margin this sequence is bounded on both its upper and lower limits by current deposited and reworked sands of non-glacigenic origin. Although this type of model cannot predict where debris flows will occur, it does give an indication as to the style of facies architecture that can be expected. The lithological information from the core indicates that current deposition and reworking are important processes but that the resulting sediments are difficult to distinguish from the glacimarine muds on seismic profiles. It also shows that thin debris flows, or the distal end of larger flows which are below seismic resolution, can be hidden within acoustically stratified units, suggesting debris flows maybe more common than is apparent from seismic records.

Application of the model to the rock record

The Macduff Slate Formation forms the uppermost section of the Southern Highland Group of the Late Proterozoic Dalradian succession of Northern Scotland (Fig. 1). It comprises a combination of psammites, semipelitic greywackes and prominent beds of gritty quartzites, forming a 2 km thick sequence of metasediments (Harris *et al.* 1994). These sediments have been interpreted as deep-water distal turbidites and sandy channel-fill deposits from a submarine fan complex (Kneller 1987; Trewin

1987). A 19 m interval near the top of an 83 m exposed section near Banff (Fig. 8) is termed the Macduff Boulder Bed. This unit consists predominantly of greywackes with dispersed clasts from granule to boulder grade.

The Boulder Bed section at Banff is described in detail by Stoker *et al.* (1999). The glacially-influenced section is dominated by thin graded siltstones and lenticular sandstones with sediment starved ripples. Dropstones with distorted laminae occur within two distinct bands and a series of thin planar and lensoid muddy diamictites were also recorded. In addition, a 1.3 m thick diamictite occurs close to the top of the section, immediately overlying a slumped unit containing disrupted bedding (the 'Allochthonous unit' of Stoker *et al.* 1999).

The origin of the Boulder Bed, and in particular the origin of the diamictite bed near the top, has been variously interpreted as a glacially-influenced sequence (Sutton & Watson 1954), a glacimarine section capped by a lodgement till deposited by grounded ice (Hambrey & Waddams 1981) and a glacially-influenced section with frozen masses of sediment deposited from floating ice (Trewin 1987). In these interpretations, the term glacially-influenced is used to indicate glacimarine deposition by ice-rafting without the presence of tills deposited by grounded ice. The section was re-interpreted by Stoker *et al.* (1999) as a glacially-influenced deep marine succession, with periodic input from glacigenic debris flows which formed the thin diamictites. This interpretation was thought to be analogous with the scenario observed in the Pleistocene deposits of the Faroe–Shetland Channel (Stoker *et al.* 1999).

In comparing the glacigenic sequences from borehole 99/3 and the Macduff Boulder Bed, there are obvious similarities. Both comprise silty muds with dropstones punctuated by muddy diamictons and are known to be from, or interpreted as, deep-water environments. The scale of the diamictite bodies seen in the Macduff section is smaller than those from the 99/3 borehole. However, the two sequences have a similar geometry and lithological character.

The comparison between the Macduff Boulder Bed and the Faeroe–Shetland Channel succession suggested the thin Macduff diamictons were the distal end of larger, unseen debris flows similar in nature to those on the West Shetland margin (Stoker *et al.* 1999). However, this original comparison lacked the lithological data from long cores that borehole 99/3 provides. The new core data allows a direct

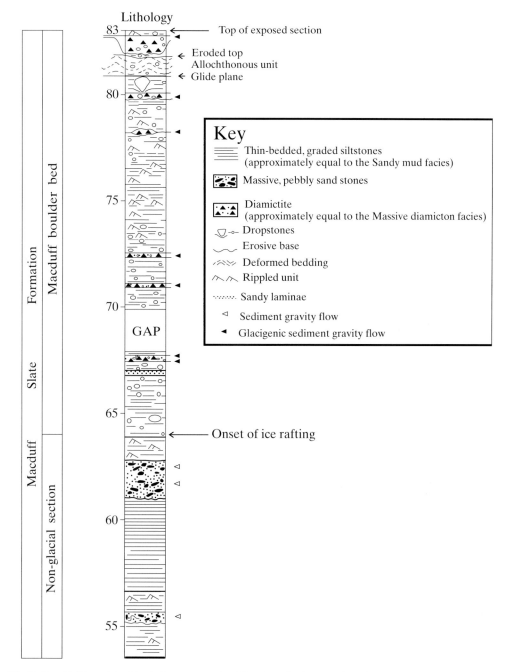

Fig. 8. Detailed log of the Macduff sequence, simplified from Stoker *et al.* (1999).

comparison of lithofacies, confirming the similarity of the Macduff rocks to glacigenic sediments from a deep-water continental margin setting. In addition, the regional seismic data and sea-bed image can now provide an insight

into how the unexposed up-slope section of the Macduff sequence may have appeared. Conversely, the distal debris flows of the Macduff Formation may give an indication as to the nature of the lithologies and smaller scale architecture of the West Shetland margin sequence, down slope and lateral to the borehole location. This is particularly useful in considering the architecture of the distal toe of debris flows that are too small to resolve on seismic records and which were only poorly sampled by the 99/3 borehole. Within the scheme of the model, the characteristics of the section from borehole 99/3 are represented in Figure 7 by the log at locality A. Locality B shows how the characteristics seen in the Macduff section fit into the model.

Conclusions

The use of sea-bed imagery, high resolution seismic surveys and borehole cores provides a powerful combination of data with which to characterize the glacially-influenced deep marine environment. The model developed from the Quaternary data can be used to infer the likely depositional scenario for an ancient sequence for which there is less regional information. In addition, the smaller scale features seen in outcrop in ancient sequences can be used to fill in the gaps in the Quaternary model and allow the development of a more accurate version which could be applied to other continental margins, both ancient and modern.

Permission to use core data from borehole 99/3 was provided by the British Geological Survey (BGS) and UK Rockall Consortium (BGS, Agip, Amerada Hess, BG International, BP Exploration, Conoco, Enterprise, ExxonMobil Exploration, Statoil, TotalFinaElf, ChevronTexaco, DTI). Sea-bed image data were provided by the Western Frontiers Association (WFA) (Agip, Amerada Hess, BP, Conoco, Enterprise, Norsk Hydro, Shell, Statoil, Texaco, TotalFinaElf). Work by S. Davison was funded by the WFA. This paper is published with the permission of the Director of the British Geological Survey (NERC).

References

ACKHURST, M A. 1991. Aspects of late Quaternary sedimentation in the Faeroe–Shetland Channel, northeast UK Continental Margin. British Geological Survey Technical Report, WB/91/2

BARKER, P. F., CAMERLENGHI, A., ACTON, G. D., ET AL. 1999. Proceedings of the Ocean Drilling Programme, Initial Reports, 178 [CD-ROM]. Ocean Drilling Programme, Texas A & M University, College Station, TX 77845–9547, USA.

BOND, G. C. & LOTTI, R. 1995. Iceberg discharges into the North Atlantic on millennial time scales during the last glaciation. Science, 267, 1005–1010.

BULAT, J. & LONG, D. 1998. Creation of seabed feature maps from 3D seismic horizon data sets.

British Geological Survey Technical Report, WB/98/38C.

CLAUSEN, L. 1998. The Southeast Greenland glaciated margin: 3D stratal architecture of shelf and deep sea. In: STOKER, M., EVANS, D. & CRAMP, A. (eds) Geological Processes on Continental Margins: Sedimentation, Mass-wasting and Stability. Geological Society, London, Special Publications, 129, 173–203.

COOPER, A. K., BARRETT, P. J., HINZ, K., TRAUBE, V., LEITCHENKOV, G. & STAGG, H. M. J. 1991. Cenozoic prograding sequences of the Antarctic continental margin: a record of glacio-eustatic and tectonic events. Marine Geology, 102, 175–213.

DOWDESWELL, J. A., ELVERHØI, A. & SPIELHAGEN, R. 1998. Glacimarine sedimentary processes and facies on the polar North Atlantic margins. Quaternary Science Reviews, 17, 243–272.

ERICSON, D. B. 1959. Coiling direction of Globigerina pachyderma as a climatic index. Science, 130, 219–220.

EYLES, N., DANIELS, J., OSTERMAN, L. E. & JANUSZCZAK, N. 2001. Ocean Drilling Program Leg 178 (Antarctic Peninsula): sedimentology of glacially influenced continental margin topsets and forsets. Marine Geology, 178, 135–156.

HAMBREY, M. J. 1994. Glacial environments. 296 pp. UCL Press Ltd, London.

HAMBREY, M. J. & WADDAMS, P. 1981. Glacigenic boulder-bearing deposits in the Upper Dalradian Macduff Slates, north-eastern Scotland. In: HAMBREY, M. J. & HARLAND, W. B. (eds) 1981. Earth's pre-Pleistocene glacial record. Cambridge University Press. 571–575.

HARRIS, A. L., HASELOCK, P. J., KENNEDY, M. J. & MENDUM, J. R. 1994. The Dalradian Supergroup in Scotland, Shetland and Ireland. In: GIBBONS, W. & HARRIS, A. L. (eds) A revised correlation of Precambrian rocks in the British Isles. Geological Society, London, Special Reports, 22, 33–53.

HITCHEN, K. (compiler) 1999. Rockall Continental Margin Project. Shallow Borehole Drilling Programme 1999. Geological Report. BGS Technical Report WB/99/21C.

HOLMES, R. 1997. Quaternary Stratigraphy: The offshore record. In: GORDON, J. E. (ed.) Reflections on the ice age in Scotland. Scottish Natural Heritage, Glasgow. 72–94.

JANSEN, E., RAYMO, M. E., BLUM, P. ET AL. 1996. In: Proceedings, Ocean Drilling Program, Initial Reports, 162, College Station, Tx (Ocean Drilling Program).

KNELLER, B. C. 1987. A geological history of Northeast Scotland. In: TREWIN, N. H., KNELLER, B. C. & GILLEN, C. (eds) Excursion guide to the geology of the Aberdeen area. Scottish academic press, Edinburgh, 1–54.

KRISTOFFERSON, Y., WINTERHALTER, N. & SOLHEIM, A. 2000. Shelf progradation on a glaciated continental margin, Queen Maud Land, Antarctica. Marine Geology, 165, 109–122.

MONCRIEFF, A. C. M. 1989. Classification of poorly sorted sedimentary rocks. Sedimentary Geology, 65, 191–194.

O'BRIEN, P. E., COOPER, A. K., RICHTER, C., *ET AL*. 2001. Proceedings of the Ocean Drilling Programme, Initial Reports, **188** [CD-ROM]. Ocean Drilling Programme, Texas A & M University, College Station, TX, 77845–9547, USA.

PIPER, D. J. W. 1988. Glaciomarine sedimentation on the continental slope off Eastern Canada. *Geoscience Canada*, **15**, 23–28.

SOLHEIM, A., FALEIDE, J., ANDERSEN, E. S., ELVERHOI, A., FORSBERG, C., VANNESTE, K., UENZELMANN-NEBEN, G. & CHANNELL, J. E. T. 1998. Late Cenozoic seismic stratigraphy and glacial geological development of the east Greenland and Svalbard-Barents sea continental margins. *Quaternary Science Reviews*, **17**, 155–184.

STEIN, A. B. & SYVITSKI, J. P. M. 1997. Glacially-influenced debris flow deposits: East Greenland slope. *In*: DAVIES, T. A., BELL, T., COOPER, A. K., JOSENHANS, H., POLYAK, L., SOLHEIM, A., STOKER, M. S. & STRAVERS, J. A. (eds) *Glaciated Continental Margins: An Atlas of Acoustic Images*. Chapman & Hall, London, 134–136.

STRAVERS, J. A. 1997. Debris flows and slumps; Overview. *In*: DAVIES, T. A., BELL, T., COOPER, A. K., JOSENHANS, H., POLYAK, L., SOLHEIM, A., STOKER, M. S. & STRAVERS, J. A. (eds) *Glaciated Continental Margins: An Atlas of Acoustic Images*. Chapman & Hall, London, 115–118.

STOKER, M. S. 1990. *Judd sheet, 60° N–06° W. Quaternary Geology*. 1: 250 000 Map Series. British Geological Survey, Nottingham.

STOKER, M. S. 1995. The influence of glacigenic sedimentation on slope-apron development on the continental margin off Northwest Britain. *In*: SCRUTTON, R. A., STOKER, M. S., SHIMMIELD, G. B. & TUDHOPE, A. W. *The Tectonics, Sedimentation and Palaeoceanography of the North Atlantic Region*. Geological Society, London, Special Publications, **90**, 159–177.

STOKER, M. S. 1999. *Stratigraphic nomenclature of the UK North West Margin. 3. Mid- to Late Cenozoic Stratigraphy*. British Geological Survey, Edinburgh.

STOKER, M. S. & HOLMES, R. 1991. Submarine end-moraines as indicators of Pleistocene ice limits off north-west Britain. *Journal of the Geological Society, London*, **148**, 431–434.

STOKER, M. S., HARLAND, R. & GRAHAM, D. K. 1991. Glacially influenced basin plain sedimentation in the southern Faeroe-Shetland Channel, north-west United Kingdom continental margin. *Marine Geology*, **100**, 185–199.

STOKER, M. S., HITCHEN, K. & GRAHAM, C. C. 1993. *United Kingdom offshore regional report: the geology of the Hebrides and West Shetland shelves, and adjacent deep-water areas*. HMSO, London, for British Geological Survey.

STOKER, M. S., HOWE, J. A. & STOKER, S. J. 1999. Late Vendian-? Cambrian glacially influenced deep-water sedimentation, Macduff Slate Formation (Dalradian), NE Scotland. *Journal of the Geological Society, London*, **156**, 55–61.

SUTTON, J. & WATSON, J. V. 1954. Ice-borne boulders in the Macduff Group of the Dalradian of Banff-shire. *Geological Magazine*, **91**, 391–398.

TREWIN, N. H. 1987. Macduff, Dalradian turbidite fan and glacial deposits. *In*: TREWIN, N. H., KNELLER, B. C. & GILLEN, C. (eds) *Excursion Guide to the Geology of the Aberdeen Area*. Scottish Academic Press, Edinburgh, 79–88.

VORREN, T. O., LABERG, J., BLAUME, F., DOWDESWELL, J. A., KENYON, N. H., MIENERT, J., RUMOHR, J. & WERNER, F. 1998. The Norwegian-Greenland Sea continental margins: Morphology and Late Quaternary sedimentary processes and environment. *Quaternary Science Reviews*, **17**, 273–302.

WILKINSON, I. P., RIDING, J. B., LEES-BURNETT, J. A. & BOWN, P. 2000. Micropalaeontological analysis of Quaternary deposits of boreholes 99/3 and 99/6. *In*: DAVISON, S. (ed.) *An Investigation of the Quaternary Sediments from Boreholes 99/3, 99/4, 99/5 and 99/6, Faroe-Shetland Channel*. BGS Commercial Report **CR/00/27**, British Geological Survey, Edinburgh.

Late Quaternary sedimentation in Kejser Franz Joseph Fjord and the continental margin of East Greenland

J. EVANS[1*], J. A. DOWDESWELL[1], H. GROBE[2], F. NIESSEN[2], R. STEIN[2],
H.-W. HUBBERTEN[3] & R. J. WHITTINGTON[4]

[1]*Scott Polar Research Institute, University of Cambridge, Lensfield Road, Cambridge,
CB2 1ER, UK*

**Present address: Geological Division, British Antarctic Survey,
High Cross, Madingley Road, Cambridge, CB3 0ET, UK (e-mail: JEV@bas.ac.uk)*

[2]*Alfred Wegener Institut fur Polar –und Meeresforschung, Columbusstrasse,
D-27568 Bremerhaven, Germany*

[3]*Alfred Wegener Institut fur Polar –und Meeresforschung, Telegrafenberg A43,
D-14473 Potsdam, Germany*

[4]*Institute of Geography and Earth Science, University of Wales, Aberystwyth, Ceredigion,
SY23 3DB, UK*

Abstract: The marine sedimentary record in Kejser Franz Joseph Fjord and on the East Greenland continental margin contains a history of Late Quaternary glaciation and sedimentation. Evidence suggests that a middle-shelf moraine represents the maximum shelfward extent of the Greenland Ice Sheet during the last glacial maximum. On the upper slope, coarse-grained sediments are derived from the release of significant quantities of iceberg-rafted debris (IRD) and subsequent remobilization by subaqueous mass-flows. The middle–lower slope is characterized by hemipelagic sedimentation with lower quantities of IRD (dropstone mud and sandy mud), punctuated episodically by deposition of diamicton and graded sand/gravel facies by subaqueous debris flows and turbidity currents derived from the mass failure of upper slope sediments. The downslope decrease of IRD reflects either the action of the East Greenland Current (EGC) confining icebergs to the upper slope, or to the more ice-proximal setting of the upper slope relative to the LGM ice margin. Sediment gravity flows on the slope are likely to have fed into the East Greenland channel system, contributing to its formation in conjunction with the cascade of dense brines down the slope following sea-ice formation across the shelf.

Deglaciation commenced after 15 300 ^{14}C years BP, as indicated by meltwater-derived light oxygen isotope ratios. An abrupt decrease in both IRD deposition and delivery of coarse-grained debris to the slope at this time supports ice recession, with icebergs confined to the shelf by the EGC. Glacier ice had abandoned the middle shelf before 13 000 ^{14}C years BP with ice loss through iceberg calving and deposition of diamicton. Continued retreat of glacier-ice from the inner shelf and through the fjord is marked by a transition from subglacial till/bedrock in acoustic records, to ice-proximal meltwater-derived laminated mud to ice-distal bioturbated mud. Ice abandoned the inner shelf before 9100 ^{14}C years BP and probably stabilized in Fosters Bugt at 10 000 ^{14}C years BP. Distinct oxygen isotope minima on the inner shelf indicate meltwater production during ice retreat. The outer fjord was free of ice before 7440 ^{14}C years BP. Glacier retreat through the mid–outer fjord was punctuated by topographically-controlled stillstands where ice-proximal sediment was fed into fjord basins. The dominance of fine-grained, commonly laminated facies during deglaciation supports ablation-controlled, ice-mass loss.

Glacimarine sedimentation within the Holocene middle–outer fjord system is dominated by sediment gravity flow and suspension settling from meltwater plumes. Suspension sediments comprise mainly mud facies indicating significant meltwater-deposition that overwhelms debris release from icebergs in this East Greenland fjord system. The relatively widespread occurrence of fine-grained lithofacies in East Greenland fjords suggests that meltwater sedimentation can be significant in polar glacimarine environments. The ice-distal continental margin is characterized by meltwater sedimentation in the inner shelf deep, iceberg scouring over shallow shelf regions, winnowing and erosion by the East Greenland Current on the middle–outer shelf, and hemipelagic sedimentation on the continental slope.

From: DOWDESWELL, J. A. & Ó COFAIGH, C. (eds) 2002. *Glacier-Influenced Sedimentation on High-Latitude Continental Margins.* Geological Society, London, Special Publications, **203**, 149–179. 0305-8719/02/$15.00
© The Geological Society of London 2002.

Marine sediments in the fjords and on the continental margin of East Greenland record a history of sedimentation associated with Quaternary glacier-fluctuations and climate change. Investigations of marine sediments in East Greenland have been augmented by detailed studies from the Scoresby Sund fjord system (Marienfeld 1991, 1992*a*, *b*; Dowdeswell *et al.* 1993, 1994*a*, *b*; Ó Cofaigh *et al.* 2001), Kangerdlugssuaq Fjord (Syvitski *et al.* 1996*a*; Andrews *et al.* 1994, 1996) and the adjacent continental margin (Mienert *et al.* 1992; Dowdeswell *et al.* 1997*b*; Nam *et al.* 1995; Stein *et al.* 1996; Nam 1996). Correlation of the marine and terrestrial sedimentary records has provided a comprehensive reconstruction of glacial history and climatic events in East Greenland (e.g. Funder *et al.* 1998).

These geological investigations indicate that glacier fluctuations in East Greenland have been relatively minor during Late Quaternary glacial–interglacial periods (Funder 1989; Funder & Hansen 1996; Funder *et al.* 1998) in comparison to major variations elsewhere in the Polar North Atlantic (e.g. Elverhøi *et al.* 1998). However, Late Quaternary sedimentation and glacial history in the more northerly fjords and continental margin of East Greenland is poorly constrained. Our understanding of the glacial history in this region has been based on only a few terrestrial geological and lake studies (Hjort 1979, 1981; Funder 1989; Wagner *et al.* 2000; Cremer *et al.* 2001). Therefore, the aim of our work is to investigate the marine sedimentary record from the northern part of East Greenland (Kejser Franz Joseph Fjord and the adjacent continental margin) in order to: (1) characterize glacimarine sedimentation along a transect from the middle–outer fjord to the continental slope; (2) to reconstruct the Late Quaternary glacial history and sedimentation for this region; and (3) to place the study within the context of sedimentation and glacier fluctuations in East Greenland and the Polar North Atlantic.

Study area

Physiography and bathymetry

Kejser Franz Joseph Fjord is located in East Greenland at $73°$ N, covering an area of 2200 km^2 and extending 220 km from fjord head to mouth (Fig. 1). Nordfjord, Geologfjord and Isfjord form tributary fjords. A prominent shallow sill characterizes the intersection of Nordfjord and Kejser Franz Joseph Fjord (Fig. 1b). The middle–outer Kejser Franz Joseph Fjord is 10–20 km wide and has water depths of up to 550 m, with the outer fjord basin subdivided into three sub-basins (Fig. 1b). Fosters Bugt forms a wide embayment at the fjord mouth with maximum water depths of 340 m. The inner continental shelf is characterized by a bathymetric deep with maximum water depths reaching 520 m. A prominent bathymetric high is located across the inner–middle shelf with water depths of 235 m (Fig. 1b). The remaining shelf is 280 to 340 m deep and extends 110 km to the shelf break.

The hinterland of the middle–outer Kejser Franz Joseph Fjord and Fosters Bugt comprises a mountainous inland that slopes down to coastal lowlands, and contrasts physiographically with the steep-walled fjord interior. Glacially abandoned valleys and cirques incise the hinterland. Glacifluvial and fluvial systems fed by melting glacier-ice or snow and precipitation dissect these coastal plains (e.g. Badlanddal and Paralleldal discharging into Fosters Bugt and the outer fjord, respectively; Fig. 1c). These systems produce fjord-margin outwash deltas or alluvial fans, and surface meltwater plumes that extend to shore-distal locations. Subaerial rock-falls form talus cones along steep fjord margins.

Glaciology and oceanography

The Greenland Ice Sheet drains through the inner coastal mountain zone via Walterhausen, Adolf Hoels, Jætte, Gerrard de Geer and Nordenskjøld Gletschers that terminate at the head of Kejser Franz Joseph Fjord and its tributary fjords (Fig. 1). Walterhausen Gletscher is the largest of these glaciers, with a terminus width of 10.2 km. The total glacier drainage-basin area exceeds 8400 km^2 compared to 50 000 km^2 for the inner Scoresby Sund fjord system (Dowdeswell *et al.* 1994*b*).

Approximately 8 km^3 of ice is calved into the fjord system each year (compared to 18 km^3 in Scoresby Sund and 7 km^3 in Dove Bugt), accounting for 3% of total iceberg production in Greenland (Reeh 1985). Observations indicate that icebergs in Kejser Franz Joseph Fjord are highly variable both in size and shape and their net drift is towards the fjord mouth. Shorefast sea ice prevents drift between October and June. Icebergs escaping the fjord system drift south along the continental shelf parallel to the coast within the East Greenland Current (Wadhams 1981). Multi-year sea-ice across the continental margin retreats to NE Greenland during mild summers and remains in East Greenland during moderate summers.

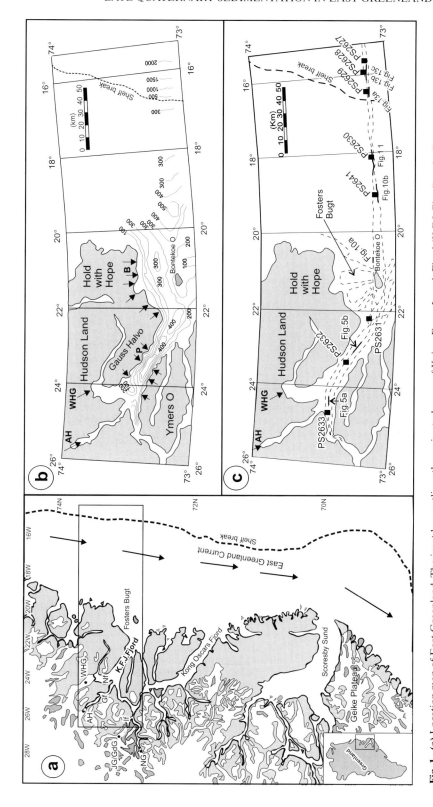

Fig. 1. (**a**) Location map of East Greenland. The inset box outlines the main study area of Kejser Franz Joseph Fjord (K.F.J. Fjord) and adjacent continental margin discussed in this paper. The fjord is fed by several tributary fjords that include: Nordfjord (Nf), Geologfjord (Gf) and Isfjord (If). White areas correspond to ice-covered landscape with outlet glaciers terminating at the head of the fjord system indicated: Walterhausen Gletscher (WHG), Adolf Hoels Gletscher (AH), Jaette Gletscher (JG), Gerrard de Geer Gletscher (GdG) and Nordenskjold Gletscher (NG). Black arrows mark the southward flowing East Greenland Current. (**b**) Detailed bathymetric map of the study area with arrows denoting the influx of meltwater to the outer fjord and Fosters Bugt. Two major fluvial systems are marked: P – Paralleldal; B – Badlandal. (**c**) Map showing cruise tracks along which Parasound acoustic data were acquired and the locations of sediment cores. Locations of Parasound records shown in this paper are illustrated.

Table 1. *Acoustic facies identified from Parasound records from Kejser Franz Joseph Fjord and the adjacent East Greenland continental margin*

Acoustic facies	Description
1a	Acoustically stratified sediment that infills fjord basins. Smooth and continuous sea floor reflector with multiple, parallel, continuous to semi-continuous sub-sea floor reflectors that can pinch out laterally. Basin fill can be down-fjord in direction.
1b	Acoustically stratified sediment comprising a smooth continuous sea floor reflector with parallel, distinct, semi-continuous to continuous reflectors that can pinch-out laterally. Facies is confined to the continental shelf and slope.
2	Acoustically transparent lens-shaped sediment bodies that are up to 20 m thick and up to 750 m wide, although widths of several km are present in fjord basins. Distinct and well-defined reflectors enclose sediment lenses. Hummocky sea floor where sediment lenses are located close to the sea floor on the upper slope.
3	Acoustically homogeneous sediment with a highly irregular sea floor reflector comprising paired ridges and an intervening trough. Crest-trough amplitude reaches 15 m and crest-to-crest width metres to tens of metres. Occurs down to water depths of 400 m. Sea floor reflector is discontinuous and diffuse, and displays high intensity irregularities with sharp ridge crests above 300 m water depths. Ridge crests are more rounded and sea floor reflector more distinct and continuous with some areas of flat sea floor in water depths between 200 and 300 m.
4	Acoustically transparent to stratified sediment (can exceed 25 m thick) with multiple, parallel continuous to semi-continuous sea floor and sub-sea floor reflectors draping underlying topography. Sea floor reflector can be slightly irregular, comprising small-scale, low intensity irregularities separated by extensive regions of smooth sea floor.
5	Acoustically semi-transparent to crudely stratified sediment with a distinct and irregular (small-scale <5m wide) top-surface reflector. Facies thickness is highly variable but generally <10 m thick. The facies is confined to the mid–outer shelf.
6	Acoustically homogeneous sediment on the upper slope. Low acoustic penetration. Sea floor reflector is continuous, semi-prolonged and generally smooth and planar, but can be hummocky in association with facies 2.

The East Greenland Current (EGC) flows south along the continental margin (Fig. 1a). It comprises cold ($-1\,^{\circ}$C), polar water to a depth of 250 m, and warm, saline return Atlantic intermediate water (RAIW) below this (Hopkins 1991). Kejser Franz Joseph Fjord comprises three water masses (Vogt *et al.* 1995), similar to other East Greenland fjords (Marienfeld 1991, 1992*b*; Syvitski *et al.* 1996*a*; Ó Cofaigh *et al.* 2001): (1) warm (>0 °C), low saline (<31 per mil) surface water (<25 m thick) that extends onto the inner shelf; (2) very cold (<0 °C), high salinity polar waters to depths of 200–300 m derived from the inflow of the EGC; and (3) warm (0–3 °C), high salinity (>34 per mil) RAIW intruding into the fjord below 300 m.

Data acquisition and methods

Geological and geophysical data were collected during the 1994 cruise of RV *Polarstern* to Kejser Franz Joseph Fjord and the East Greenland continental margin (Hubberten 1995; Fig. 1). The regional distribution of sediments was analysed using a Krupp-Atlas Parasound system (Grant & Schreiber 1990). The Parasound system adopts the parametric principle where the profiling beam is generated from non-linear interaction of two primary signals of different frequencies. The resultant profiling beam produces a footprint diameter of 7% of water depth and a width of 4°. This enables up to 100 m of sediment penetration with a vertical resolution of 0.3 m leading to better spatial resolution than with conventional 3.5 kHz systems (Kuhn & Weber 1993; Dowdeswell *et al.* 1997*a*). Six acoustic facies are identified from Parasound records collected along all ship tracks (Fig. 1c; Table 1) on the basis of sea floor and sub-sea floor reflectors and associated lateral continuity, morphology and geometry (e.g. Damuth 1978).

Eight gravity cores were recovered along the fjord–shelf–slope transect (Fig. 1c; Table 2). Cores were described both visually and with X-radiographs, and lithofacies were identified using the nomenclature of Eyles *et al.* (1983)

Table 2. *Location, water depth and recovery length of gravity cores from Kejser Franz Joseph Fjord and the adjacent continental margin of East Greenland*

Core	Latitude	Longitude	Water depth (m)	Recovery (m)
PS2633	73° 28.8 S	24° 36.8 W	283	5.85
PS2632	73° 24.4 S	23° 38.0 W	505	2.58
PS2631	73° 10.7 S	22° 11.0 W	430	7.25
PS2641	73° 09.3 S	19° 28.9 W	469	7.00
PS2630	73° 09.5 S	18° 04.1 W	287	3.02
PS2629	73° 09.5 S	16° 29.0 W	850	2.70
PS2628	73° 09.8 S	15° 58.0 W	1694	2.35
PS2627	73° 07.4 S	15° 40.9 W	2009	4.14

(Table 3). Grain size distribution was determined using wet and dry sieving and SediGraph. Mean grain size and sorting were calculated using the statistical graphical method of Folk & Ward (1957). The number of particles coarser than 2 mm were point-counted using X-radiographs, and the percentage sand and gravel >500 µm determined, as an index of iceberg-rafted debris (IRD) (cf. Grobe 1987; Elverhøi *et al.* 1995).

The chronology of the sediments was established by radiocarbon dating of carbonate shells using the accelerator mass spectrometer (AMS) at the University of Århus. AMS radiocarbon ages were determined for specific horizons using 2000–3000 shells of the planktonic foraminifera

Table 3. *Lithofacies in cores from middle–outer Kejser Franz Joseph Fjord and adjacent continental margin (after Eyles* et al. *1983)*

Lithofacies	Description
Diamicton	
Dmm	Diamicton, matrix-supported and massive. Dispersed to clustered clasts. Can form a rare sandy gravel-rich lag
Dmm(r)	Diamicton, matrix-supported and massive with dispersed clasts to locally imbricated clasts
Gravelly sand	
GSng	Gravelly sand, normally graded
Sand	
Sm	Sand, massive
Sm(d)	Sand, massive with dispersed clasts
Muddy sand	
FSng	Muddy sand, normally graded
Sandy mud	
SFng	Sandy mud, normally graded
SFm	Sandy mud, massive
SFm(d)	Sandy mud, massive with dispersed clasts
SFb(d)	Sandy mud, bioturbated with dispersed clasts
SFc(m-l)	Sandy mud, rhythmic couplets comprising sand/silt rich and clay rich units with a massive to planar parallel to cross-laminated structure, water escape structures
Mud	
Fm	Mud, massive
Fm(d)	Mud, massive with dispersed clasts
Fm(d-sl)	Mud, massive with dispersed clasts and lenses of poorly-sorted sand
Fb	Mud, bioturbated
Fb(d)	Mud, bioturbated with dispersed clasts
Fl	Mud, laminated with dispersed clasts
Fl(d)	Mud, laminated with dispersed to layered clasts

Table 4. *Radiocarbon dates for cores PS2631, PS2641, PS2630, PS2629, PS2628 and PS2627. Ages are shown in uncorrected and corrected form. The corrected ages assume a reservoir age of 550 years in East Greenland. Ages were determined on planktonic forminifera (*N. pachyderma*), gastropoda (*Buccinum hydrophanum*) and bivalvia (*Thyasira gouldi, Bathyarca glacialis or Portlandia fraterna*) species*

Core	Core Depth (cm)	Species	Uncorrected age ^{14}C years BP	Corrected age ^{14}C years BP
PS2631	99	Buccinum hydrophanum	1695 +/− 55	1145 +/− 55
	390	Thyasira gouldi	7990 +/− 210	7440 +/− 210
PS2641	375	Bathyarca glacialis	6980 +/− 130	6430 +/− 130
	413	Bathyarca glacialis	7600 +/− 70	7050 +/− 70
	535	Bathyarca glacialis	8700 +/− 75	8150 +/− 75
	554	Bathyarca glacialis	9130 +/− 80	8580 +/− 80
	565	Portlandia fraterna	9280 +/− 80	8730 +/− 80
	585	Portlandia fraterna	9560 +/− 120	9010 +/− 120
PS2630	180	N. pachyderma	13 560 +/− 130	13 010 +/− 130
PS2629	70	N. pachyderma	17 510 +/− 160	16 960 +/− 160
	130	N. pachyderma	19 500 +/− 210	18 950 +/− 210
PS2628	30	N. pachyderma	13 570 +/− 120	13 020 +/− 120
	150	N. pachyderma	15 910 +/− 160	15 360 +/− 160
	210	N. Pachyderma	19 390 +/− 190	18 840 +/− 190
PS2627	20	N. pachyderma	9300 +/− 100	8750 +/− 100
	220	N. pachyderma	15 880 +/− 120	15 330 +/− 120
	270	N. pachyderma	19 040 +/− 230	18 490 +/− 230
	330	N. pachyderma	26 350 +/− 380	25 800 +/− 380

Neogloboquadrina pachyderma sin, obtained from the 125–250 μm sand fraction in PS2641, PS2630, PS2629, PS2628 and PS2627, and gastropoda and bivalvia shells in PS2631 and PS2641 (Table 4). The ocean reservoir effect for East Greenland is 550 years (Hjort 1973) and is subtracted from the raw age to obtain the reservoir corrected radiocarbon age (^{14}C years BP). There were no ash layers within the cores to corroborate radiocarbon ages independently.

Chronology

Radiocarbon ages and sediment flux

Radiocarbon ages are presented in uncorrected and marine reservoir corrected form in Table 4. Ages indicate that core sediments extend back to the Late Weichselian glaciation. An age of 10 000 ^{14}C years BP is estimated for the surface of PS2630, as this is the last time that significant quantities of icebergs influenced the shelf to produce diamicton (see below). This interpretation is supported by an absence of Holocene diamicton on the inner shelf and in the fjords of E/NE Greenland (Stein *et al.* 1993; Nam 1996). An age of 13 000 ^{14}C years BP is assumed for the top of the sand-mud couplet facies in PS2627 as this correlates with an identical, stratigraphic position in PS2628 that is dated to this time.

Radiocarbon ages allow calculation of linear sedimentation rates (cm ka^{-1}) and bulk accumulation rates (g cm^{-2} ka^{-1}) for the cores (Fig. 2). Sediment flux decreases eastward, from a point depending on the position of the ice margin through time. Sediment delivery is generally greatest on the continental margin under full Late Weichselian glacial and deglacial conditions in response to the Greenland Ice Sheet being located on the continental shelf. Sediment flux during the Late Weichselian glaciation is 30 cm ka^{-1} and 29 to 65 g cm^{-2} ka^{-1} on the upper slope decreasing to 16 cm ka^{-1} and 16–24 g cm^{-2} ka^{-1} on the mid-lower slope (Fig. 2). Sedimentation rates of 51–79 cm ka^{-1} and 47–98 g cm^{-2} ka^{-1} are reached between 13 000 and 15 300 ^{14}C years BP on the mid-lower slope, decreasing to <4 cm ka^{-1} and <3 g cm^{-2} ka^{-1} after 13 020 ^{14}C years BP (Fig. 2). Holocene sediment flux is highest in the fjord and on the inner shelf (up to 111 cm ka^{-1} and 117 g cm^{-2} ka^{-1}) reflecting greater proximity to the ice margin during and after retreat to its present day position (Fig. 2).

Stable isotope stratigraphy

The stable isotope stratigraphy of three cores (PS2627, PS2630 and PS2641) is presented in Figure 3. Isotope stage 2 (LGM) in PS2627 is characterized by heavy $\delta^{18}O$ isotopes (>4 ‰), and isotope stage 1 by lighter $\delta^{18}O$ values (<3.5 ‰) (Fig. 3). Intense bioturbation of

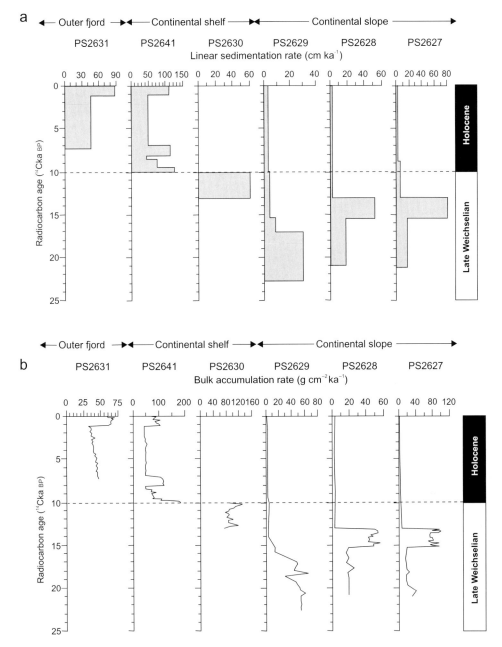

Fig. 2. (**a**) Sedimentation and (**b**) accumulation rates for cores PS2631, PS2641, PS2630, PS2629, PS2628 and PS2627, outer Kejser Franz Joseph Fjord and adjacent continental margin.

Holocene mud in PS2641 (see below) has smoothed any short-term isotopic variations that may have been present. A distinct shift of 1.67 ‰ in δ^{18}O in PS2627 occurs at the Stage 2/1 transition between 13 020 and 15 300 ^{14}C years BP (Fig. 3). Similarly, δ^{18}O minima (as low as 0.91 ‰) characterize the base of PS2641 (pre-dating 9010 ^{14}C years BP), corresponding to meltwater-derived laminated mud (Fig. 3). PS2630 is characterized by light δ^{18}O values.

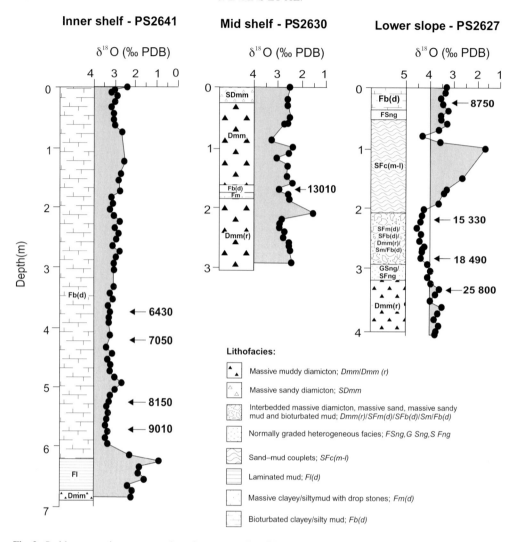

Fig. 3. Stable oxygen isotope records and corresponding lithological log of cores PS2641, PS2630 and PS2627 from the East Greenland continental margin. Core locations are shown in Figure 1c. AMS radiocarbon dates (^{14}C years BP) obtained from all cores are marked.

The distinct shift in δ^{18}O values between Stage 2 and the δ^{18}O minima exceed the 1.1–1.3 ‰ associated with the glacial–interglacial ice-volume effect (e.g. Chappell & Shackleton 1986; Shackleton 1987). The excess shift in δ^{18}O is unlikely to be a result of temperature change because the East Greenland Current is at present –1 °C, and additional cooling is unlikely during glacial periods. Instead, the δ^{18}O minima are attributed to a decrease in surface water salinity associated with a major pulse of isotopically light meltwater during the last deglaciation (cf. Jones & Keigwin 1988; Sarnthein *et al.* 1992; Stein *et al.* 1994 a& b; Elverhøi *et al.* 1995; Nam *et al.* 1995; Hald *et al.* 1996). This meltwater event influenced the slope between 15 300 and 13 020 ^{14}C years BP, the shelf after 13 020 ^{14}C years BP and terminated on the inner shelf before 9010 ^{14}C years BP. The meltwater signal in PS2627 corresponds to a sequence of thinly interbedded turbidites and hemipelagic muds (see below), but the signal is in-sequence (i.e. between Stages 2 and 1) and its timing is consistent with meltwater production in

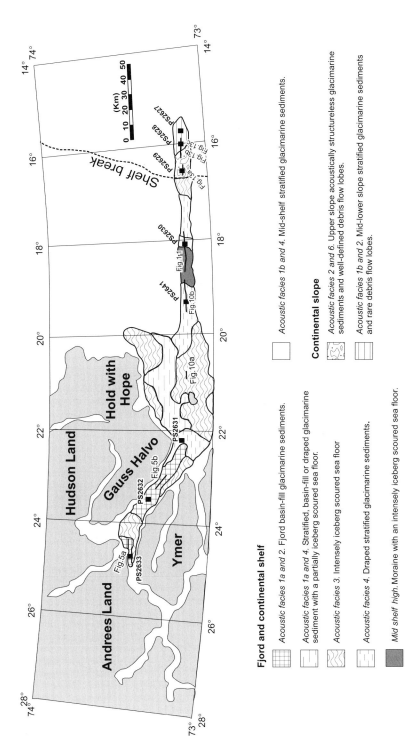

Fig. 4. Detailed map of the study area showing the distribution of acoustic facies defined from Parasound records. Locations of Parasound records shown in this paper are illustrated.

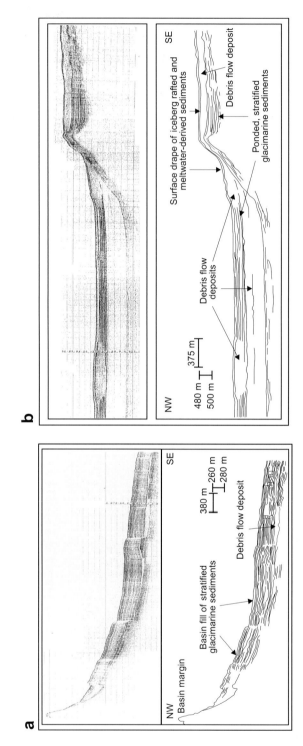

Fig. 5. Parasound records of acoustic facies within middle–outer Kejser Franz Joseph Fjord. (**a**) Ponded sediment within a deep middle fjord basin bounded by steep sills and comprising stratified sediment (facies 1a) and discontinuous sediment lobes (facies 2) that fill the basin in a down-fjord direction. Horizontal and vertical scales are shown. (**b**) Ponded sediment fill within deep outer fjord basins, consisting of acoustically homogeneous sediment lobes (facies 2), stratified sediment (facies 1a) and a thin surface sediment drape (facies 4). Note that facies 2 overspills a sill separating the basins.

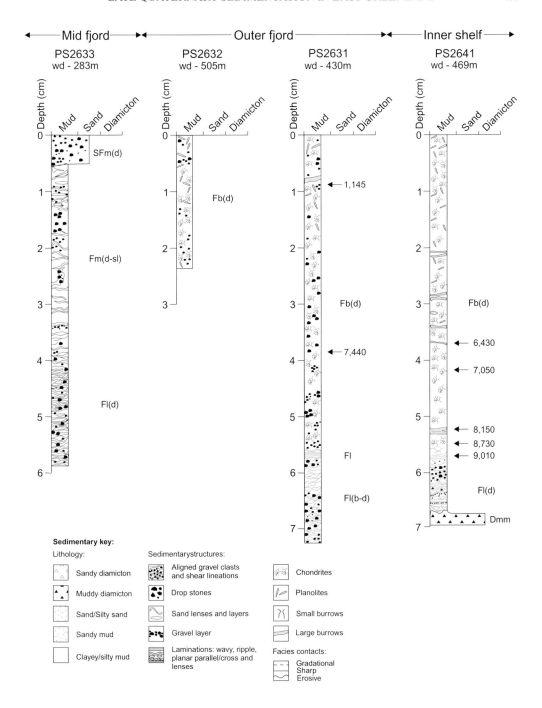

Fig. 6. Sedimentological logs of cores PS2633, PS2632, PS2631 and PS2641, middle–outer Kejser Franz Joseph Fjord and inner continental shelf. Core locations are shown in Figure 1c. AMS radiocarbon dates (^{14}C years BP) obtained from PS2631 and PS2641 are marked. Explanation of lithofacies codes is given in Table 3.

high-latitude regions. Therefore, it is likely that the reworking of foram shells by turbidity currents is minimal and immediately succeeds primary deposition.

Kejser Franz Joseph Fjord and Fosters Bugt

Sediment thickness and acoustic facies distribution

The mid-fjord basin comprises well-stratified sediment (facies 1a) interbedded with lens-shaped sediment bodies (facies 2) that fill the basin in a down-fjord direction (Figs 4 & 5a). Marginal fjord regions are draped by stratified sediment (facies 4). A prominent sill separating the mid- and outer fjord comprises a thin sediment cover with an irregular sea floor (facies 3).

The innermost sub-basin of the outer fjord comprises a sequence (>60 m thick) of stacked sediment lenses (facies 2) extending the length of the basin (several kilometres) overlain by well-stratified sediment (facies 1a) with isolated lenses of acoustically transparent sediment (<750 m wide, less than 20 m thick; facies 2) (Figs 4 & 5b). In the intermediate sub-basin

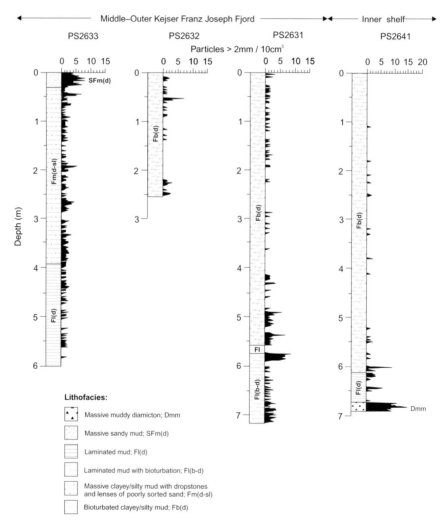

Fig. 7. Coarse-particle counts (particles >0.2 cm/10 cm³) from cores PS2633, PS2632, PS2631 and PS2641. Core locations are shown in Figure 1c.

a. PS2631

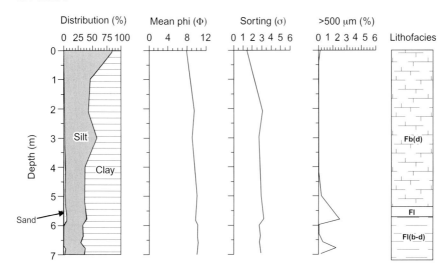

b. PS2641

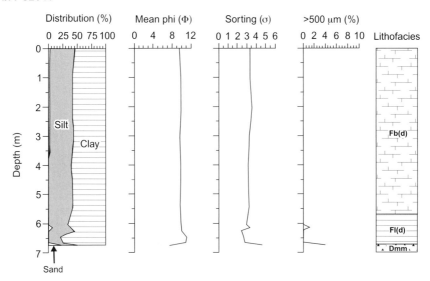

Fig. 8. Down-core grain size distribution, mean grain size, sorting and particles over 500 μm from (a) PS2631, (b) PS2641, (c) PS2630, (d) PS2629, (e) PS2628 and (f) PS2627, outer Kejser Franz Joseph Fjord and inner continental shelf. Core locations are shown in Figure 1c. Lithofacies codes are explained in Table 3.

sediment is <30 m thick and comprises acoustically opaque sediment or bedrock overlain by a drape of stratified sediment (facies 4) and a basin-length acoustically transparent sediment unit (facies 2) (Figs 4 & 5b). Transparent to stratified sediment (facies 1a, 2) is present in the outermost sub-basin (>60 m thick), with a less than 12 m thick drape of stratified sediment (facies 4) along more down-fjord marginal regions of the basin. A thin drape of sediment (<2 m) (facies 4) characterizes recent sedimentation in the outer fjord (Fig. 5b).

c. PS2630

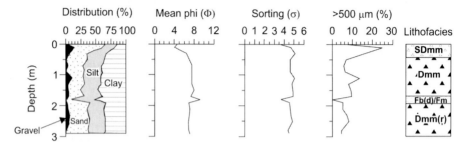

d. PS2629

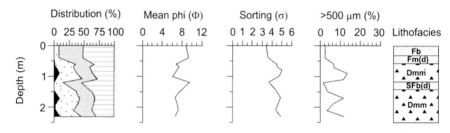

e. PS2628

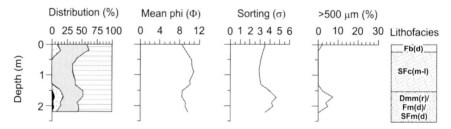

f. PS2627

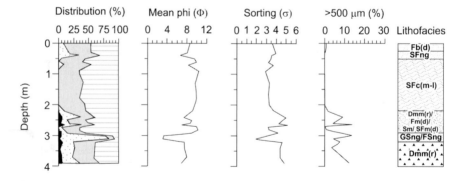

Fig. 8. continued.

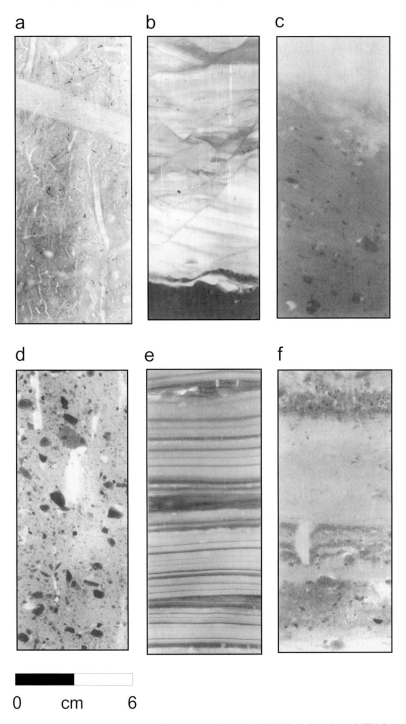

Fig. 9. Core X-radiographs of representative lithofacies in this study. (a) Bioturbated mud (Fb) from PS2641. (b) Laminated mud (Fl) and massive diamicton (Dmm) from PS2641. (c) Resedimented massive diamicton (Dmmr) from PS2630. (d) Massive diamicton (Dmm) from PS2629. (e) Mud-sand couplets (SFc) from PS2627 and PS2628. (f) Interbedded dropstone mud/sandy mud (Fmd/SFmd) and massive diamicton (Dmm) in PS2627.

Sediment within the deeper part of Fosters Bugt to the north of Bontekoe Island is up to 15 m thick and comprises a drape of stratified sediment with rare sea floor irregularities (facies 4), and rare ponded stratified sediment within some basins (facies 1a) (Fig. 4). The shallower regions of Fosters Bugt (<300 m) comprise a highly irregular sea floor (facies 3) (Fig. 4).

Core sedimentology

Core **PS2631** was recovered from a stratified sediment drape in the outermost sub-basin of the outer fjord, and core **PS2632** from the surface drape of the innermost sub-basin (facies 4) (Fig. 4). Both cores are dominated by bioturbated clay-rich mud (Fb) with only a few clasts (Figs 6, 7 & 8a). Bioturbation is characterized by pyritized *Chondrites* and rare *Planolites* burrows (Fig. 9a). The lowermost unit in PS2631 comprises laminated clay-rich mud (Fl) with diffuse and planar parallel-to-wavy laminae disturbed by bioturbation, and rare clasts (Figs 6, 7 & 8a).

Core **PS2633** was recovered from stratified sediments draping the flanks of the mid-fjord basin (facies 4) (Fig. 4). The top half of the core consists of massive clay-rich mud (Fm) and small lenses of poorly sorted sand (<10 mm) (Fig. 6). The core-top comprises massive sandy mud with abundant clasts (SFmd) (Fig. 6). Laminated mud (Fl) characterizes the lower half of the core comprising fine-to-crude scale, diffuse to well-defined, wavy-to-wispy-to-planar parallel laminae, with thin units of massive mud (<5 cm). Clasts, although rare, are relatively more abundant throughout PS2633 than in PS2631 and PS2632 (Fig. 7).

Interpretation of acoustic and core sedimentology

The ponded stratified sediments (facies 1a) within the basins of the fjord and Fosters Bugt are the result of deposition from sediment gravity flows and suspension settling (Syvitski 1989; Niessen & Whittington 1997). Acoustically transparent sediment lenses of facies 2 are consistent with debris flow deposits derived from failure of sediment outside the basin (Laberg & Vorren 1995; Dowdeswell *et al.* 1997*b*; Niessen & Whittington 1997; King *et al.* 1998). The depositional processes, producing facies 1a and 2, dominate sedimentation in the middle–outer fjord basins. Down-fjord progradation of sediment in the mid-fjord basin represents ice-proximal sedimentation derived

from a temporarily stable ice-margin (cf. Ó Cofaigh *et al.* 2001). The drape of stratified sediment (facies 4) in the outer fjord and Fosters Bugt is derived from iceberg rafting and suspension settling (Syvitski 1989; Niessen & Whittington 1997), and directly overlies acoustically opaque till or bedrock in the outermost fjord.

Cores PS2632, PS2633 and PS2631 recovered from the drape of glacimarine sediments in the middle–outer fjord (facies 4) indicate that deposition of fine-grained muds by meltwater processes greatly overwhelms the supply and release of debris by icebergs. Massive and bioturbated muds support sedimentation from meltwater under ice- or fjord margin-distal conditions (Elverhøi *et al.* 1983; Elverhøi & Solheim 1983; Cowan *et al.* 1997). In contrast, the laminated muds indicate deposition from turbid meltwater plumes under comparatively more ice-proximal conditions (Powell 1983; Mackiewicz *et al.* 1984; Cowan & Powell 1990; Cowan *et al.* 1997, 1999). Debris release from iceberg rafting is supported by the presence of small amounts of dispersed clasts (PS2631, PS2632 and PS2633), and sandy mud with clasts and lenses of poorly sorted sand (PS2633). An increase in IRD content up-fjord reflects the increasing proximity to tidewater glaciers in the inner fjords (Fig. 7). The highly irregular sea floor and acoustically homogeneous sediment (facies 3) in water depths of less than 300 m indicate that icebergs scoured the sea floor as they drifted through the fjord (cf. Dowdeswell *et al.* 1993, 1994*a*).

Continental shelf

Sediment thickness and acoustic facies distribution

The inner–middle shelf and the shelf break comprises a highly irregular sea floor and acoustically homogeneous sediment (facies 3) (Figs 4, 10a & 11). A drape of stratified sediment (<10 m thick; facies 4) overlies acoustically opaque sediment or bedrock in the inner shelf bathymetric deep (Figs 4 & 10b). A drape of surface sediment (up to 4 m thick; facies 4) extends across an irregular topography from a prominent bathymetric high on the mid-shelf (see below) to the outer shelf. The surface drape is underlain by stratified sediment (facies 1b) in the region extending 2 km from the mid-shelf bathymetric high, and acoustically transparent to crudely stratified sediment of variable extent and thickness (<10 m; facies 5) across the remaining shelf (Figs 4 & 11).

Middle shelf ridge

A wedge-shaped bathymetric high composed of acoustically homogeneous sediment (facies 3) is located on the inner–middle shelf (Figs 4 & 11). The surface of this ridge is highly irregular. The ridge is *c.* 50–60 m in height and is bounded by a low gradient inner-shelf facing slope that merges with the eastern flank of the inner-shelf bathymetric deep, and a steeper outer-shelf facing margin (3°) (Fig. 11). The lower part of the steep outer-shelf facing margin is covered by a thin drape of sediment (facies 4).

Core sedimentology

Core **PS2641** was recovered from the drape of sediment in the inner shelf deep (facies 4) (Figs 4 & 10b). The core is dominated by bioturbated mud facies (Fb) comprising pyritized *Chondrites* and *Planolites* burrows, and rare clasts (Figs 6, 8b & 9a). This facies is underlain by laminated mud (Fl) with cyclically intercalated, millimetre-scale, planar-parallel, silty mud and clayey mud laminae, and contains only rare clasts (Figs 6 & 9b). The base of the core comprises massive, matrix supported diamicton (Dmm) with a sharp and irregular upper contact (Figs 6, 8b & 9b).

Core **PS2630** was recovered from stratified sediment (facies 1b, 4) 1 km in front of the mid-shelf ridge (Figs 4 & 11b). The top 30 cm of the core comprises massive sandy gravelly diamicton (13% gravel, 63% sand), separated via an indistinct contact from a massive muddy diamicton with dispersed to clustered clasts (Figs 8c & 12a). A sharp and irregular contact separates the muddy diamicton from a thin unit of bioturbated mud (Fb) which, in turn, is underlain by massive mud (Fm) (Figs 9c & 12a). The core base comprises massive, muddy diamicton (Dmmr), with a fabric of inclined and aligned clasts in the top 10 cm of the unit, which become dispersed below (Figs 8c, 9c & 12a).

Interpretation of the acoustic record and core sedimentology

Middle-shelf ridge: The prominent wedge-shaped morphology and acoustically homogeneous internal structure of the mid-shelf ridge are consistent with an origin as an ice-contact

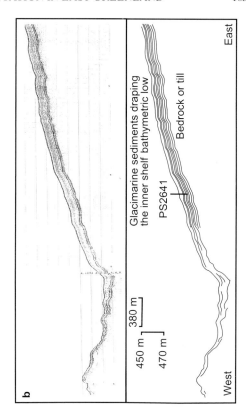

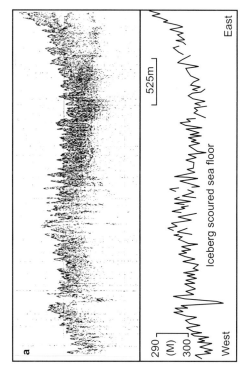

Fig. 10. Parasound records of acoustic facies from the inner continental shelf. (a) Highly irregular sea floor with no sub-sea floor sedimentary structure (facies 3). (b) Thin glacimarine sediment drape (facies 4) overlying acoustically impenetrable sediment or bedrock within the inner shelf basin.

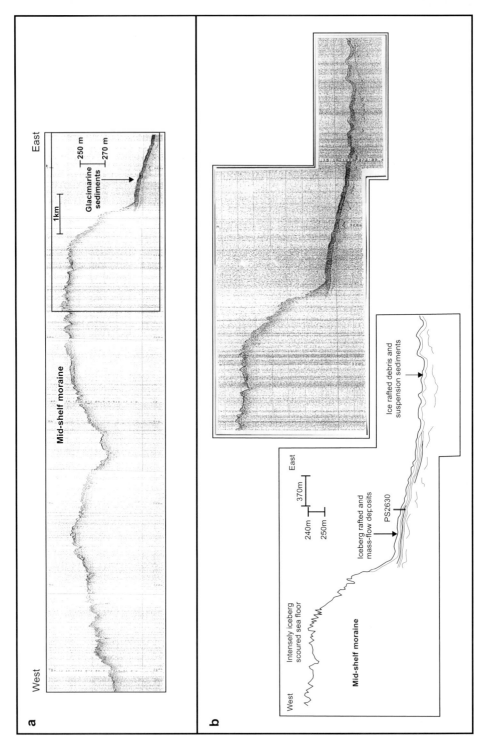

Fig. 11. Parasound record of the moraine and acoustic facies on the middle continental shelf. (a) Wide perspective of the moraine and glacimarine sediments. (b) Close-up of the eastern margin of the moraine and glacimarine sediments.

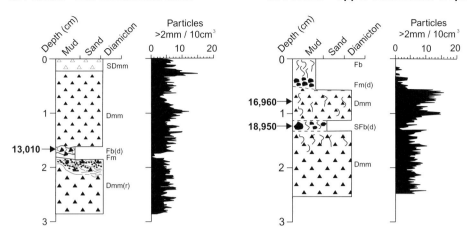

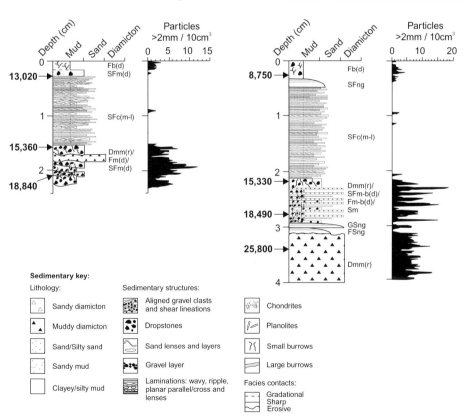

Fig. 12. Sedimentological logs and coarse-particle counts (particles >0.2 cm/10 cm^3) of (a) PS2630, (b) PS2629, (c) PS2628 and (d) PS2627, middle continental shelf and continental slope. Core locations are shown in Figure 1c. AMS radiocarbon dates (^{14}C years BP) obtained from all cores are marked. Lithofacies codes are explained in Table 3.

moraine (e.g. Syvitski et al. 1996b; Maclean 1997) (Fig. 11). The moraine is located directly east of the inner shelf deep, indicating that sediment may have been glacially excavated and redeposited from this region during the advance of grounded glacier-ice. Equally, the moraine may have formed by continual bulldozing of ice-proximal glacigenic sediments forming a morainal bank (Powell & Molnia 1989).

Acoustic and core sedimentology: The highly irregular sea floor and acoustically structureless sediment (facies 3) across the inner–middle shelf and shelf break indicates significant scouring by icebergs (Dowdeswell et al. 1993, 1994a). Core PS2641 reveals that glacimarine sediments draping acoustically opaque till or bedrock in the inner shelf deep comprise, in part, bioturbated mud deposited under ice- or fjord-distal conditions by meltwater escaping the East Greenland fjords (Elverhøi et al. 1983; Elverhøi & Solheim 1983), and/or from the settling of bottom current remobilized shelf debris. The laminated mud facies (Fl) indicates deposition from turbid-meltwater plumes with variable discharge under comparatively more ice-proximal conditions (Mackiewicz et al. 1984; Powell & Molnia 1989; Cowan & Powell 1990). Rare clasts indicate that iceberg rafting is relatively insignificant across the inner shelf. The basal massive muddy diamict is glacimarine in origin rather than a till on account of its unconsolidated nature and meltwater-derived $\delta^{18}O$ isotope minima (see above).

The acoustically stratified sediment (facies 1b, 4) deposited in front of the mid-shelf moraine is derived from ice-proximal sediment gravity flows, and rain-out/suspension settling. The upper massive diamicton in PS2630 corresponds to the drape of surface sediment (facies 4) extending from the moraine to the outer shelf, and is interpreted to result from the release of iceberg-rafted debris. Intense winnowing by the East Greenland Current has modified this sediment, producing a surface unit of sandy gravelly diamicton (cf. Mienert et al. 1992). The upper part of the underlying sequence of stratified sediment (facies 1b) comprises diamicton (PS2630), and the inclined and aligned clast fabric, sharp upper contact and fine-grained matrix supports an origin from cohesive debris flow (Walker 1992). The pinching out of reflectors in this sequence indicates the presence of a number of stacked mass-flow deposits. The origin of the acoustically transparent to crudely stratified sediment (facies 5) on the middle–outer shelf is uncertain on account of the spatially restrictive data

coverage and absence of cores, but the crudely stratified nature suggests formation by ice-distal glacimarine sedimentation.

Continental slope

Acoustic facies distribution

The uppermost continental slope down to about 1200 m is dominated by acoustically structureless sediment (facies 6; Figs 4 & 13a). Below 1200 m the mid–lower slope is characterized by stratified sediment (facies 1b) with isolated sediment lenses (facies 2) (Figs 13b & c). Sediment lenses up to 7 m thick and orientated in a downslope direction (facies 2) are present close to the sea floor at 800–1300 m between the upper slope sediment and mid/lower slope stratified sediment resulting in a hummocky sea floor (Figs 4 & 13b).

Core sedimentology

Core **PS2629** was recovered from acoustically structureless sediment (facies 6) on the upper slope (Figs 4 & 13a). The core-top comprises bioturbated clay-rich mud (Fb) and rare clasts, which are most abundant above the underlying diamicton (Figs 8d, 12b). A massive, poorly sorted diamicton facies (Dmm) dominates the core and comprises dispersed to clustered clasts (10%) and more than 30% sand (Figs 8d, 9d & 12b). The diamicton is separated into two units via gradational contacts by a biogenic-rich (>50%) bioturbated sandy mud (SFbd) with dispersed clasts (4%), and up to 15% poorly sorted sand (Figs 8d & 12b).

Cores PS2628 and PS2627 were recovered from stratified sediment (facies 1b) and sediment lenses (facies 2) on the middle–lower slope (Figs 4 & 13c). Surface facies consist of biogenic-rich (>30%) bioturbated clay-rich mud (Fb) (Fig. 12). The facies is underlain by a thick sequence of rhythmically-intercalated, sand-mud couplets (SFc). The couplets comprise a lower layer or lenses (<8 mm thick) of massive or planar/wavy parallel-to-cross laminated sandy or silty mud, and an upper layer (<20 mm thick) of massive to weakly parallel laminated clayey mud (Figs 9e, 12c & d). Climbing ripples and dewatering structures are rare. Contacts are sharp and well defined, with the basal contact of the couplet flat to undulating and in some cases erosive (Fig. 9e). A succession of thinly inter-bedded, massive muddy diamicton, massive to bioturbated sandy mud and massive mud dominate the lower half of both cores (Figs 9f, 12c & d). The muddy diamicton (Dmm)

a. Upperslope

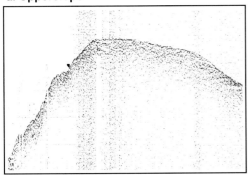
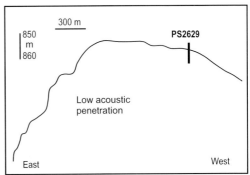

b. Upper/mid slope

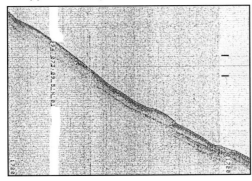
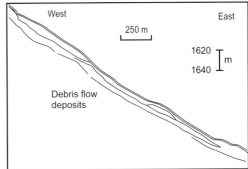

c. Mid/lower slope

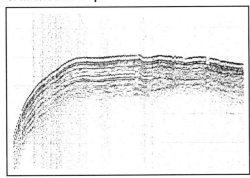
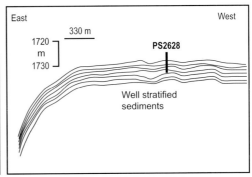

Fig. 13. Parasound records of acoustic facies from the continental slope. (a) Acoustically impenetrable sediments (facies 6) characterizing the upper slope. (b) Acoustically homogeneous sediment lobes (facies 2) overlain by a thin sediment drape (facies 4) on the upper–mid slope. (c) Acoustically stratified sediment (facies 1b) on the mid–lower slope. Horizontal and vertical scales are shown.

contains dispersed clasts and is bounded by sharp contacts. Sandy mud facies (SFmb) contain up to 20% dispersed clasts and poorly sorted sand (Figs 8e & f, 12c & d). The base of PS2627 comprises normally graded gravelly sand and muddy sand facies (GSng/SFng) and a thick, massive muddy diamicton (Dmm), all separated by sharp contacts (Figs 12c & d).

Interpretation of acoustic and core sedimentological data: Acoustically structureless sediment (facies 6) on the upper slope is interpreted to reflect both a coarse-grained sediment texture, and a steep slope gradient (cf. Kuhn & Weber 1993; Melles & Kuhn 1993). Massive muddy diamicton with large numbers of clustered to dispersed clasts, interbedded with bioturbated sandy mud via gradational contacts in PS2629 indicates that iceberg rafting is an important depositional mechanism on the upper slope (cf. Dowdeswell *et al.* 1994*a*). Sandy mud facies supports a period of reduced IRD supply coupled to an increase in hemipelagic sedimentation. Down-slope orientated sediment lenses (facies 2) on the upper-mid slope transitional region are consistent with debris flow deposits (Laberg & Vorren 1995; Dowdeswell *et al.* 1997*b*; King *et al.* 1998), supporting sediment failure and mass-flow on the upper slope. Recent sedimentation on the upper slope comprises thin hemipelagic bioturbated mud (Fb) (PS2629). A gradual decrease in the number of clasts from the diamicton through the surface mud represents a gradual cessation in IRD delivery to the upper slope (Fig. 8).

Stratified and lens-shaped sediment (facies 1a, 2) on the middle-lower slope indicates deposition from sediment gravity flow, iceberg rafting and suspension settling producing interbedded fine- and coarse-grained facies (PS2628 and PS2627). Surface bioturbated muds represent recent hemipelagic sedimentation with only limited deposition of IRD. The sedimentary characteristics of the sand–mud couplet facies are consistent with down-slope currents (distal turbidites) and intervening periods of hemipelagic sedimentation (Piper 1978; Stow & Shanmugam 1980; Hill 1984; Yoon *et al.* 1991; Anderson *et al.* 1996). The lower sections of PS2628 and PS2627 are sedimentologically more variable. Bioturbated mud and sandy muds contain a high biogenic content and rare to common clasts, indicating hemipelagic sedimentation and the release of low but variable amounts of IRD. Hemipelagic sedimentation is punctuated by episodic deposition of massive diamicton by cohesive debris flows (Hampton 1972; Middleton & Hampton 1976; Laberg & Vorren 1995; King *et al.* 1998). Normally graded sandy/gravelly facies indicate further deposition by turbidity currents (Bouma 1962; Walker 1992). A debris flow origin for the diamicton is confirmed by the correspondence of thicker units to sediment lenses (facies 2) within acoustic records.

Discussion: Late Quaternary sedimentary record

Late Weichselian ice-sheet extent (LGM)

A thin veneer of Holocene and Late Weichselian glacimarine sediment overlying acoustically impenetrable till or bedrock on the inner shelf, Fosters Bugt and outer Kejser Franz Joseph Fjord indicates that active, grounded glacier-ice of the Greenland Ice Sheet occupied the fjord and extended onto the inner shelf during the LGM, removing pre-existing sediment cover (Fig. 14). The existence of a floating ice shelf within the Kejser Franz Joseph Fjord and across the inner shelf can therefore be ruled out, as this would be incapable of eroding sediment. However, our data are inconclusive in terms of whether the ice-sheet margin was floating or grounded on the middle continental shelf, although Funder *et al.* (1998) suggest that East Greenland ice masses formed ice shelves.

The prominent mid-shelf moraine consisting of unlithified sediment marks the margin of the grounded palaeo-Greenland ice sheet on the shelf. This moraine is directly overlain by a thin iceberg-rafted diamicton unit that dates 13 000 ^{14}C years BP (PS2630; Figs 11 &12a), indicating that the moraine is probably Late Weichselian in age. The moraine therefore represents either the maximum ice-sheet extent during the LGM or marks a recessional position during Late Weichselian deglaciation. Although it is conceivable that the moraine could mark a recessional position of an ice margin retreating from the shelf break, current evidence indicates that the moraine is more likely to represent the outermost limit of the LGM ice sheet for the following related reasons. (1) Terrestrial geological evidence indicates that the Greenland Ice Sheet reached the inner-middle shelf during the LGM in this region (Hjort 1981; Funder 1989; Funder & Hansen 1996; Funder *et al.* 1998). (2) There is an apparent absence of ice-contact features on acoustic records from the outer shelf, suggesting glacier-ice probably terminated inshore of the shelf break. (3) There is an apparent absence of major debris flows, trough mouth fans and large-scale sliding in this region of the East Greenland continental slope (Mienert *et al.* 1993, 1995). Such features are characteristic of regions around the Polar North Atlantic where ice sheets extended to the shelf break during glacial maxima (Laberg & Vorren 1995; Dowdeswell *et al.* 1996, 1998, 2002; Vorren *et al.* 1998).

Glacier extent during the LGM in East and NE Greenland appears to have been more restricted (Hjort 1981; Funder 1989; Funder & Hansen 1996) when compared to the region south of Scoresby Sund where glacier-ice extended to the shelf break during the LGM (e.g. Mienert *et al.* 1992; Andrews *et al.* 1996). This contrast probably reflects an increase in aridity north of Scoresby Sund in response to cyclonic drift tracks delivering precipitation to SE Greenland prior to moving out into the Polar North Atlantic away from the East Greenland coast (Funder & Hansen 1996; Funder *et al.* 1998).

Continental slope sedimentation during the Late Weichselian glaciation

The continental slope is characterized by significant sediment flux and deposition of coarse-grained lithofacies in response to the advance of the Greenland Ice Sheet onto the continental margin (Fig. 2). Sediment flux is highest on the upper slope (29–65 g cm^{-2} ka^{-1}; 30 cm ka^{-1}), reflecting the more proximal location relative to the palaeo-ice sheet margin on the continental shelf, and decreases on the middle–lower slope (16–24 g cm^{-2} ka^{-1}; 16–17 cm ka^{-1}). The upper slope is characterized by coarse-grained sediment, and core PS2629 indicates that this sediment is composed, in part, of iceberg-rafted diamicton with high concentrations of IRD. This suggests that sedimentation on the upper slope during the Late Weichselian was dominated by the release of debris from a significant number of icebergs calved from the Greenland Ice Sheet with a further contribution from marine sources and distal meltwater. The coarse-grained content of the diamicton exceeds 30% suggesting modification by bottom currents associated with the East Greenland Current (Mienert *et al.* 1992). The presence of dropstone sandy mud and mud-rich facies with low amounts of IRD on the mid to lower slope (PS2628 and PS2627) indicates that the release of IRD is greatly reduced and hemipelagic sedimentation dominates. This spatial difference in the amount of IRD between the upper slope and the mid–lower slope reflects either: (1) a downslope gradient in the number of icebergs traversing the slope, possibly in response to the southward flowing EGC confining most of the icebergs to the upper slope, or (2) the more ice-proximal setting of the upper slope relative to the LGM ice margin where a large amount of debris is deposited from icebergs.

Sediment gravity flows redistribute sediment down the East Greenland slope. The acoustically impenetrable sediment characterizing the upper slope reflects the coarse grain size of ice-rafted diamicts and subaqueous mass-flow deposits in this region (cf. Kuhn & Weber 1993; Melles & Kuhn 1993). Debris flow lobes on the upper–mid slope region indicate that mass-flows were derived from upper slope sediment instability and failure of rapidly accumulating, iceberg rafted sediment. Debris flows transport sediment over variable distances down the slope where it can cross from the upper slope region of acoustically impenetrable sediment into the area of acoustically stratified sediment on the mid- to lower-slope. On the mid- to lower slope, episodic debris flow activity punctuated intervals of more quiescent, hemipelagic sedimentation, resulting in a sequence of diamicts (cm- to metre-scale thickness) interbedded with dropstone sandy mud and laminated, massive or stratified mud facies. Turbidites are also present on the mid–lower slope, forming graded sand and gravel facies produced from episodic turbidity currents that are triggered by debris flow activity further up slope (cf. Hampton 1972).

Large-scale sedimentation on the East Greenland continental margin

Spatial differences in the style of large-scale sedimentation around the margins of the Polar North Atlantic have been attributed to contrasts in ice-sheet dynamics, notably with respect to ice-sheet extent and the rate at which sediment is delivered to the continental slope (Laberg & Vorren 1995; Dowdeswell *et al.* 1996, 1998; Vorren *et al.* 1998). The sedimentary record in this study covers only a limited area of the East Greenland slope and is, therefore, placed within a regional context using published GLORIA imagery and geological data from the slope and the abyssal plain of the Greenland Sea (cf. Mienert *et al.* 1993, 1995; Dowdeswell *et al.* 2002; Ó Cofaigh *et al.* 2002).

Sedimentation is greatest on the central East Greenland slope during full-glacial periods in response to ice-sheet advance across the continental shelf. LGM sedimentation rates of up to 30 cm ka^{-1} are recorded on the slope and contrast with Holocene rates of less than 4 cm ka^{-1}. Geological evidence presented in this study and from other work in East Greenland (e.g. Dowdeswell *et al.* 1994*b*; Funder 1989; Funder *et al.* 1998) indicates that the Greenland Ice Sheet has exhibited relatively minor fluctuations between Late Quaternary glacial and interglacial periods. These fluctuations limited the amount of sediment transferred to the East Greenland slope, resulting in a sediment-starved environment relative to other margins of the Polar North Atlantic where ice sheets

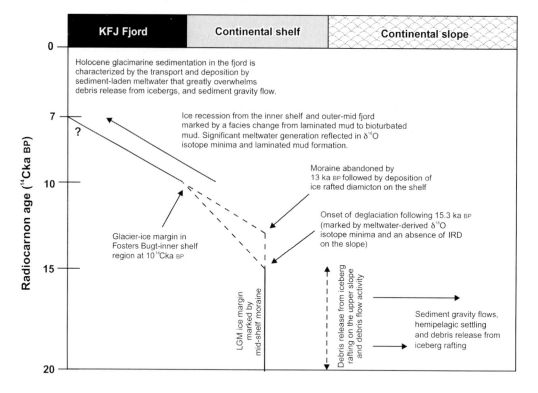

Fig. 14. Summary time–distance diagram of the Late Weichselian and Holocene glacial and sedimentation history of Kejser Franz Joseph Fjord and adjacent continental margin.

reached the shelf break and delivered sediment directly to the slope at rates up to, and possibly exceeding, 170 cm ka^{-1} (Laberg & Vorren 1996; Vorren *et al.* 1998; Dowdeswell *et al.* 1996, 1998; Solheim *et al.* 1998).

The East Greenland margin north of 72°N is characterized by a network of submarine channels that extend from the upper continental slope to the abyssal plain of the Greenland Sea, and the apparent absence of major debris flow lobes or a trough-mouth fan (Mienert *et al.* 1993, 1995; Dowdeswell *et al.* 2002). This is confirmed by our core and acoustic records showing that debris flows are confined mainly to the upper slope and do not form the main architectural sediment body so typical of slopes characterized by trough-mouth fans (e.g. Laberg & Vorren 1995; Dowdeswell *et al.* 1997*b*; King *et al.* 1998). Sediment cores and acoustic records in this study were recovered from immediately upslope of the channels and also from the upper regions of the channel system itself. Diamicton, sand and mud facies in cores, and sediment lobes in the acoustic records, indicate episodic down-slope transport of coarse-grained sediment by debris flows and turbidity currents, derived from mass failure of upper slope sediments.

Sediment gravity flows sourced from the upper slope are likely to have fed into the deep-water channel systems. The passage of turbidity currents through the channel system is indicated by the presence of sandy and muddy turbidite facies in cores recovered from the channel system further downslope (Ó Cofaigh *et al.* 2002). This evidence suggests that sediment gravity flows contributed to the formation of the channel system, possibly in conjunction with dense brines cascading down the slope following their rejection during the formation of sea ice across the East Greenland shelf (Dowdeswell *et al.* 1996, 1998, 2002).

Onset of Late Weichselian deglaciation

The initial onset of deglaciation of the Greenland Ice Sheet is indicated by a distinct $\delta^{18}O$ isotope minima after 15 300 ^{14}C years BP on the continental slope (PS2627) attributed to a major pulse of low-saline meltwater (Fig. 14). A meltwater event following 15 800 years BP is also recorded in oxygen isotope records from further south along the East Greenland continental margin (Nam *et al.* 1995; Nam 1996). The similarity in timing indicates that meltwater production was a regional phenomenon and that the onset of deglaciation in East Greenland was broadly synchronous. The timing of deglaciation is supported by oxygen isotope data from the Renland Ice Core (Scoresby Sund region), which shows termination of the last glacial period at *c.* 15 000 ^{14}C years BP (Johnsen *et al.* 1992). Meltwater spikes in isotope records from the Fram Strait and Norwegian Sea indicate that the deglaciation of ice sheets surrounding these regions occurred slightly later, beginning after 15 000 ^{14}C years BP (e.g. Jones & Keigwin 1988; Sarnthein *et al.* 1992; Elverhøi *et al.* 1995; Hald *et al.* 1996; Hebbeln *et al.* 1998). Initial deglaciation in the Eurasian Arctic occurred earlier at 15 800 years BP (Stein *et al.* 1994a,b).

Deglaciation of the Greenland Ice Sheet further to the north in East and NE Greenland may have sourced these meltwaters with subsequent transport south in the East Greenland Current (cf. Stein *et al.* 1996). Meltwaters influencing the East Greenland continental margin may be associated, in part, with the decay of the Russian Arctic and Svalbard–Barents Sea ice sheets (Stein *et al.* 1994a,b; Nam *et al.* 1995). The influence of meltwater across the continental slope had terminated by 13 000 ^{14}C years BP, probably in response to continued retreat of the ice-sheet margin.

There is an apparent absence of diagnostic sedimentary and acoustic facies on the middle–outer shelf that could provide information on the nature (continual or staggered retreat), mechanism (iceberg calving versus melting) and rate of ice sheet recession in East Greenland. However, a thin unit of iceberg-rafted diamicton drapes the front of the middle-shelf moraine indicating that the ice sheet had abandoned the middle shelf before 13 000 ^{14}C years BP (Fig. 14).

Sedimentation on the continental shelf and slope during Late Weichselian deglaciation

An abrupt termination in both IRD delivery and deposition of coarse-grained sediment gravity flow deposits across the East Greenland slope occurs concomitantly with the onset of deglaciation after 15 300 ^{14}C years BP. Both parameters are consistent with the slope becoming an increasingly ice-distal environment, and with an associated decrease in the flux of glacial sediment to the slope, concomitant with ice-sheet retreat. Reduced sediment flux would result in greater sediment stability on the slope, thereby preventing sediment gravity flows and the deposition of coarse-grained lithofacies. The East Greenland Current may have confined icebergs to the shelf and uppermost slope (possibly in response to an increase in current velocity), thereby preventing both the drift of icebergs over the slope and deposition of IRD.

Hemipelagic sedimentation dominates the upper slope during Late Weichselian deglaciation with deposition of dropstone sandy mud (PS2629). The mid- to lower slope between 15 300 and 13 000 ^{14}C years BP is characterized by the deposition of finely interbedded sand-mud couplets (47–98 g cm^{-2} ka^{-1}; 51–79 cm ka^{-1}) by a combination of downslope current activity and hemipelagic settling. This lithofacies has been observed on the slope off western Spitsbergen where it is associated with the growth and decay of the Barents Sea Ice Sheet (Andersen *et al.* 1996). These currents may have been derived from relatively small-scale sediment failure on the upper slope. Alternatively, downslope currents could be associated with movement of dense brines down the continental slope following their rejection during increased seasonal sea-ice formation (cf. Dowdeswell *et al.* 1998). Hemipelagic settling dominates sedimentation after 13 000 ^{14}C years BP with reduced sediment flux (<3 g cm^{-2} ka^{-1}; <4 cm ka^{-1}).

Sedimentation proximal to the middle-shelf moraine before 13 000 years BP is dominated by debris flow diamicton (PS2630) but it is uncertain how far this mass-flow activity extends back in the Late Weichselian. Iceberg-rafted diamicton is deposited across the shelf in regions proximal to the moraine between 13 000 and 10 000 ^{14}C years BP and may extend across the outer shelf. This facies post-dates the moraine and indicates that the ice sheet had retreated from the middle-shelf by 13 000 ^{14}C years BP, and that ice-mass loss was by iceberg calving. Very light δ^{18}O values corresponding to this diamicton drape indicate significant meltwater production associated with either melting of icebergs or the ice sheet itself.

Late Weichselian–Early Holocene deglaciation of the inner shelf and fjord system

Deglaciation of the inner continental shelf and fjord is marked by an up-sequence facies change (PS2633, PS2631 and PS2641) from laminated mud to bioturbated mud, representing meltwater deposition and a progressive shift from ice-proximal to ice-distal conditions during ice-sheet recession (cf. Svendsen *et al.* 1992). These sediments directly overlie acoustically opaque till or bedrock on the inner shelf and in the outer fjord, and they represent grounded ice-sheet conditions as opposed to an ice shelf. Significant production of low-saline meltwater during deglaciation of the inner shelf and outer fjord is confirmed by a prominent δ^{18}O-spike in isotope records corresponding to the laminated mud facies on the inner shelf (PS2641). The abrupt termination of both the δ^{18}O-minima and deposition of laminated mud supports rapid deglaciation and establishment of ice-distal conditions. Deposition of fine-grained litho-facies dominates ice recession, indicating that ice mass-loss is controlled by ablation with significant production of meltwater. The low concentrations of IRD within these lithofacies indicate that the supply of IRD was over-whelmed by meltwater-derived sediment. However, iceberg scouring of the inner shelf and Fosters Bugt support iceberg production during deglaciation and the low IRD content could alternatively reflect the influence of polar water of the EGC that prevented iceberg melt and debris release.

Radiocarbon dates from sediments directly overlying the meltwater-derived laminated mud facies indicate that ice had abandoned the inner shelf before 9100 ^{14}C years BP and the outer fjord before 7440 ^{14}C years BP (Fig. 14). The thin nature of sediment cover on the shelf indicates that ice abandoned the inner shelf and established ice-distal conditions rapidly, but the timing of this is unknown. Terrestrial geological data from Kejser Franz Joseph Fjord point to stabilization of the ice margin in Fosters Bugt at 10 000 ^{14}C years BP (Hjort 1979, 1981; Funder 1989; Funder & Hansen 1996; Funder *et al.* 1998) (Fig. 14). The shallow Fosters Bugt and Bontekoe Ø would have formed natural pinning points, thereby facilitating stabilization of the retreating ice margin during the Younger Dryas. Marine geological evidence in the form of moraines and ice-proximal sediment depocentres, such as that associated with the Milne Land stadial ice front in Scoresby Sund (e.g. Dowdeswell *et al.* 1994*b*), are absent. This may reflect Holocene iceberg scouring which has destroyed a significant proportion of the marine record in Fosters Bugt and on the inner shelf. Alternatively, the dominance of melt-water sediments in Fosters Bugt may reflect the continental retreat of the ice sheet margin from a position immediately east of Bontekoe Ø after 10 000 ^{14}C years BP.

There is no chronological data on deglacia-tion and the rate of glacier retreat from the middle fjord, but evidence from other East Greenland fjords suggests that glacier-ice had attained its present-day position by 6–7000 ^{14}C years BP (Funder 1978, 1989) (Fig. 14). A drape of stratified glacimarine sediment directly overlying till or bedrock, coupled to the absence of ponded sediment depocentres suggests that glacier retreat through the outer fjord before 7440 ^{14}C years BP was fairly continuous. In contrast, glacier retreat through middle Kejser Franz Joseph Fjord was punctuated by still-stands in response to numerous pinning points. Glacier-ice stabilized in these topographically favourable positions, and fed sediment into ice-proximal basins (prograding down-fjord) by subaqueous sediment gravity flow and suspen-sion settling.

Holocene ice-distal sedimentation in the middle–outer fjord and continental margin

Ice-distal conditions became established in the middle–outer fjord during the Early Holocene. In this setting, a significant volume of sediment is transferred to the deep basins of the middle-outer fjord by subaqueous debris flows and turbidity currents, producing interbedded, homogeneous sediment lobes and stratified sediment (Fig. 5; cf. Niessen & Whittington 1997). Mass-flows are derived from sediments deposited along the margins of the middle–outer fjord by rock fall, deltas, alluvial fans and melt-water-processes. These sediments are prone to failure due to rapid sedimentation and the irregular and steep bathymetry of the fjord. Sediment-laden meltwater discharged into the middle–outer fjord and Fosters Bugt via a number of glacifluvial and fluvial systems (e.g. Badlandal and Paralleldal; Fig. 1c) and from glacier-fed turbid overflow plumes escaping the inner fjord, producing thick sequences of biotur-bated and massive muds (PS2633, PS2632 and PS231). Sediment flux can reach 90 cm ka^{-1}, reflecting the narrow fjord physiography and proximity of meltwater sources, and the high volume of sediment transferred to this fjord by meltwater. The low amount of IRD indicates that meltwater sedimentation is dominant over iceberg rafting in this East Greenland fjord. Meltwater-derived muds are abundant close to fast-flowing outlet glaciers in fjords further to the south in East Greenland, in both Scoresby Sund and Kangerlussuaq Fjord (Smith & Andrews 2000; Ó Cofaigh *et al.* 2001), as well as in fjords in the Canadian Arctic (Stewart 1991; Syvitski & Hein 1991; Hein & Syvitski 1992). Therefore, the relatively widespread occurrence of fine-grained lithofacies suggests that meltwater sedimentation can be relatively significant in polar glacimarine environments. Icebergs that do traverse the fjord system and Fosters Bugt actively scour the sea floor above 300 m water depth.

Meltwater escaping the middle–outer fjord and fjords further to the north in East Greenland also contribute significantly to the deposition of thick bioturbated mud facies in the inner shelf bathymetric deep. The high flux of sediment to the inner shelf (up to 117 g cm^{-2} ka^{-1}) reflects the high volume of meltwater-derived sediment escaping the fjord systems, and is probably a function of the proximity of fluvial and glacifluvial systems that drain into the nearby outer fjord system. The middle–outer shelf is subject to intense erosion and winnowing by the East Greenland Current resulting in a lag of sandy gravelly diamicton (PS2630) and an apparent absence of fine-grained sediment at the sea floor. The continental slope is comparatively sediment starved (<4 cm ka^{-1}) where hemipelagic sedimentation produces mud facies.

Conclusions

1. Geophysical and geological evidence indicates that during the LGM the Greenland Ice Sheet extended as far as the mid-continental shelf where its maximum grounded extent is marked by a prominent moraine.
2. Sediment flux to the continental slope during the LGM was high (15–30 cm ka^{-1}; 16–65 g cm^{-2} ka^{-1}) in response to ice-sheet advance onto the shelf. Sedimentation across the upper slope was characterized by the release of significant quantities of iceberg-rafted debris (diamicton facies) with subsequent downslope remobilization of this sediment by mass-flow (typically debris flow). On the middle–lower slope iceberg rafting and hemipelagic sedimentation (forming dropstone mud and sandy mud) were punctuated by deposition of diamicton and graded sand/gravel facies by mass flows derived from sediment failure on the upper slope. A downslope decrease in IRD reflects either the influence of the East Greenland Current confining icebergs to the upper slope, or progressive distance from the ice-sheet margin. Sediment gravity flows on the slope are likely to have fed into the East Greenland channel system, contributing to its formation.
3. Deglaciation commenced after 15 300 ^{14}C years BP as indicated by δ^{18}O isotope minima. Ice had abandoned the middle-shelf moraine before 13 000 ^{14}C years BP. The

presence of iceberg-rafted diamicton on the middle–outer shelf supports ice-mass loss through calving after 13 000 years BP.
4. Iceberg rafting and deposition across the slope ceased during deglaciation reflecting increasingly ice-distal conditions and, possibly, confinement of icebergs to the shelf by the EGC. Fine-grained deposition by downslope currents dominated slope sedimentation at 15 300 and 13 000 ^{14}C years BP, and may have been linked to an increase in brine rejection on the shelf.
5. Ice abandoned the inner shelf before 9100 ^{14}C years BP and stabilized in Fosters Bugt at 10 000 ^{14}C years BP. The outer fjord was deglaciated before 7440 ^{14}C years BP. Sedimentologically, deglaciation is marked by an upward stratigraphic transition from acoustically opaque till or bedrock to laminated mud and bioturbated mud representing increasingly ice-distal conditions. Distinct δ^{18}O isotope minima on the inner shelf indicate major meltwater production during deglaciation. Ice retreat through the middle–outer fjord was punctuated by topographically-controlled still-stands and ice-proximal sedimentation within the mid-fjord basin.
6. Holocene sediments on the middle–outer continental shelf are winnowed and eroded by the EGC, and the shelf is iceberg scoured. Sediment gravity flows transfer sediment to the deep basins of the Holocene ice-distal middle–outer fjord, producing interbedded acoustically transparent sediment lobes and stratified sediment. Sediment-laden meltwaters transfer high volumes of sediment at high fluxes (up to 111 cm ka^{-1} and 117 g cm^{-2} ka^{-1}) to the outer fjord, Fosters Bugt and inner shelf, and produce thick sequences of bioturbated and massive mud. Meltwater sedimentation overwhelms iceberg rafting in this East Greenland fjord and shelf system. The relatively widespread occurrence of fine-grained lithofacies in East Greenland fjords suggests that meltwater sedimentation can be significant in polar glacimarine environments.

We thank the officers and crew of RV *Polarstern* during ARK X cruise in 1994, the shipboard scientific and technical party for assistance, and the technical support at AWI Bremerhaven and University of Wales, Aberystwyth. We are grateful to C. J. Pudsey (British Antarctic Survey) for the review of an earlier version of the manuscript. We also thank S. I. Nam and C. Vogt for discussions on the work. Data are available through the information system PANGAEA (http://www.pangaea.de). We thank Juergen Mienert and Antoon Kuijpers for formal reviews of the manuscript, and Colm Ó Cofaigh for scientific editing.

References

ANDERSEN, E. S., DOKKEN, T. M., ELVERHØI, A., SOLHEIM, A. & FOSSEN, I. 1996. Late Quaternary sedimentation and glacial history of the western Svalbard continental margin. *Marine Geology*, **133**, 123–156.

ANDREWS, J. T., MILLIMAN, J. D., JENNINGS, A. E., RYNES, N. & DWYER, J. 1994. Sediment thicknesses and Holocene glacial marine sedimentation rates in three East Greenland fjords (c. 68°N). *The Journal of Geology*, **102**, 669–683.

ANDREWS, J. T., JENNINGS, A. E., COOPER, T., WILLIAMS, K. M. & MIENERT, J. 1996. Late Quaternary sedimentation along a fjord to shelf (trough) transect, East Greenland (*c.* 68 N). *In*: ANDREWS, J. T., AUSTIN, W. E. N., BERGSTEN, H. & JENNINGS, A. E. (eds) *Late Quaternary palaeoceanography of the North Atlantic margins.* Geological Society, London, Special Publication, **111**, 153–166.

BOUMA, A. H. 1962. *Sedimentology of flysch deposits. A graphic approach to facies interpretation.* Elsevier, pp168.

CHAPPELL, J. & SHACKLETON, N. J. 1986. Oxygen isotopes and sea level. *Nature*, **324**, 137–140.

COWAN, E. A. & POWELL, R. D. 1990. Suspended sediment transport and deposition of cyclically interlaminated sediment in a temperate glacial fjord, Alaska, U.S.A. *In*: DOWDESWELL, J. A. & SCOURSE, J. D. (eds) *Glacimarine environments: Processes and sediments.* Geological Society, London, Special Publication, **53**, 75–89.

COWAN, E. A., CAI, J., POWELL, R. D., CLARK, J. D. & PITCHER, J. N. 1997. Temperate glacimarine varves: An example from Disenchantment Bay, Southern Alaska. *Journal of Sedimentary Research*, **67**, 536–549.

COWAN, E. A., SERAMUR, K. C., CAI, J. & POWELL, R. D. 1999. Cyclic sedimentation produced by fluctuations in meltwater discharge, tides and marine productivity in an Alaskan fjord. *Sedimentology*, **46**, 1109–1126.

CREMER, H., WAGNER, B., MELLES, M. & HUBBERTEN, H. W. 2001. The postglacial environmental development of Raffles Sø, East Greenland. *Journal of Paleolimnology*, **26**, 67–87.

DAMUTH, J. E. 1978. Echo character of the Norwegian-Greenland Sea: relationship to Quaternary sedimentation. *Marine Geology*, **28**, 1–36.

DOWDESWELL, J. A., VILLINGER, H., WHITTINGTON, R. J. & MARIENFELD, P. 1993. Iceberg scouring in Scoresby Sund and on the East Greenland continental shelf. *Marine Geology*, **111**, 37–53.

DOWDESWELL, J. A., WHITTINGTON, R. J. & MARIENFELD, P. 1994a. Origin of massive diamicton facies by iceberg rafting and scouring, Scoresby Sund, East Greenland. *Sedimentology*, **41**, 21–35.

DOWDESWELL, J. A., UENZELMANN-NEBEN, G., WHITTINGTON, R. J. & MARIENFELD, P. 1994b. The Late Quaternary sedimentary record in Scoresby Sund, East Greenland. *Boreas*, **23**, 294–310.

DOWDESWELL, J. A., KENYON, N. H., ELVERHØI, A., LABERG, J. S., HOLLENDER, F.-J., MIENERT, J. & SIEGERT, M. J. 1996. Large-scale sedimentation on the glacier-influenced polar North Atlantic margins: long-range side-scan sonar evidence. *Geophysical Research Letters*, **23**, 3535–3538.

DOWDESWELL, J. A., WHITTINGTON, R. J. & VILLINGER, H. 1997a. Iceberg scours: records from broad and narrow-beam acoustic systems. *In*: DAVIES, T. A., BELL, T., COOPER, A. K., JOSENHANS, H., POLYAK, L., SOLHEIM, A., STOKER, M. S. & STRAVERS, J. A. (eds) *Glaciated continental margins: An atlas of acoustic images,* Chapman and Hall, London, 27–29.

DOWDESWELL, J. A., KENYON, N. H. & LABERG, J. S. 1997b. The glacier-influenced Scoresby Sund Fan, East Greenland continental margin: evidence from GLORIA and 3.5 kHz records. *Marine Geology*, **143**, 207–221.

DOWDESWELL, J. A., ELVERHØI, A. & SPIELHAGEN, R. 1998. Glacimarine sedimentary processes and facies on the Polar North Atlantic margins. *Quaternary Science Reviews*, **17**, 243–272.

DOWDESWELL, J. A., Ó COFAIGH, C., TAYLOR, J., KENYON, N. H. & MIENERT, J. 2002. On the architecture of high-latitude continental margins: the influence of ice sheet and sea ice processes. *In*: DOWDESWELL, J. A. & Ó COFAIGH, C. (eds) *Glacier-Influenced Sedimentation on High-Latitude Continental Margins.* Geological Society, London, Special Publication, **203**, 33–54.

ELVERHØI, A. & SOLHEIM, A. 1983. The Barents Sea ice sheet – a sedimentological discussion. *Polar Research*, **1**, 23–42.

ELVERHØI, A., LONNE, O. & SELAND, R. 1983. Glacimarine sedimentation in a modern fjord environment. *Polar Research*, **1**, 127–149.

ELVERHØI, A., ANDERSEN, E. S., DOKKEN, T. M., HEBBELN, D., SPIELHAGEN, R., SVENDSEN, J. L., SORFLATEN, M., RORNES, A., HALD, M. & FORSBERG, C. F. 1995. The growth and decay of the Late Weichselian Ice Sheet in Western Svalbard and adjacent areas based on provenance studies of marine sediments. *Quaternary Research*, **44**, 303–316.

ELVERHØI, A., DOWDESWELL, J. A., FUNDER, S., MANGERUD, J. & STEIN, R. 1998. Glacial and oceanic history of the Polar North Atlantic Margins: an overview. *Quaternary Science Reviews*, **17**, 1–10.

EYLES, N., EYLES, C. H. & MIALL, A. D. 1983. Lithofacies types and vertical profile models; an alternative approach to the description and environmental interpretation of glacial diamict and diamictite sequences. *Sedimentology*, **30**, 393–410.

FOLK, R. L. & WARD, W. C. 1957. Brazos river bar, a study in the significance of grain size parameters. *Journal of Sedimentary Petrology*, **27**, 34–59.

FUNDER, S. 1978. Holocene stratigraphy and vegetation history in the Scoresby Sund area, East Greenland. *Grønlands geologiske Undersogeise,* Bulletin **129**, pp 66.

FUNDER, S. 1989. Quaternary geology of the ice-free areas and adjacent shelves of Greenland. *In*: FULTON, R. J. (ed.) *Quaternary geology of Canada and Greenland.* Geological Survey of Canada, Geology of Canada, **1**, 743–792.

FUNDER, S. & HANSEN, L. 1996. The Greenland Ice Sheet – a model for its culmination and decay during and after the last glacial maximum. *Bulletin of the Geological Society of Denmark*, **42**, 137–152.

FUNDER, S., HJORT, C., LANDVIK, J. Y., NAM, S.-I., REEH, N. & STEIN, R. 1998. History of a stable ice margin – East Greenland during the middle and upper Pleistocene. *Quaternary Science Reviews*, **17**, 77–123.

GRANT, J. A. & SCHREIBER, R. 1990. Modern swath sounding and sub-bottom profiling technology for research applications: The Atlas Hydrosweep and Parasound system. *Marine Geophysical Research*, **12**, 9–19.

GROBE, H. 1987. A simple method for the determination of ice rafted debris in sediment cores. *Polarforschung*, **57**, 123–126.

HALD, M., DOKKEN, T. & HAGEN, S. 1996. Palaeoceanography of the European Arctic margin during the last deglaciation. In: ANDREWS, J. T., AUSTIN, W. E. N., BERGSTEN, H. & JENNINGS, A. E. (eds) *Late Quaternary palaeoceanography of the North Atlantic margins*. Geological Society, London, Special Publications, **111**, 275–288.

HAMPTON, M. A. 1972. The role of subaqueous debris flow in generating turbidity currents. *Journal of Sedimentary Petrology*, **42**, 775–793.

HEBBELN, D., HENRICH, R. & BAUMANN, K.-H. 1998. Paleoceanography of the last interglacial/glacial cycle in the Polar North Atlantic. *Quaternary Science Reviews*, **17**, 125–153.

HEIN, F. A. & SYVITSKI, J. P. M. 1992. Sedimentary environments and facies in an Arctic basin, Itirbilung Fiord, Baffin Island, Canada. *Sedimentary Geology*, **81**, 17–45.

HILL, P. R. 1984. Sedimentary facies of the Nova Scotian upper and middle continental slope, offshore eastern Canada. *Sedimentology*, **31**, 293–309.

HJORT, C. 1973. A sea correction for East Greenland, *Geologiska Forengungen i Stockholm Forhandlungar*, **95**, 132–134.

HJORT, C. 1979. Glaciation in northern East Greenland during the Late Weichselian and Early Flanderian. *Boreas*, **8**, 281–296.

HJORT, C. 1981. A glacial chronology for northern East Greenland. *Boreas*, **10**, 259–274.

HOPKINS, T. S. 1991. The GIN Sea – A synthesis of its physical oceanography and literature reviews 1972–1985. *Earth Science Reviews*, **30**, 175–318.

HUBBERTEN, H. W. (ed.) 1995. The expedition ARK-X/2 with RV "Polarstern" 1994. *Berichte zur Polarforschung*, **174**, Alfred Wegener Institut, Bremerhaven, p186.

JOHNSEN, S. J., CLAUSEN, H. B., DANSGAARD, W., GUNDERSTRUP, N. S., HANSSON, M., JONSSON, P., STEFFENSEN, J. P. & SVEINBJØRNSDOTTIR, A. E. 1992. A 'deep' ice core from East Greenland. *Meddelelser om Grønland*, Geoscience **29**, 22 pp.

JONES, G. A. & KEIGWIN, L. D. 1988. Evidence from Fram Strait (78°N) for early deglaciation. *Nature*, **336**, 56–59.

KING, E. L., HAFLIDASON, H., SEJRUP, H. P. & LØVLIE,

R. 1998. Glacigenic debris flows on the North Sea Trough Mouth Fan during ice-stream maxima. *Marine Geology*, **152**, 217–246.

KUHN, G. & WEBER, M. E. 1993. Acoustic characterization of sediments by *Parasound* and 3.5 kHz systems: related sedimentary processes on the southeastern Weddell Sea continental slope, Antarctica. *Marine Geology*, **113**, 201–217.

LABERG, J. S. & VORREN, T. O. 1995. Late Weichselian submarine debris flow deposits on the Bear Island Trough Mouth Fan. *Marine Geology*, **127**, 45–72.

LABERG, J. S. & VORREN, T. O. 1996. The middle and late Pleistocene evolution of the Bear Island Trough Mouth Fan. *Global and Planetary Change*, **12**, 309–330.

MACKIEWICZ, N. E., POWELL, R. D., CARLSON, P. R. & MOLNIA, B. F. 1984. Interlaminated ice-proximal glacimarine sediments in Muir Inlet, Alaska. *Marine Geology*, **57**, 113–147.

MACLEAN, B. 1997. Submarine lateral moraine in the South Central Region of Hudson Strait, Canada. In: DAVIES, T. A., BELL, T., COOPER, A. K., JOSENHANS, H., POLYAK, L., SOLHEIM, A., STOKER, M. S. & STRAVERS, J. A. *Glaciated continental margins: An atlas of acoustic images*, Chapman and Hall, London, 86–87.

MARIENFELD, P. 1991. Holozäne sedimentationsentwicklung im Scoresby Sund, Ost-Grönland. *Berichte zur Polarforschung*, **96**, Alfred Wegener Institut, Bremerhaven, 166 p.

MARIENFELD, P. 1992*a*. Postglacial sedimentary history of Scoresby Sund, East Greenland. *Polarforschung*, **60**, 181–195.

MARIENFELD, P. 1992*b*. Recent sedimentary processes in Scoresby Sund, East Greenland. *Boreas*, **21**, 169–186.

MELLES, M. & KUHN, G. 1993. Sub-bottom profiling and sedimentological studies in the southern Weddell Sea, Antarctica: evidence for large-scale erosional/depositional processes. *Deep-Sea Research*, **40**, 739–760.

MIDDLETON, G. V. & HAMPTON, M. A. 1976. Subaqueous sediment transport and deposition by sediment gravity flows. In: STANLEY, D. J. & SWIFT, D. J. P. (eds) *Marine sediment transport and environmental management*. John Wiley, New York, 197–218.

MIENERT, J., ANDREWS, J. T. & MILLIMAN, J. D. 1992. The East Greenland continental margin (65°N) since the last deglaciation: changes in seafloor properties and ocean circulation. *Marine Geology*, **106**, 217–238.

MIENERT, J., KENYON, N. H., THIEDE, J. & HOLLENDER, F. J. 1993. Polar continental margins: studies off East Greenland. *EOS*, **74**, 225, 234, 236.

MIENERT, J., HOLLENDER, F. J. & KENYON, N. H. 1995. GLORIA survey of the East Greenland margin: 70°N to 80°N. In: CRANE, K. & SOLHEIM, A. (eds) *Seafloor atlas of the northern Norwegian–Greenland Sea*. Norsk Polarinstitut Meddeleler, **137**, 150–151.

NAM, S.-I. 1996. *Late Quaternary glacial history and palaeoceanographic reconstructions along the*

East Greenland continental margin: Evidence from high resolution records of stable isotopes and ice rafted debris. PhD thesis, University of Bremen, 158 p.

NAM, S.-I., STEIN, R., GROBE, H. & HUBBERTEN, H.-W. 1995. Late Quaternary glacial/interglacial changes in sediment composition at the East Greenland continental margin and their palaeoceanographic implications. *Marine Geology,* **122,** 243–262.

NIESSEN, F. & WHITTINGTON, R. J. 1997. Typical sections along a transect of a fjord in East Greenland. *In:* DAVIES, T. A., BELL, T., COOPER, A. K., JOSENHANS, H., POLYAK, L., SOLHEIM, A., STOKER, M. S. & STRAVERS, J. A. (eds) *Glaciated continental margins: An atlas of acoustic images.* Chapman and Hall, London, 182–186.

Ó COFAIGH, C., DOWDESWELL, J. A. & GROBE, H. 2001. Holocene glacimarine sedimentation, inner Scoresby Sund, East Greenland: the influence of fast-flowing ice-sheet outlet glaciers. *Marine Geology,* **175,** 103–129.

Ó COFAIGH, C., TAYLOR, J., DOWDESWELL, J. A., ROSELL-MELÉ, A., KENYON, N. H., EVANS, J. & MIENERT, J. 2002. Sediment reworking on high-latitude continental margins and its implications for palaeoceanographic studies: insights from the Norwegian-Greenland Sea. *In:* DOWDESWELL, J. A. & Ó COFAIGH, C. (eds) *Glacier-Influenced Sedimentation on High-Latitude Continental Margins.* Geological Society, London, Special Publications, **203,** 325–348.

PIPER, D. J. W. 1978. Turbidite muds and silts on deep-sea fans and abyssal plains. *In:* STANLEY, D. J. & KELLING, G. (eds) *Sedimentation in Submarine Canyons, Fans and Trenches.* 163–176.

POWELL, R. D. 1983. Glacial-marine sedimentation processes and lithofacies of temperate tidewater glaciers, Glacier Bay, Alaska. *In:* MOLNIA, B. F. (ed.) *Glacial-marine sedimentation.* Plenum, 185–232.

POWELL, R. D. & MOLNIA, B. F. 1989. Glacimarine sedimentary processes, facies and morphology of the south-southeast Alaska shelf and fjords. *Marine Geology,* **85,** 359–390.

REEH, N. 1985. Greenland Ice-Sheet mass balance and sea-level change. *In:* Report DOE/EV/60235-1 *Glaciers, ice sheets and sea level: Effect of a CO_2-induced climatic change.* U.S. Department of Energy, Washington DC, 155–171.

SARNTHEIN, M., JANSEN, E., ARNOLD, M., DUPLESSY, J.-C., ERLENKEUSER, H., FLATOY, A., VEUM, T., VOGELSANG, E. & WEINELT, M. S. 1992. $\delta^{18}O$ time-slice reconstructions of meltwater anomalies at Termination I in the North Atlantic between 50 and 80 N. *In:* BARD, E. & BROECKER, W. S. (eds) *The last deglaciation: Absolute and radiocarbon chronologies.* NATO ASI Series, **I2,** Springer-Verlag Berlin Heidelberg, 184–200.

SHACKLETON, N. J. 1987. Oxygen isotopes, ice volume and sea level. *Quaternary Science Reviews,* **6,** 183–190.

SMITH, L. M. & ANDREWS, J. T. 2000. Sediment characteristics in iceberg dominated fjords,

Kangerlussuaq region, East Greenland. *Sedimentary Geology,* **130,** 11–25.

SOLHEIM, A., FALEIDE, J. I., ANDERSEN, E. S., ELVERHØI, A., FORSBERG, C. F., VANNESTE, K. & NENZELMANN-NEBEN, G. 1998. Late Cenozoic seismic stratigraphy and glacial geological development of the East Greenland and Svalbard-Barents Sea continental margins. *Quaternary Science Reviews,* **17,** 155–184.

STEIN, R., GROBE, H., HUBBERTEN, H.-W., MARIENFELD, P. & NAM, S.-I. 1993. Latest Pleistocene to Holocene changes in glaciomarine sedimentation in Scoresby Sund and along the East Greenland continental margin: Preliminary results. *Geo-Marine Letters,* **13,** 9–16.

STEIN, R., NAM, S.-I., SCHUBERT, C., VOGT, C., FÜTTERER, D. & HEINEMEIER, J. 1994*a.* The last deglaciation event in the Eastern Central Arctic Ocean. *Science,* **264,** 692–695.

STEIN, R., SCHUBERT, C., VOGT, C. & FÜTTERER, D. 1994*b.* Stable isotope stratigraphy, sedimentation rates, and salinity changes in the latest Pleistocene to the Holocene eastern central Arctic Ocean. *Marine Geology,* **119,** 333–355.

STEIN, R., NAM, S.-I., GROBE, H. & HUBBERTEN, H.-W. 1996. Late Quaternary glacial history and short term ice rafted debris fluctuations along the East Greenland continental margin. *In:* ANDREWS, J. T., AUSTIN, W. E. N., BERGSTEN, H. & JENNINGS, A. E. (eds) *Late Quaternary palaeoceanography of the North Atlantic margins.* Geological Society, London, Special Publication **111,** 135–151.

STEWART, T. G. 1991. Glacial marine sedimentation from tidewater glaciers in the Canadian High Arctic. *Geological Society of America,* Special Paper **261,** 95–104.

STOW, D. A. V. & SHANMUGAM, G. 1980. Sequence of structures in fine-grained turbidites: comparison of recent deep-sea and ancient flysch sediments. *Sedimentary Geology,* **25,** 23–42.

SVENDSEN, J. I., MANGERUD, J., ELVERHØI, A., SOLHEIM, A. & SCHUTTERHELM, R. T. E. 1992. The Late Weichselian glacial maximum on western Spitsbergen inferred from offshore sediment cores. *Marine Geology,* **104,** 1–17.

SYVITSKI, J. 1989. On the deposition of sediment within glacier-influenced fjords: oceanographic controls. *Marine Geology,* **85,** 301–329.

SYVITSKI, J. P. M. & HEIN, F. J. 1991. Sedimentology of an Arctic Basin: Itirbilung Fiord, Baffin Island, Northwest Territories. *Geological Survey of Canada,* **91.**

SYVITSKI, J. P. M., ANDREWS, J. T. & DOWDESWELL, J. A. 1996*a.* Sediment deposition in an iceberg-dominated glacimarine environment, East Greenland: basin fill implications. *Global and Planetary Change,* **12,** 251–270.

SYVITSKI, J. P. M., LEWIS, C. F. M. & PIPER, D. J. W. 1996*b.* Palaeoceanographic information derived from acoustic surveys of glaciated continental margins: examples from eastern Canada. *In:* ANDREWS, J. T., AUSTIN, W. E. N., BERGSTEN, H. & JENNINGS, A. E. (eds) *Late Quaternary* *palaeoceanography of the North Atlantic margins*, Geological Society, London, Special Publications, **111**, 51–76.

VOGT. C., MATTHIESSEN, J., HUBBERTEN, H.-W. & MONK, J. 1995. Water column investigations: Hydrography. *In*: HUBBERTEN, H.-W. (ed) *The expedition ARKTIS-X/2 of R.V. 'Polarstern' in 1994.* Berichte zur Polarforschung, **174**, 89–94.

VORREN, T. O., LABERG, J. S., BLAUME, F., DOWDESWELL, J. A., KENYON, N. H., MIENERT, J., RUMOHR, J. & WERNER, F. 1998. The Norwegian–Greenland Sea continental margins: morphology and Late Quaternary sedimentary processes and environment. *Quaternary Science Reviews*, **17**, 273–302.

WADHAMS, P. 1981. The ice cover in the Greenland and Norwegian Seas. *Reviews of Geophysics and Space Physics*, **19**, 345–393.

WAGNER, B., MELLES, M., HAHNE, J., NIESSEN, F. & HUBBERTEN, H. W. 2000. Holocene deglaciation and climate history on Geographical Society Island, East Greenland – evidence from lake sediments. *Palaeogeography, Palaeoclimatology, Palaeoecology*, **160**, 45–68.

WALKER, R. G. 1992. Turbidites and submarine fans. *In*: WALKER, R. G. & JAMES, N. P. (eds) *Facies models: Response to sea level change.* Geological Association of Canada, 239–263.

YOON, S. H., CHOUGH, S. K., THIEDE, J. & WERNER, F. 1991. Late Pleistocene sedimentation on the Norwegian continental slope between 67° and 71° N. *Marine Geology*, **99**, 187–207.

Contrasting glacial sedimentation processes and sea-level changes in two adjacent basins on the Pacific margin of Canada

J. VAUGHN BARRIE & KIM W. CONWAY

Geological Survey of Canada – Pacific Institute of Ocean Sciences, PO Box 6000, Sidney, BC V8L 4B2, Canada (e-mail: barrie@pgc-gsc.nrcan.gs.ca)

Abstract: During the late Wisconsin Fraser Glaciation on the Pacific Margin of Canada, ice moved offshore from the Coast Mountains of the Canadian Cordillera and south into the Strait of Georgia, reaching a maximum extent at about 14 000 ^{14}C BP. Most of the strait was ice-free by 11 300 ^{14}C BP. Deglaciation was very rapid with regional downwasting and widespread stagnation. This resulted in a stratigraphy of thick till (30–60 m), overlain by ice-proximal glacimarine sediments and a thin and discontinuous ice-distal glacimarine unit. Glaciation of Queen Charlotte Basin reached a maximum sometime after 21 000 ^{14}C BP. Deglaciation in this region began sometime after 16 000 to 15 000 ^{14}C BP and ice had retreated fully onto mainland British Columbia by 13 500 ^{14}C BP. Deglaciation was rapid, with the eastward retreat of an ice shelf. This resulted in a stratigraphy of a till up to 50 m in thickness, usually turbated by iceberg scour and overlain in some areas by thin, ice-proximal glacimarine sediments and much thicker (20 m) widespread ice-distal glacimarine sediments. A significant difference between these two regions is the deglacial relative sea-level history. Rapid regression of the outer Queen Charlotte Islands shelf occurred between approximately 14 600 and 12 500 ^{14}C BP, primarily due to rapid isostatic rebound and contemporaneous with deglaciation of the continental shelf. Sea-level reached a maximum lowstand of greater than 150 m and remained low until approximately 12 400 ^{14}C BP. In Georgia Basin, sea-level was at a relative high stand of 50 to 200 m during initial deglaciation, falling to between 0 to 50 m below present sometime after 10 000 ^{14}C BP. We suggest that rapid emergence on the northern margin of the outer shelf was due to forebulge effects. Further, the very limited extent of glacial ice on the Queen Charlotte Islands and the exposure to the open Pacific forced the retreat of the Cordilleran ice-sheet margin eastwards thereby resulting in dominantly ice-distal glacimarine sedimentation. In contrast, the initial relative sea-level highstand during deglaciation between the Vancouver Island and Cordilleran glaciers in the Strait of Georgia resulted in significant ice-proximal deposition and limited ice-distal glacimarine deposition.

The style of late Wisconsin glaciation off the west coast of Canada was strongly influenced by the existence of Vancouver Island and the Queen Charlotte Islands, which are separated by semi-enclosed shelf basins (Figs 1 and 2). The relationship between the advance and retreat of glaciers that developed on these islands and the Cordilleran ice sheet is key to understanding the sea-level history of this region and, in turn, the sea-level history has influenced glacier dynamics and glacial sedimentation. Our objective in this paper is to investigate the distribution and style of glacimarine sedimentation and the known late Pleistocene to Holocene sea-level history for the two areas, the Queen Charlotte Basin and Georgia Basin (Figs 1 and 2). Based on these analyses, the primary mechanisms determining glacial sedimentation for this region can be understood better (e.g. Syvitski 1993).

Data for Queen Charlotte Basin are derived from previously published work, which includes an extensive collection of high resolution sub-bottom profiles, sidescan sonograms and sediment cores collected over the past 15 years by the Geological Survey of Canada (e.g. Luternauer *et al.* 1989*a*; Barrie *et al.* 1991; Josenhans *et al.* 1995; Barrie & Conway 1999, 2002). These results are compared to recent investigations undertaken within the Strait of Georgia.

Methods

For the Strait of Georgia, a regional grid of high resolution Huntec DTS sub-bottom profiles, Simrad sidescan sonar and airgun seismic lines were collected during scientific programs in 1996, 1997 and 2000 with a regional grid spacing of 6 km. From these field programs, 38 vibrocores and 34 piston cores were retrieved (Fig. 3). Another seven piston cores were collected in 1992 along the BC Hydro electrical cable

From: DOWDESWELL, J. A. & Ó COFAIGH, C. (eds) 2002. *Glacier-Influenced Sedimentation on High-Latitude Continental Margins.* Geological Society, London, Special Publications, **203**, 181–194. 0305-8719/02/$15.00

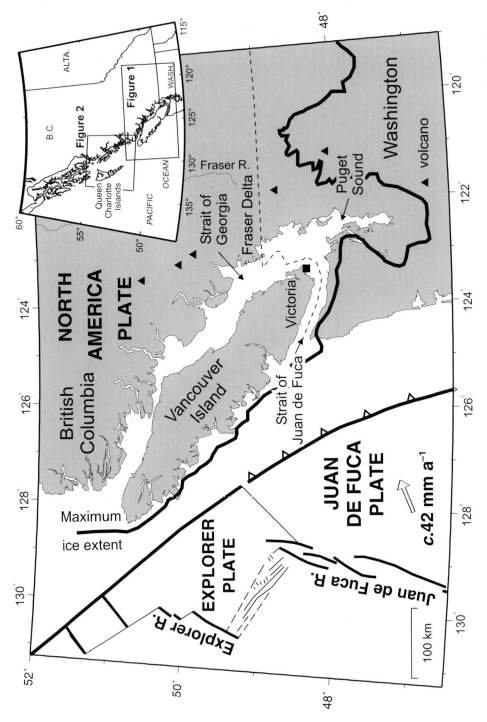

Fig. 1. The regional and tectonic setting of Georgia Basin, including the Strait of Georgia, Strait of Juan de Fuca and Puget Sound. The heavy line shows the maximum extent of late Wisconsin ice (after James *et al.* 2000).

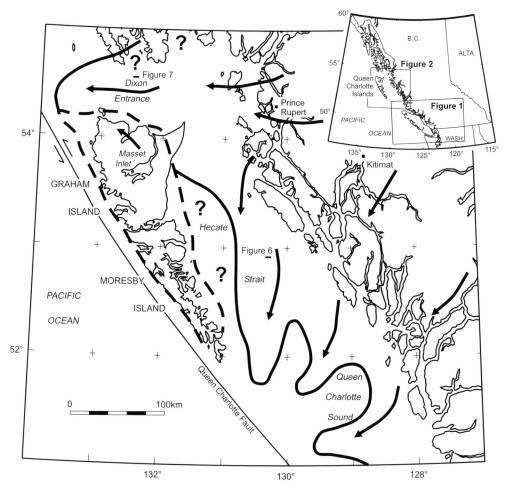

Fig. 2. The Queen Charlotte Basin showing the extent and flow direction of the late Wisconsin ice (heavy line) during glacial maximum. Locations of Figures 6 and 7 are shown.

corridor and are also used by this study (Fig. 3). Cores were split in the laboratory, photographed and sampled for textural analyses, radiocarbon dating, foraminiferal, pollen and diatom analyses. The results of the foraminiferal, pollen and diatom analyses will be discussed in a future publication.

Regional setting

Georgia Basin

Georgia Basin consists of three inland water bodies, the Straits of Georgia and Juan de Fuca and Puget Sound, surrounded by the British Columbia mainland, Washington State and Vancouver Island (Fig. 1). Only the Strait of

Juan de Fuca leads to the Pacific Ocean. This forearc basin lies between southern British Columbia and Vancouver Island, with subsidence that began in the late Cretaceous (85 million years ago). The three Straits consist of a series of structural depressions, over-deepened by Tertiary fluvial erosion and Quaternary glaciation, and partially infilled by glacial and post-glacial sediments. The North American plate is overriding the oceanic Juan de Fuca plate (Fig. 1) at a rate of about 45 mm yr^{-1} (Riddihough & Hyndman 1991).

Queen Charlotte Basin

The present morphology of the Queen Charlotte Islands and surrounding seabed (Queen

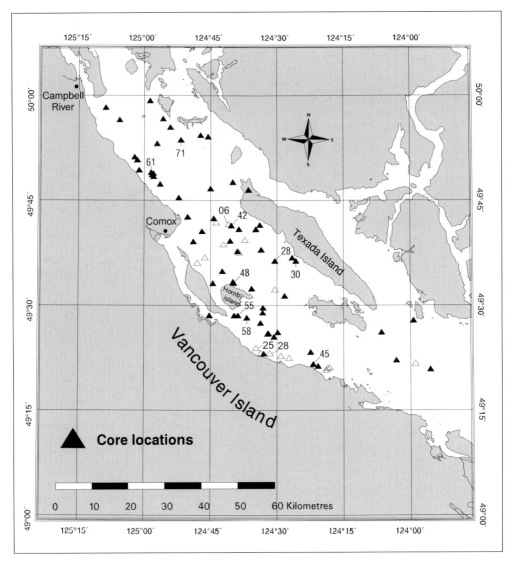

Fig. 3. Distribution of sediment cores collected in the central and northern Strait of Georgia. Cores with a number designation refer to the last two digits of cores listed in Table 1 and shown in Figures 4 and 5 (solid triangles for TUL92A and TUL97B and open triangles for TUL00A).

Charlotte Basin), which includes Dixon Entrance, Hecate Strait and Queen Charlotte Sound (Fig. 2), is a product of glaciation, tectonism, sea-level change and a dynamic oceanography over the Quaternary. Shelf physiography is controlled by the transform fault boundary between the Pacific and North American lithospheric plates along the west coast of the Queen Charlotte Islands (Fig. 2). Relative movement along the fault has been calculated at 50–60

mm yr^{-1} (Riddihough 1988) with Canada's largest earthquake (magnitude 8.1) occurring off the northwestern tip of the Queen Charlotte Islands in 1949.

Late Quaternary glacial stratigraphy of Georgia Basin

Based on the interpretation of seismostratigraphical and sedimentological core data,

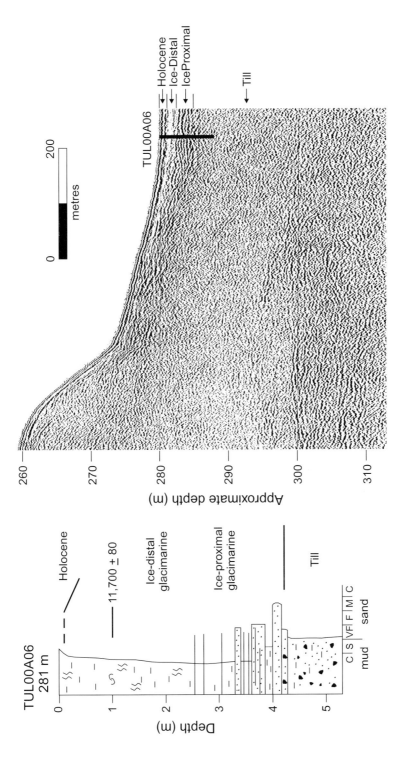

Fig. 4. Huntec DTS sub-bottom profile and sediment core (TUL00A06) from central Strait of Georgia (Fig. 3) illustrating the glacial stratigraphy from till, through ice-proximal sediments into 2.4 m of ice-distal glacimarine sediments and a very thin Holocene cover. Notice the lack of iceberg scours on the surface of the till.

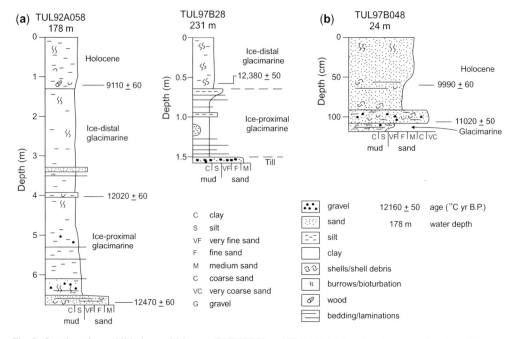

Fig. 5. Stratigraphy and lithology of (**a**) cores TUL97B58 and TUL97B28 showing the succession from till, through ice-proximal glacimarine sediments and ice-distal glacimarine sediments into a thin Holocene unit; and (**b**) a core recovered in a nearshore environment (24 m water depth) off Hornby Island. Locations of the cores are shown in Figure 3.

Table 1. *Radiocarbon dates obtained from cores recovered in central and northern Strait of Georgia. Shell dates are corrected for an 800 year reservoir effect based on the corrections of Southon* et al. *(1990)*

Lab number	Core	Water depth (m)	Sample depth (cm)	Dated specimen	Radiocarbon date (yr BP)	Lithological unit
33804*	T92A55	103	60	Yoldia amygdalea	12 100±60	ID
33933*	T92A58	178	122	wood	9110±60	H
33802*	T92A58	178	402	Rhabdus rectius	12 020±60	ID
33803*	T92A58	178	670	Nuculana fossa	12 470±60	IP
51003*	T97B28	231	55	Yoldia thraciaeformis	12 380±50	IP
51004*	T97B30	170	19	Yoldia martyria	12 340±50	IP
50333*	T97B42	206	280	shell	12 400±50	IP
51006*	T97B42	206	252	Yoldia thraciaeformis	12 160±50	IP
51007*	T97B45	28	132	Macoma lipara	10 130±50	H
51008*	T97B48	24	59	Compsomyax subdiaphana	9990±60	H
52959*	T97B48	24	106	Panomya arctica	11 120±50	ID
52086*	T97B61	104	77	Nuculana fossa	11 270±50	ID
52091*	T97B71	179	118	Macoma lipara	12 310±60	IP
T0-9315	T00A06	281	104	shell	11 700±80	ID
T0-9321	T00A25	41	81	Nuculana minuta	11 770±90	ID
T0-9325	T00A28	71	97	shell	12 090±100	ID
T0-9326	T00A28	71	151	Nuculana minuta	12 340±90	IP

*Radiocarbon analysis undertaken at the Centre for Accelerator Mass Spectrometry, Lawrence Livermore National Laboratory.
Lithological units (ID, ice-distal glacimarine; H, holocene; IP, ice-proximal glacimarine) are discussed in the text and the core locations are shown in Figure 3.

glacial deposits occur throughout the Strait of Georgia. Till is defined from seismic profiling by its uniform unstratified character and high internal backscatter in sub-bottom profiles (Fig. 4). The unit varies in thickness from a few metres to over 60 m. Cores recovered from the till consist of massive, poorly sorted gravelly muddy sands (15% gravel, 45% sand and 40% mud) with striated and faceted pebbles up to 10 cm in diameter (Figs 4 and 5) suggesting that these were deposited by grounded ice. These sedimentological characteristics are consistent in all cores that penetrated the till (15 total) and appear to be very similar to the sandy till of the adjacent coastal lowland (Fyles 1963).

Overlying the till is a generally thin unit (less than 10 m) that is acoustically stratified with discontinuous reflectors in the acoustic records (Fig. 4). In core, these sediments are primarily laminated grey clays with thin silt laminations and ice-rafted pebbles, interbedded with well-sorted sand layers of variable thickness (Fig. 5). The thickness and spacing of the silt laminations is very irregular and no bioturbation is observed. The contact between the till and the laminated grey clays is conformable, suggesting that the latter was deposited by rapid suspension as ice lifted off the bed in an ice-proximal environment, similar to the deglacial sedimentology of the Franz Victoria Trough in the Barents Sea (Lubinski et al. 1996). Radiocarbon dates obtained from this unit consistently give ages of 12 400 ^{14}C BP (Table 1) and indicate very high sedimentation rates (see Core TUL9258; Fig. 5), further supporting the interpretation of an ice-proximal setting (e.g. Syvitski 1993; Cai et al. 1997). The timing of deposition of the unit is consistent with the modelled relative sea-level observations for the collapse of the Cordilleran ice sheet (James et al. 2000) and eustatic meltwater pulse 1A (Fairbanks 1989).

Thin (< 20 m), discontinuous glacimarine mud overlies the ice proximal sediments, and in a few areas, the till with no intervening ice-proximal facies. The mud consists of bioturbated sediments that contain in decreasing order clay (45%), silt (35%), sand (15%) and ice-rafted gravel (5%). Acoustically the facies is weakly reflective, and well-stratified. Six dates from shells in this unit vary from 12 100 to 11 300 ^{14}C BP (Table 1), and are consistent with the interpretation of the end of deglaciation (Clague 1981). This facies compares closely with the ice-distal glacimarine muds found in Queen Charlotte Basin (see below) but is consistently finer.

There is no evidence of iceberg scours anywhere in the Georgia Basin. Recent multi-beam swath bathymetry of the southern half of the Strait of Georgia displays no evidence of iceberg scour. Recent studies from the Strait of Juan de Fuca by Hewitt & Mosher (2001) also show no evidence for iceberg scours in this area.

Late Quaternary glacial stratigraphy of Queen Charlotte Basin

Distribution of glacial ice on the continental shelf surrounding the Queen Charlotte Islands (Fig. 2) is inferred from seismostratigraphic investigations which suggest ice-contact sediments (tills), up to 50 m in thickness, occurring to the shelf edge within the cross-shelf troughs (Josenhans et al. 1995) and Dixon Entrance (Barrie & Conway 1999). Though these acoustic characteristics are normally associated with till, it is difficult to differentiate genetically between massive diamict facies using acoustic data alone (Syvitski et al. 1997).

Extensive glacimarine mud, up to 20 m thick, overlies thin, ice-proximal sediments (Fig. 6) or more usually a till, over most of the Queen Charlotte Basin in water depths generally greater than 200 m (Luternauer et al. 1989a; Barrie et al. 1991; Barrie & Conway 1999). The muds contain approximately equal proportions of sand, silt and clay with ice-rafted debris and they are bioturbated. The unit is interpreted to have been ice-distal, deposited possibly by iceberg rafting and from floating sea ice, similar to present day conditions on the Labrador margin of eastern Canada (Gilbert & Barrie 1985).

Overlying the glacimarine mud in the troughs is a sedimentary sequence, designated Unit B, up to 20 m thick. The sequence was sub-divided by Luternauer et al. (1989a) into two primary sub-units based on the radiocarbon age and texture of the sediments. B_1, the lowermost in the package dates between 13 000 and 12 000 ^{14}C BP in age. This mud unit is thought to have developed as sea-levels were falling on the shelf and, unlike the glacimarine mud, contains no ice-rafted debris.

Unit B_2 overlies B_1 or, where B_1 sediments are not found, overlies the glacimarine mud. This sandy mud unit, 0.01 to 4 m in thickness is found within the shelf troughs and represents a lag formed between 12 900 ^{14}C BP and 10 500 to 10 200 ^{14}C BP. It is thought to have been deposited over a period of several hundred years when sea-level was lower than at present (Luternauer et al. 1989b; Barrie et al. 1991; Barrie & Conway 1999).

Iceberg furrows are ubiquitous in the troughs of the basin between 110 and 350 m depth

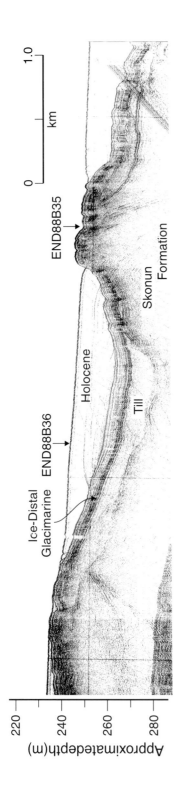

Fig. 6a. Huntec DTS sub-bottom profile and sediment cores (END88B35 and END88B36) from central Hecate Strait (Fig. 2), illustrating the glacial stratigraphy from till, through thin ice-proximal glacimarine sediments into the pervasive ice-distal glacimarine sediments (6 m thick) all overlain by Holocene mud. The strong reflector at the base of the ice-distal glacimarine section is the thin ice-proximal glacimarine unit (after Barrie *et al.* 1991).

(Luternauer & Murray 1983; Barrie & Bornhold 1989; Barrie & Conway 1999). The surface of the till is usually heavily reworked by the action of icebergs (pits and scours) whether exposed at the seafloor (Fig. 7) or overlain by glacimarine and hemipelagic Holocene sediments. The curvilinear scours have incision depths of up to 7 m but are mostly less than 3 m deep. They typically display a preferred orientation in the direction of the trough.

Chronology of the Late Wisconsin glaciation

Glaciation affected the Pacific margin of Canada many times, although extensive evidence has been found for only the youngest glacial episode over much of the area (Clague 1989). The Fraser Glaciation began approximately 25 000–30 000 years ago (Clague 1977, 1981). During the early stages of the Fraser Glaciation, thick, well-sorted sand deposits (Quadra Sand) were deposited in front of, and possibly along the margins of, glaciers moving down the Strait of Georgia as distal outwash aprons (Clague 1976, 1977, 1994). Ice moving south from the Coast Mountains of the Canadian Cordillera and Vancouver Island progressively coalesced, over-rode and eroded these deposits (Fig. 1). This large glacier advanced down the Strait of Georgia to the south end of the Puget Lowland (Waitt & Thorson 1983), and reached its maximum extent (Fig. 1) about 14 000 ^{14}C BP (Porter & Swanson 1998). It deposited a till of variable thickness throughout most of the basin.

In the Queen Charlotte Basin, a glacier from the massive Cordilleran ice sheet extended across northern Hecate Strait and through Dixon Entrance and coalesced with ice from the Queen Charlotte Islands, deflecting it westward within Dixon Entrance (Sutherland-Brown 1968; Barrie & Conway 1999). This coalescence was probably short-lived (Clague 1989). Ice also moved south down the central trough in Hecate Strait (Barrie & Bornhold 1989) and coalesced with ice flowing through the troughs of Queen Charlotte Sound to the shelfbreak (Luternauer & Murray 1983; Luternauer *et al.* 1989a; Hicock & Fuller 1995; Josenhans *et al.* 1995; Josenhans 1997) (Fig. 2). Glaciation reached its maximum extent sometime after 21 000 ^{14}C BP (Blaise *et al.*

1990), and therefore, earlier than in Georgia Basin.

On the Queen Charlotte Islands, small ice caps and piedmont glaciers, up to 500 m thick, developed that were independent of the Cordilleran Ice Sheet (Clague *et al.* 1982b;

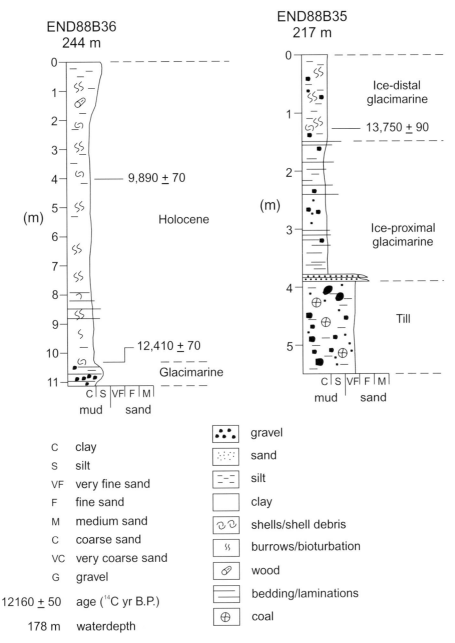

Fig. 6b.

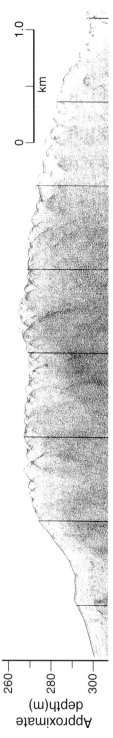

Fig. 7. Huntec DTS sub-bottom profile of the heavily iceberg-scoured surface of a till in central Dixon Entrance (Fig. 2).

Clague 1983). There may have been ice-free areas on the islands (Fig. 2), and on the coastal lowlands of Graham Island glaciation was minimal and of short duration (Clague *et al.* 1982*b*; Clague 1989). The limited size and extent of the Queen Charlotte mountain source areas and the proximity of deep water of the open Pacific Ocean, Dixon Entrance and Queen Charlotte Sound limited expansion of Queen Charlotte ice (Clague 1981, 1989; Warner *et al.* 1982; Barrie *et al.* 1993; Barrie & Conway 1999). In contrast, on Vancouver Island, ice thickness exceeded 1500 m at the height of glaciation (Clague 1983), and ice remained on the northern island until after 13 000 ^{14}C BP (Howes 1983).

Glacial retreat in the Puget Lowland of Georgia Basin began at 13 900 ^{14}C BP (Porter & Swanson 1998) and 14 000 ^{14}C BP within the Strait of Georgia (Clague 1981, 1994). Deglaciation appears to have been rapid, with marine incursion of most of the Strait by 12 400 ^{14}C BP. Ice-distal glacimarine sedimentation continued until about 11 300 ^{14}C BP. Offshore glacial retreat in the Queen Charlotte Basin began somewhat earlier, at 15 000 ^{14}C BP with eastward retreat. Ice had completely withdrawn from tidewater in the Queen Charlotte region by 13 500 to 13 000 ^{14}C BP (Barrie & Conway 1999).

Sea-level history

Georgia Basin

At the maximum extent of the Fraser Glaciation, the entire Strait of Georgia region was isostatically depressed with the greatest vertical displacement exceeding 250 m (Clague 1983). Isostatic uplift followed deglaciation which commenced at approximately 12 500 ^{14}C BP. Uplift exceeded eustatic sea-level rise and resulted in a relative sea-level fall of between 50 to 200 m, depending on locality (Clague 1994). The uplift also occurred at different times during deglaciation due to diachronous retreat and melting of the ice sheet (Clague 1983). For example, in the Fraser Lowland, data from lake and bog cores show that sea-level fell from above 180 m to about 80 m between 12 500 and 12 000 ^{14}C BP (James *et al.* 2002). This rate of sea-level fall (approximately 0.8 m a⁻¹) may have briefly slowed, but it had dropped to 10 m asl by 10 000 ^{14}C BP (Fig. 8a). Near Victoria, however, the sea-level fall from 60 m asl to below present sea-level occurred between 12 500 and 11 000 ^{14}C BP (James *et al.* 2002). Sea-level fell below its present position by 10 m or greater between 12 000 and 10 000 ^{14}C BP

(Fig. 8a) in the southern reaches of the Strait of Georgia (near Victoria) and in the Strait of Juan de Fuca (Clague *et al.* 1982*a*; Clague 1983; Linden & Schurer 1988). A marine transgression occurred in these southern regions until 5 000 ^{14}C BP (Clague *et al.* 1982*a*), due primarily to eustatic rise (Fig. 8a). In the central and northern Strait of Georgia it appears that isostatic adjustment was balanced by eustatic effects resulting in minimal variation in sea-level once it had fallen to close to its present position (Fig. 8a). No evidence was found within the cores from the central and northern Strait of a sea-level low stand. One core (TUL97B48, Fig. 5), taken in 24 m water depth, shows continuous deposition from ice-proximal glaci-marine conditions to present nearshore sands with no indication of subaerial exposure. Another core (TUL97B45), in 28 m of water (Fig. 3), has a similar stratigraphy.

Queen Charlotte Basin

A regional sea-level regression in western Hecate Strait and Dixon Entrance of the northern Pacific margin of Canada began soon after the late Wisconsin glacial maximum and continued throughout deglaciation (Fig. 8b). The oldest date on marine conditions is 14 380 ^{14}C BP (Fig. 8b) from the present-day

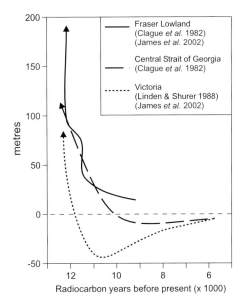

Fig. 8. Generalized sea-level curves for eight locations ranging from (a) central Strait of Georgia, Fraser Lowlands and Victoria for the Georgia Basin and (b) the mainland of British Columbia to the Queen Charlotte Islands in Queen Charlotte Basin. Also shown is the eustatic sea-level curve of Fairbanks (1989).

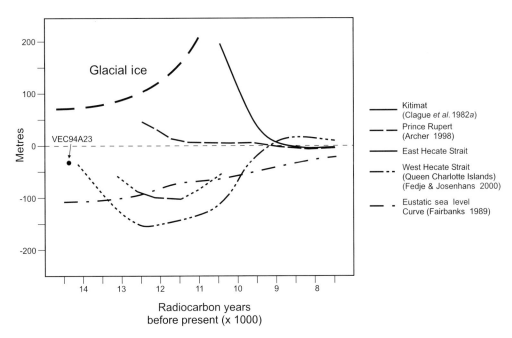

Fig. 8b.

water depth of 37 m in northern Hecate Strait (Barrie & Conway 1999). Relative sea-level had reached a maximum lowering after 13 000 ^{14}C BP in western Hecate Strait (Fig. 2) and remained low until approximately 12 400 ^{14}C BP (Josenhans et al. 1997; Barrie & Conway 1999). Isostatic rebound and the development of a glacioisostatic forebulge accounts for the relative sea-level change during this period of regression. Eustatic sea-level rise, coupled with collapse of a glacioisostatic fore-bulge resulted in relative sea-level rise (Clague 1983; Luternauer et al. 1989b). Present sea-level was reached on the Queen Charlotte Islands by about 9100 ^{14}C BP and reached a maximum of 13 to 16 m asl by 8900 ^{14}C BP, returning to the present level by 2000 ^{14}C BP (Clague et al. 1982a; Clague 1983; Josenhans et al. 1995; 1997; Fedje & Josenhans 2000).

Relative sea-level change has been influenced by isostatic crustal depression and rebound, the raising and lowering of eustatic sea-level, and local tectonic crustal adjustments. The Queen Charlotte Islands archipelago is located at the western margin of the North American litho-spheric plate (Fig. 2) where high heat-flow, evidenced by low mantle viscosity, and a rela-tively thin lithosphere implies rapid crustal response to changes in surface load and a short wave-length (James et al. 2000). The amount of change varies dramatically from east to west (Fig. 8b) in response to loading and rebound from the Cordilleran ice advance (Clague 1983; Barrie & Conway 1999). As the eastern ice load retreated the crust responded, with the maximum flexure occurring nearest the area of maximum change in the crustal load. For example, at 13 000 ^{14}C BP when ice had retreated to near the British Columbia mainland, a contemporaneous shelf tilt existed across the northern Pacific margin of Canada ranging from submergence at the ice edge (Fig. 8b), to greater than 100 m of emerg-ence on the western edge of the Queen Charlotte Islands (Barrie & Conway 2002).

Discussion

The glacial stratigraphy and relative sea-level history between these two regions of the western Canadian continental shelf are quite different, and result from differing deglacial histories. In the Queen Charlotte Basin, iso-static rebound occurred immediately after initial deglaciation at the western extreme, while close to the thick ice load of the Cordilleran ice sheet the crust remained depressed. The development of shelf tilt and the intrusion of Pacific waters could have been

factors in the eastward retreat of ice. Retreat appears to have been rapid (Hicock and Fuller 1995). Sedimentation changes from deposition of till to hemipelagic sedimentation and iceberg rafting, over a very short time period (Barrie & Conway 1999). Ice distal glaciamarine sediments are primarily of ice-rafted origin (e.g. Osterman & Andrews 1983) with the ice source being well removed from the depositional basin.

Guilbault et al. (1997) suggest that open marine conditions and vigorous circulation occurred shortly after 13 000 ^{14}C BP, when water temperatures changed from glacial to tran-sitional in Dixon Entrance. The intrusion of warmer Pacific waters against the ice front probably enhanced ice melt and the calving of icebergs. Iceberg production must have been significant considering the occurrence of a highly scoured seafloor in water depths greater than 300 m (Fig. 7).

In contrast, sea-level in Georgia Basin remained high during initial deglaciation. Thick ice in the Strait of Juan de Fuca (Hewitt & Mosher 2001), along with the basin physiogra-phy, prevented early intrusion of warmer Pacific waters. The high Coast Mountains of the British Columbia mainland, the Olympic Mountains of Washington State and the Vancouver Island Mountains all slowed down ice retreat in the basin. The glacial stratigraphy suggests that once deglaciation began ice-proximal sedimentation rates were high (Fig. 4) over a short period of time near 12 400 ^{14}C BP. Within approximately 300 years no ice was left within the present marine areas, except to the north between the two basins, where ice may have remained (Howes 1983). Ice-distal glaciamarine sedimen-tation continued in some areas for the next 800 years. However, the coarse-grained component is far less than that found in the Queen Charlotte Basin, suggesting reduced ice rafting. Over this period, sea-level fell to near its present level. There is no evidence of any iceberg scouring in the shallowing basin, implying that there was little or no iceberg calving or movement of icebergs across the basin.

Regional downwasting and ice stagnation appears to have been both rapid and primarily in-situ based on coastal mapping (Clague 1981). Our data suggest that ice-proximal sedimen-tation began sometime after 13 000 ^{14}C BP until the ice had disappeared. This interpretation is based on the high sedimentation rates and lack of bioturbation within this unit between 12 400 and to 12 100 ^{14}C BP. Subsequent localized ice-distal glaciamarine sedimention between 12 100 to 11 300 ^{14}C BP was controlled by local varia-tions in basin physiography.

Summary

This study demonstrates that sea-level change can dramatically influence sedimentary processes on high-latitude continental shelves (Barrie & Conway 2002). As shown here, the difference in sedimentation processes at the onset of deglaciation on the continental shelf between two regions of western Canada, was influenced by differences in relative sea-level during deglaciation. The differential cross-shelf isostatic response of a thin crust along the transform plate boundary drove rapid ice retreat eastwards in Queen Charlotte Basin. However, in Georgia Basin, a thicker ice load in a more enclosed basin of a subducting margin resulted in a slower isostatic response. Consequently, deglaciation was primarily manifested as an *in-situ* downwasting of the ice.

We would like to thank the Captains and crews of *CCGS John P. Tully* and *CCGS Vector* for their support in the collection of the field data. Technical support for all field operations were provided by B. Hill, I. Frydecky and B. Macdonald. R. Franklin produced the graphics and the manuscript was improved and enhanced by the critical revision of H. Josenhans, D. Evans, J. Dowdeswell and C. Ó Cofaigh. Support for this work was through the Georgia Basin Geohazards Initiative of the Geological Survey of Canada. This is Geological Survey of Canada Publication 2001037.

References

ARCHER, D. J. W. 1998. *Early Holocene Landscapes on the North Coast of BC*. 31st Annual Meeting of the Canadian Archaeological Association, Victoria, British Columbia.

BARRIE, J. V. & BORNHOLD, B. D. 1989. Surficial geology of Hecate Strait, British Columbia continental shelf. *Canadian Journal of Earth Sciences*, **26**, 1241–1254.

BARRIE, J. V. & CONWAY, K. W. 1999. Late Quaternary glaciation and postglacial stratigraphy of the northern Pacific margin of Canada. *Quaternary Research*, **51**, 113–123.

BARRIE, J. V. & CONWAY, K. W. 2002. Rapid sea level change and coastal evolution on the Pacific margin of Canada. *Sedimentary Geology*, **150**, 171–183.

BARRIE, J. V., BORNHOLD, B. D., CONWAY, K. W. & LUTERNAUER, J. L. 1991. Surficial geology of the northwestern Canadian continental shelf. *Continental Shelf Research*, **11**, 701–715.

BARRIE, J. V., CONWAY, K. W., MATHEWES, R. W., JOSENHANS, H. W. & JOHNS, M. J. 1993. Submerged Late Quaternary terrestrial deposits and paleoenvironment of northern Hecate Strait, British Columbia continental shelf, Canada. *Quaternary International*, **20**, 123–129.

BLAISE, B., CLAGUE, J. J. & MATHEWES, R. W. 1990.

Time of maximum Late Wisconsin glaciation, west coast of Canada. *Quaternary Research*, **34**, 282–295.

CAI, J., POWELL, R. D., COWAN, E. A. & CARLSON, P. R. 1997. Lithofacies and seismic-reflection interpretation of temperate glacimarine sedimentation in Tarr Inlet – Glacier Bay, Alaska. *Marine Geology*, **143**, 5–37.

CLAGUE, J. J. 1976. Quadra Sand and its relation to the late Wisconsin glaciation of southwest British Columbia. *Canadian Journal of Earth Sciences*, **13**, 803–815.

CLAGUE, J. J. 1977. *Quadra Sand: a study of the late Pleistocene geology and geomorphic history of coastal southwest British Columbia*. Geological Survey of Canada, Paper 77–17.

CLAGUE, J. J. 1981. *Late Quaternary geology and geochronology of British Columbia. Part 2: Summary and discussion of radiocarbon-dated Quaternary history*. Geological Survey of Canada, Paper 80–35.

CLAGUE, J. J. 1983. Glacio-isostatic effects of the Cordilleran Ice Sheet, British Columbia, Canada. *In*: SMITH, D. E. & DAWSON, A. G. (eds) *Shorelines and isostasy*. Institute of British Geographers Special Publication, **16**, 321–343.

CLAGUE, J. J. 1989. Quaternary geology of the Queen Charlotte Islands. *In*: SCUDDER, G. G. E. & GESSLER, N. (eds) *The outer shores*. Queen Charlotte Islands Museum Press, Skidegate, Queen Charlotte Islands, 65–74.

CLAGUE, J. J. 1994. Quaternary stratigraphy and history of south-coastal British Columbia; *In*: MONGER, J. W. H. (ed.) *Geology and geological hazards of the Vancouver region, southwestern British Columbia*. Geological Survey of Canada, Bulletin, **481**, 181–192.

CLAGUE, J. J., HARPER, J. R., HEBDA, R. J. & HOWES, D. E. 1982a. Late Quaternary sea levels and crustal movements, coastal British Columbia. *Canadian Journal of Earth Sciences*, **19**, 597–618.

CLAGUE, J. J., MATHEWES, R. W. & WARNER, B. G. 1982b. Late Quaternary geology of eastern Graham Island, Queen Charlotte Islands, British Columbia. *Canadian Journal of Earth Sciences*, **19**, 1786–1795.

FAIRBANKS, R. G. 1989. A 17 000-year glacio-eustatic sea level record: Influence of glacial melting rates on the Younger Dryas event and deep-ocean circulation. *Nature*, **352**, 637–642.

FEDJE, D. W. & JOSENHANS, H. 2000. Drowned forests and archaeology on the continental shelf of British Columbia, Canada. *Geology*, **28**, 99–102.

FYLES, J. G. 1963. *Surficial geology of Horne Lake and Parksville map-areas, Vancouver Island, British Columbia*. Geological Survey of Canada, Memoir **318**.

GILBERT, G. R. & BARRIE, J. V. 1985. Provenance and sedimentary processes of ice-scoured surficial sediments, Labrador Shelf. *Canadian Journal of Earth Sciences*, **22**, 1066–1079.

GUILBAULT, J-P., PATTERSON, R. T., THOMSON, R. E., BARRIE, J. V. & CONWAY, K. W. 1997. Late Quaternary paleoceanographic changes in Dixon

Entrance, northwest British Columbia, Canada: Evidence from the foraminiferal faunal succession. *Journal of Foraminiferal Research*, **27**, 151–174.

HEWITT, A. T. & MOSHER, D. C. 2001. Late Quaternary stratigraphy and seafloor geology of the eastern Juan de Fuca Strait, British Columbia and Washington. *Marine Geology*, **177**, 295–316.

HICOCK, S. R. & FULLER, E. A. 1995. Lobal interactions, rheologic superposition, and implications for a Pleistocene ice stream on the continental shelf of British Columbia. *Geomorphology*, **14**, 167–184.

HOWES, D. E. 1983. Late Quaternary sediments and geomorphic history of northern Vancouver Island, British Columbia. *Canadian Journal of Earth Sciences*, 20, 57–65.

JAMES, T. S., CLAGUE, J. J., WANG, K. & HUTCHINSON, I. 2000. Postglacial rebound at the northern Cascadia subduction zone. *Quaternary Science Reviews*, **19**, 1527–1541.

JAMES, T. S., HUTCHINSON, I. & CLAGUE, J. J. 2002. Improved relative sea-level histories for Victoria and Vancouver from isolation basin coring. Geological Survey of Canada, *Current Research*, 2002-A16, 1–7.

JOSENHANS, H. 1997. Glacially overdeepened troughs and ice retreat 'till tongue' deposits in Queen Charlotte Sound, British Columbia, Canada. *In*: DAVIS, T. A., BELL, T., COOPER, A. K., JOSENHANS, H., POLYAK, L., SOLHEIM, A, STOKER, M. S. & STRAVERS, J. A. (eds) *Glaciated continental margins: An atlas of acoustic images*. Chapman & Hall, London, 40–41.

JOSENHANS, H. W., FEDJE, D. W., CONWAY, K. W. & BARRIE, J. V. 1995. Post glacial sea-levels on the western Canadian continental shelf: Evidence for rapid change, extensive subaerial exposure, and early human habitation. *Marine Geology*, **125**, 73–94.

JOSENHANS, H., FEDJE, D., PIENITZ, R. & SOUTHON, J. 1997. Early humans and rapidly changing sea levels in the Queen Charlotte Islands – Hecate Strait, British Columbia. *Science*, **277**, 71–74.

LUBINSKI, D. J., KORSUN, S., POLYAK, L., FORMAN, S. L., LEHMAN, S. J., HERLIHY, F. A. & MILLER, G. H. 1996. The last deglaciation of the Franz Victoria Trough, northern Barents Sea. *Boreas*, **25**, 89–100.

LINDEN, R. H. & SCHURER, P. J. 1988. Sediment characteristics and sea-level history of Royal roads Anchorage, Victoria, British Columbia. *Canadian Journal of Earth Sciences*, **25**, 1800–1810.

LUTERNAUER, J. L. & MURRAY, J. W. 1983. *Late Quaternary morphologic development and sedimentation, central British Columbia continental shelf*. Geological Survey of Canada, Paper **83–21**.

LUTERNAUER, J. L., CONWAY, K. W., CLAGUE, J. J. & BLAISE, B. 1989a. Late Quaternary geology and geochronology of the central continental shelf of western Canada. *Marine Geology*, **89**, 57–68.

LUTERNAUER, J. L., CLAGUE, J. J., CONWAY, K. W., BARRIE, J. V., BLAISE, B. & MATHEWES, R. W. 1989b. Late Pleistocene terrestrial deposits on the continental shelf of western Canada: evidence for rapid sea-level change at the end of the last glaciation. *Geology*, **17**, 357–360.

OSTERMAN, L. E. & ANDREWS, J. T. 1983. Changes in glacial-marine sedimentation in core HU77-159, Frobisher Bay, Baffin Island, NWT: A record of proximal, distal, and ice-rafting glacial-marine sediments. *In*: MOLNIA, B. F. (ed.) *Glacial-marine sedimentation*. Plenum Press, New York, 451–493.

PORTER, S. C. & SWANSON, T. W. 1998. Radiocarbon age constraints on rates of advance and retreat of the Puget lobe of the Cordilleran ice sheet during the last glaciation. *Quaternary Research*, **50**, 205–213.

RIDDIHOUGH, R. P. 1988. The northeast Pacific Ocean and margin. *In*: NAIRN, A. E. M., STEHLI, F. W. & UYEDA, S. (eds) *The Ocean Basins and Margins*. Vol. 7B, *The Pacific Ocean*. Plenum, New York, 85–118.

RIDDIHOUGH, R. P. & HYNDMAN, R. D. 1991. Modern plate tectonic regime of the continental margin of western Canada. Chapter 13 *In*: GABRIELSE, H. & YORATH, C. J. (eds) *Geology of the Cordilleran Orogen in Canada*. Geological Survey of Canada, Geology of Canada, **No 4**, 435–455.

SOUTHON, J. R., NELSON, D. E. & VOGEL, J. S. 1990. A record of past ocean-atmosphere radiocarbon differences from the northeast Pacific. *Paleoceanography*, **5**, 197–206.

SUTHERLAND-BROWN, A. 1968. *Geology of Queen Charlotte Islands, British Columbia*. British Columbia Department of Mines and Petroleum Resources, Bulletin **54**.

SYVITSKI, J. P. M. 1993. Glaciomarine environments in Canada: an overview. *Canadian Journal of Earth Sciences*, **30**, 354–371.

SYVITSKI, J. P. M., STOKER, M. S. & COOPER, A. K. 1997. Seismic facies of glacial deposits from marine and lacustrine environments. *Marine Geology*, **143**, 1–4.

WAITT, R. B., JR. & THORSON, R. M. 1983. The Cordilleran ice sheet in Washington, Idaho, and Montana. *In*: PORTER, S.C. (ed) *Late-Quaternary environments of the United States, Volume 1, the Late Pleistocene*. University of Minnesota Press, Minneapolis, Minnesota, 53–70.

WARNER, B. G., MATHEWES, R. W. & CLAGUE, J. J. 1982. Ice free conditions on the Queen Charlotte Islands, British Columbia, at the height of late Wisconsin glaciation. *Science*, **218**, 678–684.

Developing high-resolution chronologies in glacimarine sediments: examples from southeastern Alaska

JOHN M. JAEGER

Department of Geological Sciences, University of Florida, PO Box 112120, Gainesville, FL 32611–2120, USA (e-mail: jaeger@geology.ufl.edu)

Abstract: Glacial systems release sediment to the marine environment over a range of time periods, including short-term seasonal, tidal, and diurnal scales. Often, the sedimentary record providing the highest temporal resolutions of short-term processes is found in relatively inaccessible areas such as proximal to glacier termini or below floating ice sheets. To assess the importance of short-term glacial and oceanographic processes in creating glacimarine strata, it is necessary to evaluate strata production over similar time scales. Time-series coring, sediment traps, or bathymetric profiles may be difficult to perform in these harsh settings. Coupling the observations of sedimentary structures seen in core X-ray radiographs with sound chronologies allows for the evaluation of short-term glacimarine sedimentation. For seasonal time-scales, the highly particle-reactive radio-isotope ^{234}Th (24 day half-life) can be used to measure the rates of strata production under both steady and non-steady sediment deposition. To create appropriate age–depth relationships, two approaches are used: a rigorous and exacting mathematical model developed from a steady-state transport-reaction equation; and the less exacting CIC (constant initial concentration) and CRS (constant rate of supply or constant flux) point transformations. These methods are used for developing chronologies of cores collected in 1995 in Icy Bay, Alaska at 4, 12, and 32 km from the tidewater terminus of the Guyot Glacier. Examples are given of the boundary conditions that must be satisfied to use each approach in developing age–depth relationships. For the two cores collected closest to the terminus, age–depth relationships can be generated using all three approaches, whereas the presence of bioturbation in the most ice-distal sample severely complicates matters. The high (~0.5 cm d^{-1}) but non-steady sedimentation rates at the ice-proximal station create a high-resolution sedimentary record. The controls on sedimentation at this location are evaluated by examining sedimentary structures (alternating light and dark laminae and beds) seen in digitized X-ray radiographs. The CRS method is used to convert depth to time in X-ray radiographs. The down-core pixel intensities of the grey-scale X-ray radiograph positives are used to create a time series of sedimentation in proxy data (the bulk density of the sediments). By using Blackman-Tukey spectral analyses and relatively new wavelet techniques to evaluate periodic processes, sedimentation at this location can be related to seasonal variability in meltwater production, fortnightly tidal influences on particle settling rates, and episodic precipitation-induced sediment deposition.

Numerous processes are responsible for the creation of strata in glacimarine environments having characteristic time-scales ranging from hours (tidal) to millennia (Milankovitch cycles). A reliable chronology, developed for the time-scales over which the strata are observed is required to interpret the processes responsible for the formation of sedimentary strata in a particular setting. Once a chronology is developed, the role of individual sedimentary processes in strata formation can be addressed. Although many glacimarine studies have focused on strata production by slower processes related to ice-margin advances and retreats throughout the Quaternary (Andersen *et al.* 1996; Elverhøi *et al.* 1998), there also is need to evaluate strata production on shorter time-scales. Strong seasonal fluctuations in sediment and water discharge occur from temperate and sub-polar glaciers due to increased solar insolation in the summer (Cowan & Powell 1990; Syvitski 1989, 1993). This in turn leads to very high short-term sediment deposition rates during this period (Cowan & Powell 1991; Syvitski 1989). In ice-proximal glacimarine environments, a myriad of glacial and oceanographic processes influence strata production on seasonal time-scales, and in many cases similar types of deposits (i.e. laminated sediments) can be formed by processes occurring over different time-scales and in different climatic regimes (Syvitski & Shaw 1995). For example, rhythmically inter-laminated sediments have been observed in

From: DOWDESWELL, J. A. & Ó COFAIGH, C. (eds) 2002. *Glacier-Influenced Sedimentation on High-Latitude Continental Margins.* Geological Society, London, Special Publications, **203**, 195–213. 0305-8719/02/$15.00
© The Geological Society of London 2002.

Antarctic fjords, possibly forming sub-glacially due to tidal processes (Domack 1990), whereas nearly identical deposits form in temperate Alaska due also to tidal processes, but additionally due to annual meltwater/oceanographic controls (Mackiewicz *et al.* 1984; Cowan & Powell 1990; Jaeger & Nittrouer 1999).

In many glacimarine settings, it is difficult to evaluate seasonal strata production because of sampling difficulties or lack of access due to ice cover. Often in proglacial settings, the highest deposition rates are within a few kilometres of calving ice fronts or even under floating ice sheets (Syvitski 1989). Deploying sediment traps or collection of time-series cores or bathymetric/seismic surveys can be difficult due to the presence of icebergs or glacial ice sheets. These methods, in addition, have their limitations. Although care is taken to construct the most efficient traps, numerous biases (both over- and under-estimates) of sediment fluxes occur in sediment-trap measurements (Knauer & Asper 1989). Reliable comparisons may not exist between what is observed in the water column and sediment traps and what is deposited on the seabed, where subsequent erosion by gravity and tidal currents will affect strata formation without necessarily producing equal deposition/erosion as reflected in traps. Deposition-rate measurements made using seismic and bathymetric methods average rates over many months to years and cannot resolve seasonal or high-frequency dynamics. The use of short-lived radioisotopic measurements coupled with X-ray radiography in glacimarine settings can eliminate problems associated with these two methods, because potentially they allow for rate measurements over a range of time-scales (days to years) within the seabed itself.

The objective of this paper is to describe the use of a short-lived radioisotope geochronometer (^{234}Th) useful for documenting glacimarine strata production over seasonal time-scales. Although ^{234}Th has been used in other marine settings, it has not been used widely in glacimarine environments to evaluate strata production. Radioisotopic data can be used in conjunction with X-ray radiography to measure sedimentation from daily to annual time periods, allowing for better correlation between observed lithostratigraphies and the glacial processes leading to their formation especially when the events occur over short time-scales. Radioisotopes can be powerful tools in creating sediment chronologies, but can be misapplied providing meaningless results if the boundary conditions for their applications are not satisfied (Boudreau 1986; Robbins & Herche 1993).

Because there are numerous processes that can create identical tracer profiles in sediments, X-ray radiography of cores is necessary to establish the appropriate boundary conditions. In addition, X-ray radiographs of laminated glacimarine strata can be analysed using image- and time-series analysis techniques to provide detailed information about the relative magnitude and timing of sedimentary events. Specifically, this paper will describe boundary conditions under which a short-lived (seasonal) radioisotope (^{234}Th) can be used to measure the rates of strata production under both steady and non-steady sediment deposition. This paper also demonstrates that by developing appropriate age–depth transformations using ^{234}Th and through the use of time-series techniques, sedimentary structures observed in X-ray radiographs can be quantitatively related to oceanographic and glacial processes.

Establishing sedimentary chronologies of seasonal glacimarine processes

To develop chronologies of sedimentary processes occurring over short time-scales, a decaying tracer must have a characteristic half-life commensurate with the processes. Thus to measure processes occurring over a seasonal (i.e. 3 month) time-scale, the appropriate tracer would decay at a rate that would allow for measurement of time over a four-month period. In the case of radioisotopes, current alpha and gamma spectroscopic measurements allow for accurate measurement of the isotope of interest back 4–5 half lives (Mann *et al.* 1988). After five half lives (i.e. 5 $t_{1/2}$), only about 3% of the original amount of activity exists, which is usually within the analytical precision of the measurements (approximately 5%). The advent of ICP-MS measurements of isotopes, with analytical precisions of less than 1% (Becker & Dietze 2000) may allow for measurements back 6–7 half-lives, as is the situation in AMS[14]C dating (Tuniz 2001). Therefore, using common gamma spectroscopic measurements, a tracer of seasonal processes (about 100 days) should have a half-life of 20 days (100 d/5 half-lives). The most commonly used radioisotope tracer for this purpose is ^{234}Th, which has a half-life of 24.1 days and allows for measurements of sediment mixing and deposition over time-scales of 100–120 days (Aller *et al.* 1980; McKee *et al.* 1984). In marine settings, suspended sediment is supplied with ^{234}Th activity by only one source, decay from the parent ^{238}U, dissolved within the water column, and whose activity is conservative with salinity (Borole *et al.* 1982).

An additional radioisotope that allows for measurements over similar seasonal time-scales is ^{7}Be (half life of 53 days), a cosmogenic tracer that is the atmospheric product of cosmic-ray spallation of nitrogen and other gases (Olsen *et al.* 1986). ^{7}Be is a difficult isotope to use in glacimarine environments because the activity cannot be absorbed by englacial sediment. Also, any activity directly deposited by precipitation into the marine environment is rapidly diluted with open ocean water resulting in particulate ^{7}Be activities below the detection limit of most gamma spectroscopy systems.

Evaluating sedimentary dynamics with mathematical models and point transformations

A mathematical approach

Numerous theoretical models have been developed to measure rates of sediment mixing, deposition, and accumulation using radioisotopes or geochemical tracers. The non-steady modes and timing of sediment production from glaciers (i.e. sub-glacial discharge, iceberg melting, strong seasonality of melting) warrant an overview of how these theoretical models can be applied to the glacimarine environment. The ability to generate reliable rate measurements depends on the use of an appropriate model, and chronologies created by such models depend on the recognition of the numerous processes that affect the delivery and behaviour of the tracer over time (Robbins & Herche 1993). The sediment profile of a tracer in a glacimarine environment is the combined result of: atmospheric, oceanic, and glacial delivery of the tracer to the marine environment; scavenging behaviour in the water column by suspended sediment; depositional and erosional processes active at the seabed surface; lateral transport and focused deposition of sediment; post-depositional mixing by physical and biological processes; and post-depositional adsorption/desorption reactions and porewater diffusion and advection of the tracer (Aller & Cochran 1976; Boudreau 1986; Cochran 1992). Attempts to model any one of these processes using sedimentary tracers require a quantitative method that can address all of these controls.

Two approaches can be used to quantify sedimentary processes using a time-decaying tracer. The most rigorous and exacting model is a mathematical one developed from a steady-state transport-reaction equation (Benninger *et al.* 1979):

$$\frac{\delta A}{\delta t} = D_b \frac{\delta^2 A}{\delta z^2} - S \frac{\delta A}{\delta z} - \lambda A, \qquad (1)$$

where A = the activity of the tracer, t = time, z = depth in sediment (assuming there is no depth variations in porosity), λ = the first-order decay constant of the tracer, D_b = the sediment mixing coefficient [assuming mixing is a diffusional process; (Boudreau 1986)], and S = the sediment accumulation or deposition rate (length/time). For short-lived tracers that are only found in the upper 10 cm of the seabed where porosity variations are significant (Berner 1980), the shape of the tracer profile may be affected strongly by porosity changes. To overcome this obstacle, the activity of the tracer (dpm g^{-1} or Bq kg^{-1} dry sediment; where dpm=disintegrations per minute and Bq= Becquerel, SI unit of activity, disintegrations per second) can be plotted against the cumulative dry weight of the sediment (g cm^{-2}) determined from porosity or bulk density measurements made when drying the sediment for analyses (Robbins & Herche 1993).

To calculate a steady-state solution to Equation 1, it is necessary to assume that $\delta A/\delta t$ = 0, meaning that the delivery rate of excess (unsupported) activity (dpm cm^{-2} time^{-1}) to the sediment is constant over the period of interest (100 days for ^{234}Th). A steady-state solution to Equation (1) also requires that D_b is constant spatially (with depth) and temporally (over 100 d). Boudreau (1986) believes that a spatially constant D_b is the only one that can be justified given the measurement errors and lack of resolution in tracer data. Without time-series data on mixing rates, the choice of a temporally constant D_b is difficult to justify (Gerino *et al.* 1998). Important controls on biological sediment mixing rates in coastal sediments are bottom-water temperature and organic-carbon flux (Gerino *et al.* 1998). Bottom water temperatures are usually constant in many deeper glacimarine settings but there can be distinct seasonality to organic-carbon fluxes (Syvitski *et al.* 1987). If there is little evidence of biological mixing as seen in cores or core X-ray radiographs, the value of D_b is zero (constant, see below), otherwise a time-varying function of D_b is necessary (Boudreau 1986). Likewise, a steady-state solution to Equation (1) also requires that S is constant temporally (100 days for ^{234}Th). If $\delta S/\delta t \neq 0$, then an alternative approach is taken to estimate S (see the following section on point transformations).

If the above assumptions (i.e. $\delta A/\delta t$ = 0, $\delta S/\delta t$

$= 0$, $\delta D_b/\delta t = 0$) are satisfied, the solution to Equation 1 is:

$$A(z) = A_0 e^{\left(\dfrac{S - \sqrt{S^2 + 4\lambda D_B}}{2D_B}\right)z} + A(s) \quad (2)$$

for activity versus depth (z), or:

$$A(g) = A_0 e^{\left(\dfrac{S - \sqrt{S^2 + 4\lambda D_B}}{2D_B}\right)g} + A(s) \quad (3)$$

for activity versus cumulative mass (g cm^{-2}).

For Equations 2 and 3, $A(s)$ represents the background activity produced by decay of the

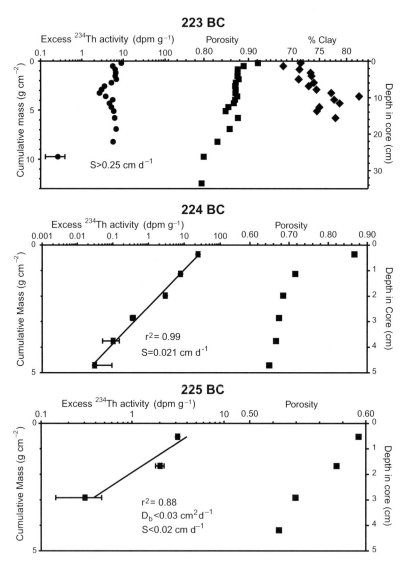

Fig. 1. Excess ^{234}Th, porosity and percent clay (for core 223 BC) data from box cores collected in July 1995. (**a**) Depth-varying excess activity profiles in an upper-fjord station (core 223 BC) were not due to grain-size changes, but reflected non-steady sedimentation. (**b**) Steady sedimentation (as suggested by laminated X-ray radiographs) occurred at a mid-fjord station (core 224 BC). (**c**) The excess ^{234}Th activity profile at a lower-fjord station was influenced by biological mixing. Sedimentation rates (S) and/or sediment mixing coefficients (D_b) were calculated from the best-fit regression line for cores 224 and 225.

parent radioisotope (e.g. ^{238}U for ^{234}Th). If post-depositional mixing can be shown to be absent (i.e. $D_b = 0$), sedimentation or mass accumulation rates can be determined from profiles of excess activity of the radioisotope as shown below:

$$A_{xs}(z) = A_0 \exp\left[-\left(\frac{\lambda}{S}\right)z\right] - A(s), \qquad (4)$$

for sedimentation rate, S (cm d^{-1}) or:

$$A_{xs}(g) = A_0 \exp\left[-\left(\frac{\lambda}{R}\right)g\right] - A(s), \qquad (5)$$

for mass accumulation rate, R (g cm^{-2} d^{-1}) or:

$$S = \frac{-\lambda}{m} \text{ or } R = \frac{-\lambda}{m}, \qquad (6)$$

where A_{xs} = the excess radioisotope activity at depth z or g, A_0 = the initial excess radioisotope activity at depth $z = 0$ or $g = 0$, λ = radioisotope decay constant (= ln(2)/ $t_{1/2}$), and m = the slope of the regression line fit to the natural log of excess radioisotope activity (Fig. 1). Excess activity in sediments is used to calculate accumulation rates because it represents disequilibrium between the *in situ* activity of the parent/daughter isotopes (e.g. ^{238}U/^{234}Th) resulting from the allochthonous input of activity to the sediments.

A mean sedimentation rate, $\bar{S}$, over the depth of rapidly changing porosity near the surface of the core is calculated as:

$$\bar{S}(\text{cm d}^{-1}) = \frac{R}{\rho_d}, \qquad (7)$$

where ρ_d = the mean dry bulk density (g cm^{-3}) over the depth interval of excess activity. A modelled age for any depth, z, or cumulative mass, g, can be determined from Equations 6 or 7, as shown below:

$$\text{Model age } (z) = \frac{z}{\bar{S}} \text{ or } \text{Model age } (g) = \frac{g}{R}. \quad (8)$$

A point transformation approach to quantifying sedimentary dynamics

The second approach to deriving age–depth relationships from a sedimentary tracer relies upon estimating a value of time for any given depth, z, or cumulative mass, g, in the sediments (Robbins & Herche 1993). Technically called a *point transformation* because each depth point is transformed into an age value. This approach has the disadvantage over the mathematical method outlined above because the results of the point transformation cannot be confirmed (Robbins & Herche 1993). The transformations will always produce increasing ages with depth in the sediments, and there is no predicated profile (such as that produced by Equations 2 and 3) to which the results can be compared statistically.

Two transformations have seen widespread use in creating age–depth relationships: the Constant Initial Concentration (or Constant Initial Activity when dealing with radioisotopes) (CIC method) and Constant Rate of Supply (CRS method) (Appleby & Oldfield 1992; Robbins 1978). The CIC method is identical to the mathematical model above with the notable exception that S can vary with depth in the sediments (Robbins & Herche 1993). This depth-varying S would be indicated by a profile that differs from a 'classic' exponential radiotracer profile by creating breaks in the profile. In fact, many glacimarine environments experience rapid depositional events (i.e. gravity flows) that create intervals with irregular or constant activity in the zone of exponentially decreasing activity (Jaeger et al. 1998). Because the sedimentation rate is not constant with depth or time, Equations 1–3 cannot strictly be applied; however, more complicated multi-layer mathematical models identical to Equation 1 can be developed and used to test the validity of the model fit to the measured data (Benninger et al. 1979).

The second approach to measuring sedimentation rates is the constant rate of supply (CRS) method. The most rigorous application of this method assumes that the flux of radionuclide to the sediments has been constant with time and that there have been 4–5 half-lives of deposition so that steady state is achieved (Robbins 1978). The best way of assuring a constant flux to the sediment–water interface would be if the radioisotope of interest (e.g. ^{234}Th) were quantitatively removed from the water column, which would be apparent as very low dissolved and particulate radioisotope activities in the water column relative to the parent (i.e. ^{238}U). It is then possible to assert with some confidence that changes in the specific activity of recently deposited sediment were due to changes in sediment flux and not to changes in both sediment *and* specific activity flux.

Two techniques can be used to measure whether the radioisotope has been completely removed from the water column. The most exacting and laborious technique is to measure the dissolved and particulate activities of the isotope in the water column (Buesseler et al. 1992). If the water column is stratified with respect to suspended sediment concentrations or salinity, multiple measurements of activity are required through the water column (McKee et al. 1984). A less rigorous technique applicable to short-lived radioisotopes like ^{234}Th requires the measurement of suspended sediment concentrations and water column salinity. In highly turbid waters where suspended sediment concentrations are above 100 mg l^{-1}, essentially all of the ^{234}Th activity is adsorbed to particles (McKee et al. 1984). The residence time of sediment in the water column in fjords is probably less than one day (Hill et al. 1998). Therefore, in temperate fjords, it is likely that all of the ^{234}Th activity is adsorbed to particles that are deposited within a day's time, thus meeting the main criteria for using the CRS method.

If complete removal of the radioisotope from the water column cannot be established, then it is necessary to assume that the flux was constant. However, if the sediment flux has changed with time, as is common in most temperate and sub-polar glacial environments where meltwater and sediment release increase over the summer and there is a large reservoir of dissolved radioisotope to be scavenged, then the movement of radioisotope to the sediments has probaly changed as well and the CRS method would not be a valid chronometric approach.

To calculate deposition (or accumulation) rates using the CRS method, sediment inventories, defined as depth-integrated excess radioisotope activity, must be calculated using the formula:

$$I = \sum_{i=1}^{n} (A_i \rho_i X_i), \qquad (9)$$

where A_i is the excess activity in the sediments (dpm g^{-1}), ρ_i is the bulk density of the sediment (g$_{dry\ sediment}$ cm^{-3}), and X_i is the sampling interval thickness (cm). The depth-varying sedimentation rate, $S(z)$ (cm d^{-1}) at a given depth, z (cm), in the sediment is calculated using the formula (Robbins 1978):

$$t = \frac{-1}{\lambda} \ln \left[1 - \frac{I(z)}{I(\infty)} \right] \qquad (10)$$

and

$$S(z) = \frac{z}{t} \qquad (11)$$

where z is any given depth in the excess activity profile (cm), t is the age of that depth, $I(z)$ is the excess activity inventory integrated to depth z, and $I(\infty)$ is the total depth integrated inventory. When sedimentation occurs as a steady process, the value of S calculated from Equations 7 and 11 should agree (Appleby & Oldfield 1992; Robbins 1978).

Quantifying seasonal glacimarine strata formation

Sediment deposition and mixing rates (100 day average) were determined from the profiles of excess ^{234}Th (Fig. 1) for a proglacial fjord in southern Alaska during the mid-summer of 1995, which is the period of rising to maximum sediment input from glacial melting. Details of the glacimarine sedimentary processes of this fjord are described in Jaeger & Nittrouer (1999). Proglacial strata production was examined in Icy Bay, a coastal fjord opening into the Gulf of Alaska (Fig. 2), which has existed since 1904. Prior to that time, the Guyot Glacier filled Icy Bay and formed a terminal moraine that shoals to a present depth of 11 m. This fjord experiences strong down-fjord gradients in glacial and oceanographic forces, and the resultant strata reflect these two dominant controls (Jaeger & Nittrouer 1999). Because of the exponential decrease in sediment deposition rates away from glacier termini, the resultant tracer behaviour and sediment activity profiles are controlled by the balance between time-varying deposition rates, biological mixing rates, and radiotracer fluxes.

Excess ^{234}Th was measured in three box cores collected during 1995 in the upper fjord (223 BC: 4 km from ice front, 145 m water depth), mid-fjord (224 BC: 24 km, 145 m water depth), and lower fjord (225 BC: 54 km, 70 water depth) (Fig. 2). Sediment activities of ^{234}Th were measured on sediment dried, ground, and counted on a low-energy germanium detector. The 63.3 keV photopeak was examined to calculate ^{234}Th activities using the method of Cutshall et al. (1983) to convert from counts per minute (cpm) to disintegrations per minute (dpm). Excess ^{234}Th activities were determined by subtracting supported ^{234}Th activities (which were determined by recounting the samples after 110 days) from total ^{234}Th activities.

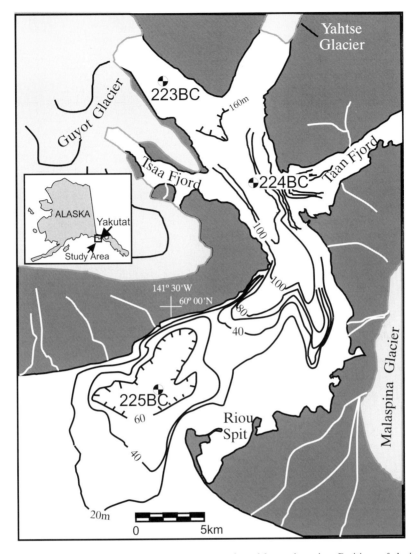

Fig. 2. Map of Icy Bay, Alaska. Locations of box cores are plotted for each station. Positions of glacier termini shown on map correspond to 1994, as measured from ship's radar. Location of Yakutat shown in inset.

Because porosity changed significantly over the upper 10 cm in all cores, excess ^{234}Th activity is plotted against cumulative mass, however a separate equivalent centimetre-scale depth axis is provided. The 'classic' exponential decrease in activity with depth (linear decrease on a log-linear plot) was observed in core 224 BC and partially in 225 BC (Fig. 1). Core 223 BC did not have steady sedimentation as reflected in the depth-varying activity profile, and estimations of sedimentation rates in this core will be discussed below. X-ray radiographs of the cores

224 BC and 225 BC show that the same shape profile was generated by two different processes: sediment deposition only at site 224 and sediment deposition and bioturbation at site 225 (Fig. 3). The surface of the core 224 BC contained a faintly laminated mud layer several centimetres thick. No biogenic structures were observed and organisms were not seen during subsampling. The surface of core 225 BC consisted of a bioturbated muddy sand found to a depth of about 7 cm. This upper sand bed contained some faint laminae and a few small

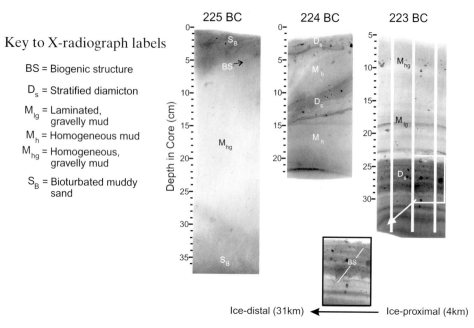

Fig. 3. X-ray radiograph positives from box cores collected in Icy Bay. The lower-fjord core (225 BC) is bioturbated and contains abundant sand layers. Upper-fjord cores 223 BC and 224 BC contain homogeneous and laminated mud beds and diamictons. The high-porosity upper 5 cm of cores 223 BC and 224 BC was disturbed during sampling and not used in subsequent time-series analysis. Biogenic structures are observed in the diamicton bed in core 223 BC. The three white stripes in 223 BC are the 2 mm wide transects (not to scale) used to create the pixel intensity values (see Fig. 6).

clasts of iceberg rafted debris (IBRD). Because the upper 3–5 cm in both cores contained excess ^{234}Th activity, the sediments had been deposited or reworked within the past 100 days, revealing that both bioturbation and sediment deposition were active during this time period.

To constrain the rates at which these two processes were acting, Equation 1 was applied to the excess ^{234}Th activity data. In both cases, it was assumed that the activity flux to the seabed was constant ($\delta A/\delta t = 0$, $\delta S/\delta t = 0$). This assumption can be checked by also applying the CRS method to the data in core 224 BC (see below). The CRS method is not applicable when bioturbation is active (Robbins 1978). For core 224 BC, the lack of observed biogenic structures in X-ray radiographs allowed for the establishment that $D_b = 0$, whereas for core 225 BC, the simplest case was assumed that $\delta D_b/\delta t = 0$. The two most important controls on biological sediment mixing rates in coastal sediments (bottom-water temperature and organic-carbon flux) do not change substantially during the summer in the Gulf of Alaska region (Reed & Schumacher 1987; Sambrotto & Lorenzen

1987). Therefore, Equations 3 and 5 were applied to activity profiles in cores 225 BC and 224 BC, respectively. Depth-averaged sedimentation rates were determined using Equation 7.

At site 224, sedimentation appeared to be constant at 0.02±0.002 cm d^{-1} based on the solution to Equation 5 (Fig. 1). To check this result, the CRS method was applied to the data. The activity flux to the seabed needed to be constant over the 100 days prior to core collection to meet the strict boundary conditions of the CRS method. Dissolved and particulate ^{234}Th activity data were not measured directly. Therefore, the relationship between suspended sediment concentration and particulate ^{234}Th activity was used (McKee *et al.* 1984). CTD casts were collected at each coring station and suspended sediment concentrations (SSC) were qualitatively described using an attached 25 cm pathlength SeaTech transmissometer. The raw voltage output of the transmissometer was converted to beam attenuation coefficient (BAC) using the method outlined in Puig & Palanques (1998). BAC values measured in this study were correlated by means of least-squares

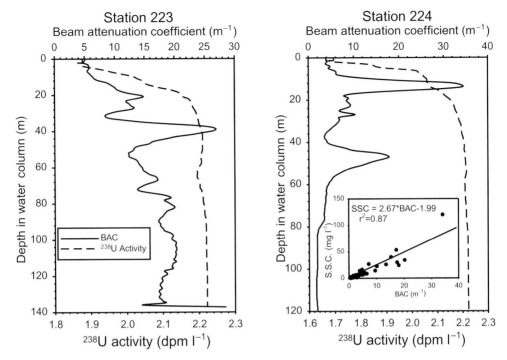

Fig. 4. Transmissometer beam attenuation coefficients and water-column [238]U total activity profiles for stations 223 BC and 224BC. Suspended sediment concentration positively correlates with beam attenuation coefficients (inset). [238]U activities were determined from salinity measurements using the method of Borole *et al.* (1982). At both stations, maximum beam attenuation coefficients were 25–30 m[-1] (~80 ± 20 mg l[-1]), indicating that all dissolved [234]Th produced by decay from the parent [238]U is adsorbed to particles, and the activity flux of [234]Th to the seabed is controlled by differential sedimentation rates and not changes in the rate of activity scavenging.

linear regression with the SSC obtained from filtered water samples for transmissometer data calibration. The relation between the SSC and the BAC for all water samples ($n = 108$), showed a significant (at 95% confidence) correlation R^2 = 0.87 (Fig. 4). Thus, the BAC values can be considered as a semi-quantitative parameter to determine the variations of suspended sediment concentration. Beam attenuation coefficients at stations 223 and 224 both averaged 5–15 m[-1] (about 10–30 mg l[-1]), however, distinct peaks of 25–30 m[-1] (about 80 ± 20 mg l[-1]) were observed (Fig. 4). Because of the limited transport of oceanic water into the fjord, decay from dissolved [238]U in fjord waters was the only source of dissolved [234]Th activity available for particle scavenging. Consequently, as particles sourced from glacial meltwater discharge settled through the fjord, all dissolved [234]Th activity was eventually scavenged onto particles, satisfying the main condition of the CRS method.

At station 224, the CRS method also produced a mean deposition rate of 0.021 cm d[-1] (Fig. 5), corroborating the assumption of steady-state activity flux and sediment deposition. At sedimentation rates of less than 1 mm d[-1], the resultant lithofacies are likely to be thin silt laminae that are too fine in detail to be resolved in X-ray radiographs (Fig. 3).

For the most-distal station (225 BC), the presence of biogenic structures in the upper few centimetres of X-ray radiographs suggests that the behaviour of [234]Th within this core was strongly influenced by bioturbation. Therefore, only Equations 2 or 3 could be employed. Given that only one tracer was used, there are an infinite number of solutions for S and D_b. Thus, only maximum estimates for S and D_b could be determined, which are 0.025±0.01 cm d[-1] and 0.021±0.01 cm[2] d[-1] (7.7±3.5 cm[2] y[-1]), respectively. Because it appears that bioturbation was the dominant sedimentary process at this

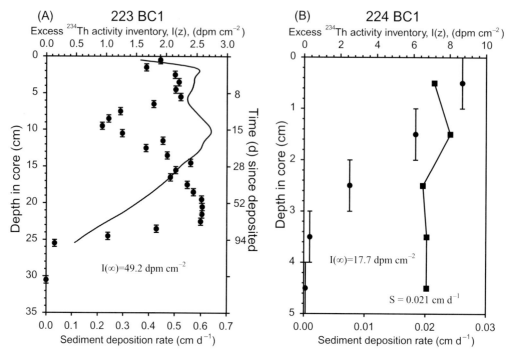

Fig. 5. Application of the constant rate of supply (CRS) method to excess ^{234}Th activity profiles. Deposition rates at a given depth (z), are calculated in Equations 10 & 11, and can be determined by dividing cumulative inventories of excess ^{234}Th activity, $I(z)$ by the total inventory, $I(\infty)$. Plotted on the depth axis is the depth in the core (cm) and the corresponding age (d) of that interval, as calculated from Equation 10. Excess activities and bulk density measurements in core 223 BC were interpolated for points not measured below 16 cm. (**a**) Depth-varying deposition rates are plotted for 223 BC, and demonstrate relatively constant deposition rates of 0.6 cm d^{-1} above 10 cm depth. Deposition rates had increased steadily prior to that. (**b**) Deposition rates calculated for 224 BC using the CRS method. Comparison with the deposition rate calculated using Equation 1 (Fig. 1) shows very good agreement, demonstrating that deposition is steady at this location and all dissolved ^{234}Th activity is quantitatively removed from the water column.

location, realistic deposition rates were assumed to be at least an order of magnitude lower than the maximum estimate.

For the most ice-proximal station (223 BC), excess ^{234}Th was measured to 25 cm (Fig. 1). The profile shows fluctuating activity with depth. Fluctuating activities with depth in core of particle-reactive radioisotopes can be attributed to changes in grain size (i.e. percent clay-size sediment); coarser sediments have lower activities (Jaeger & Nittrouer 1995) due to the reduced surface area available for scavenging. However, variable percentages of clay-size sediment were not responsible for the fluctuating excess ^{234}Th activity in core 223 BC because the region of lower excess ^{234}Th activity (6–11 cm) did not consist of coarser material. The lack of bioturbation in the upper 25 cm of core 223 BC indicates that the behaviour of

excess ^{234}Th activity was due solely to sedimentation. The fluctuating excess ^{234}Th activity with depth observed in this core indicates that one of the main boundary conditions of Equation 1 was violated ($\delta A/\delta t \neq 0$) and a steady-state deposition rate could not be calculated with Equation 7. Only a mean sedimentation rate of more than 0.25 cm d^{-1} (25 cm/100 d) could be calculated based on the disappearance of excess ^{234}Th activity at approximately 25 cm depth.

However, the successful use of the CRS method at site 224 and the higher suspended sediment concentrations at site 223 suggest that the CRS method was applicable at this location. The results of this method are shown in Figure 5 and reveal that the sediment deposition rate was variable at station 223 over the previous 100 days, ranging from 0.5 to 0.6 cm d^{-1} in the upper 11 cm. The decrease in rates in the upper 2 cm

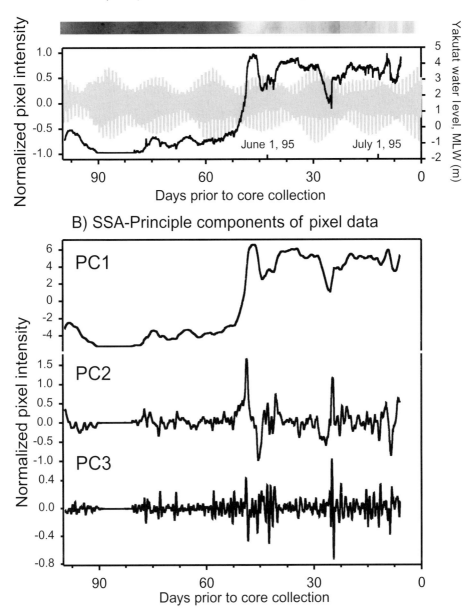

Fig. 6. Grey-scale pixel intensity values from core 223 BC X-ray radiograph positive and deconvolution of signal into its first three principal components. (**a**) Unprocessed pixel intensity data, normalized from 0–255 shades of grey to values between –1(darkest) and 1 (lightest). Representative slice from the X-ray radiograph has been distorted to correspond to time versus depth in core. Also plotted is water level measured by NOAA tide gauge in Yakutat, Alaska. Strong seasonality is observed in the unprocessed pixel intensity data, with secondary fluctuations associated with spring tidal periods. (**b**) Deconvolution of the normalized pixel intensity data into its first three principal components using the singular spectrum analysis (SSA) technique (Vautard & Ghil 1989). The strong seasonal signal is contained within the first principal component. This component was removed from the time-series providing a signal without strong seasonality or a significant trend, creating the time-series used for subsequent analyses (Figs 7 & 8).

may be an induced artefact due to poor sub-sampling of the highly fluid core surface. Below about 10 cm, the linear trend in rates suggests that deposition had undergone a relatively steady increase over that period, corresponding to the observation that sediment discharge from Alaskan tidewater glaciers increases during the spring and early summer (Cowan & Powell 1991). The daily to seasonal lithofacies resulting from such high sedimentation rates is addressed in a following section.

Radioisotope calibration of X-ray radiographs and application of time-series analysis techniques

Quantifying X-ray radiographs

The use of image analysis for the examination of sedimentary structures is well documented (Cooper 1997). The advent of image processing applications for use on personal computers, and the availability of high resolution digital cameras and scanners, have made these techniques available to most researchers. These methods are useful in counting laminae in order to construct a chronology, or to compile time series from lamina thickness or composition for comparison with those derived from other data sets. The basic approach is to perform a grey-scale digitation of a core photograph or X-ray radiograph. The pixel intensity (0–255 grey-scales, 8-bit) is related to the density of sediment (Cooper 1997). Because the X-ray radiographs in this study were taken of a constant-thickness slab (2.5 cm thick), any difference in pixel intensity (i.e. X-ray attenuation) is due to changes in wet bulk density of the sediments not thickness of the slab. Rapid sediment deposition inhibits the dewatering of sediment, maintaining higher porosity sediments and lower wet bulk densities, which appear as lighter regions in X-ray radiograph positives (Fig. 3). Also, coarser sediments (i.e. coarse silts and sands) have higher wet bulk densities, leading to darker regions in X-ray radiographs. Because the bulk density of sediment sampled in this slab-technique is due solely to depositional processes, it is possible to use the entire grey-scale spectrum of an X-ray radiograph to glean information about such processes.

The technique of Cooper (1997) was used to prepare and digitize the X-ray radiograph positive of core 223 BC. It was scanned at 300 dpi, as higher resolutions only added to the noise in the resultant conversion to pixel intensity. Three 2-mm wide down-X-ray radiograph scans were made of pixel intensity using the public domain NIH Image program (developed at the US National Institutes of Health and available on the Internet at http://rsb.info.nih.gov/nih-image/). Three scans were made of regions where large dropstones that could have skewed intensity values were mostly absent. Averaging the three scans reduces the influence of smaller dropstones on overall trends of pixel intensity. Pixel intensity (0–255) was normalized to span from –1 (darkest, most dense) to 1 (lightest, most watery) (Archer 1994; Fig. 6).

By using the CRS method, it is possible to transform the length scale in X-ray radiograph scans to age thereby providing an additional tool to relate sedimentary structures directly to active glacimarine processes (Fig. 5). However, because the CRS technique results in an estimate of the true age–depth relationship, propagation of the 5–10% error in radioisotope activities resulting from gamma spectroscopy creates age errors of up to one day near the upper part of the core and about five days deeper in the core.

The resultant age–pixel intensity profile contains apparent cyclicity at a number of frequencies that can be quantified by performing spectral analysis on the normalized pixel-intensity values (Archer 1994). To facilitate the spectral analysis, normalized pixel-intensity values were resampled to provide pixel values at half-day intervals and the record was padded with zeros to create a record 2^8 (256) values long. The CRS model and error on the ^{234}Th activity data do not create age–depth relationships with temporal resolutions better than one day. Therefore, spectral analysis results with a frequency higher than one day are ignored as noise. However, no significant peaks at higher frequencies were observed (see below).

Spectral analysis of apparent cyclicity in sedimentary sections may leave some uncertainties because it cannot be excluded that part of the signal may be lost due to periods of non-deposition or erosion. In addition, because most spectral analyses require equally spaced, continuous data, the resultant spectral density will be in error. The fundamental assumption of the CRS method is that sedimentation is continuous, but not necessarily constant. The resampled pixel versus time data provide equally spaced data points and therefore the spectral analysis is applicable in this case.

Trends in the series and any seasonality must be removed to prepare a data set for time-series analysis. (Chatfield 1975). Examination of the 223BC pixel intensity data revealed no

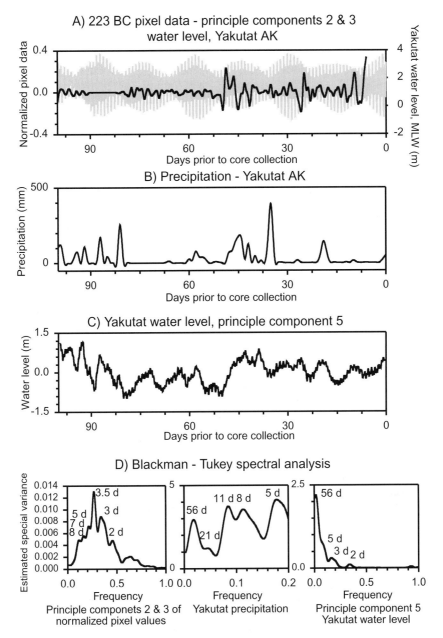

Fig. 7. Time-series analysis techniques applied to 223 BC pixel intensity data and oceanographic and meteorological forces. (**a**) Plot of combined second and third principal components of the normalized pixel intensity versus time established from CRS method. Also plotted is the water level at Yakutat Alaska during same time period. (**b**) Daily precipitation in Yakutat. (**c**) The non-tidal fifth principal component of the Yakutat water level time series established with the SSA technique (Fig. 6). The fifth component results from the onshore movement of water by low-pressure storm systems (positive water level is associated with rainfall events). (**d**) Blackman-Tukey spectral analyses performed on time-series 7a–c. The second and third principal components of pixel intensity series contains significant peaks at around 2, 3, 5, 7, and 8 days. The Yakutat precipitation series contains significant peaks at about 5, 8, 11, 21, and 56 days. Likewise, the non-tidal component of the Yakutat water level record contains significant peaks at 2, 3, 5, and 56 days. The pixel intensity time series appears to be a combination of tidal signals (7 days) and non-tidal water level (2, 3, 5, 8 days) associated with storm events.

significant trends, but there was strong seasonality (Fig. 6a). A dominant seasonality of about 60 days results from seasonal sedimentation creating the darker, denser stratified diamicton in the winter–spring period changing over to increased sedimentation from late-summer meltwater production (Cowan *et al.* 1997; Jaeger & Nittrouer 1999). Several methods can be applied to remove this seasonality, such as a moving average or high-pass filter (Chatfield 1975). However, these techniques have distinct disadvantages. One shortcoming of the running mean is that oscillations at high frequencies can make interpretation of fluctuations in a filtered time series difficult (Chatfield 1975). A disadvantage of the high-pass filter method is that there is no *a priori* knowledge of where to place the frequency cut-off, and valuable signals may be lost.

A powerful noise filter through its ability to separate self-coherent features from random ones is the singular spectrum analysis (SSA) technique (Vautard & Ghil 1989). The singular spectral analysis method is not truly a spectral analysis method. Rather, it performs an empirical orthogonal function (EOF) analysis in the time domain, and thus represents the signal as a sum of principal components that are not necessarily oscillations, but more general, data adaptive functions (Paillard *et al.* 1996).

The SSA routine in the AnalySeries program was used to identify the principal components of the pixel intensity time series (Fig. 6b). The importance of the various components can be judged qualitatively by noting which components contribute significantly more variance relative to the noise background within a SSA eigenvalue spectrum. For the 223 BC pixel data, the first principal component contained the seasonal signal and constituted most of the variance (*c.* 80%). The remaining principal components contributed decreasing amounts of variance to the total, so it was decided to use only the combined second and third components in constructing a non-seasonal time series (Fig. 7a).

An estimate of the power spectrum of the pixel-intensity time series was created on the second and third principal components using the Blackman-Tukey method (Blackman & Tukey 1958), a classical method for spectral analysis. The algorithm first computes the autocovariance of the data, then applies a Tukey window, and finally Fourier-transforms it to compute the spectrum. It is a very robust method, unlikely to present spurious spectral features (Paillard *et al.* 1996). The Blackman-Tukey routine in the AnalySeries program

(Paillard *et al.* 1996) was used, and showed cyclical deposition occurring over various time-scales (Fig. 7d). The dominant period occurred at 3.45 days, with minor higher-frequency peaks at around 1, 2, and 3 days lower-frequency peaks at 5, 7, and 8 days (Fig. 7d). With the exception of the 7 day peak (neap-spring tides), none of the other peaks appear to be tidal, although tidal processes can influence deposition on non-tidal time-scales in deep water Alaskan fjords (see below; Cowan *et al.* 1998). Given the dating error associated with the CRS method, dominant periodicities of half a day should be rounded to the nearest integer (e.g. 3.45d = 3 d). Possible forcing mechanisms of these periodicities are discussed below.

One of the limitations of spectral analysis techniques such as Fourier and Blackman-Tukey transforms is that they are unable to resolve localized variations of power within a time series. For example, is 3-, 5-, and 8-day periodicity in the pixel intensity constant over the period of interest or is it localized in time? Wavelet analysis is becoming a common tool for providing this temporal information in spectral analyses. By decomposing a time series into time–frequency space, one is able to determine both the dominant modes of variability and how those modes vary in time. The wavelet transform has been used for numerous studies in geophysics, such as the El Niño–Southern Oscillation (Wang & Wang 1996). A clear description of wavelet theory and its application to geophysical studies is given in Torrence & Compo (1998). The wavelet analyses performed with these data were accomplished using wavelet software provided by Torrence & Compo (1998), and is available at http://paos.colorado.edu/research/wavelets/.

Wavelet analysis of the normalized pixel intensity showed a dramatic rise in overall wavelet power after day 60 (*c.* 5/15/95) (Fig. 8b). This indicates that the controls on sedimentation were interacting with the increased meltwater discharge of the summer period to modulate sedimentation rates. Similarly, Cowan *et al.* (1998, 1999) document seasonal variability in both the number and thickness of laminations deposited in Muir Inlet, Glacier Bay. During the summer in Icy Bay, diurnal and tidal periodicities (14, 28 day) not apparent in the Blackman-Tukey method become more prevalent (at 90% confidence interval, above dashed line of global wavelet, Fig. 8c). The cyclicities of about 3, 5, and 8 days also become more pronounced as seen in the global wavelet. The global wavelet (Figs 8 c, e, & g) is equivalent to a Blackman-Tukey spectral variance plot and highlights the

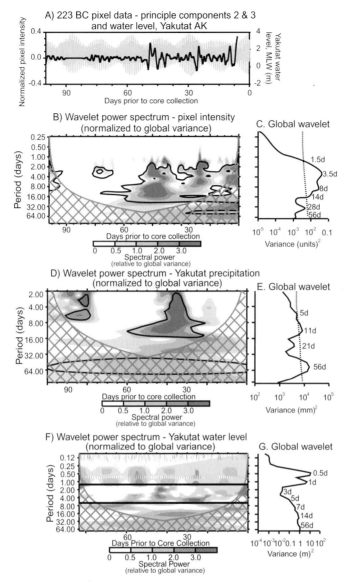

Fig. 8. The wavelet power spectra of 223BC X-ray radiograph pixel intensity, daily precipitation, and water level in Yakutat Alaska. For all wavelet spectra, the power has been scaled by the global wavelet spectrum (at right of each spectra). The global wavelet power spectrum is roughly equivalent to the Blackman-Tukey power spectra. The dashed line is the significance for the global wavelet spectrum, assuming the same significance level and background spectrum as in wavelet spectra. The crosshatched region is the cone of influence, where zero padding has reduced the variance. Heavy black contours are the 90% significance level, using a white-noise background spectrum. (**a**) The second and third principal components of the 223BC pixel intensity used for the wavelet analysis, and water level at Yakutat Alaska. (**b**) The wavelet power spectra of (a) reveals a dramatic rise in spectral power after day 60, corresponding to the onset of summertime glacial melting, a spring tide, and a large rainfall event that together may have initiated summer meltwater discharge. The global wavelet (**c**) shows the dominant periodicities within the wavelet power spectra and contains the same peaks as the Blackman-Tukey spectra (Fig. 7d). With the exception of the fortnightly tidal signal at 14 and 28 days, the power spectra (**b**) shows that about 3 and 8 day periodicities are associated with a combination of rainfall events (**d, e**) and periods of high wind-induced water level (**f** –power density shown between black lines and **g**), although they appear to lag the storm events by about 7 days, the transit time of rainfall through meltwater conduits.

dominant periodicities within the wavelet power spectra.

Evaluation of short-term tidal and storm-induced sedimentation processes

For Alaskan and many other tidewater glaciers, the dominant controls on sedimentation proximal to glacier termini are driven by diurnal fluctuations in meltwater discharge tied to daily maxima in solar insolation (Phillips et al. 1991), tidal controls on meltwater discharge and suspended sediment settling rates (Domack & Williams 1990; Cowan & Powell 1991), and episodic rainfall events (Cowan et al. 1988). Tidal processes can play an especially important role in controlling sedimentation rates. Mackiewicz et al. (1984) hypothesized that settling of suspended sediment from hyperpycnal overflow plumes resulted in the formation of faintly laminated and cyclically interlaminated mud, possibly formed through tidal processes. This facies consisted of rhythmite couplets of sand/mud (cyclopsams) or silt/mud (cyclopels), about 1 cm thick. Cyclopsams were deposited very proximal to the glaciers (less than 100 m) and cyclopels farther from the ice front. Cowan & Powell (1990) proposed that tidally modulated, differential settling resulted in the formation of these deposits. During periods of reduced tidal energy (daily slack tides; fortnightly neap periods), there is a greater tendency to deposit sediment from turbid overflow plumes due to a decrease in tidal mixing and turbulence. The increases in tidally induced turbulence during spring tides may aid in the distal transport of coarser sediments as well as accentuating meltwater discharge rates (Cowan & Powell 1990; Domack & Williams 1990). In addition, episodic events (e.g. storms, gravity flows) may influence sedimentation rates (Syvitski 1989). Cowan et al. (1988) observed increased sedimentation (up to 8 mm/hr) 1–2 days after a large rainfall event near the McBride Glacier, Alaska.

The pronounced periodicity in X-ray radiograph pixel intensity is tied to the summation of all these periodic and episodic processes. Therefore, time series of water level (as a measure of the tidal conditions) and precipitation at Yakutat, Alaska (the nearest station with high-resolution data) were analysed in the same manner as the pixel intensity data to explore relationships between sedimentation processes and the likely driving forces (Figs 7 & 8).

In general, there is little evidence of daily processes in influencing the pixel intensity series (Figs 7 & 8). This could be due to the poor temporal resolution (about one day) resulting from the CRS age–depth model or to the fact that this site was 4 km from the terminus and daily depositional processes were not strongly active (Cowan & Powell 1990). However, Cowan et al. (1998, 1999) have demonstrated that in the deep waters (>240 m) of Muir Inlet in Glacier Bay, there are on average 7.4 cyclic rhythmites (cyclopels) produced during a 14-day spring-neap period. In this deep-water fjord, deposition from lower-amplitude high tides is 'filtered' out, depositing fewer couplets than predicted (14 for diurnal, 28 for mixed semi-diurnal). The strong two-day periodicity (7–8 laminations per 14 day period, Fig. 7a) observed in this record may be the result of similar water column processes that preclude higher frequency tidal deposition.

There is a signal in the pixel intensity series tied to the fortnightly tidal cycle. In general, darker, denser portions of the X-radiograph were deposited towards the end of spring tidal periods, with the darkest laminae formed during spring tides with the highest tidal ranges (Fig. 6a; ~day 25). These darker laminae may be the result of the increased water-column turbulence during springs that allows for more distal transport of coarser sediments. The 14- and 28-day periodicity is not apparent prior to the onset of summer-time increases in meltwater discharge, so the formation of these fortnightly deposited laminae may be tied to the suspended transport of sediment from discharge plumes.

In addition to the fortnightly tidal controls, there is a strong 3 day periodicity. Evaluation of this signal in the pixel data using the wavelet method reveals that this periodicity is most pronounced during the later 3–5 days of each spring tide (Fig. 8b). The most likely explanation for this correlation is that the spring tides are able to flush more sediment from the discharge conduits and the increased water-column turbulence is able to keep the material in suspension. As the tidal energy falls during the spring-neap transition, suspended sediment is allowed to settle over a 3 to 5 day period (Cowan & Powell 1990).

Both the Blackman-Tukey power spectra and the wavelet analyses of the pixel data show significant (>90% confidence level) 5-, 8- and 56-day periodicities in laminae deposition suggestive of storm or gravity flow events (Figs 7d & 8b,c). In general, the contacts between coarser laminae and finer muds are not sharp but tend to be gradational, suggesting that there was no erosion of sediment prior to deposition, as would be the case for a gravity-flow deposit. At this same location, several centimetre-thick graded beds bounded by sharp contacts are observed in longer cores and are the likely product of gravity flows (Jaeger & Nittrouer 1999). Also, it would be expected in this deep-water setting that gravity-flow deposition would be a stochastic, non-cyclical process, in contrast to the shallow-water/intertidal cyclic gravity flow deposition observed in macrotidal Muir Inlet (Smith *et al.* 1990). Therefore, storms are a more likely cause of the sedimentation at these periodicities. To test the hypothesis that these events were rainfall-induced, hourly precipitation measurements (integrated to produce daily amounts) from Yakutat, Alaska (about 150 km from the Guyot Glacier) were analysed using the Blackman-Tukey method (Fig. 7b,d). Power spectra reveal that during the period of interest (4/01/95–7/12/95), there were pronounced about 56 day (seasonal), 21 day, 11 day, and 5–8 day periodicities in precipitation (Fig. 7d). This suggests that some of the cyclicity observed in the pixel intensity spectrum at the 5–8 day periodicity can be attributed to changes in precipitation.

In addition, NOAA tidal gauge data at Yakutat showed that there were non-tidal periodicities in the water-level record. To evaluate these periodicities, the SSA routine was used on the water-level record. The first five principal components accounted for more than 95% of the variance, and the first four components were related to tidal periodicities. However, the fifth component contained the underlying non-tidal signal (Fig. 7c). Assessment of this time series using the Blackman-Tukey routine revealed prominent 56-, 5-, 3-, and 2-day periodicities (Fig. 7) nearly identical to the precipitation time-series. Positive water-level anomalies seem to be strongly correlated with rainfall events. The proximity of Yakutat to the Gulf of Alaska suggests that the increase in water level was due to general low-barometric pressure stormy periods (i.e. onshore southerly winds, lower barometric pressure). Given the scale of water level variability (~ 2 m) associated with these storms, the likely cause for the fluctuations in water level can be attributed to onshore winds. Icy Bay also opens onto the Gulf of Alaska, so the same increase in water level due to southerly winds probably also occurred in this fjord.

Because of the lack of information regarding the time required for a rainfall event to transit the Guyot Glacier meltwater conduits, it is difficult to relate individual precipitation events to strata evident in X-ray radiographs (Fig. 7a & b). Yet, such events may create interlaminated silt and mud layers (Cowan *et al.* 1988), which at this location would be difficult to distinguish

from tidally formed cyclopels. For the smaller McBride Glacier, there was a 1–2 day lag between rainfall and meltwater increases (Cowan *et al.* 1988). So, for the larger Guyot Glacier, a 2–7 day lag may be possible. Evaluation of the wavelet power spectra for all three time series reveals that the 5 and 8 day periodicities in the pixel data can be explained by rainfall events, but lagged about a week.

One interesting result of the wavelet analysis is the correlation of the onset of summertime sedimentation with a strong precipitation event around day 47 (Figs 7 & 8). A dramatic rise in water level corresponding to the onset of a rain event correlates with the deposition of a thick, low-density mud layer, after which time diamicton sedimentation ceased. This interesting correlation compares well with observations of Cowan *et al.* (1988), who observed 2-month seasonal cyclicity in the thickness of spring-neap bundles in Muir Inlet. The early months of the summer melt season (May–June) contained relatively thicker bundles, which they suggest could be the result of initial flushing of winter-stored sediment and a peak in snowfall melt during this period. For the Guyot Glacier, this rainfall event may have been an early-season trigger leading to the reopening of meltwater discharge conduits and flushing of built-up rock flour accumulating during low-flow winter months.

Conclusions

Glacial systems release sediment to the marine environment over a range of time-scales. Regardless of the geographical setting, sediment release from melting glaciers can occur on short (daily to tidal) time-scales. Also, the magnitude of processes occurring on shorter time-scales can be strongly augmented by weather-controlled precipitation, which is ultimately driven by climate changes. Because sedimentation rates are highest within a few kilometres of temperate glacier termini or under floating ice shelves in polar settings, the highest-resolution sedimentary record is found in this difficult-to-sample location. The high-particle concentrations found in these ice-proximal environments coupled with the short half-life of ^{234}Th permits the application of certain geochronological methods to measure sedimentation rates over very short time-scales. Time-series and wavelet analyses of digital X-ray radiographic pixel-intensity values enable the evaluation of various sedimentary processes over a range of time-scales using proxy data present in core X-ray radiographs. In this example, sediment bulk density related to both grain-size and sedimentation rates created variations in X-ray radiograph pixel intensities. At a deep-water site 4 km from the terminus of the temperate Guyot Glacier, seasonal meltwater cyclicity modified by tidal processes and precipitation-augmented meltwater release controlled sedimentation. Because most tidewater glaciers terminate in relatively saline waters providing ample ^{234}Th for creating chronologies of short-term processes, such as the deposition of laminated sediments, the techniques outlined herein are applicable to most glacimarine environments allowing for additional means of evaluating seasonal and short-term sedimentary processes.

The author appreciates the assistance provided by the fellow scientists and students involved in this project, especially Chuck Nittrouer, as well as John Milliman, Sue Henrichs, and the students from SUNY Stony Brook, Virginia Institute of Marine Science, and the University of Alaska. The author acknowledges the thoughtful reviews and insight provided by Robert Aller, J. Kirk Cochran, Glenn Lopez, Eugene Domack, and Ellen Cowan. Additional thanks is given to Yohan Guyodo for assistance with time-series analysis techniques. The successful completion of this study was greatly facilitated by the crew of the R/V Alpha Helix. The research was funded by National Science Foundation grant OCE9223114 to Charles Nittrouer.

References

ALLER, R. C., BENNINGER, L. K. & COCHRAN, J. K. 1980. Tracking particle-associated processes in nearshore environments by use of 234Th/238U disequilibrium. *Earth and Planetary Science Letters*, **47**, 161–175.

ALLER, R. C. & COCHRAN, J. K. 1976. 234Th/238U disequilibrium in nearshore sediment: particle reworking and diagenetic time-scales. *Earth and Planetary Science Letters*, **26**, 37–50.

ANDERSEN, E. S., DOKKEN, T. M., ELVERHØI, A., SOLHEIM, A. & FOSSEN, I. 1996. Late Quaternary sedimentation and glacial history of the western Svalbard continental margin. *Marine Geology*, **133**, 123–156.

APPLEBY, P. G. & OLDFIELD, F. 1992. Application of lead-210 to sedimentation studies, *In:* M., I. & HARMON, R. S. (eds) *Uranium-series Disequilibrium: Applications to Earth, Marine, and Environmental Sciences*. Oxford, Clarendon Press, 731–778.

ARCHER, A. W. 1994. Extraction of sedimentological information via computer-based image analyses of shales in carboniferous coal-bearing sections of Indiana and Kansas, USA. *Mathematical Geology*, **26**, 47–65.

BECKER, J. S. & DIETZE, H. J. 2000. Precise and accurate isotope ratio measurements by ICP-MS. *Fresenius Journal of Analytical Chemistry*, **368**, 23–30.

BENNINGER, L. K., ALLER, R. C., COCHRAN, J. K. & TUREKIAN, K. K. 1979. Effects of biological sediment mixing on the 210Pb chronology and trace metal distribution in a Long Island sediment core. *Earth and Planetary Science Letters*, **43**, 241–259.

BERNER, R. A. 1980. *Early Diagenesis: A Theoretical Approach*. Princeton Series in Geochemistry Princeton University Press, Princeton, N.J.

BLACKMAN, R. B. & TUKEY, J. W. 1958. *The Measurement of Power Spectra from the Point of View of Communication Engineering*. Dover Publications, New York.

BOROLE, D. V., KRISHNASWAMI, S. & SOMAYAJULU, B. L. K. 1982. Uranium isotopes in rivers, estuaries and adjacent coastal sediments of western India: Their weathering, transport and oceanic budget. *Geochimica et Cosmochimica Acta*, **46**, 125–137.

BOUDREAU, B. P. 1986. Mathematics of tracer mixing in sediments: I Spatially-dependent, diffusive mixing. *American Journal of Science*, **286**, 161–198.

BUESSELER, K. O., COCHRAN, J. K., BACON, M. P., LIVINGSTON, H. D., CASSO, S. A., HIRSCHBERG, D., HARTMAN, M. C. & FLEER, A. P. 1992. Determination of thorium isotopes in seawater by nondestructive and radiochemical procedures. *Deep-Sea Research Part a-Oceanographic Research Papers*, **39**, 1103–1114.

CHATFIELD, C. 1975. *The Analysis of Time Series: Theory and Practice*. Chapman and Hall, London.

COCHRAN, J. K. 1992. The oceanic chemistry of the uranium- and thorium-series nuclides, *In:* M., I. & HARMON, R. S. (eds) *Uranium-series Disequilibrium: Applications to Earth, Marine, and Environmental Sciences*. Oxford, Clarendon Press, 334–395.

COOPER, M. C. 1997. The use of digital image analysis in the study of laminated sediments: *Journal of Paleolimnology*, **19**, 33–40.

COWAN, E. A. & POWELL, R. D. 1990. Suspended sediment transport and deposition of cyclically interlaminated sediment in a temperate glacial fjord, Alaska, U.S.A. *In:* DOWDESWELL, J. A. & SCOURSE, J. D. (eds) *Glacimarine Environments: Processes and Sediments*. Geological Society, London, Special Publications, **53**, 75–89.

COWAN, E. A. & POWELL, R. D. 1991. Ice-proximal sediment accumulation rates in a temperate glacial fjord, southeastern Alaska, *In:* ANDERSON, J. B. & ASHLEY, G. M. (eds) *Glacial Marine Sedimentation: Paleoclimatic Significance*. Geological Society of America, Boulder, CO, Special Paper, **261**, 61–73.

COWAN, E. A., POWELL, R. D. & SMITH, N. 1988. Rain-storm-induced event sedimentation at the tidewater front of a temperate glacier. *Geology*, **16**, 409–412.

COWAN, E. A., CAI, J., POWELL, R. D., CLARK, J. D. & PITCHER, J. N. 1997. Temperate glacimarine varves: An example from Disenchantment Bay, Alaska. *Journal of Sedimentary Research*, **67**, 536–549.

COWAN, E. A., CAI, J., POWELL, R. D., SERAMUR, K. C. & SPURGEON, V. L. 1998. Modern tidal rhythmites deposited in a deep-water estuary. *Geo-Marine Letters*, **18**, 40–48.

COWAN, E. A., SERAMUR, K. C., CAI, J. K. & POWELL, R. D. 1999. Cyclic sedimentation produced by fluctuations in meltwater discharge, tides and marine productivity in an Alaskan fjord. *Sedimentology*, **46**, 1109–1126.

CUTSHALL, N. H., LARSEN, I. L. & OLSEN, C. R. 1983. Direct analysis of 210Pb in sediment samples: Self-adsorption corrections: *Nuclear Instrumentation and Methods*, **206**, 309–312.

DOMACK, E. W. 1990. Laminated terrigenous sediments from the Antarctic Peninsula; the role of subglacial and marine processes, *In:* DOWDESWELL, J. A. & SCOURSE, J. D. (eds) *Glacimarine Environments: Processes and Sediments*. Geological Society, London, Special Publications, **53**, 91–103.

DOMACK, E. W. & WILLIAMS, C. R. 1990. Fine-structure and suspended sediment transport in three Antarctic fjords. *Antarctic Research Series*, **50**, 71–89.

ELVERHØI, A., DOWDESWELL, J. A., FUNDER, S., MANGERUD, J. & STEIN, R. 1998. Glacial and oceanic history of the Polar North Atlantic Margins: An overview. *Quaternary Science Reviews*, **17**, 1–10.

GERINO, M., ALLER, R. C., LEE, C., COCHRAN, J. K., ALLER, J., GREEN, M. & HIRSCHBERG, D. 1998. Comparison of different tracers and methods. *Coastal Estuarine Shelf Science*, **46**, 531–547.

HILL, P. S., SYVITSKI, J. P., COWAN, E. A. & POWELL, R. D. 1998. In situ observations of floc settling velocities in Glacier Bay. *Marine Geology*, **145**, 85–94.

JAEGER, J. M. & NITTROUER, C. A. 1995. Tidal controls on the formation of fine-scale sedimentary strata near the Amazon River mouth. *Marine Geology*, **125**, 259–281.

JAEGER, J. M. & NITTROUER, C. A. 1999. Sediment deposition in an Alaskan fjord: Controls on the formation and preservation of sedimentary structures in Icy Bay. *Journal of Sedimentary Research*, **69**, 1011–1026.

JAEGER, J. M., NITTROUER, C. A., SCOTT, N. D. & MILLIMAN, J. D. 1998. Sediment accumulation along a glacially impacted mountainous coastline. Northeast Gulf of Alaska. *Basin Research*, **10**, 155–173.

KNAUER, G. & ASPER, V. 1989. *Sediment Trap Technology and Sampling*. US GOFS (Global Ocean Flux Study) Planning Report, **10**, 94.

MACKIEWICZ, N. E., POWELL, R. D., CARLSON, P. R. & MOLNIA, B. F. 1984. Interlaminated ice-proximal glacimarine sediments in Muir Inlet, Alaska. *Marine Geology*, **57**, 113–147.

MANN, W. B., RYTZ, A. & SPERNOL, A. 1988. Radioactivity Measurements – Principles and Practice. *Applied Radiation and Isotopes*, **39**, 717–937.

MCKEE, B. A., DEMASTER, D. J. & NITTROUER, C. A. 1984. The use of 234Th/238U disequilibrium to examine the fate of particle-reactive species on the Yangtze continental shelf. *Earth and Planetary Science Letters*, **68**, 431–442.

OLSEN, C. R., LARSEN, I. L., LOWRY, P. D. & CUTSHALL, N. H. 1986. Geochemistry and deposition of Be-7 in river-estuarine and coastal waters. *Journal of Geophysical Research*, **91**, 896–908.

PAILLARD, D., LABEYRIE, L. & YIOU, P. 1996. Macintosh Program Performs Time-Series Analysis. *Eos, Transactions, American Geophysical Union*, **77**, 379.

PHILLIPS, A. C., SMITH, N. D. & POWELL, R. D. 1991. Laminated sediments in prodeltaic deposits, Glacier Bay, Alaska. *In:* ANDERSON, J. B. & ASHLEY, G. M. (eds) *Glacial marine sedimentation: Paleoclimatic significance*. Boulder, CO, Geological Society of America, Special Paper, **261**, 51–60.

PUIG, P. & PALANQUES, A. 1998. Nepheloid structure and hydrographic control on the Barcelona continental margin, northwestern Mediterranean. *Marine Geology*, **149**, 39–54.

REED, R. K. & SCHUMACHER, J. D. 1987. Physical oceanography. *In:* HOOD, D. W. & ZIMMERMAN, S. T. (eds) *The Gulf of Alaska: Physical Environment and Biological Resources*. US Dep. Commerce, Jul-76. Washington, D.C., National Oceanic and Atmospheric Administration.

ROBBINS, J. A. 1978. Geochemical and geophysical applications of radioactive lead. *In:* NRIAGO, J. P. (ed.) *Biochemistry of Lead in the Environment*. Amsterdam, Elsevier, 285–393.

ROBBINS, J. A. & HERCHE, L. R. 1993. Models and uncertainty in Pb-210 dating of sediments. *Verhandlungen der Internationalen Vereinigung für Theoretische und Angewandte Limnologie*, **25**, 217–222.

SAMBROTTO, R. N. & LORENZEN, C. J. 1987. Phytoplankton and primary productivity, *In:* HOOD, D. W. & ZIMMERMAN, S. T. (eds) *The Gulf of Alaska: Physical environment and biological resources*. Washington, D.C., National Oceanic and Atmospheric Administration, U.S. Dep. Commerce, 249–284.

SMITH, N. D., PHILLIPS, A. C. & POWELL, R. D. 1990. Tidal drawdown: A mechanism for producing cyclic sediment laminations in glaciomarine deltas. *Geology*, **18**, 10–13.

SYVITSKI, J. P. M. 1989. On the deposition of sediment within glacier-influenced fjords; oceanographic controls. *Marine Geology*, **85**, 301–329.

SYVITSKI, J. P. M. 1993. Glaciomarine environments in Canada; an overview. *Canadian Journal of Earth Sciences*, **30**, 354–371.

SYVITSKI, J. P. M., BURRELL, D. C. & SKEI, J. M. 1987. *Fjords: Processes and products*. Springer Verlag, New York.

SYVITSKI, J. P. M. & SHAW, J. 1995. Sedimentology and geomorphology of fjords, *In:* PERILLO, G. M. E. (ed.) *Geomorphology and sedimentology of estuaries: Developments in sedimentology*. New York, Elsevier, 113–178.

TORRENCE, C. & COMPO, G. P. 1998. A practical guide to wavelet analysis. *Bulletin of the American Meteorological Society*, **79**, 61–78.

TUNIZ, C. 2001. Accelerator mass spectrometry: ultra-sensitive analysis for global science. *Radiation Physics and Chemistry*, **61**, 317–322.

VAUTARD, R. & GHIL, M. 1989. Singular spectrum analysis in nonlinear dynamics, with applications to paleoclimatic time series. *Physica D*, **35**, 395–424.

WANG, B. & WANG, Y. 1996. Temporal structure of the Southern Oscillation as revealed by waveform and wavelet analysis. *Journal of Climate*, **9**, 1586–1598.

A glacial sequence stratigraphic model for temperate, glaciated continental shelves

ROSS D. POWELL & JAMES M. COOPER

Department of Geology and Environmental Geosciences, Northern Illinois University, DeKalb, IL 60115, USA (e-mail: ross@geol.nin.edu)

Abstract: Temperate, glaciated continental shelves are by nature complex basins in that they not only have typical low-latitude siliciclastic processes acting to produce a sedimentary record and depositional architecture, but they also have the consequences of glacial action superimposed. For these settings glacial systems tracts are defined and related to glacial advance and retreat signatures, which can then be evaluated relative to changes in other external forces. We define glacial maximum (GMaST), retreat (GRST), minimum (GMiST) and advance (GAST) systems tracts separated by bounding disconformities, which are respectively, the grounding-line retreat surface (GRS), the maximum (glacial) retreat surface (MRS) and the glacial advance surface (GAS). Each glacial advance and retreat sequence is bounded by a regionally significant unconformity, a glacial erosion surface (GES), or its equivalent conformity. Parasequence motifs vary across the shelf but include facies dominated by the movement of grounding lines. Facies motifs may have subglacial till above a GES and this is the main facies of the GMaST; however, tills are commonly absent. Although diamictic debrites are often associated with grounding-line deposystems, the GRST succession is dominated by sorted deposits of gravel, rubble and poorly sorted sands due to the dominance of glacial meltwater. These deposits commonly have the geometry of banks or wedges/fans. They form the offlap-break at the outer continental shelf and also form a retrogradational stacking of bank systems on the shelf. Banks are capped by glacimarine rhythmites including thin debrites, turbidites, cyclopsams and cyclopels and perhaps iceberg-rafted varvites in fan to sheet-like geometries. Above these are draped sheets of bergstone muds that grade into paraglacial muds. The GMiST occurs with glacial retreat onto land or into fjords. The GMiST is represented in nearshore areas by progradational deposits of paralic systems dominated by deltaic and siliciclastic shelf systems, and in offshore areas by condensed sections. The GAST is represented by the inverse of the GRST facies succession, but is also the most likely interval to be eroded during readvance.

Temperate, glaciated continental shelves are an extreme end-member of glacimarine environments, typically occurring at the low-latitude end of the spectrum, and as such, their glaciers rely on active tectonism and very dynamic climatic systems to maintain their presence at sea level. Alternatively, they are at the extreme limit of large ice sheets as they expand into lower latitudes, or they may occur at high latitudes during major climatic transitions, such as into and out of icehouse and hot(green)house periods. These forcing mechanisms dictate an extreme condition of many important external factors that interact to build the resulting stratigraphic architecture and succession motifs.

Here we attempt to integrate data from modern to Miocene age records in Alaska and detailed facies and facies sequences recovered from the Oligocene–Miocene record in Cape Roberts Project drill cores, McMurdo Sound, Antarctica, when that region is thought to have been under much milder climatic regimes (e.g. Hambrey *et al.* 1992; Powell *et al.* 1998, 2000, in press). Alaska data include those from (1) modern fjords where facies and depositional system architecture from coring and seismic reflection data can be related to quantified process data from modern monitoring programmes; (2) seismic reflection records of Quaternary age sedimentary packages on the southern Alaskan continental shelf; and (3) continental shelf and upper slope deposits of the Yakataga Formation, uplifted and exposed in coastal outcrops, which are Miocene to late Pleistocene in age. The data are integrated in order to establish long time-series facies changes in a three-dimensional architecture, and which can be interpreted using documented modern processes. We use these data as a basis for constructing a sequence stratigraphic model that is specifically related to glacial fluctuations, and which may occur independently from

From: DOWDESWELL, J. A. & Ó COFAIGH, C. (eds) 2002. *Glacier-Influenced Sedimentation on High-Latitude Continental Margins.* Geological Society, London, Special Publications, **203**, 215–244. 0305-8719/02/$15.00

Table 1. *Lithofacies and depositional systems from Alaskan fjords*

Lithofacies	Features	Origin	Depositional systems
Gravel (G)	clast- to matrix-supported, rounded and angular clasts, locally openwork, sharp and graded/ amalgamated contacts	marine outwash, rockfall, iceberg dumping, fluviodeltaic	ice-contact grounding line, ice-proximal iceberg zone, delta/shoreline
Diamicton – massive (Dm)	diamict texture may be over-consolidated	subglacial till, local debrites	subglacial
Diamicton – stratified (Ds)	diamict texture, weakly stratified in matrix or by clasts	iceberg rainout + hemipelagite, local debrites	ice-contact grounding line, ice-proximal iceberg zone
Sand – stratified (Ss)	very fine to medium sand, often fines upward, mainly parallel laminated, some cross-laminated, sharp base, gradational top	turbidites, fluvial, shoreline	ice-proximal iceberg zone, ice-distal iceberg zone, deltaic, beach
Sand – massive (Sm)	fine to medium sand, sharp top and bottom contacts, either grain supported (clean) or matrix supported (muddy – increases up)	sediment gravity flow deposits, fluvial, shoreline	ice-proximal iceberg zone, ice-contact grounding line, ice-distal iceberg zone, deltaic, beach
Silt – stratified (Zs)	sharp top and bottom contacts, horizontal laminae, some mud inter-laminae, clean (no clay)	turbidites	ice-distal iceberg zone, ice-proximal iceberg zone, deltaic
Silt – massive (Zm)	sharp top and bottom contacts, clean (no clay)	turbidites	ice-distal iceberg zone, ice-proximal iceberg zone, deltaic
Mud – laminated (Ms)	Thick mud lamina coarsening down to sand or silt lamina (some one particle thick), rhythmic lamination, isolated gravel and sand	cyclopsams and cyclopels, tidalites	ice-proximal iceberg zone, ice-distal iceberg zone, tidal mud flat
Mud – massive (Mm)	structureless, bioturbation increases as isolated gravel decreases	hemipelagites, tidalites	ice-distal iceberg zone, ice-proximal iceberg zone, ice-contact grounding line, tidal mud flat

changes in relative sea-level. Hence the term 'glacial sequence stratigraphy' is introduced to distinguish it from the original 'sequence stratigraphy' derived from low latitude margins. In concept, this adaptation is similar to that used to interpret fluvial successions, but differs greatly in details.

Compared with non-glacial successions, sea level on glaciated continental shelves is not necessarily the only control on base level, the base of the glacier is considered a control as well. The glacier has the ability to erode major regional unconformities well below sea level,

and the glacier bed is also a primary source of large volumes of sediment introduced below sea level. The sediment produced at grounding lines accumulates as aggradational, progradational or retrogradational packages that are only secondarily dependent on relative sea-level changes for controlling accommodation space. Furthermore, at a parasequence-scale, changes in relative sea-level are also a function of glacial isostasy in addition to the more common isostatic forces (water and sediment loading), tectonism, and local water-depth controls of local erosion and sediment

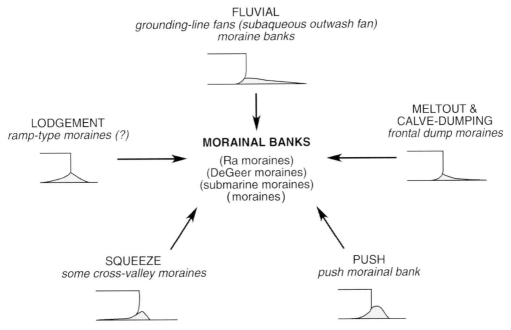

Fig. 1. End-member processes that contribute to forming grounding-line systems in the form of morainal banks (not to scale). Beneath each process are terms that have been used to describe the deposit in the literature. Here it is recognized that morainal banks can form by an end-member process or any combination of those processes. Alternative terms for morainal banks in the literature are also provided (after Powell & Domack 1995).

accumulation rates. These are all important distinctions to be made relative to the original sequence stratigraphic concepts. Facies changes, such as large-scale fining and coarsening trends, may occur simply by a glacial advance and retreat and can be independent of these other variables.

Therefore, glacial systems tracts (GST) are defined here in relation to glacial advance and retreat signatures, which can then be evaluated relative to changes in other external forces. We use glacial maximum (GMaST), retreat (GRST), minimum (GMiST) and advance (GAST) systems tracts separated by glacial bounding surfaces (GBS); each sequence being separated by a glacial erosion surface (GES) or its equivalent conformity. Parasequences of these systems tracts vary in character across the shelf and slope but include facies dominated by the movement of grounding lines rather than necessarily being tied to relative sea-level. There may be the equivalent of low-latitude sequence boundaries, but they need not coincide with a GES. Often a GES is produced at a time when relative sea-level is high due to glacial isostatic loading during glacial advance.

Facies, facies associations and depositional systems

Fjord data

A comprehensive review of facies and seismic architecture of depositional systems in fjords of southern Alaska has been provided by Cai *et al.* (1997). Major lithofacies types include gravel (G), structureless diamicton (Dm), weakly stratified diamicton (Ds), stratified sand (Ss), structureless sand (Sm), stratified silt (Zs), structureless silt (Zm), laminated mud (Ms) and structureless mud (Mm). Processes that generate these facies and their spatial distribution within modern settings have been used to characterize different modern environments (Table 1). Equivalent depositional systems have then been defined from seismic reflection profiles and from cores. These provide an understanding of the 3-D geometry of a system, which is related to glacial movements. Aspects of the different depositional systems are discussed below.

Subglacial depositional systems are difficult to characterize in modern settings because of

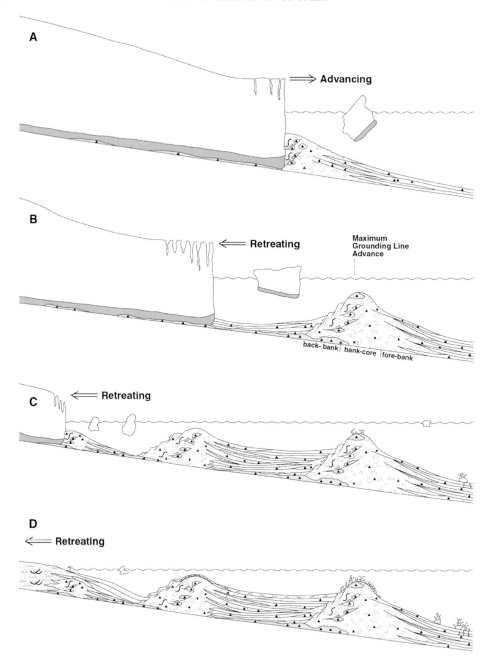

Fig. 2. An hypothetical glacial advance, retreat and readvance sequence that depicts how facies associations and sequences are thought to have been generated to produce the succession recovered by the Cape Roberts Project drilling in Antarctica. Scale varies in order to allow pertinent features to be shown: vertical scale is *c.* 1:15 000 and the horizontal scale varies between *c.* 1:140 000 and > *c.* 1:40 000 for morainal banks and non-bank areas respectively. The facies sequences are constructed to emphasize variations relative to glacial activity, so water depth does not vary, instead the effects of sea level, isostasy and tectonics are not considered. However, a particular preserved stratigraphic record will vary depending on relative lengths of time a glacier spends within different stages of its advance and retreat modes, the amount of erosion (glacial mainly) and the other variables not considered in relative water depth changes and their rates and their timing relative to the glacial advance and retreat cycle. (After Powell *et al.* 2000.)

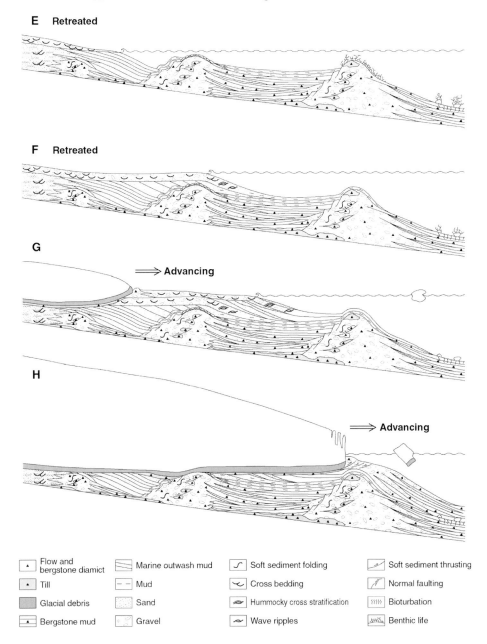

Fig. 2. continued.

the problem of sampling accessibility. Most studies assume that the subglacial processes and deposits in marine environments are similar to those of terrestrial environments both in terms of diamicton and sorted sediment. A window into temperate marine subglacial systems is provided by icebergs that calve from a submarine location near the base of the glacier and then turn over. As they calve either pore pressure in the till must drop rapidly causing freezing to the iceberg sole (Powell & Molnia 1989) or the till must be previously frozen to the

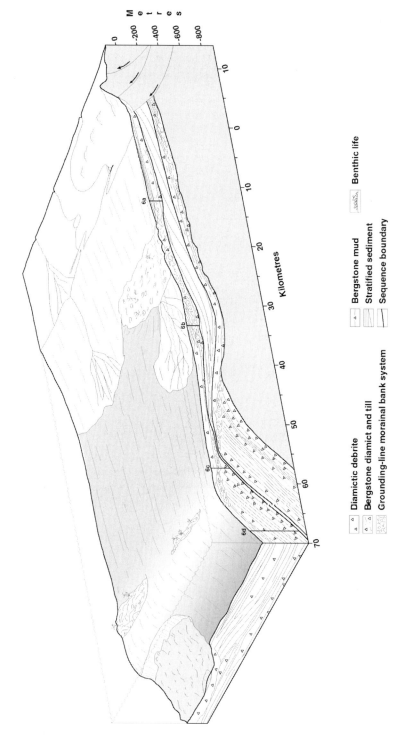

Fig. 3. A three-dimensional conceptual model of three sequences on a temperate glaciated shelf, where a Type IIAK occurs between two Type IAK sequences. Each of the Type IAK sequences has four glacial systems tracts (GAST, GMaST, GRST and GMiST), and the Type IIAK is dominated by a progradational system that represents a partial glacial advance. The model is set at the GMiST phase of the youngest Type IAK sequence and scaling is approximate but is taken from the modern Alaskan shelf (Fig. 5). Only general facies are shown; detailed facies sequences within the model are more clearly demonstrated in individual columns from different parts of the continental shelf and slope (Fig. 6), their locations being shown on this figure.

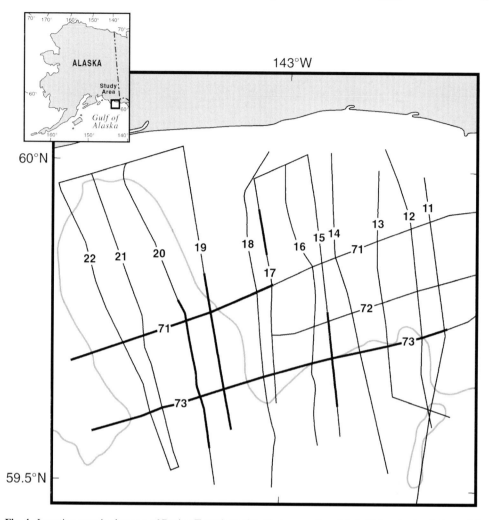

Fig. 4. Location map in the area of Bering Trough on the Alaskan continental shelf showing track lines of the seismic reflection profiles used in this study. All track lines were examined, the bold line segments being considered the most critical. The bold line segments of lines 15 and 19 are shown in Figure 5.

glacier sole with the decollement plane along which the glacier is moving, being within the till. A till from the back of a bank in Glacier Bay has also been sampled (Powell 1983). These submarine tills are similar to those of terrestrial tills (mainly Dm, rarely Ds) and their genetic processes appear similar. The processes of subpolar (Glasser & Hambrey 2001) and polar (Anderson 1999; Domack *et al.* 1999; Shipp *et al.* 1999) subglacial systems also substantiate the interpretation that subglacial processes in marine environments are similar to those described in terrestrial environments. Seismic facies analysis of seismic reflection records from

the fjords, show that these tills are rare in the temperate marine record (Cai *et al.* 1997; Seramur *et al.* 1997) and are modelled to have a patchy distribution at the base of a glacimarine succession (Powell 1981).

Ice-contact and ice-proximal grounding-line depositional systems contain complex facies and geometries produced by sedimentation, re-sedimentation and glacitectonism. Facies are gravels, rubble and sands, or less commonly, diamicton. They can exhibit significant soft-sediment deformation structures due to grounding-line movement, and iceberg dump structures may occur locally. Geometrically

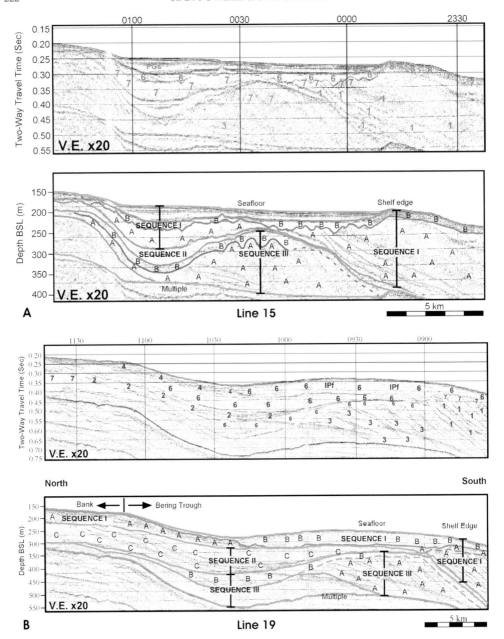

Fig. 5. Interpreted seismic reflection profiles, in mainly dip orientation, from the area of Bering Trough on the Alaskan continental shelf. Seismic facies are shown as numbers that refer to those described in Table 4, and the capital letters show locations of the seismic facies associations (SFA) described in the text. These are mini-sparker (400–800 J, 100–900 Hz) profiles recovered by the US Geological Survey (P. Carlson pers. comm.).

these systems can be divided into various types including banks, fans and ramps, and have been collectively grouped under the term 'morainal bank' (Fig. 1; Powell & Domack 1995). Each morainal bank may be subdivided into back, core and fore bank systems with slightly different facies associations and geometries (see labels on Stage B, Figure 2; following Cai *et al.* 1997). Back-bank systems include most of the till within grounding-line systems and can show collapse geometries from the removal of support following grounding-line retreat. Bank-core systems are a poorly-sorted mix of gravel, sand and diamicton facies, with rare mud, and may include major compressional structures from glacier push and over-riding. Fore-bank systems comprise mainly sediment gravity flow sands, rarer diamictons, and some interstratified muds which may be laminated. Grounding-line fan systems originating from the point source of a subglacial stream conduit, comprise stratified sands and gravels of marine outwash and sediment gravity flow deposits.

Temperate iceberg-zone depositional systems are dominated by laminated silt and mud with rarer sand, interstratified sand and mud, structureless mud, and rainout debris from icebergs (iceberg-rafted debris – IBRD); generally all being grouped within a bergstone mud facies association. Fine-grained stratified sediment is dominated by cyclopels and cyclopsams, although the latter are more common in fore-bank areas. Diamictons are either thin beds of iceberg-rafted debris (bergstone diamictons) or thicker debrites. Clast nests or clusters and lonestones from iceberg dumping and rainout are commonly scattered throughout.

Temperate ice-distal marine systems are dominated by structureless bergstone mud made massive primarily by bioturbation. Rare laminated mud intervals, thin turbidite sands and iceberg-rafted lonestones also occur.

After glacial retreat from fjords, the so-called marine outwash fjords have major deltaic systems that prograde over the earlier glacial succession at fjord heads. Structureless paraglacial hemipelagic mud, lacking IBRD lonestones, is produced distally. Where tidal forces dominate the fluvial system on the delta plain, tidal flat (mainly mud flat) systems are produced (Powell 1983; Powell & Molnia 1989). These paraglacial deposits are incorporated with the glacial fjord successions to complete the full model for fjord sedimentation (Powell 1981, 1983); a similar representation is shown here as Stage F, Figure 2. All of these data have been synthesized to construct an idealized stratigraphic architecture (Powell 1981; Fig. 3) and facies sequences (Powell 1984; Fig. 4) for a glacial advance and retreat scenario, as is depicted in a similar way for this paper (see below and in Figs 3 & 5).

These original facies models have been verified and elaborated by detailed analyses of older uplifted Quaternary examples and high-resolution seismic reflection profiles from other locations (e.g. McCabe et al. 1984; McCabe 1986; Belknap & Shipp 1991; Stewart 1991; Hart & Roberts 1994; Lønne 1995; McCabe & Ó Cofaigh 1995; Benn & Evans 1996; Hunter et al. 1996; Lønne & Lauritsen 1996; Lønne & Syvitski 1997; Plink-Björklund & Ronnert 1999). The advantage of these Quaternary deposits is that they allow a full documentation of lithofacies that in modern settings are depicted primarily as seismic facies with local samples from cores of their surfaces. They also allow verification of lateral facies changes and geometries, large-scale deformation structures, glacitectonic structures and deformation tills produced from the over-riding of glacimarine sediments, and of detailed facies successions, especially in coarser-grained deposits, which are difficult to sample in modern environments.

In general, the facies and geometries of Quaternary examples are the same as those of modern settings, but details of some facies and the complexities of interfaces geometries are better defined. For example, small- and large-scale deformation structures associated with grounding-line oscillations and glacial over-riding are better defined (e.g. McCabe et al. 1984; McCabe 1986; Lønne 1995, 2001; Hunter et al. 1996; Lønne & Lauritsen 1996; Lønne & Syvitski 1997; Bennett et al. 1999; Boulton et al. 1999; Plink-Björklund & Ronnert 1999). Varieties of gravel and rubble facies associated with grounding-line systems, especially grounding-line fans, are well documented (e.g. Powell 1990; Plink-Björklund & Ronnert 1999), as are structures generated by iceberg rafting (e.g. Thomas & Connell 1985; Powell 1990).

Continental shelf and slope data

The southern Alaskan continental margin has approximately a 7-km thickness of the late-Miocene to Pleistocene Yakataga Formation, which primarily consists of glacimarine diamicton, sand, and mud (Zellers & Lagoe 1992). Approximately 2-km of Yakataga Formation sediments have been uplifted along the coast of southern Alaska, with exposures in Yakataga Reef, the Robinson Mountains near Icy Bay, and Middleton Island (Eyles & Lagoe 1990; Eyles et al. 1991). Lithofacies distributions within these outcrops provide data on the type and style of long-term temperate glacial sequences and insights into the influence of temperate glaciation on continental margin development (Eyles & Lagoe 1990; Eyles et al. 1991) (see Table 2). This latter aspect of margin development is relevant because temperate glacial systems may commonly occur in these tectonic settings where active uplift along coastal mountains is associated with very high

Table 2. *Lithofacies of the onshore exposures of Yakataga Formation (after Eyles & Lagoe 1990)*

Lithofacies	Water depth	Genesis	Ice-sheet Stage
1. Diamict			
a. Massive	Outer shelf/upper slope	Rainout	Extended
		Suspension settling and iceberg rafting	Intermediate
b. Stratified	Outer shelf/upper slope	Winnowing by traction currents	Intermediate
		Subaqueous mass movement of glacigenic deposits	Intermediate
2. Boulder Pavement	20 to 150 m	Winnowing of diamict by currents followed by abrasion beneath partially floating ice sheet	Extended
3. Coquina Bed	20 to 50 m	Current winnowing	Intermediate
4. Gravel	Outer shelf/upper slope	Sediment gravity flow; stacked channel fill	Extended
5. Sand	150 to 250 m	Turbidity current	Intermediate
6. Fine-grained			
a. Laminated Silt	Shallow marine	Turbidity current; upper channel fill	Intermediate
b. Massive Mud	150 to 250 m	Suspension settling within quiescent basin ice sheet	Withdrawn

snow accumulation rates that can drive glaciers into the sea and maintain very high erosion rates and sediment fluxes (Powell 1991; Hallet *et al.* 1996). This style of glaciation may well be important in the development of broad continental margins, relative to those convergent margin settings that occur at lower latitudes, due to the high sediment fluxes. An important result of this analysis relative to the younger record is that the older parts of the Yakataga Formation comprise slope deposits, whereas younger sections consist of temperate shelf successions formed after the margin had become wide enough to allow accumulation of shallower water deposits.

The lowest Yakataga Formation is defined by the first appearance of IBRD within the section. Facies of the package are primarily turbidites and debrites, and palaeobathymetric interpretations from foraminiferal and facies analysis, indicate that the continental shelf was narrow and the margin was a deep basin at the onset of late-Miocene glaciation (Eyles & Lagoe 1990).

Sediments of the middle Yakataga Formation are exposed in the Robinson Mountains around Icy Bay (Eyles & Lagoe 1990). This section consists primarily of glacimarine diamictites from a combination of suspension settling, rainout and high-density sediment gravity flows, and turbidites. However, shallow marine sandstones are also present and record the

middle-Pliocene warm event. This section exhibits rapid progradation of the continental shelf by sediment gravity-flow processes down the slope with high sediment fluxes being supplied by temperate glaciers.

Outer shelf–upper slope deposits comprise the upper Yakataga Formation and are recorded on Middleton Island, which is located near the shelf edge (Eyles & Lagoe 1990). These deposits are between 1.8 and 0.7 Ma old. Major deposits include rainout diamictite, channel-filling gravels, coquinas and boulder lags. Sediment gravity-flow deposits are a major constituent of the shelf-break sedimentary package (Eyles & Lagoe 1990). This section records transition to a shallow shelf environment.

Rainout (bergstone) diamictites are here inferred to be continental shelf equivalents of the bergstone mud described above, having distinctive polymodal rather than bimodal size distributions. This distinction represents one facies contrast between fjord and shelf settings. Consequently, on the shelf, sediment plumes originating from stream discharges are unconfined relative to fjord systems. The plumes then spread over greater areas and thus decrease local hemipelagic accumulation rates. Therefore, in ice proximal areas, the IBRD component is increased relative to that in fjord settings to result in a diamict texture on the shelf.

Megachannels within the middle Yakataga Formation have been described from the Robinson Mountains and Icy Bay region (Armentrout 1983; Eyles & Lagoe 1998). Megachannel development is hypothesized to occur by sediment slumping on the outer shelf–upper slope where sediment accumulation rates are high. Megachannels are quickly infilled by gravelly and sandy turbidites. Frequent and voluminous sediment failure along the southern Alaska shelf is due to both high sedimentation rates and strong seismicity. The Gulf of Alaska is one of the most seismically active areas on earth, where frequent earthquakes liquify under-consolidated sediments and trigger mass wasting on the over-steepened upper slope (Plafker & Addicott 1976; Carlson & Molnia 1977). These earthquakes also cause major resedimentation events of morainal bank systems on low slopes such as fjord floors (e.g. Powell 1991) and the continental shelf.

In addition to these lithofacies associations just described, Eyles *et al.* (1992) also have suggested typical ichnofacies for the succession, these include: (1) Cruziana ichnofacies assemblage for muddy shelves that may be associated with common escape structures (fungichnia), mass bivalve mortality and perhaps dwarf feeding traces in areas of high sedimentation and high meltwater (brackish conditions) delivery; (2) Glossifungites ichnofacies assemblage of burrowed firm grounds experiencing erosion in shallow water and which may be associated with coquinas or winnowed boulder lags; and (3) diverse Cruziana ichnofacies assemblage on the slope due to food transport down slope by regular sediment redepositional processes.

The continental margin of southern Alaska, because it is in a compressional/transpressional tectonic regime, has experienced significant compressional stresses associated with under-thrusting of the Yakutat terrane. This compression has resulted in substantial deformation and shortening of the shelf (Bruns & Schwab 1983). The intensity of deformation varies spatially and temporally across the shelf, but locally deformation rates are sufficiently fast that they influence glaciation and depositional geometries of a glacial cycle. As deformation progressed, anticlines subsided while others were activated. Intervening basins were infilled with sediment, burying subsided anticlines within a glacial cycle. This tectonic subsidence is thought to have provided significant accommodation space for the glacimarine succession on the shelf such that it was not exposed during eustatic sea

level lowstands and has also allowed local preservation of older successions during younger glacial advances.

Recently, a 1500-m-thick succession on the Antarctic continental shelf has been described and interpreted as a temperate to polythermal glacially-influenced, early Cenozoic glacial record (Powell *et al.* 1998, 2000, 2001). The facies used to interpret the succession are very similar to modern glacial and glacimarine facies from Alaskan fjords and shelf (Table 3), as is the palaeo-environmental reconstruction (Fig. 2). These have been used to infer numerous glacial fluctuations through about 17 million years of Antarctic glaciation and thus provide an excellent database with which to analyse facies and facies associations. They are also integrated into the database used here for constructing the sequence stratigraphic model.

Facies successions and motifs

Facies successions for separate depositional systems in Alaskan fjords have been established (summarized by Cai *et al.* 1997; see above) and combined into simplified facies sequences representing hypothesized successions from a full glacial advance and retreat cycle, which also include inferences for the continental shelves (Powell 1981, 1984). Advance facies are rarely preserved in temperate fjord settings due to the efficiency of erosion during the advance stage, although isolated pockets may be preserved (Cai *et al.* 1997). Pre-advance sediment may be eroded to varying depths depending on several factors such as subglacial regime, length of glacial cycle, and geographic position of the site relative to maximum glacial advance position. In some circumstances, if erosion of older strata above bedrock is incomplete, a definite erosion surface may be absent. Instead, pre-existing sediment may be remoulded by glacitectonic processes into deformed subglacial till, and depending on the type of older sediment and the extent of deformation, the section may lack an exact 'erosion' surface.

A facies motif has also been suggested for sequences of the Antarctic Cenozoic shelf succession described above (Fielding *et al.* 2001; Naish *et al.* 2001). In the facies motif, Fielding *et al.* (2001) and Naish *et al.* (2001) combined relative sea-level changes with the inferred glacial cycles, where, for example, lowstands of relative sea-level were interpreted to occur at a glacial maximum and coincide with a GES. Lower parts of the facies motif have a broad fining upwards trend from pebbly sandstone, conglomerate and diamictite, to poorly sorted

Table 3. *Summary of facies characteristics and their interpretations, Cape Roberts cores (Powell et al. 1998, 2000, 2001)*

Facies number and name	Key sedimentological characteristics	Depositional process interpretation	Key interpretation criteria
1. Mudstone	- massive, often sandy - local laminae - common lonestones - locally brecciated - marine macro- and microfossils	- hemipelagic suspension settling - rainout from ice rafting - may be modified by other processes - brecciated by tectonism or glacial tectonism	- fine-grained character - isolated clasts - marine fossils
2. Interstratified sandstone and mudstone	- sandstones on sharp contacts - sandstones grade up, often to mudstones - massive and amalgamated beds - planar stratified - local ripple cross-lamination - some normal, local reverse grading - dispersed to abundant clasts - marine macro- and microfossils	- range of marine processes: low- to moderate-density sediment gravity flow deposition; combined wave and current action - rapid deposition and resedimentation	- sandstone/mudstone association - style of internal stratification and grading - marine fossils
3. Poorly-sorted (muddy) very fine to coarse sandstone	- various poorly sorted sandstones - locally massive and amalgamated - locally planar laminated and bedded - normal grading, local reverse - local ripple cross-lamination - local soft-sediment deformation, boudinage - local dispersed clasts grading to matrix-supported conglomerate - marine macro- and microfossils	- medium- to high-density sediment gravity flow deposition - very fine to fine sandstones may be from settling from turbid plumes with high sediment concentrations - may be massive due to depositional processes or mixing by bioturbation, freeze/thaw, loading	- style of internal stratification and grading - degree of sorting - marine fossils
4. Moderately to well sorted, stratified fine sandstone	- local low angle cross-bedding and cross-lamination - locally planar, thin bedded to laminated - possible HCS - quartz rich, local coal laminae - locally with dark mudstone, bituminous - penecontemporaneous soft-sediment deformation - marine macro- and microfossils	- dilute tractional currents (within or about wave base to shoreface)	- style of internal stratification - particle size and sorting - marine fossils

Table 3. *continued.*

Facies number and name	Key sedimentological characteristics	Depositional process interpretation	Key interpretation criteria
5. Moderately to well sorted, stratified or massive, fine to coarse sandstone	- mostly medium-grained, locally fine or coarse - planar- to cross-stratified - locally massive and amalgamated - dispersed to abundant clasts - local gravelly layers at base - weak to moderate bioturbation - marine fossils	- marine currents/wave influence (perhaps shoreface) - local erosion with hiatuses - rainout from iceberg rafting	- particle size and sorting - style of internal stratification - bioturbation - marine fossils
6. Stratified diamictite	- clast-rich to clast-poor, sandy or muddy - a-axes locally aligned with stratification outsized clasts - stratification by: mudstone, siltstone, very fine to very coarse sandstone laminae; change in mean size in matrix sand; varying proportions of mud - commonly grade with massive diamictite - commonly interbedded with conglomerates, diamictites, sandstones and mudstones - locally strong soft-sediment deformation - locally includes marine macrofossils	- amalgamated or single debris-flow deposition - rainout with currents - subglacial deposition	- particle size and sorting - style of internal stratification and grading - style of contacts - associated facies - marine fossils
7. Massive diamictite	- clast-rich to clast-poor, sandy or muddy - graded contacts with conglomerate, sandstone and mudstone - or lower contact sharp (loaded and deformed) - rarely a-axes apparent preferred sub-horizontal orientation - locally includes marine macrofossils and lapilli	- subglacial deposition - amalgamated or single debris-flow deposition - rainout with currents	- particle size and sorting - style of contacts - associated facies - marine fossils

Table 3. *continued.*

Facies number and name	Key sedimentological characteristics	Depositional process interpretation	Key interpretation criteria
8. Rhythmically interstratified sandstone and siltstone	- very fine and fine sandstone sharply interstratified with mudstone - mudstone with discrete siltstone laminae - lonestones, dropstones and outsized clasts - often with Facies 2, 6 and 7	- suspension settling from turbid plumes - may include low density turbidity current deposition	- one-grain-thick lamina style of cyclopels and cyclopsams - graded style of turbidites
9. Clast-supported conglomerate	- massive, poorly sorted, locally graded - no clast orientation, some clasts angular - sharp lower contacts - gradational up into matrix-supported conglomerate to sandstone	- settling from submarine jet from subglacial streams - fluvial/shallow marine deposition - may include rainout of ice-rafted debris - redeposition of conglomerate by mass flow	- clast-support style - clast features - style of contacts
10. Matrix-supported conglomerate	- massive, very poorly sorted - angular clasts quite common - gradational into clast-supported conglomerate or sandstone	- high-density mass flow deposit - hyperconcentrated flows from submarine jet of subglacial stream - very local mass flow redeposition of fluvial sediment	- clast-support style - clast features - particle sorting - style of contacts
11. Mudstone breccia	- massive - intraclasts of mudstone, angular to subrounded - clast-supported - within soft-sediment deformed sequences	- suspension settling and rainout - mass flow deposition	- clast features - clast-support style - facies sequence

muddy sandstone and interstratified mudstone and sandstone, and laminated mudstone with decreasing glacial influence up-section (Fielding *et al.* 2001; Naish *et al.* 2001). The transition to fine-grained facies is gradational but rapid, in most cases occurring over a 1-m-thick interval. In the middle interval of the facies motif are relatively ice-distal, offshore shelf mudstones. Upper parts of the motif were described as a coarsening-upwards, regressive lithofacies succession from mudstone, to sandy mudstone, to well-sorted sandstone with increasing glacial influence. The succession was interpreted as being nearshore marine deposits primarily controlled by relative sea-level change, but with increasing glacial influence up-section (Fielding *et al.* 2001; Naish *et al.* 2001).

Variations in particle size and lithofacies up an entire sequence were thought to primarily reflect oscillations in depositional energy that were thought to be controlled by the combination of changes in shoreline and glacial proximity (Fielding *et al.* 2001; Naish *et al.* 2001). Typically, these sequences were interpreted as representing: (1) erosion and minor deposition during glacial advance followed by a fining upwards interval deposited during retreat and bathymetric deepening assigned to the lowstand/transgressive systems tract; (2) an interval of relatively ice-free, open marine pelagic sediment assigned to the highstand systems tract; and (3) a coarsening-upwards regressive succession representing glacial re-advance into a shallow marine setting, in association with bathymetric shoaling assigned to the regressive systems tract (Fielding *et al.* 2001; Naish *et al.* 2001). Although such interpretations may be a reasonable first approximation and appropriate for some successions, they do not necessarily represent all sequences. We prefer to establish a glacial sequence stratigraphic interpretation independent of eustatic sea-level changes.

A stratigraphic framework

General principles

Establishing a stratigraphic framework on which to interpret temperate glaciated continental margins is more complex than for lower latitude margins. Two major problems are: (1) the number of forces controlling accommodation space, and (2) documenting the causes of water depth variations. Previous studies (e.g. Powell 1984; Bednarski 1988; Boulton 1990; Powell & Alley 1997) have outlined the complex interaction of glacial advance and retreat,

eustatic sea level changes, glacial and sediment isostatic loading, rates and styles of sediment delivery to the sea, marine dispersal and redepositional processes, local tectonic movements, continental shelf morphology, and type of glacial terminus in controlling sediment accommodation space on temperate glaciated shelves. The complexity created by differences in rates, magnitudes and relative timing of changes in each of these factors is enhanced when trying to interpret a stratigraphic record in which palaeo-depth indicators are commonly few. Furthermore, changes in facies driven by glacial proximity v. relative sea-level changes are often debatable (e.g. Powell *et al.* 1998, 2000, 2001). This section attempts to establish that cycles of glacial advance and retreat probably act as a more important control on sedimentation and genesis of particular systems tracts than other forces such as relative sea-level. Specific points are:

(1) The timing of glacial fluctuations relative to eustatic sea-level changes are perhaps more critical in temperate glacial systems than for others because local glaciers may commonly fluctuate independently of major global events (e.g. Powell 1991). Furthermore, ice streams within major ice sheets may behave independently of eustatic changes (e.g. Hudson Bay ice stream producing Heinrich layer debris; e.g. MacAyeal 1993; Alley & MacAyeal 1994; Dowdeswell *et al.* 1995, 1999).

(2) Rapid sediment accumulation at temperate glacial grounding-lines is known to enhance their stability and reduce water depth without a change in relative sea-level or the location of the glacier terminus (Brown *et al.* 1982; Powell 1991; Warren *et al.* 1995; van der Veen 1996; Fischer & Powell 1998). Because the terminus is a vertical tidewater cliff, the calving line coincides with the grounding-line, and depositional systems can aggrade to and above sea level (Powell & Molnia 1989; Powell 1990; Plink-Björklund & Ronnert 1999). Furthermore, glacial advances are partly controlled by sediment accumulation and may occur independently of major eustatic changes (Powell *et al.* 1995), especially with surging glaciers (Solheim 1991).

(3) Bednarski (1988) and Boulton (1990; modified by Eyles & Eyles 1992) have evaluated the influence of crustal loading and rebound on marine sequences within a few hundred kilometres of a grounding line. They summarize what many workers have found by describing Quaternary coastal exposures in the northern hemisphere, that is, facies changes vary as a function of water depth changes due to a combination of the magnitude of glacial loading, and

the relative timing of glacial fluctuation v. relative sea-level change.

(4) Temperate glaciation on shelves can often be created by active tectonism, which causes vertical motions of significant rates and magnitude to influence sedimentation during an icehouse period as well as over shorter terms during glacial events (Meigs & Sauber 2000; Jaeger *et al.* 2001).

(5) On temperate glaciated shelves, major cross-shelf erosional surfaces may not be due to exposure during a lowstand in relative sea level, but rather record subglacial erosion during glacial advance. The glacial advance may coincide with global eustatic lowering but that need not be the case for every such surface. Furthermore, isostatic loading by a glacial advance may well be larger than the eustatic sea-level fall, thus not allowing a sufficient relative sea-level fall to erode a sequence boundary equivalent to a lower-latitude shelf.

Alaskan shelf architecture

Two distinctive styles of seismic sequence architecture are observed within Sparker records of the youngest Quaternary section of the Yakataga Formation from the southern Alaskan continental shelf. The two styles are designated Type IAK and Type IIAK (Figs 4 & 5), in which, AK stands for Alaska (following the style of Cooper *et al.* 1991). The AK suffix is used to distinguish these architectures from the classic lower-latitude sequences formed by eustatic sea-level oscillations, which are distinctly different.

A typical Type IAK sequence has a lower bounding surface that is erosional on the continental shelf and conformable down the continental slope (Figs 4 & 5; Table 4). At the shelf edge, an oblique to complex-sigmoid, slope-prograding package is overlain by semi-transparent topsets. Locally, clinoform reflections change from aggrading to prograding without an intervening unconformity. Architecture of the seismic facies association (SFA) above the lower bounding surface on the shelf, SFA-A, varies subtly from glacially-eroded cross-shelf troughs to adjacent shallow shelf banks (Fig. 5). On the banks, prograding clinoform reflections terminate up-dip within a chaotic to reflection-free hummocky facies which may thicken shoreward and form a semi-transparent, tabular sheet on the mid-shelf. Within troughs, clinoform reflections terminate within thin linear ridges of a semi-transparent seismic facies that pinches out on the outer shelf. Where the hummocky facies occurs, its upper surface is preserved as a hummocky upper bounding surface of this SFA-A.

The superjacent seismic facies association (SFA-B) also varies from banks to troughs. On banks, semi-transparent seismic facies form a fill within shelf basins and have prograding clinoform reflections; however, elsewhere within troughs, the semi-transparent facies may have a subtle mound morphology. This SFA is bounded above by a regional unconformity which truncates semi-transparent successions on the inner shelf. A seaward-thinning wedge of stratified seismic facies occurs above this unconformity on the inner to middle shelf.

This Alaskan shelf succession is interpreted as glacial advance/retreat cycles across the continental shelf, some to the shelf edge. In any one area, part or all of a succession may be eroded by the next glacial advance, depending on various scenarios of water depth on the continental shelf and glacial erosion processes. Glacial advance is represented on the shelf either as an unconformity or as a thin layer of till. Most sediment produced during advance is deposited at the shelf break, and consequently, that is the time of significant movement of the offlap break with progradation and/or aggradation. These changes in style represent changes in sediment delivery between shelf-basin infilling and sediment by-pass directly to the slope, indicating either a fall in eustatic sea level, tectonic uplift, sedimentary filling of accommodation space, or some combination of these factors prior to ice-sheet expansion onto the continental shelf.

An upper hummocky surface in seismic profiles represents the top of the retrogradational package of back-stacked, ice-contact morainal bank systems (SFA-A) formed at a retreating grounding line. That retrogradational package rests on the grounding-line retreat surface (GRS) above patchy subglacial till of the GAST. The hummocky surface is overlain by ice-proximal deposition (SFA-B) during full glacial retreat, interpreted as hemipelagic bergstone mud and diamicton from turbid plume sediments and iceberg-rafted debris deposited away from the grounding line (GRST). These are overlain by a wedge of paraglacial shelf sediments primarily derived from glacially-fed fluvial systems on land (GMiST). Nearshore the wedge coarsens upward from hemipelagic mud to prograding shoreline/deltaic sands that downlap onto the mud. The downlapped surface represents a time of maximum glacial retreat (maximum retreat surface; MRS), and is equivalent in time to being at or just following an offshore condensed section. The nearshore

Table 4. *Seismic facies characteristics and their interpretations as lithofacies*

Seismic facies No.	Acoustic stratification	Internal configuration	Reflection		External form	Boundary condition	Position & distribution	Interpretation (lithofacies)
			Continuity	Amplitude				
1	Stratified	Oblique-tangential	High to discontinuous	Medium–high	Slope fan	Sharp to diffuse erosional top	Continental slope	Debris-flow diamicton (DFo)
2	Stratified	Sigmoid	Medium–high	High	Channel fill	Sharp truncating top, downlap on to sharp bottom	Within troughs	Interbedded glacimarine and flow deposits (IGFs)
3	Stratified	Complex sigmoid-oblique	Medium to discontinuous	Medium–high	Slope fan	Sharp to diffuse top, erosional and toplap; no bottom	Continental slope	Combination of IGFs and DFo
4	Stratified	Hummocky clinoform	Low to medium	Medium	Lens	Gradational top and bottom	Trough walls	Debris flow deposit (DFh)
5	Stratified	Dipping-parallel	Medium	Medium	Sheet	Sharp top and bottom	Continental slope	Sediment gravity flow deposit (SGFs)
6	Unstratified	Semi-transparent	Locally chaotic, occasional discontinuous reflections, point-source diffractions		Tabular to hummocky sheets	Sharp hummocky top, sharp to diffuse bottom	Mid-shelf to outer shelf	Ice-marginal diamicton (IMf)
			Locally discontinuous reflections, point-source diffractions		Basin fill	Sharp top, sharp hummocky bottom	Mid-shelf basins above SF	Ice-proximal diamicton (IPf)
7	Unstratified	Chaotic	Local discontinuous, hummocky reflections		Lens	Sharp top and diffuse bottom	Forms sea floor shoal on outer shelf	Winnowed lag deposit (Wlc)

Note: The capital letters in the abbreviations refer to the lithofacies; and the lower case letters refer to their seismic facies.

sands may become subaerial and be truncated by fluvial sands and gravels. Locally, some GAST deposits may be preserved during the next glacial advance. This entire sequence is very similar to the sequence established for advance and retreat of temperate valley glaciers in fjords of the same region (Powell 1981, 1983, 1984) as discussed above, but the fjord successions are more likely to be eroded during younger glacial advances.

Type IIAK sequences consist of a sigmoidal, prograding facies association (SFA-C) on the continental shelf (Figs 4 & 5). This sequence is characterized by sigmoidal prograding reflections that converge to form a fan and overlap the lower sequence bounding surface. The top of the prograding facies may be truncated by the base of an overlying sequence. This sequence type may be preserved best within shelf troughs; however, it may also occur at the shelf edge. The sigmoidal prograding facies within shelf troughs grades conformably into discontinuous and contorted stratified facies on adjacent shelf banks where internal reflections are strongly aggrading. Internal high-amplitude reflectors within the discontinuous and contorted stratified facies on banks may have a mounded morphology. SFA-C is either overlain by a semi-transparent facies, which is then truncated by an erosion surface or SFA-C is directly truncated by that erosion surface.

Type IIAK sequences may be produced during minor ice-sheet expansion that covers part of a shelf. The resulting depositional systems are overlain by grounding-line systems as grounded glaciers advance over part of the inner shelf. During a period of terminus stability at the most advanced glacier position, effluxes from subglacial streams build an extensive fan of suspension settling and sediment gravity flow (mainly turbidite) deposits in cross-shelf troughs and ice-contact morainal banks and submarine outwash deposits on shelf banks. The youngest semi-transparent seismic facies, represents remnants of the retreat and para-glacial systems. Such sequences may be preserved in inner shelf areas when tectonic subsidence provides accommodation space prior to the next glacial advance. A succession that looks seismically similar to a Type IIAK may be produced by normal-marine coastal processes during a glacial minimum. That succession would be a combination of the last stages of a GMiST and first stages of the next GAST if there is no glacial advance into the sea. In that case, it would not be a separate sequence and is distinguished from the Type IIAK by the latter having ice-proximal and perhaps

ice-contact glacimarine systems (although not necessarily a GES) as part of the full package of systems tracts within it. Detailed facies analysis can distinguish these two cases.

Thus, data from the younger Yakataga Formation in continental shelf seismic profiles can be integrated with those from the older shelf/slope successions within Yakataga Formation in outcrop, data from modern fjords, uplifted Pleistocene sections, and a thick older Cenozoic core record from Antarctica. The result is a comprehensive temporal and spatial picture of temperate glacimarine sedimentation from which it is possible to create an idealized sequence stratigraphic framework for temperate glaciated shelves, which is described below.

An integrated model

3-D model

A three-dimensional conceptual model is presented that describes a Type IIAK sequence between two Type IAK sequences (Fig. 3) and is sketched at the GMiST phase of the youngest sequence. Each of the Type IAK sequences has the four glacial systems tracts, and the Type IIAK sequence is dominated by a progradational system which represents a partial glacial advance. Scaling is approximate but is taken from the modern Alaskan shelf. Only general facies are shown in this 3-D sketch: diamictons, grounding-line morainal bank systems, bergstone muds, and stratified deposits. Sequences are unconformable and conformable on the shelf and conformable on the upper slope. Facies motifs within the model are more clearly demonstrated in individual columns from different parts of the continental shelf and slope rather than in the 3-D model (Fig. 6). A full

Fig. 6. Selected individual facies motifs from different locations on the 3-D conceptual model for a Type IAK sequence (Fig. 3). They show a similar motif across the continental shelf but vary in style mainly due to the relative thickness of deposits and presence of some local depositional systems (e.g. coastal deltaic/shoreface system). The lithofacies columns were taken from the nearshore (**a**), mid-shelf (**b**), outer-shelf (**c**) and upper slope (**d**) areas. Lithofacies are depicted using particle size variations with major facies types and sedimentary structures as described in the key. Associated columns are a descriptive summary for a facies, an inferred depositional environment presented with name as well as a continuum curve, the major glacial systems tracts and bounding surfaces, and an inferred sediment accumulation rate.

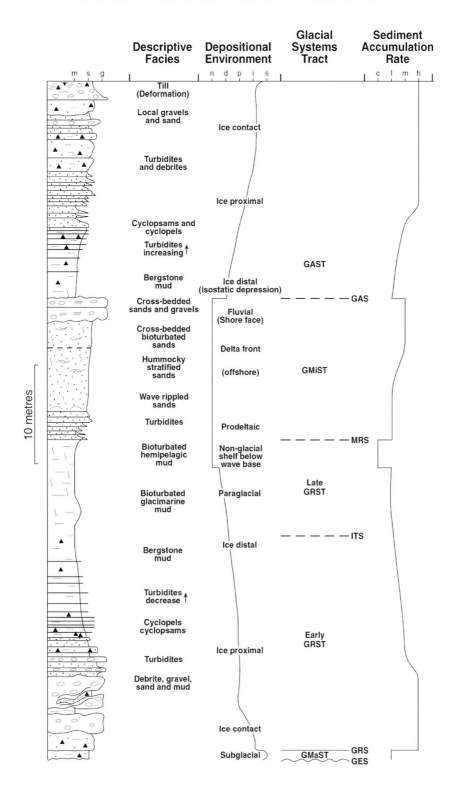

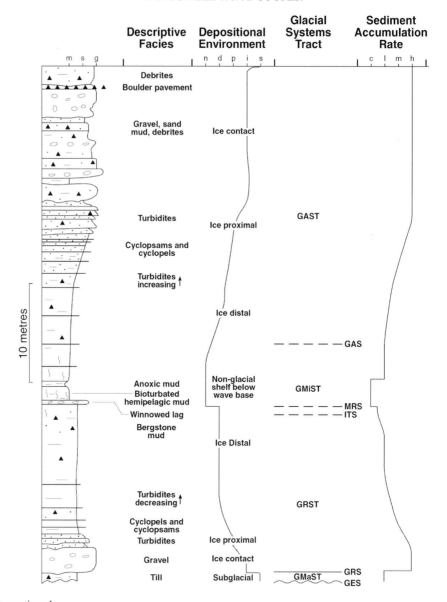

Fig. 6. continued.

glacial sequence with designated glacial systems tracts is described for each location of the inner, middle and outer shelf and the upper slope (Fig. 6), and will be discussed in more detail in later sections.

Cross-shelf troughs and marine fans are not included in the 3-D model for simplicity, but they do deserve brief comment. Troughs or sea valleys cutting across continental shelves are common on high latitude continental margins

and they are often associated with fan systems (e.g. Stevenson & Embley 1987; Vorren *et al.* 1989, 1998; Eittriem *et al.* 1995; Laberg & Vorren 1995; Dowdeswell *et al.* 1996, 1998; Vorren & Laberg 1997; Elverhøi *et al.* 1998*a*; Solheim *et al.* 1998; Anderson 1999; O'Brien *et al.* 1999). Troughs are thought to be sites of streaming ice under convergent flow, during stages of glacial advance, maximum and retreat over the shelf. Ice elsewhere on the continental

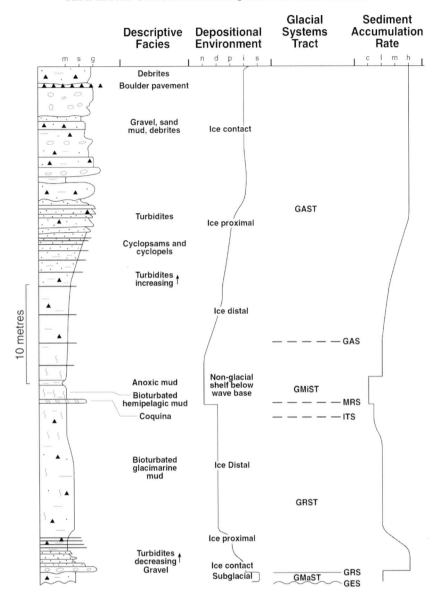

Fig. 6. continued.

shelf would be flowing more slowly, mostly under divergent flow. Fan systems on continental slopes are interpreted to be primarily active during glacial periods where a glacier can deliver sediment to them most directly as it is sitting at continental shelf-slope break. This is also the time of major progradation of the offlap break by sediment released directly at the grounding line.

Two varieties of marine fans have been described within the Pleistocene glacimarine record: (1) the common style of submarine fans being mainly composed of turbidites occurring at the toe of the continental slope and on the continental rise; and (2) trough-mouth fans, made of mainly debris flow deposits and occurring on the upper continental slope at the mouths of the cross-shelf troughs. The submarine fan types are dominant on the active tectonic margin of the Gulf of Alaska; the best

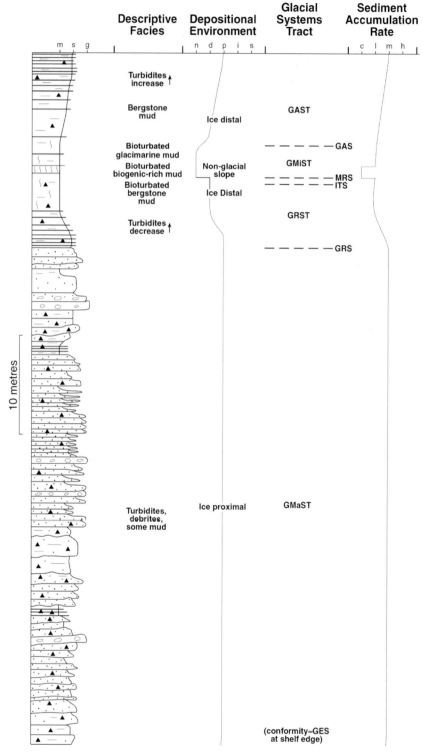

Fig. 6. continued.

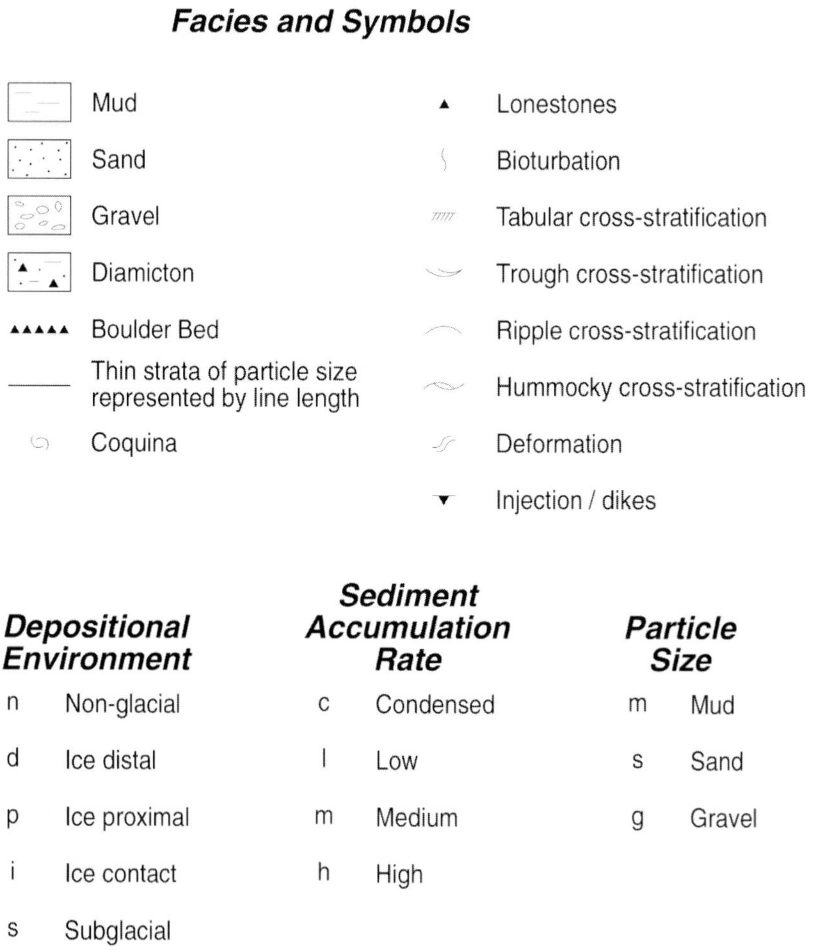

Facies and Symbols

Mud	
Sand	
Gravel	
Diamicton	
▲▲▲▲▲ Boulder Bed	
—— Thin strata of particle size represented by line length	
◡ Coquina	

▲	Lonestones
⌇	Bioturbation
⊞⊞	Tabular cross-stratification
⌣	Trough cross-stratification
⌒	Ripple cross-stratification
⌇	Hummocky cross-stratification
ʃ	Deformation
▼	Injection / dikes

Depositional Environment		**Sediment Accumulation Rate**		**Particle Size**	
n	Non-glacial	c	Condensed	m	Mud
d	Ice distal	l	Low	s	Sand
p	Ice proximal	m	Medium	g	Gravel
i	Ice contact	h	High		
s	Subglacial				

Fig. 6. Legend.

example being the Surveyor Fan. This huge, low relief fan was tectonically moved along the transform-faulted border of the Pacific Plate and as it passed glacier-filled drainage basins, the fan channels constantly changed on a large scale as they encountered each new sediment source (Stevenson & Embley 1987; Dobson *et al.* 1998). Glacial fluctuations superimposed another variable on sediment flux to the fan with glacial advance stages increasing fan growth. DSDP cores from the fan contain iceberg-rafted pebbles and mud interbedded with thin silt and sand turbidites (Kulm *et al.* 1973). Piston cores near and within the Surveyor Channel also contain turbidites (Ness & Kulm 1973). Glacial regime along the Gulf of Alaska coast during the period of fan formation is thought to have been similar to today, that is,

temperate with large volumes of meltwater producing a dominance of sorted glacimarine sediment (e.g. Powell 1984). In that regard, this modern system may be seen as representing similar deposits to part of the Yakataga Formation described above. Similar, but smaller, fans are being formed in some fjords today (Carlson *et al.* 1999), and a similar system exists off Hudson Strait as the NW Atlantic Mid-Ocean Channel (NAMOC) (Hesse *et al.* 1996, 1997, 1999; Klaucke & Hesse 1996; Wang & Hesse 1996; Hesse & Khodabakhsh 1998).

The upper slope successions evaluated in this study, both in seismic records of modern systems and in the inferred upper slope deposits of the Yakataga Formation, have successions of diamicton interpreted both as rainout and debris flow deposits. The latter debrites are the most common component facies of the second type of Pleistocene marine fan associated with glaciated margins, termed trough-mouth fans. These types of fans are mainly made of debris-flow diamictons fed from a subglacial till system, probably a deforming bed of a fast flowing or surging ice stream (e.g. the Wilkes Land margin, Antarctica (Eittreim *et al.* 1995) and Spitsbergen and East Greenland (Laberg & Vorren 1995, 2000; Dowdeswell *et al.* 1996, 1998; Vorren & Laberg 1997; Vorren *et al.* 1998; Elverhøi *et al.* 1998*a*; Dimakis *et al.* 2000)).

Based on their facies, the two varieties of fans are inferred to be controlled by glacial regime and bed conditions in terms of the location and volume of meltwater (Powell & Alley 1997). In polar, and some subpolar, situations where large ice sheets extend over continental shelves, much of the subglacial system appears to lack free-flowing water in major conduits and is retained as pore water within subglacial sediment. In that case, most sediment is transported by subglacial deformation and it has a diamict texture, a till. In contrast, temperate glaciers such as those in Alaska, have an abundance of free-flowing water in subglacial conduits and they discharge sorted sediment from point sources at a terminus to produce a majority of glacimarine deposits (cf. Hunter *et al.* 1996). This is not to say that on occasions, conduit flow does not occur under polar ice sheets, nor that they cannot produce turbidite fans. In fact, where polar ice sheets rest on bedrock and ground-water storage in till is not possible, conduit systems have been described, as have turbidite fans (e.g. Anderson 1999; Shipp *et al.* 1999). Similarly, temperate systems can have diamicton facies with deforming beds. However, it is a matter of relative dominance of the lithofacies produced under each regime. In fact, some of the glaciers that produced trough-mouth fans during advance appear to have changed their style of deposition on retreat to produce morainal-bank style grounding-line deposits on the continental shelf (e.g. King *et al.* 1991). We suggest these bank deposits could be used to indicate a warmer glacial regime associated with the retreat phase.

The upper slope deposits of our model are more sheet-like than fan-like, and stratified successions are interbedded with diamictons. During some intervals, stream discharges produce stratified deposits, whereas on other occasions, the area is dominated by a sheet of deforming subglacial till issuing along a grounding line acting as a line-source for debris flows. These alternations may represent changing bed conditions, at least locally through the period

when the grounding line was at the shelf edge. In the model, stratified facies dominate with time.

Facies sequences

Individual facies sequences (Fig. 6) have a similar motif across the continental shelf but vary in style mainly due to the relative thickness of deposits and presence of some local depositional systems (e.g. coastal deltaic/shoreface system). A GES is taken as the sequence boundary at the base of each facies motif. A GES may well occur within a diamicton succession and not necessarily at its base (e.g. Fig. 6a), or there may be no subglacial till above a GES as occurs in Alaskan fjords and on the shelf. Furthermore, a GES may be absent at the base of any succession on the shelf if the glacier does not fully advance across the shelf, which creates a Type IIAK sequence boundary as shown in Figure 3. That situation has also been documented elsewhere such as for 'till tongues' or grounding-line wedges on the shelf; King *et al.*, 1991; Powell and Alley 1997; Anderson 1999; Shipp *et al*, 1999). A slope succession also most likely lacks a GES; such a succession having a continuous record at the same time a GES forms on the shelf.

Above the GES is the GMaST, which in temperate glacial systems is commonly thin or absent. Most commonly the GMaST is represented by subglacial till, which is mainly aggradational in character, although it may be partly progradational at the maximum glacial extent (Fig. 3). The thickest succession of GMaST occurs on the upper slope where interstratified turbidites, debrites and bergstone mud or diamicton prograde as clinoforms (Figs 3 & 6), often creating significant seaward movement of the offlap break. Note that this seaward movement of the offlap break is not necessarily tied to sea level.

Above the GMaST is the GRS, which if till is absent, may be indistinguishable from the GES. A GRS may be equivalent to a drowning (flooding or transgressive) surface in lower latitude sequences with one important distinction, there is no abrupt deepening across the surface apparent in facies. However, an abrupt facies dislocation may occur on the slope (Fig. 6d) where the system starts to become starved of coarser sediment, which is now being stored on the shelf, and becomes more mud-rich.

The GRST rests on the GRS and includes an overall fining upwards succession of retrogradationally stacked grounding-line deposystems below aggradational ice-proximal to ice-distal systems. Although diamictic debrites occur in grounding-line deposystems, the early phase GRST succession is dominated by sorted deposits of breccias, conglomerates, and poorly sorted sandstones due to the dominance of glacial meltwater. These grounding-line systems may also include prominent glacitectonic deformation structures and have a characteristic sharp, hummocky upper surface. The ice-proximal deposits that lie on this surface include glacimarine rhythmites of thin debrites, turbidites, cyclopsams and cyclopels and perhaps iceberg-rafted varves in fan to sheet geometries. Above these are draped sheets of bergstone muds. The thickness of all of these deposits depends on sediment fluxes, but it is also a function of the rate of grounding-line retreat, i.e. packages are thinner during faster retreat. Furthermore, the degree of bioturbation follows the same trends where it is more intense with low fluxes or fast retreat. If the terminus has extended still-stands during retreat, the grounding-line systems can aggrade significantly, and can be dominated by fluvial processes and fluxes. As retreat is renewed and these grounding-line systems are abandoned, they stand high above the rest of the shelf and due to their hardground character, they are sites of epibenthic colonization. Other GRST sediment may pond between the banks, but eventually they may be overlain sharply by a condensed section of thin bergstone mud and paraglacial mud. In shallow water sites, or with isostatic rebound, these sites, may be reworked into lag deposits (e.g. Smith 1982; Bednarski 1988; Retelle & Bither 1989; Hunter et al. 1996) or be the source for re-deposited coquina beds.

Iceberg rafting continues to contribute IBRD to the succession until the glacier terminus is so far away that all bergs are melted before reaching a location or the terminus has retreated onto land. The decrease in IBRD may be especially affected when glaciers retreat into fjords where the larger bergs can be trapped easily behind entrance sills. The last occurrence of IBRD can be recognized in outcrops, in cores and in core logs, and is defined as the iceberg-rafting termination surface (ITS). In offshore successions this surface may not carry strong significance and is not used as a bounding surface of systems tracts. In these offshore areas the ITS could indicate lateral changes in circulation systems on the shelf in terms of altering the directions of berg drift. However, the major utility of recognizing the ITS occurs in nearshore sections (Fig. 6a) where it can be used to indicate the time at which the terminus receded from tidewater onto land.

In nearshore settings the late phase GRST above the ITS is represented by paraglacial mudstones that lack IBRD and may be more intensely bioturbated (Fig. 6a). It is the interval of time at these nearshore sites when sedimentation rates are at their lowest. With isostatic rebound, this same time interval may be represented in offshore sections by winnowed lag deposits or coquinas (Fig. 6b, c), or if sites remain in deeper water, condensed sections of intensely bioturbated hemipelagic mud occur.

The GRST terminates at the MRS, which is often sharp nearshore and sharp or gradational offshore. Nearshore an abrupt facies change occurs across the MRS from hemipelagite to prodeltaic turbidites, which downlap onto the surface as part of the lower section of prograding clinoforms of paralic deltaic or siliciclastic shoreface systems (Fig. 6a). Offshore, this is a period of low sedimentation rates, being represented by condensed sections with possible anoxia between high bank forms, depending on water depth and wave climate (Fig. 6b, c & d). These facies motifs represent the GMiST when glaciers are landward of the shelf or onshore from the head of fjords in their minimum phase.

The glacial advance surface (GAS) nearshore may be sharp and represented by another abrupt facies dislocation with drowning due to isostatic depression by glacial re-loading. The coarser-grained deposits of the GMiST wedge are overlain abruptly by bergstone muds. In some circumstances the initial indication of glacial advance may be recorded in the progradational wedge where the rate of progradation may increase due to increasing sediment flux as the glacier nears the coast. If crustal loading by the glacier occurs during this phase, and there is sufficient sediment supply, the wedge then may show an aggradation of facies. Offshore the change is more gradational and is represented by an increase in hemipelagic mud sedimentation. However, offshore the strongest evidence of advance is the first appearance of IBRD, and this lowest occurrence is taken as the GAS.

Above the GAS is the GAST, being represented by a coarsening-upward succession that mirrors the earliest phase of the GRST. The succession grades from distal bergstone mud, to ice-proximal turbidites, cyclopels and cyclopsams that coarsen and thicken, and finally include increasing diamictic debrites. This package changes upward from aggradational to progradational.

Although the full section is presented in these idealized motifs (Fig. 6), the GAST, as well as

Table 5. *Characteristics of temperate glaciated continental margins compared with polar and subpolar (polythermal) glaciated shelves, and non-glaciated shelves*

Temperate glaciated shelves	Non-temperate glaciated shelves	Low latitude non-glaciated shelves
Active tectonism	Active tectonism not required	Active tectonism not required
Very dynamic climatic systems	Climatic systems need not be dynamic	Climatic systems need not be dynamic
Debris fluxes and erosion rates are very high	Debris fluxes and erosion rates lower	Debris fluxes and erosion rates can be much lower
Rapid growth of continental margin	Slower growth of continental margin	Can be much slower growth of continental margin
Deposition of thick, rapidly deposited sequences	Sequences more slowly deposited	Sequences more slowly deposited
Erosional feedback enhancing isostatic uplift of mountains	Locally may have feedback	Locally may have feedback
Enhanced local relief by over-deepening	Enhanced local relief by over-deepening	Less relief creation by fluvial systems over same time interval
Enhanced accommodation space by glacial isostasy (plus water and sediment loading), and tectonics	Accommodation space by glacial isostasy (plus water and sediment loading), not necessarily tectonics	Accommodation space by isostatic water and sediment loading, and tectonics
Sequences maybe preserved on shelf due to enhanced accommodation	Sediment preservation potential may not be as high	Sediment preservation potential independent of significant glacial erosion
Facies changes may occur by glacial advance and retreat	Facies changes may occur by glacial advance and retreat	Facies changes occur due to changes in relative sea-level
Thick interglacial wedges on the shelf	May not have thick interglacial wedges on the shelf	Not applicable
Upper continental slope includes sorted sediment	Debris flow diamicts dominate upper continental slope	Not applicable

parts of the older systems tracts, may well be eroded at the next GES. Preservation of the full sequence is enhanced by rapid tectonic subsidence of the shelf to take it away from the area of intense glacial erosion; that is, beyond the depth of glacial buoyancy. Rapid tectonic subsidence also enhances the accommodation space for the sequences. The upper-most sediments preserved below the GMaST of the next sequence may exhibit soft-sediment deformation due to glacial over-riding. It then becomes a question as to where the exact GES is, because many include the glacially deformed sediment as subglacial till, which places the GES at the base of the subglacially deformed unit.

Conclusion and perspective

Temperate, glaciated continental shelves typically occur at low-latitudes, with glaciers relying on active tectonism and very dynamic climatic systems to maintain their presence at sea level. Under these conditions, debris fluxes and inferred erosion rates are very high (Powell 1991; Hallet *et al.* 1996; Elverhøi *et al.* 1998*b*) which provide for a significant source of sediment for growth of a continental margin and deposition of thick stratigraphic sequences. In addition, these settings provide for positive feedbacks to the whole continental margin system in that the high erosion rates may well enhance isostatic uplift of the mountains (Meigs & Sauber 2000; Jaeger *et al.* 2001) and the glaciers may enhance local relief by increasing erosion at lower elevations as shown by over-deepened fjords and a higher debris flux with larger drainage basins (Powell 1991). The tectonic regime of either compressional/transpressional or extensional/transtensional can also provide significant space in which to accommodate the large volumes of sediment produced rapidly from high sediment fluxes and high accumulation rates. Sediment can accumulate to hundreds of metres thick over very short time periods of tens (locally) to hundreds-to-thousands of years. Although the shelf is eroded or by-passed during advance, with most sediment originating in fjords and foreland areas to then be redeposited on the slope, older sequences may still be preserved on the shelf. However, the shelf preserves most of its sediment during the subsequent retreat.

Furthermore, sea level on glaciated continental shelves is not necessarily the only control on base level, the glacier sole is as well. On temperate glaciated shelves, changes in relative sea-level are a function of glacial isostasy as well as the more common isostatic forces (water and sediment loading), tectonism, and local water-depth controls of local erosion and sediment accumulation rates. Within these controlling forces, facies changes may occur simply by a glacial advance and retreat that is independent of other variables. Even with a significant water depth on the shelf provided by tectonic forces, a glacier can supply sediment to fill that space, and even aggrade to above sea level if the terminus stabilizes for some time, during one glacial advance into the sea without any change in eustatic sea level.

These characteristics make temperate glaciated margins not only different from non-glaciated shelves, but also different to polar and subpolar (polythermal) glaciated shelves (Table 5). The active tectonic setting is not a necessary condition for continental shelf glaciation in other glacial regimes, and thus the same sediment fluxes and accommodation space factors need not be as significant in those regimes. Glacial regimes other than temperate may not have thick interglacial wedges on the shelf. Furthermore, if active tectonic subsidence is not necessary for continental shelf glaciation, then sediment preservation need not be as high, because previously deposited sediment is not taken out of the main erosion depth of a readvancing glacier as rapidly. In addition, in the case of non-temperate margins, upper continental slope sediments may be dominated more by debris flow diamictons than sorted sediment.

This stratigraphic sequence model may be used to compare direct glacial records of ice volume changes through time with more distal proxy records such as global eustatic sea-level curves and oxygen isotopes. After all, on high latitude margins it is the glacial signal that is strongest and may be compared with the more distal proxy records of global ice volume changes. Eustatic sea level history is probably more easily established from non-glaciated shelves. However, if, after establishing a glacial sequence stratigraphy, it is possible to also establish temporal changes in local water depth and their forcings, then it would be possible to infer both glacial fluctuations and relative sea level change, and that could be ultimately the definitive case.

This paper is an out-growth of research studies in Alaska that have been made possible with the help of many students and colleagues, and supported by the US National Science Foundation. Paul Carlson of the US Geological Survey is recognized for his continued support over many years in providing geophysical equipment and data that have been used in those Alaskan studies. The Cape Roberts Project research in Antarctica was also supported by the US National Science Foundation and discussions of glacial sequences with project scientists, especially Tim Naish, have stimulated the development of ideas presented herein. The manuscript has been improved by comments made by reviewers Owen Sutcliff, Colm Ó Cofaigh and John Anderson.

References

ALLEY, R. B. & MACAYEAL, D. R. 1994. Ice-rafted debris associated with binge-purge oscillations of the Laurentide Ice Sheet. *Paleoceanography*, **9**, 503–511.

ANDERSON, J. B. 1999. *Antarctic Marine Geology*. Cambridge University Press, Cambridge.

ARMENTROUT, J. M. 1983. Glacial lithofacies of the Neogene Yakataga Formation, Robinson Mountains, southern Alaska Coast Range, Alaska. *In*: MOLNIA, B. F. (ed.) *Glacial-Marine Sedimentation*. Plenum Publishing Co., New York, 629–666.

BEDNARSKI, J. 1988. The geomorphology of glaciomarine sediments in a high arctic fjord. *Géographie Physique et Quaternaire*, **42**, 65–74.

BELKNAP, D. F. & SHIPP, R. C. 1991. Seismic stratigraphy of glacial marine units, Maine inner shelf. *In*: ANDERSON, J. B. & ASHLEY, G. M. (eds) *Glacial Marine Sedimentation – Paleoclimatic Significance*. Geological Society of America, Special Paper, **261**, 137–158.

BENN, D. I. & EVANS, D. J. A. 1996. The interpretation and classification of subglacially-deformed materials. *Quaternary Science Reviews*, **15**, 23–52.

BENNETT, M. R., HAMBREY, M. J., HUDDART, D., GLASSER, N. F. & CRAWFORD, K. 1999. The landform and sediment assemblage produced by a tidewater glacier surge in Kongsfjorden, Svalbard. *Quaternary Science Reviews*, **18**, 1213–1246.

BOULTON, G. S. 1990. Sedimentary and sea level changes during glacial cycles and their control on glacimarine facies architecture. *In*: DOWDESWELL, J. A. & SCOURSE, J. D. (eds) *Glacimarine Environments: Processes and Sediments*. Geological Society, London, Special Publications, **53**, 15–52.

BOULTON, G. S., VAN DER MEER, J. J. M., BEETS, D. J. & RUEGG, G. H. J. 1999. The sedimentary and structural evolution of a recent push moraine complex; Holmstrombreen, Spitsbergen. *Quaternary Science Reviews*, **18**, 339–371.

BROWN, C. S., MEIER, M. F. & POST, A. 1982. *The calving relation of Alaskan tidewater glaciers with application to Columbia Glacier.* US Geological Survey Professional Paper **1258-C**.

BRUNS, T. R. & SCHWAB, W. C. 1983. *Structure maps and seismic stratigraphy of the Yakataga segment of the continental margin, northern Gulf of Alaska.* US Geological Survey Miscellaneous Field Studies Map **MF-1424**, scale 1: 250 000.

CAI, J., POWELL, R. D., COWAN, E. A. & CARLSON, P. R. 1997. Lithofacies and seismic reflection interpretations of temperate glacimarine sedimentation in Tarr Inlet, Glacier Bay, Alaska. *Marine Geology*, **145**, 5–37.

CARLSON, P. R. & MOLNIA, B. F. 1977. Submarine faults and slides on the continental shelf, northern Gulf of Alaska. *Marine Geotechnology*, **2**, 275–290.

CARLSON, P. R., COWAN, E. A., POWELL, R. D. & CAI, J. 1999. Growth of a post Little-Ice-Age submarine fan, Glacier Bay, Alaska. *Geo-marine Letters*, **19**, 227–236.

COOPER, A. K., BARRETT, P. J., HINZ, K., TRAUBE, V., LEITCHENKOV, G. & STAGG, H. M. J. 1991. Cenozoic prograding sequences of the Antarctic continental margin: a record of glacio-eustatic and tectonic events. *Marine Geology*, **102**, 175–213.

DIMAKIS, P., ELVERHØI, A., HOEG, K., SOLHEIM, A., HARBITZ, C., LABERG, J. S., VORREN, T. O. & MARR, J. 2000. Submarine slope stability on high-latitude glaciated Svalbard-Barents Sea margin. *Marine Geology*, **162**, 303–316.

DOBSON, M. R., O'LEARY, D. & VEART, M. 1998. Sediment delivery to the Gulf of Alaska: source mechanisms along a glaciated transform margin. *In*: STOKER, M. S., EVANS, D. & CRAMP, A. (eds) *Geological Processes on Continental Margins: Sedimentation, Mass-Wasting and Stability*. Geological Society, London, Special Publications, **129**, 43–66.

DOMACK, E. W., JACOBSON, E. A., SHIPP, S. & ANDERSON, J. B. 1999. Late Pleistocene-Holocene retreat of the West Antarctic ice-sheet system in the Ross Sea, Part 2, Sedimentologic and stratigraphic signature. *Bulletin of the Geological Society of America*, **111**, 1517–1536.

DOWDESWELL, J. A., ELVERHØI, A. & SPIELHAGEN, R. 1998. Glacimarine sedimentary processes and facies on the polar North Atlantic margins. *Quaternary Science Reviews*, **17**, 243–272.

DOWDESWELL, J. A., MASLIN, M. A., ANDREWS, J. T. & McCAVE, I. N. 1995. Iceberg production, debris rafting, and the extent and thickness of Heinrich layers (H1, H-2) in North Atlantic sediments. *Geology*, **23**, 301–304.

DOWDESWELL, J. A., KENYON, N., ELVERHØI, A., LABERG, J. S., HOLLANDER, F.-J., MIENERT, J. & SIEGERT, M. J. 1996. Large-scale sedimentation on glacier-influenced polar North Atlantic margins: long-range side-scan sonar evidence. *Geophysical Research Letters*, **23**, 3535–3538.

DOWDESWELL, J. A., ELVERHØI, A., ANDREWS, J. T. & HEBBELN, D. 1999. Asynchronous deposition of ice-rafted layers in the Nordic seas and North Atlantic Ocean. *Nature*, **400**, 348–351.

EITTREIM, S. L., COOPER, A. K. & WANNESSON, J. 1995. Seismic stratigraphic evidence of ice-sheet advances on the Wilkes Land margin of Antarctica. *Sedimentary Geology*, **96**, 131–156.

ELVERHØI, A., DOWDESWELL, J. A., FUNDER, S., MANGERUND, J. & STEIN, R. 1998*a*. Glacial and oceanic history of the polar North Atlantic margins: an overview. *Quaternary Science Reviews*, **17**, 1–10.

ELVERHØI, A., HOOKE, R. LeB. & SOLHEIM, A. 1998*b*. Late Cenozoic erosion and sediment yield from Svalbard-Barents Sea region: implications for understanding erosion of glacierized basins. *Quaternary Science Reviews*, **17**, 209–241.

EYLES, N. & EYLES, C. H. 1992. Glacial Depositional Systems. *In*: WALKER, R. G. & JAMES, M. P. (eds) *Facies Models: Response to Sea Level Change*. Geological Association of Canada, 73–100.

EYLES, C. H. & LAGOE, M. B. 1990. Sedimentation patterns and facies geometries on a temperate glacially-influenced continental shelf: the Yakataga Formation, Middleton Island, Alaska. *In*: DOWDESWELL, J. A. & SCOURSE, J. D. (eds) *Glacimarine Environments: Processes and Sediments*. Geological Society, London, Special Publications, **53**, 363–386.

EYLES, C. H. & LAGOE, M. B. 1998. Slump-generated megachannels in the Pliocene-Pleistocene glacimarine Yakataga Formation, Gulf of Alaska. *Bulletin of the Geological Society of America*, **110**, 395–408.

EYLES, C. H., EYLES, N. & LAGOE, M. B. 1991. The Yakataga Formation; A late Miocene to Pleistocene record of temperate glacial marine sedimentation in the Gulf of Alaska. *In*: ANDERSON, J. B. & ASHLEY, G. M. (eds) *Glacial Marine Sedimentation—Paleoclimatic Significance*. Geological Society of America, Special Paper, **261**, 159–180.

EYLES, N., VOSSLER, S. M. & LAGOE, M. B. 1992. Ichnology of a glacially influenced continental shelf and slope: the Late Cenozoic Gulf of Alaska (Yakataga Formation). *Palaeogeography, Palaeoecology, Palaeoclimatology*, **94**, 193–221.

FISCHER, M. P. & POWELL, R. D. 1998. A simple model for the influence of push morainal banks on the calving and stability of glacial tidewater termini. *Journal of Glaciology*, **44**, 31–41.

FIELDING, C. R., NAISH, T. R. & WOOLFE, K. J. 2001. Facies architecture of the CRP-3 drillhole, Victoria Land Basin, Antarctica. *Terra Antarctica*, **8**, 217–224.

GLASSER, N. F. & HAMBREY, M. J. 2001. Tidewater glacier beds: insights from iceberg debris in Kongsfjorden, Svalbard. *Journal of Glaciology*, **47**, 295–302.

HALLET, B., HUNTER, L. & BOGEN, J. 1996. Rates of erosion and sediment evacuation by glaciers: A review of field data and their implications. *Global and Planetary Change*, **12**, 213–235.

HAMBREY, M. J., BARRETT, P. J., EHRMANN, W. U. & LARSEN, B. 1992. Cenozoic sedimentary processes on the Antarctic continental shelf: the record from deep drilling. *Zeitschrift für Geomorphologie.*, Suppl. Bd., **86**, 73–99.

HART, J. K. & ROBERTS, D. H. 1994. Criteria to distinguish between subglacial glaciotectonic and glaciomarine sedimentation, I. Deformation styles and sedimentology. *Sedimentary Geology*, **91**, 191–213.

HESSE, R. & KHODABAKHSH, S. 1998. Depositional facies of late Pleistocene Heinrich events in the Labrador Sea. *Geology*, **26**, 103–106.

HESSE, R., KLAUCKE, I., RYAN, W. B. F., EDWARDS, M. B. & PIPER, D. J. W. 1996. Imaging Laurentide ice sheet drainage into the deep sea; impact on sediments and bottom water. *GSA Today*, **6** (9), 3–9.

HESSE, R., KLAUCKE, I., RYAN, W. B. F. & PIPER, D. J. W. 1997. Ice-sheet sourced juxtaposed turbidite systems in Labrador Sea. *Geoscience Canada*, **24** (1), 3–12.

HESSE, R., KLAUCKE, I., KHODABAKHSH, S. & PIPER, D. 1999. Continental slope sedimentation adjacent to an ice margin; III, The upper Labrador Slope. *Marine Geology*, **155**, 249–276.

HUNTER, L. E., POWELL, R. D. & SMITH, G. W. 1996. Facies architecture and grounding-line fan processes of morainal banks during the deglaciation of coastal Maine. *Bulletin of the Geological Society of America*, **108**, 1022–1038.

JAEGER, J., HALLET, B., PAVLIS, T., SAUBER, J., LAWSON, D., MILLIMAN, J., POWELL, R., ANDERSON, S. & ANDERSON, R. 2001. Orogenic and glacial research in pristine southern Alaska. *Eos*, **82**(19), 213–216.

KING, L. H., ROKOENGEN, K., FADER, G. B. J. & GUNLEIKSRUD, T. 1991. Till-tongue stratigraphy. *Bulletin of the Geological Society of America*, **103**, 637–659.

KLAUCKE, I. & HESSE, R. 1996. Fluvial features in the deep-sea: new insights from the glacigenic submarine drainage system of the Northwest Atlantic mid-ocean channel in the Labrador Sea. *Sedimentary Geology*, **106**, 223–234.

KULM, L. D., VON HUENE, R., DUNCAN, J. R. ET AL. 1973. *Initial Reports Deep Sea Drilling Project 18.* U.S. Government Printing Office, Washington D.C., 1077 pp.

LABERG, J. S. & VORREN, T. O. 1995. Late Weichselian submarine debris flow deposits on the Bear Island Trough mouth fan. *Marine Geology*, **127**, 45–72.

LABERG, J. S. & VORREN, T. O. 2000. Flow behaviour of the submarine glacigenic debris flows on the Bear Island Trough Mouth Fan, western Barents Sea. *Sedimentology*, **47**, 1105–1117.

LØNNE, I. 1995. Sedimentary facies and depositional architecture of ice-contact glaciomarine systems. *Sedimentary Geology*, **98**, 13–43.

LØNNE, I. 2001. Dynamics of marine glacier termini read from moraine architecture. *Geology*, **29**, 199–202.

LØNNE, I. & LAURITSEN, T. 1996. The architecture of a modern push-moraine at Svalbard as inferred from ground-penetrating radar measurements. *Arctic and Alpine Research*, **28**, 488–495.

LØNNE, I. & SYVITSKI, J. P. M. 1997. Effects of the readvance of an ice margin on the seismic character of the underlying sediment. *Marine Geology*, **143**, 81–102.

MACAYEAL, D. R. 1993. Binge/purge oscillations of the Laurentide Ice Sheet as a cause of the North Atlantic's Heinrich events. *Paleoceanography*, **8**, 775–784.

MCCABE, A. M. 1986. Glaciomarine facies deposited by retreating tidewater glaciers; an example from the late Pleistocene of Northern Ireland. *Journal of Sedimentary Petrology*, **56**, 880–894.

MCCABE, A. M. & Ó COFAIGH, C., 1995. Late Pleistocene morainal bank facies at Greystones, eastern Ireland: an example of sedimentation during ice-marginal re-equilibration in an isostatically depressed basin. *Sedimentology*, **42**, 647–664.

MCCABE, A. M., DARDIS, G. F. & HANVEY, P. M. 1984. Sedimentology of a late Pleistocene submarine-moraine complex, County Down, Northern Ireland. *Journal of Sedimentary Petrology*, **54**, 716–730.

MEIGS, A. & SAUBER, J. 2000. Southern Alaska as an example of the long-term consequences of mountain building under the influence of glaciers. *Quaternary Science Reviews*, **19**, 1543–1562.

NAISH, T. R., BARRETT, P. J., DUNBAR, G. P., WOOLFE, K. J., DUNN, A. G., HENRYS, S. A., CLAPS, M., POWELL, R. D. & FIELDING, C. R. 2001. Sedimentary cyclicity in CRP drillcore, Victoria Land Basin, Antarctica. *Terra Antartica*, **8**, 225–244.

NESS, G. E. & KULM, L. D. 1973. Origin and development of the Surveyor deep-sea channel. *Bulletin of the Geological Society of America*, **84**, 3339–3354.

O'BRIEN, P. E., DE SANTIS, L., HARRIS, P. T., DOMACK, E. & QUILTY, P. G. 1999. Ice shelf grounding zone features of western Prydz Bay, Antarctica; sedimentary processes from seismic and sidescan images. *Antarctic Science*, **11**, 78–91.

PLAFKER, G. & ADDICOTT, W. O. 1976. Glaciomarine deposits of Miocene through Holocene age in the Yakataga Formation along the Gulf of Alaska margin, Alaska. *In*: MILLER, T. P. (ed.) *Recent and Ancient Sedimentary Environments in Alaska.* Proceedings of the Alaska Geological Symposium, **Q1-Q23**.

PLINK-BJÖRKLUND, P. & RONNERT, L. 1999. Depositional processes and internal architecture of Late Weichselian ice-margin submarine fan and delta settings. *Sedimentology*, **46**, 215–234.

POWELL, R. D. 1981. A model for sedimentation by tidewater glaciers. *Annals of Glaciology*, **2**, 129–134.

POWELL, R. D. 1983. Glacial-marine sedimentation processes and lithofacies of temperate tidewater glaciers, Glacier Bay, Alaska. *In*: MOLNIA, B. F. (ed.) *Glacial-Marine Sedimentation.* Plenum Publishing Co., 185–232.

POWELL, R. D. 1984. Glacimarine processes and inductive lithofacies modelling of ice shelf and tidewater glacier sediments based on Quaternary examples. *Marine Geology*, **57**, 1–52.

POWELL, R. D. 1990. Grounding-line fans and their growth to ice-contact deltas. *In*: DOWDESWELL,

J. A. & SCOURSE, J. D. (eds) *Glacimarine Environments: Processes and Sediments*. Geological Society, London, Special Publications, **53**, 53–73.

POWELL, R. D. 1991. Grounding-line systems as second order controls on fluctuations of temperate tidewater termini. *In*: ANDERSON, J. B. & ASHLEY, G. M. (eds) *Glacial Marine Sedimentation – Paleoclimatic Significance*. Geological Society of America, Special Paper, **261**, 75–94.

POWELL, R. D. & ALLEY, R. B. 1997 Grounding-line systems: Processes, glaciological inferences and the stratigraphic record. *In*: BARKER, P. F. & COOPER, A. C. (eds) *Geology and Seismic Stratigraphy of the Antarctic Margin, 2*. Antarctic Research Series, AGU, Washington, DC, **71**, 169–187.

POWELL, R. D. & DOMACK, E. W. 1995. Glaciomarine Environments, Chapter 13. *In*: Menzies, J. (ed.) *Glacial Environments – Processes, Sediments and Landforms*. Butterworth-Heinemann, Boston, 445–486.

POWELL, R. D. & MOLNIA, B. F. 1989. Glacimarine sedimentary processes, facies, and morphology on the south-southeast Alaska shelf and fjords. *Marine Geology*, **85**, 359–390.

POWELL, R. D., CAI, J. & COWAN, E. A. 1995. Marine record from a major Little Ice Age advance of a marine-ending temperate ice stream, Alaska. *Eos* (Fall AGU), **76**(46), 290.

POWELL, R. D., HAMBREY, M. J. & KRISSEK, L. A. 1998. Quaternary and Miocene glacial and climatic history of the Cape Roberts drill site region, western Ross Sea, Antarctica. *Terra Antartica*, **5**, 341–351.

POWELL, R. D., KRISSEK, L. A. & VAN DER MEER, J. J. M. 2000. Depositional environments of Cape Roberts 2/2A drill core, Antarctica: Palaeoglaciological and palaeoclimatic inferences. *Terra Antartica*, **7**, 313–322.

POWELL R. D., LAIRD, M. G., NAISH, T. R., FIELDING, C. R., KRISSEK, L. A. & VAN DER MEER, J. J. M. 2001. Depositional environments for strata cored in CRP-3 (Cape Roberts Project), Victoria Land Basin Antarctica: palaeoglaciological and palaeoclimatological inferences. *Terra Antartica*, **8**, 207–216.

RETELLE, M. J. & BITHER, K. M. 1989. Late Wisconsinan glacial and glaciomarine sedimentary facies in the lower Androscoggin Valley, Topsham, Maine. *In*: TUCKER, R. D. & MARVINNEY, R. G. (eds) *Maine Geological Survey, Studies in Maine Geology*. **6**, 33–51.

SERAMUR, K. C., POWELL, R. D. & CARLSON, P. R. 1997. Evaluation of conditions along the grounding line of temperate marine glaciers: an example from Muir Inlet, Glacier Bay, Alaska. *Marine Geology*, **140**, 307–327.

SHIPP, S., ANDERSON, J. B. & DOMACK, E. W. 1999. Late Pleistocene-Holocene retreat of the West Antarctic ice-sheet system in the Ross Sea; Part 1, Geophysical results. *Bulletin of the Geological Society of America*, **111**, 1486–1516.

SMITH, G. W. 1982. End moraines and the patterns of last ice retreat from central and southern coastal Maine. *In*: LARSON, G. J. & STONE, B. D. (eds) *Late Wisconsin glaciation of New England*. Kendall/Hunt Publishing Co., Dubuque, Iowa, 195–209.

SOLHEIM, A. 1991. The depositional environment of surging glaciers. *Norsk Polarinstiutt Skrifter*, **194**, 1–97.

SOLHEIM, A., FALEIDE, J. I., ANDERSEN, E. S., ELVERHØI, A., FORSBERG, C. F., VANNESTE, K., UENZELMANN-NEBEN, G. & CHANNEL, J. E. T. 1998. Late Cenozoic seismic stratigraphy and glacial geological development of the East Greenland and Svalbard-Barents Sea continental margins. *Quaternary Science Reviews*, **17**, 155–184.

STEVENSON, A. J. & EMBLEY, R. 1987. Deep-sea fan bodies, terrigenous turbidite sedimentation, and petroleum geology, Gulf of Alaska. *In*: SCHOLL, D. W., GRANTZ, A. & VEDDER, J. G. (eds) *Geology and Resource Potential of the Continental Margin of Western North America and Adjacent Ocean Basins – Beaufort Sea to Baja California*. Circum-Pacific Council for Energy and Mineral Resources, Earth Science Series, Houston, Texas, **6**, 503–522.

STEWART, T. G. 1991. Glacial marine sedimentation from tidewater glaciers in the Canadian high arctic. *In*: ANDERSON, J. B. & ASHLEY, G. M. (eds) *Glacial Marine Sedimentation – Paleoclimatic Significance*. Geological Society of America, Special Paper, **261**, 95–105.

THOMAS, G. S. P. & CONNELL, R. J. 1985. Iceberg drop, dump and grounding structures from Pleistocene glaciolacustrine sediments. *Journal of Sedimentary Petrology*, **55**, 243–249.

VAN DER VEEN, C. J. 1996. Tidewater calving. *Journal of Glaciology*, **42**, 375–385.

VORREN, T. O. & LABERG, J. S. 1997. Trough mouth fans; palaeoclimate and ice-sheet monitors. *Quaternary Science Reviews*, **16**, 865–881.

VORREN, T. O., LEBESBYE, E., ANDREASSEN, K. & LARSEN, K. B. 1989. Glacigenic sediments on a passive continental margin as exemplified by the Barents Sea. *Marine Geology*, **85**, 251–272.

VORREN, T. O., LABERG, J. S., BLAUME, F., DOWDESWELL, J. A., KENYON, N. H., MIENERT, J., RUMOHR, J. & WERNER, F. 1998. The Norwegian-Greenland Sea continental margins: morphology and late Quaternary sedimentary processes and environment. *Quaternary Science Reviews*, **17**, 273–302.

WANG, D. & HESSE, R. 1996. Continental slope sedimentation adjacent to an ice-margin; II, Glaciomarine depositional facies on the Labrador slope and glacial cycles, *Marine Geology*, **135**, 65–96.

WARREN, C. R., GLASSER, N. F., HARRISON, S., WINCHESTER, V., KERR, A. R. & RIVERA, A. 1995. Characteristics of tidewater calving at Glaciar San Rafael. *Journal of Glaciology*, **41**, 273–289.

ZELLERS, S. D. & LAGOE, M. B. 1992. Stratigraphic and seismic analyses of offshore Yakataga Formation sections, northeastern Gulf of Alaska. *In*: *Proceedings of the International Conference on Arctic Margins, Anchorage*. OCS Study, MMS 94–0040, 111–116.

Large-scale morphological evidence for past ice-stream flow on the mid-Norwegian continental margin

D. OTTESEN,[1] J. A. DOWDESWELL,[2] L. RISE,[1] K. ROKOENGEN[3] &
S. HENRIKSEN[4]

[1]*Geological Survey of Norway, N-7491 Trondheim, Norway*
(e-mail: dag.ottesen@ngu.no)
[2]*Scott Polar Research Institute & Department of Geography, University of Cambridge,
Cambridge CB2 1ER, UK*
[3]*Norwegian University of Science and Technology (NTNU), N-7491 Trondheim, Norway*
[4]*Statoil Research Centre, Rotvoll N-7005 Trondheim, Norway*

Abstract: The Norwegian continental margin has been affected by several cycles of growth and decay of the Fennoscandinavian Ice Sheet. Evidence for this former ice-sheet activity is found in the seismic stratigraphy of the shelf and slope, and in the morphology of sea-floor sediments. The regional bathymetry of the mid-Norwegian shelf (63°N to 68°N) comprises a series of cross-shelf troughs, separated by shallower banks. In Trænadjupet and Suladjupet, streamlined, elongate sedimentary bedforms (known as mega-scale lineations) are found, aligned along trough long-axes. Spacing between ridge-tops is 400–500 m, and ridge width and height are about 250 m and less than 10 m, respectively. Streamlined bedforms are not present on the intervening shallow banks, and terminate abruptly at the trough margins, marking the former boundary of fast glacier flow. On northwestern Trænabanken and on either side of Trænadjupet, lateral ridges are inferred to mark the shear margin of ice streams in Sklinnadjupet and Trænadjupet. The Skjoldryggen Moraine records the seaward limit of Late Weichselian ice-sheet growth, and ridges inshore result from deposition in still-stands during ice retreat. The streamlined bedforms in Trænadjupet and Suladjupet are similar in morphology and scale to streamlined Antarctic bedforms, linked to the former presence of fast-flowing ice streams on the continental shelves of Antarctica. Several geomorphological criteria, identified as diagnostic of past ice-stream flow, are observed in our geophysical studies of the mid-Norwegian shelf. Using these criteria, we identify several fast-flowing ice streams on the western margin of the former Fennoscandinavian Ice Sheet. Numerical ice-sheet model predictions of fast-flowing ice coincide closely with the inferred locations of past ice streams based on the distribution of such suites of diagnostic bedforms.

Introduction

Fast-flowing ice streams are the main pathways by which mass is lost from large ice sheets. Satellite observations of modern ice masses, ranging in dimensions from the 10^6 km^2 Antarctic and Greenland ice sheets to small Arctic ice caps of 10^3 km^2 (e.g. Bentley 1987; Dowdeswell & Collin 1990; Bamber *et al.* 2000), show that ice streams from a few kilometres to some tens of kilometres in width and tens to hundreds of kilometres in length are a typical feature of ice-mass dynamics (Fig. 1). These fast-flowing units are separated from slower-flowing ice by marked shear zones of high velocity gradient (e.g. Bindschadler *et al.* 1996). The presence of fast-flowing ice streams within former ice sheets has been inferred from observations of large-scale streamlined sedimentary features using,

first, satellite imagery of the land surface (e.g. Boulton & Clark 1990; Stokes & Clark 2001) and, more recently, marine geophysical observations of the sea floor (e.g. Longva & Thorsnes 1997; Anderson 1999; Shipp *et al.* 1999; Canals *et al.* 2000; Ó Cofaigh *et al.* 2002).

The Norwegian continental margin has been affected by a number of cycles of growth and decay of the ice sheet which covered Scandinavia intermittently during the Late Cenozoic (e.g. Holtedahl 1993; Mangerud *et al.* 1996; Sejrup *et al.* 2000). This ice sheet, in turn, formed a major dome of the larger Eurasian Ice Sheet during some glacial cycles (e.g. Siegert *et al.* 1999). Evidence for this former ice-sheet activity is found in the seismic stratigraphy of the Norwegian shelf and slope, and also in the morphology of the sediments making up the present sea floor (e.g. Gunleiksrud & Rokoengen 1980; Rise &

From: DOWDESWELL, J. A. & Ó COFAIGH, C. (eds) 2002. *Glacier-Influenced Sedimentation on High-Latitude Continental Margins.* Geological Society, London, Special Publications, **203**, 245–258. 0305-8719/02/$15.00

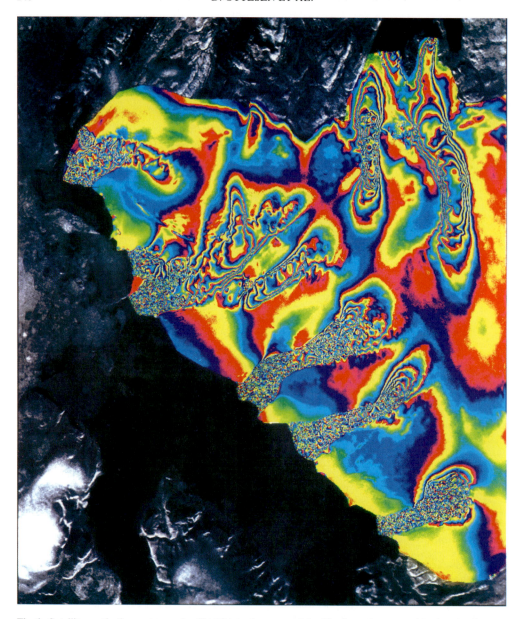

Fig. 1. Satellite synthetic aperture radar (SAR) interferogram of the Vestfonna ice cap on Nordaustlandet, Svalbard. The image is 55 km across. The 2500 km² ice cap is made up of fast-flowing ice streams (indicated by closely-spaced interference fringes), separated by slower-moving ice where interference fringes are widely spaced (Dowdeswell 1986; Dowdeswell & Collin 1990). The image was produced from a pair of ERS-1 SAR scenes acquired on 2 and 5 March 1994. Although the interferogram is 'mixed', containing information on both ice-cap motion and surface topography, the 1.08 m baseline (the distance between the satellite locations at each scene acquisition) means that it contains information mainly on ice flow in the look-direction of the satellite (from bottom to top).

Rokoengen 1984; Rokoengen *et al.* 1995; Ottesen *et al.* 2001). The likely presence of a former ice stream in the Norwegian Channel, immediately south of our study area, has also been reported by several authors (e.g. King *et al.* 1996, 1998; Sejrup *et al.* 1998; Larsen *et al.* 2000).

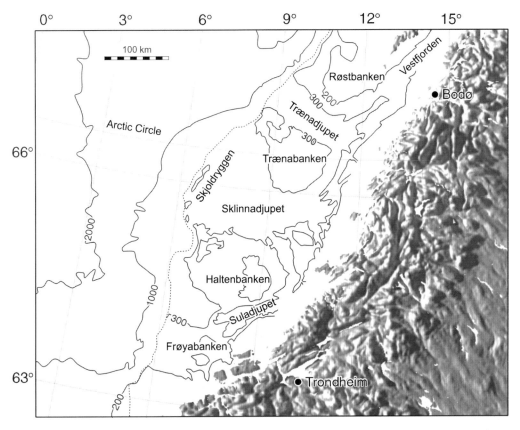

Fig. 2. Location map of the mid-Norwegian continental shelf between 63° and 68°N, showing the names of main offshore features on the shelf. Stippled line represents the shelf edge.

In this paper, our aim is to investigate the former flow dynamics of the past ice sheet draining westward from the crest of the Scandinavian Ice Sheet to the Norwegian continental margin. We use several marine geophysical datasets to describe the morphology of sea-floor sediments on the Norwegian shelf (Fig. 2). Streamlined features are observed over large areas, especially in cross-shelf bathymetric troughs. These features are then interpreted in the context of past ice flow across the shelf and discussed in relation to numerical models of the form and flow of the Late Weichselian ice sheet (Dowdeswell & Siegert 1999; Payne & Baldwin 1999; Siegert et al. 2001; Siegert & Dowdeswell 2002).

Background: seismic stratigraphy and geological setting

At the present edge of the Norwegian continental shelf, a sedimentary wedge reaching a maximum thickness of about 1500 m is typically found on seismic reflection lines (e.g. Rokoengen et al. 1995; Ottesen et al. 2001). Exploration wells on the mid-Norwegian shelf indicate that the oldest parts of this sedimentary wedge are Late Pliocene in age (Eidvin et al. 1998). The onset of glacial sedimentation has been linked to the first occurrence of ice-rafted debris in the Nordic Seas, which is interpreted to record the inception of major glaciation on Scandinavia (Jansen & Sjøholm 1991).

The seismic stratigraphy of this sedimentary wedge across Haltenbanken is shown schematically in Figure 3. The sequence is divided by an angular regional unconformity at the irregular base of unit D (Fig. 3). The units below this are complex and show a strongly prograding sequence. They are thought to be predominantly of glacial origin and were deposited through the Late Pliocene and much of the Pleistocene. Large-scale clinoforms typically prograde towards the NW, and have built out

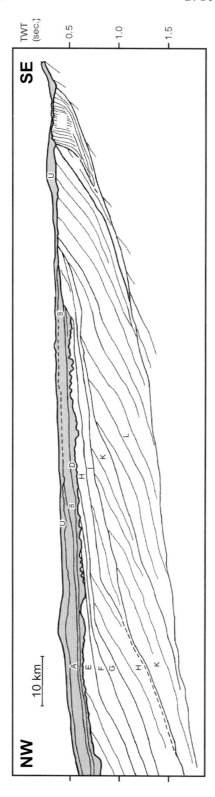

the shelf edge through time. Above the regional unconformity, the upper units (D, B, A and U in Figure 3) are sub-horizontal and exhibit both progradation and aggradation (King *et al.* 1987; Rokoengen *et al.* 1995). The dating of the units is still uncertain, but those above the regional unconformity are interpreted to represent two or three glacial–interglacial cycles. Eemian sediments are found below Unit A in the outer part of the shelf. The age of units B and D remains uncertain. The two youngest till sub-units within unit U were probably deposited about 15 000 and 13 500 years BP, respectively (Rokoengen & Frengstad 1999).

Ice appears to have reached the continental shelf on several occasions during the Weichselian, with the last two events reaching the outer shelf. From terrestrial records, Mangerud *et al.* (1996) summarize evidence for at least three, and possibly four advances of Weichselian ice to or beyond the coast, with the Late Weichselian maximum being the largest of these events. Later investigations based on extensive radiocarbon dating show that the last glacial maximum comprises two major glacial expansion phases dated at 25–20 000 and 19–15 000 years BP, respectively (Olsen 1997; Sejrup *et al.* 2000; Olsen *et al.* 2001). These two phases are tentatively correlated with seismic units A and U (Ottesen *et al.* 2001).

Data acquisition and methods

Several datasets provide information on the morphology of the Norwegian continental shelf and upper slope, in some cases down to about 1000 m water depth. First, 100 kHz single-beam echo-sounder data collected by the Norwegian Hydrographic Service (NHS) between 1965 and 1985 have been used to produce a regional bathymetry of the Norwegian shelf south of 68°N. The echo-sounder data have an average line spacing of 500 m and, accordingly, bathymetric values were gridded at a cell size of 500 m. These data can be plotted as coloured contour maps (5 m contour interval) or as shaded relief maps. Although absolute navigational accuracy using Decca Main Chain was usually better than

Fig. 3. Composite seismic profile showing the Upper Cenozoic stratigraphy across the mid-Norwegian shelf (modified from Rokoengen *et al.* 1995). Units K–E represent Upper Pliocene-Pleistocene sediments and units D, B, A and U, above the upper angular unconformity, are interpreted to represent the last two or three interglacial–glacial cycle. The seismic profile is located in Figure 4.

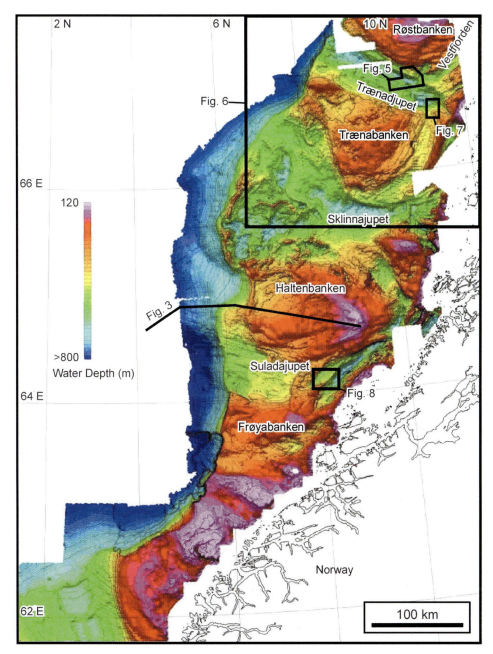

Fig. 4. Bathymetry of the mid-Norwegian shelf illustrated as a shaded relief map with 50 m depth contours. The data were collected by the Norwegian Hydrographic Service with a single beam echo-sounder. The positions of Figures 3, 5, 6, 7 and 8 are located.

100 m, the relative accuracy is much higher and thus the dataset is very well suited to the identification and description of sea-floor morphological features (Ottesen *et al.* 2001).

Three further datasets are available for some areas of the Norwegian shelf. These comprise swath bathymetric and 3-dimensional seismic imagery. The navigational control on these

Fig. 5. Shaded relief image of the sea bottom of the inner part of Trænadjupet based on 3D seismic data. Note the extensive glacial lineations (interpreted to be) caused by fast ice-stream flow out of Trænadjupet. See Figures 4 and 6 for location.

datasets is a few metres, being derived from differential GPS. Swath bathymetric data from the Trænadjupet area were acquired by the NHS using a Kongsberg Simrad EM-1002 system during a cooperative test cruise with the Norwegian Geological Survey (NGU) in 1999. The cruise data mainly cover a corridor along shipping lanes which is normally 1 km wide (dependent on water depth). The multibeam bathymetry data from a 200 km^2 area covering Sularyggen were collected by NHS during 1999. The data density varies between 3 m and 10 m and the data were gridded with a cell size of 5 m. Three-dimensional seismic data were collected for Statoil ASA in 1992, covering an area of approximately 1000 km^2 with a line spacing of 25 m. These data were gridded with a cell size of 25 m.

Geophysical observations on the mid-Norwegian shelf

Regional bathymetry

The regional bathymetry of the mid-Norwegian shelf (62°N to 68°N) is that of a series of cross-shelf troughs (indicated by the suffix 'djupet' in the Norwegian names), separated by shallower banks (Figs 2 and 4). The continental shelf is narrowest in the south of the study area, at slightly less than 100 km wide. It reaches up to 250 km in width at 66°N, before narrowing to the north (Fig. 4). The main morphological features are the three troughs, known as Trænadjupet, Sklinnadjupet and Suladjupet, and the bordering banks of Røstbanken,

Trænabanken, Sklinnabanken, Haltenbanken and Frøyabanken (Fig. 2). The cross-shelf troughs reach to depths between about 300 m and 500 m, and the upper parts of the banks are usually less than 200 m deep (Fig. 4).

The wide line-spacing (500 m) of single-beam equipment used in construction of the regional bathmetry means that multi-beam swath bathymetric and 3-dimensional seismic datasets are required in order to resolve details within the cross-shelf troughs in particular. Data from these latter sources are now presented.

Trænadjupet

Geophysical data from the floor of the 50 to 90 km-wide Trænadjupet (67°N) show a series of streamlined ridges and depressions (Fig. 5). The spacing between ridge-tops varies between 400 m and 500 m, and the width of the individual ridges is on average about 250 m. The maximum height of the ridges is 10 m, but normally they are less than 5 m high (Fig. 5). These ridges and depressions are aligned along the long-axis of the Trænadjupet trough and seismic data show that the bedforms are developed in Quaternary sediments indicating that they are unconsolidated sedimentary bedforms and are not bedrock (e.g. King *et al.* 1987). These streamlined bedforms occur along most of Trænadjupet. Similar streamlined and elongate features are not found on the neighbouring shallow banks of Røstbanken and Trænabanken, which form the northern and southern boundaries of the trough, respectively (Fig. 2). However, lateral ridges are found on

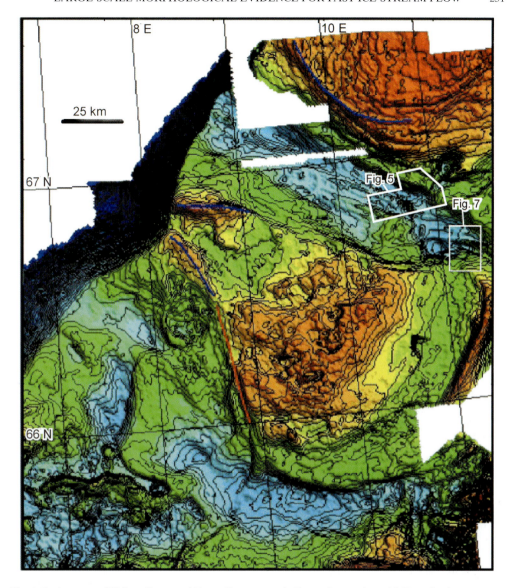

Fig. 6. Bathymetry of Sklinnadjupet and Trænadjupet cross-shelf troughs together with Trænabanken as a shaded relief map. Lateral ridges are on the SW margin of Trænabanken, and on either side of Trænadjupet, and are shown by solid lines. Part of the Skjoldryggen moraine-ridge complex is also shown in the bottom left of the image. Image located in Figure 4.

both the northern and southern margins of Trænadjupet (Fig. 6). The ridge on the south side is approximately 35 km long, up to 50 m high and 6 km wide. On the northern side, two partly overlapping ridges are present with a total length of more than 100 km. These ridges are up to 2 km wide and 20 m high.

In the inner part of Trænadjupet, about 110 km from the shelf break, the cross-shelf trough divides into two parts which diverge to align NE–SW (Vestfjorden) and SE–NW (Fig. 4). The elongate sedimentary bedforms also form two sets, aligning along each of the deeper troughs within the inner shelf (Fig. 7). The area imaged in Figure 7 shows the region in which the two sets of streamlined bedforms coalesce. Cold-water coral mounds can also be seen on the image.

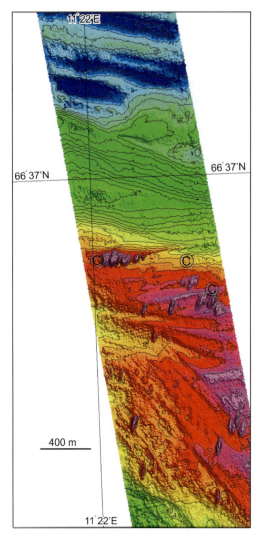

Fig. 7. Large-scale streamlined sedimentary features on the floor of Trænadjupet (emphasized by black lines), indicating former ice flow from two directions from 3-dimensional seismic data. Note the cold-water coral mounds (C). Image is located in Figure 6.

Skjoldryggen moraine system and Sklinnadjupet

A second large trough crossing the mid-Norwegian shelf, at almost 66°N where the shelf break is in excess of 200 km from the coast, is that of Sklinnadjupet (Figs 4 and 6). The trough is well-defined on the inner shelf at about 20 km wide. On the outer shelf, the water is often deeper than about 400 m across a width of over 120 km, and the trough becomes less clearly defined as the outer part of Trænabanken in particular curves away northwards (Fig. 4). On the southwestern side of Trænabanken, there is a lateral ridge up to about 50 m high and 10 km wide. It follows the edge of the bank at a water depth of 250–300 m. This feature is, in total, about 100 km long (Fig. 6).

The Skjoldryggen moraine system is an easily-identified set of morphological features on the regional bathymetric map (Fig. 4). A ridge almost 200 km long, up to 150 m high and 10 km wide is present at the continental shelf break, centred on about 65°30'N (Fig. 4). East and inshore of this well-defined moraine is a further series of depressions and ridges (Figs 4 and 6). The Skjoldryggen Moraine itself has been interpreted to be the limit of latest Weichselian ice-sheet growth on this part of the Norwegian margin. The ridges inshore are ascribed to deposition during still-stands or slow-downs during ice retreat from this maximum position. From seismic profiles, we infer that active ice in the Skjoldryggen area deposited many till packages (tongues) during glacial episodes (King *et al.* 1987).

Suladjupet

Suladjupet, located between Halten- and Frøyabanken, has a more complex morphology than Traenadjupet. The Frøyryggen sill, which rises above the 300 m bathymetric contour (Fig. 4), separates the inner and outer parts of the trough. Suladjupet is located east of Frøyryggen, and reaches a depth of more than 500 m due to extensive glacial erosion into Jurassic claystones (Rokoengen *et al.* 1988). The outer trough is about 35 km wide and reaches a water depth of about 400 m (Fig. 4).

A swath bathymetric image covering the Sularyggen along the northern flank of Suladjupet demonstrates that elongate and streamlined bedforms are present. The lineations cross the ridge 20° offset towards the WSW–ENE (70°–250°) (Fig. 8). The bedforms are orientated along the long-axis of the trough, in a similar fashion to those in Traenadjupet (Figs 5–7). The image also shows that cold-water coral mounds are mainly developed on top of the lineations on the Sularyggen and only occasionally on the sides of the iceberg scour marks that are also present (Fig. 8). West of Frøyryggen, the large-scale form of the ridges and depressions indicate that the ice is deflected and continues its movement towards the WNW (Fig. 4). Haltenbanken and Frøyabanken do not show evidence of large-scale streamlined bedforms.

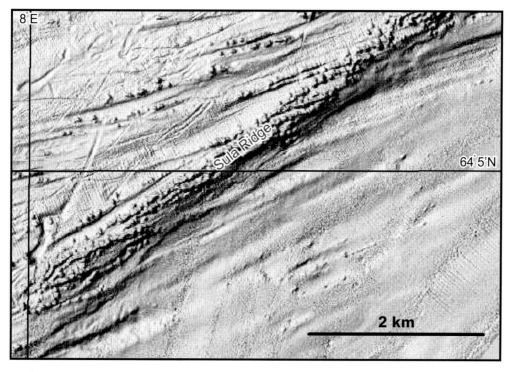

Fig. 8. Swath bathymetric image of the Sula Reef area in the Suladjupet cross-shelf trough. The image shows extensive large-scale flutes crossing the Sula Ridge where the Sula deep-water Lophelia reef is located. Note also the smaller trawl and iceberg-scour marks. Image located in Figure 4.

Interpretation: the locations of former ice streams

The streamlined, elongate ridges and depressions recorded on geophysical datasets from the sea floor in the Traenadjupet and Suladjupet troughs are similar in morphology and in scale to the sets of linear bedforms described from troughs across the Antarctic Peninsula shelf by Canals *et al.* (2000) and Ó Cofaigh *et al.* (2002), and to the streamlined forms identified by Shipp *et al.* (1999) from the Ross Sea margin of Antarctica. These bedforms have each been linked to the former presence of fast-flowing ice streams on the continental shelves of Antarctica.

A set of geomorphological criteria has been identified by Stokes & Clark (1999, 2001) as diagnostic of past ice-stream flow. It is pointed out by Stokes & Clark that it is not the presence of single features, but rather the suite of geomorphological evidence taken together, that provides the strongest indication of palaeo-ice stream locations. We identify several of these criteria from our geophysical studies of the mid-Norwegian shelf.

- *Characteristic shape and dimensions*: sets of elongate, streamlined bedforms (Figs 5 and 7) occur within cross-shelf troughs and, in some cases, extend over 100 km along trough axes (Trænadjupet). In each case, along-trough length is considerably greater than the width of the suite of features.

- *Highly convergent flow patterns*: the coalescence of sets of streamlined features in the inner Trænadjupet trough (Fig. 7) clearly suggests the presence of convergent flow within the past ice sheet. The eastern part of Sklinnadjupet has also acted as a confluence area for the drainage of ice (Fig. 6).

- *Highly attenuated bedforms*: the streamlined and elongate bedforms, illustrated from the Trænadjupet and Suladjupet troughs (Figs 5, 7 and 8), are good examples of this type of feature, referred to as mega-scale glacial lineations by Stokes & Clark (1999, 2001).

- *Abrupt lateral margins*: the streamlined bedforms observed in the troughs of the mid-Norwegian shelf are not present on the intervening shallow banks, and terminate abruptly at the trough margins. This change

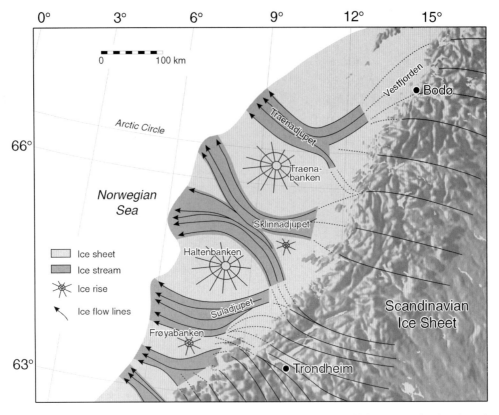

Fig. 9. Inferred ice-stream flow lines and ridges during the Late Weichselian with ice streams flowing along the main offshore depressions/troughs (from Ottesen *et al.* 2001).

in bedform pattern is interpreted to mark the former boundary of fast glacier flow (Stokes & Clark 1999).

- *Lateral ridges*: On the outer part of Træna-banken, a lateral moraine inferred to be formed at the shear margin of an ice stream flowing out of Sklinnadjupet is identified (Fig. 6). The ice stream may have been deflected northwards to impinge on Trænae-banken by slow-moving or stagnant deglacial ice to the south, which formed the Skjoldryggen moraine system. Sediment ridges on both sides of Trænadjupet are also interpreted as ice-stream lateral moraines (Fig. 6). Such ice-stream marginal moraines have rarely been described previously (cf. Dyke & Morris 1988; Shipp *et al.* 1999).

There is little data of sufficiently high resolution to provide evidence for the presence or otherwise of elongate, streamlined bedforms from the area of the Skjoldryggen moraine ridge.

However, seismic-reflection profiles from this area indicate significant progradation of the shelf. More than 400 m of sediment has been deposited during the last three glacial stages. This indicates rapid sediment transport to the outer shelf and shelf edge, implying that fast-flowing ice streams are likely to have been active in this cross-shelf trough too.

On the basis of our marine geophysical evidence, we have identified several suites of features consistent with the above geomorphological criteria associated with past ice-stream flow. The locations of these westward flowing palaeo-ice streams are shown schematically in Figure 9, together with the ice-surface ridges which separated each of the fast-flowing elements.

Several earlier studies have investigated the geomorphological record deposited by the Fennoscandian Ice Sheet during the Weichselian in order to reconstruct past glacier flow (e.g. Punkari 1997; Kleman *et al.* 1997; Boulton

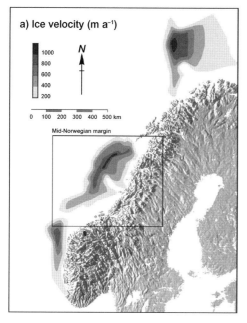 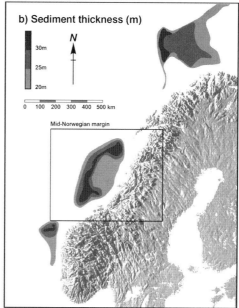

Fig. 10. Numerical-model reconstructions of part of the Late Weichselian Eurasian Ice Sheet. (a) Ice velocities (m a⁻¹) during the Late Weichselian glacial maximum. (b) Sediment accumulation (m) over the Late Weichselian glacial period from 30 000 to 10 000 years ago (from Siegert & Dowdeswell, 2002). The smaller box indicates the location of the mid-Norwegian margin shown in Figures 2 and 9.

et al. 2001). These studies used the presence of elongate, streamlined landforms at the kilometre-scale, such as drumlins and mega-flutes, to infer the locations of past ice streams. However, relatively few ice streams were identified flowing westward to the Norwegian shelf, compared with eastward flow towards and across the Baltic. This is probably because the datasets used by Punkari (1997) and Kleman *et al.* (1997), in the form of satellite imagery, aerial photographs and field observations, were based on the analysis of exposed land that lacks till cover in many places. Our marine geophysical datasets have enabled similar suites of streamlined landforms to be identified beneath modern sea level, resulting in the identification of additional fast-flowing ice streams on the western margin of the former Fennoscandian Ice Sheet (Fig. 9).

Discussion: past ice-sheet flow

A study of the geophysical observations of the sea floor in the context of numerical ice-sheet modelling work, aids the understanding of the locations and nature of former ice flow across the mid-Norwegian shelf. The submarine morphological features, described and interpreted above, imply fast ice flow in several of the troughs on the mid-Norwegian shelf (Fig. 9). Several recent numerical ice-sheet models of the Fennoscandian and Eurasian ice sheets predict that fast glacier flow has taken place along parts of the former ice sheet covering Scandinavia and the adjacent Norwegian shelf (Dowdeswell & Siegert 1999; Payne & Baldwin 1999; Siegert *et al.* 1999, 2002).

The three-dimensional thermo-mechanical model used by Payne & Baldwin (1999) was run in the form of numerical experiments to test whether ice streams would develop within the Fennoscandian Ice Sheet. There was no attempt to reproduce the geologically-inferred extent and chronology of the ice sheet but, because the model treated ice-sheet thermal structure in a relatively sophisticated way, areas where basal melting developed were predicted. By inference, this is where fast glacier flow would be expected. The model suggests that several ice streams are present on the mid-Norwegian shelf, including the Trænadjupet, Sklinnadjupet and Suladjupet troughs. This modelling work provides independent support for our interpretation of the streamlined bedforms identified in geophysical records, particularly since no geological data in the form of past ice extent or chronology were incorporated in these modelling experiments.

Other ice-sheet modelling reconstructions of the Eurasian Ice Sheet, including Fennoscandia, have used an inverse approach to fit the predicted dimensions of the ice sheet to the geological evidence for ice extent at the Late Weichselian glacial maximum (e.g. Dowdeswell & Siegert 1999; Siegert *et al.* 1999). Fast glacier flow is again predicted over much of the mid-Norwegian shelf, along with rapid motion in most of the main cross-shelf troughs on both the western and northern margins of this former ice sheet (Fig. 10a). In addition, high rates of sediment delivery to the continental margin are also predicted by the model for the full-glacial ice streams terminating at the shelf break (Fig. 10b) (Dowdeswell & Siegert 1999). Henriksen & Vorren (1996) report a prograding sedimentary wedge of Late Pliocene–Pleistocene age beyond the mouth of Trænadjupet, which they ascribe to the presence of glaciers at the shelf break on up to 12 occasions. They also suggest that the sequence is similar in form to the Bear Island and Storfjorden fans to the north, offering further support to the notion of fast glacier flow and high rates of sediment delivery to the margin (cf. Dowdeswell & Siegert 1999).

In the models of Dowdeswell & Siegert (1999) and Payne & Baldwin (1999), predictions of the locations of fast-flowing ice derived from time-dependent numerical ice-sheet modelling coincide closely with the inferred locations of past ice streams based on the distribution of suites of diagnostic bedforms identified from marine geophysical datasets (Fig. 9). However, some of the individual troughs are not well resolved in the bathymetric datasets used in numerical modelling, and this relatively coarse spatial resolution restricts the ability of the models to resolve all the ice streams individually. In addition, the model of Dowdeswell & Siegert (1999) uses simple assumptions concerning the thickness of deformation till mobilized beneath ice streams. It is the pattern, rather than the absolute magnitude, of sediment delivery to the mid-Norwegian margin that is of principal significance.

Conclusions

- The Norwegian continental margin has been affected by several cycles of ice-sheet growth and decay during the Late Cenozoic, as recorded in the seismic stratigraphy of the Norwegian shelf and slope (Fig. 3).

Fast-flowing ice streams are often the main routes by which mass is lost from ice sheets (Fig. 1), and we have used several marine geophysical datasets to describe the morphology of sea-floor sediments on the mid-Norwegian shelf, in order to investigate whether there is evidence for the presence of palaeo-ice streams on the Norwegian margin.

- The regional bathymetry of the mid-Norwegian shelf is that of a series of cross-shelf bathymetric troughs, separated by shallower banks (Figs 3 and 4). Streamlined sedimentary bedforms are observed in these troughs (Fig 5, 7 and 8), comprising ridges and depressions that are aligned along the trough long-axes (typical wavelengths are 500 m and amplitudes 5–10 m). These features are not present on neighbouring shallow banks.

- The streamlined, elongate ridges and depressions recorded on geophysical datasets from the sea floor in the Trænadjupet and Suladjupet troughs on the mid-Norwegian shelf are similar in morphology and in scale to the set of linear features described from the continental margins of Antarctica and referred to as mega-scale glacial lineations. These bedforms have each been linked to the former presence of fast-flowing ice streams on the continental shelves of Antarctica (e.g. Shipp *et al.* 1999; Canals *et al.* 2000; Ó Cofaigh *et al.* 2002).

- The Skjoldryggen moraine system, easily identified on regional bathymetric maps (Figs 4 and 6), represents the seaward limit of Late Weichselian ice-sheet growth on the mid-Norwegian margin. Ridges inshore are related to deposition during still-stands or slow-downs during ice retreat from this maximum position.

- Geomorphological criteria have been identified as diagnostic of past ice-stream flow (Stokes & Clark 1999, 2001), and several of these criteria are observed in our geophysical studies of the mid-Norwegian shelf. These features include: elongate, streamlined bedforms interpreted to represent fast ice flow (Figs 5, 7 and 8); convergent patterns of streamlined features, indicating flow convergence within the past ice sheet (Fig. 7); lateral ridges and abrupt lateral margins to sets of streamlined bedforms terminating at trough margins, both marking the former boundary of fast glacier flow (Fig. 6) (cf. Dyke & Morris 1988).

- On the basis of this marine geophysical evidence, consistent with geomorphological criteria associated with past ice-stream flow

(Stokes & Clark 1999, 2001), we identify the locations on the mid-Norwegian shelf of westward flowing palaeo-ice streams of the Late Weichselian Fennoscandian Ice Sheet (Fig. 9).

- Model predictions of the locations of fast-flowing ice derived from time-dependent numerical ice-sheet modelling coincide closely with the inferred locations of past ice streams based on the distribution of suites of diagnostic bedforms identified from marine geophysical datasets (Figs 9 and 10).

We are grateful to the Norwegian Hydrographic Service for the regional bathymetric data, to Sintef Petroleum Research for access to seismic data, and to Statoil ASA for releasing datasets for publication. We thank Chris Clark and Jon Landvik for reviewing the manuscript.

References

ANDERSON, J. B. 1999. *Antarctic Marine Geology.* Cambridge University Press, Cambridge, 289 pp.

BAMBER, J. L., VAUGHAN, D. G. & JOUGHIN, I. 2000. Widespread complex flow in the interior of the Antarctic Ice Sheet. *Science*, **287**, 1248–1250.

BENTLEY, C. R. 1987. Antarctic ice streams: a review. *Journal of Geophysical Research*, **92**, 8843–8858.

BINDSCHADLER, R., VORNBERGER, P., BLANKENSHIP, D., SCAMBOS, T. & JACOBEL, R. 1996. Surface velocity and mass balance of Ice Streams D and E, West Antarctica. *Journal of Glaciology*, **42**, 461–475.

BOULTON, G. S. & CLARK, C. D. 1990. A highly mobile Laurentide ice sheet revealed by satellite images of glacial lineations. *Nature*, **346**, 813–817.

BOULTON, G. S., DONGLEMANS, P. PUNKARI, M. & BROADGATE, M. 2001. Palaeoglaciology of an ice sheet through a glacial cycle: the European ice sheet through the Weichselian. *Quaternary Science Reviews*, **20**, 591–625.

CANALS, M., URGELES, R. & CALAFAT, A. M. 2000. Deep sea-floor evidence of past ice streams off the Antarctic Peninsula. *Geology*, **28**, 31–34.

DOWDESWELL, J. A. 1986. Drainage-basin characteristics of Nordaustlandet ice caps, Svalbard. *Journal of Glaciology*, **32**, 31–38.

DOWDESWELL, J. A. & COLLIN, R. L. 1990. Fast-flowing outlet glaciers on Svalbard ice caps. *Geology*, **18**, 778–781.

DOWDESWELL, J. A. & SIEGERT, M. J. 1999. Ice-sheet numerical modeling and marine geophysical measurements of glacier-derived sedimentation on the Eurasian Arctic continental margins. *Geological Society of America, Bulletin*, **111**, 1080–1097.

DYKE, A. S. & MORRIS, T. F. 1988. Drumlin fields, dispersal trains, and ice streams in Arctic Canada. *Canadian Geographer*, **32**, 86–90.

EIDVIN, T., BREKKE, H., RIIS, F. & RENSHAW, D. 1998. Cenozoic stratigraphy of the Norwegian Sea continental shelf, 64°–68°N. *Norsk Geologisk Tidsskrift*, **78**, 125–151.

GUNLEIKSRUD, T. & ROKOENGEN, K. 1980. Regional geological mapping of the Norwegian continental shelf with examples of engineering applications. *In:* D. A. ARDUS (ed.) *Offshore site investigations.* Graham and Trotman, London, 23–35.

HENRIKSEN, S. & VORREN, T. 1996. Late Cenozoic sedimentation and uplift history on the mid-Norwegian continental shelf. *Global and Planetary Change*, **12**, 171–199.

HOLTEDAHL, H. 1993. Marine geology of the Norwegian continental margin. *Norges Geologiske Undersøkelse Special Publication*, **6**, 150 pp.

JANSEN, E. & SJØHOLM, J. 1991. Reconstruction of glaciation over the past 6 Myr from ice-borne deposits in the Norwegian Sea. *Nature*, **349**, 600–603.

KING, L., ROKOENGEN, K. & GUNLEIKSRUD, T. 1987. Quaternary seismostratigraphy of the Mid Norwegian Shelf, 65°–67°30'N. A till tongue stratigraphy. *Continental Shelf Institute (IKU) Publication*, **114**, 58 pp.

KING, E. L., SEJRUP, H. P., HAFLIDASON, H., ELVERHØI, A. & AARSETH, I. 1996. Quaternary seismic stratigraphy of the North Sea Fan: glacially-fed gravity flow aprons, hemipelagic sediments, and large submarine slides. *Marine Geology*, **130**, 293–315.

KING, E. L., HAFLIDASON, H., SEJRUP, H. P. & LØVLIE, R. 1998. Glacigenic debris flows on the North Sea Trough Mouth Fan during ice stream maxima. *Marine Geology*, **152**, 217–246.

KLEMAN, J., HÄTTESTRAND, C., BORGSTRÖM, I. & STROEVEN, A. 1997. Fennoscandian palaeo-glaciology reconstructed using a glacial geological inversion model. *Journal of Glaciology*, **43**, 283–299.

LARSEN, E., SEJRUP, H. P., JANOCKO, J., LANDVIK, J., STALSBERG, K. & STEINSUND, P. I. 2000. Recurrent interaction between the Norwegian Channel Ice Stream and terrestrial-based ice across southwest Norway. *Boreas*, **29**, 185–203.

LONGVA, O. & THORSNES, T. (eds) 1997. Skagerrak in the past and at the present – an integrated study of geology, chemistry, hydrography and micro-fossil ecology. *Norges Geologiske Undersøkelse Special Publication*, **8**, 100.

MANGERUD, J., JANSEN, E. & LANDVIK, J. Y. 1996. Late Cenozoic history of the Scandinavian and Barents Sea ice sheets. *Global and Planetary Change*, **12**, 11–26.

Ó COFAIGH, C., PUDSEY, C. J., DOWDESWELL, J. A. & MORRIS, P. 2002. Evolution of subglacial bedforms along a paleo-ice stream, Antarctic Peninsula continental shelf. *Geophysical Research Letters*, **29** (8), 10.1029/2001GL014488.

OLSEN, L. 1997. Rapid shifts in glacial extension characterise a new conceptual model for glacial variations during the Mid and Late Weichselian in Norway. *Norges Geologiske Undersøkelse Bulletin*, **433**, 54–55.

OLSEN, L., SVEIAN, H. & BERGSTRØM, B. 2001. Rapid adjustments of the western part of the Scandinavian Ice Sheet during the Mid and Late Weichselian – a new model. *Norsk Geologisk Tidsskrift*, **81**, 93–118.

OTTESEN, D., RISE, L., ROKOENGEN, K. & SÆTTEM, J. 2001. Glacial processes and large-scale morphology on the mid-Norwegian continental shelf. *In:* MARTINSEN, O. J. & DREYER, T. (eds) *Sedimentary Environments Offshore Norway – Palaeozoic to Recent.* Norwegian Petroleum Society, Special Publication, **10**, 441–449.

PAYNE, A. J. & BALDWIN, D. J. 1999. Thermomechanical modelling of the Scandinavian ice sheet: implications for ice-stream formation. *Annals of Glaciology*, **28**, 83–89.

PUNKARI, M. 1997. Glacial and glaciofluvial deposits in the interlobate areas of the Scandinavian Ice Sheet. *Quaternary Science Reviews*, **16**, 741–753.

RISE, L. & ROKOENGEN, K. 1984. Surficial sediments in the Norwegian sector of the North Sea between 60°30'N and 62°N. *Marine Geology*, **56**, 287–317.

ROKOENGEN, K. & FRENGSTAD, B. 1999. Radiocarbon and seismic evidence of ice-sheet extent and the last deglaciation on the mid-Norwegian continental shelf. *Norsk Geologisk Tidsskrift*, **79**, 129–132.

ROKOENGEN, K., RISE, L., BUGGE, T. & SÆTTEM, J. 1988. *Berggrunnsgeologi på midtnorsk kontinentalsokkel. Kart i målestokk 1:1000 000.* Continental Shelf Institute (IKU), Publication, **118**.

ROKOENGEN, K., RISE, L., BRYN, P., FRENGSTAD, B., GUSTAVSEN, B., NYGAARD, E. & SÆTTEM, J. 1995. Upper Cenozoic stratigraphy on the mid-Norwegian continental shelf. *Norsk Geologisk Tidsskrift*, **75**, 88–104.

SEJRUP, H. P., LANDVIK, J., LARSEN, E., JANOCKO, J., EIRIKSSON, J. & KING, E. 1998. The Jæren area, a border zone of the Norwegian Channel Ice Stream. *Quaternary Science Reviews*, **17**, 801–812.

SEJRUP, H. P., LARSEN, E. LANDVIK, J. KING, E. L., HAFLIDASON, H. & NESJE, A. 2000. Quaternary glaciations in southern Fennoscandia: evidence from southwestern Norway and the northern North Sea region. *Quaternary Science Reviews*, **19**, 667–685.

SHIPP, S., ANDERSON, J. & DOMACK, E. 1999. Late Pleistocene-Holocene retreat of the West Antarctic Ice-Sheet system in the Ross Sea: Part 1 – Geophysical results. *Geological Society of America Bulletin*, **111**, 1486–1516.

SIEGERT, M. J. & DOWDESWELL, J. A. 2002. Late Weichselian iceberg, meltwater and sediment production from the Eurasian Ice Sheet: results from numerical ice-sheet modelling. *Marine Geology*, **188**, 109–127.

SIEGERT, M. J., DOWDESWELL, J. A. & MELLES, M. 1999. Late Weichselian glaciation of the Russian High Arctic. *Quaternary Research*, **52**, 273–285.

SIEGERT, M. J., DOWDESWELL, J. A., HALD, M. & SVENDSEN, J. I. 2001. Modelling the Eurasian Ice Sheet through a full (Weichselian) glacial cycle. *Global and Planetary Change*, **31**, 367–385.

STOKES, C. R. & CLARK, C. D. 1999. Geomorphological criteria for identifying Pleistocene ice streams. *Annals of Glaciology*, **28**, 67–74.

STOKES, C. R. & CLARK, C. D. 2001. Palaeo-ice streams. *Quaternary Science Reviews*, **20**, 147–1457.

SÆTTEM, J., RISE, L., ROKOENGEN, K. & BY, T. 1996. Soil investigations, offshore mid Norway: a study of glacial influence on geotechnical properties. *Global and Planetary Change*, **12**, 271–285.

Geomorphology of buried glacigenic horizons in the Barents Sea from three-dimensional seismic data

B. RAFAELSEN[1], K. ANDREASSEN[1], L. W. KUILMAN[2], E. LEBESBYE[1], K. HOGSTAD[3] & M. MIDTBØ[4]

[1]*Department of Geology, University of Tromsø, N-9037 Tromsø, Norway*
(e-mail: bjarne@ibg.uit.no)
[2]*Norsk Hydro ASA, 0246 Oslo, Norway*
[3]*Norsk Hydro ASA, 9480 Harstad, Norway*
[4]*Norsk Hydro ASA, Pb. 7190, 5020 Bergen, Norway*

Abstract: The glacigenic sequence of the southwestern Barents Sea shelf has for the first time been studied using 3-D seismic data. The close spacing of 3-D lines and powerful computer workstation interpretation techniques have allowed detailed mapping of the observed features. Several generations of subglacial lineations observed on four different palaeo-surfaces are interpreted to reflect the flow patterns of palaeo-ice sheets. To our knowledge, this is the first time that multiple levels of subglacial lineations have been observed. The lineations are 2.5–8 m in relief, 50 to 180 m wide and 0.5 to 20 km long. All four surfaces show a main lineation pattern comprising lineations with a N–S trend, suggesting that the dominant ice flow was directed northwards across the Barents Sea shelf at least four times during the last 0.8 Ma. Two of the surfaces display semi-circular to oblong depressions trending mainly in the same direction as the sub-parallel lineations. These depressions are 9–53 m in relief, 1.25–3.2 km wide and 1.9–9 km long. In contrast to the buried surfaces, the sea floor is dominated by 2.5–25 m deep cross-cutting iceberg plough-marks from the deglaciation phase of the last Barents Sea ice sheet. The 3-D seismic data are conventional industry data. Despite relatively low seismic frequencies and, hence, limited vertical resolution of seismic profiles, time slices and sub-horizontal time maps are of high spatial resolution, providing detailed images of different stages of buried Quaternary glacial geomorphology.

The Barents Sea is a large epicontinental sea, where bathymetry is characterized by an east–west orientated channel, Bjørnøyrenna, which reaches a depth of 500 m, with shallower banks in the north and south (Fig. 1a). An upper regional unconformity (URU) separates dipping well-bedded pre-glacial sedimentary rocks of Tertiary age and older (Solheim & Kristoffersen 1984) from the upper, sub-horizontal layers of glacigenic sediments (Fig. 1b). The upper regional unconformity is a time-transgressive (diachronous) surface (Lebesbye 2000) which represents mainly the incipient formation of ice sheets in the Barents Sea (Richardsen *et al.* 1991). Although its age is disputed, it has tentatively been estimated at around 0.8 million years old (Table 1; Vorren *et al.* 1988, 1991; Richardsen *et al.* 1991; Knutsen *et al.* 1992). Biostratigraphic evidence on the timing together with the magnitude of erosion (Eidvin *et al.* 1993, 1998) indicates that the present upper regional unconformity landscape formed during the late Pliocene–Pleistocene, since about 2.5 Ma. The intensity of erosion probably varied considerably, both in time and

geographically, thus explaining the diachronous nature of the surface (Lebesbye 2000). According to Vorren *et al.* (1991), Bjørnøyrenna was formed in the mid-Oligocene as a fluvial surface, prior to the onset of late Cenozoic glaciations. At least seven glacial episodes have occurred on the Barents Shelf since the onset of the glaciations (Solheim & Kristoffersen 1984) and regional ice sheets have reached the shelf edge at least five times during the last 0.8 Ma (Vorren *et al.* 1991). The stratigraphy of the glacigenic sequence in the Barents Sea, known as the Barents Sea Synthem (Vorren *et al.* 1989), has been described and classified by several authors (Solheim & Kristoffersen 1984; Vorren *et al.* 1990; Sættem *et al.* 1992a). A large proportion of the sediments above the upper regional unconformity have been interpreted either as sediment deposited directly from ice sheets as till or as proximal glacimarine sediment (Solheim & Kristoffersen 1984; Vorren *et al.* 1990).

Previous seismic investigations of the glacigenic sequence of the southwestern Barents Sea have been based on 2-D data (Solheim & Kristoffersen 1984; Vorren &

From: DOWDESWELL, J. A. & Ó COFAIGH, C. (eds) 2002. *Glacier-Influenced Sedimentation on High-Latitude Continental Margins.* Geological Society, London, Special Publications, **203**, 259–276. 0305-8719/02/$15.00
© The Geological Society of London 2002.

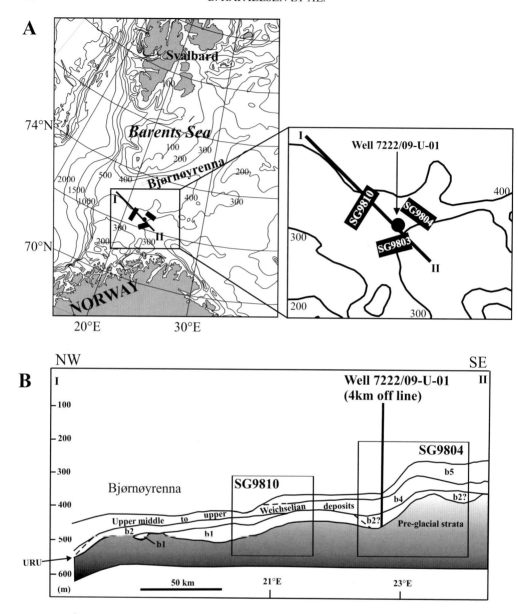

Fig. 1. (**a**) Location and bathymetry of the southwestern Barents Sea. The areas with studied 3-D seismic data are indicated by black rectangles. (**b**) Generalized geo-seismic section, location indicated in (a), outlining the stratigraphy of glacigenic sediments in the study area. Rectangles show the approximate location of two of the 3-D surveys. Modified from Sættem *et al.* (1992*b*).

Kristoffersen 1986; Vorren *et al.* 1986, 1988, 1989, 1990; Sættem 1992*b*, 1994; Sættem *et al.* 1994). This paper presents results from the first use of 3-D seismic data to study the glacigenic sequence of the Barents Sea. The study area, covering 2870 km², is located immediately south

of Bjørnøyrenna at water depths ranging from 250 to 450 m (Fig. 1). Powerful computer workstation interpretation techniques have allowed detailed 3-D mapping and visualization of areas and features of particular interest (Rafaelsen *et al.* 2000). Detailed interpretation of the 3-D

Table 1. *Relationship between the present seismo-stratigraphic units defined from 3-D data and earlier work based on 2-D data*

Horizons / units	Units in Sættem *et al.* (1992a)		Corresponding units and their ages
⌐bE	E	b5	5E	22–18 ka (Vorren *et al.* 1989, 1990)
	D	b4	4E	*c.* 28 ka (Vorren *et al.* 1989, 1990)
			b4	< 27 320 ^{14}C years BP (Hald *et al.* 1990)
			I	Solheim & Kristoffersen (1984)
⌐bD	C	b2?	D_2	200–130 ka (Sættem *et al.* 1992b)
⌐bC	B			
⌐bB	A	b1		Younger than URU (< *c.* 800 ka)
⌐bA				

seismic dataset suggests the potential contribution of industry 3-D seismic data to studies of the shallow sediments in the Barents Sea.

The objectives of this paper are, first, to emphasize the value of conventional 3-D seismic data in order to derive detailed morphological information of good horizontal resolution from shallow buried horizons; and, secondly, to focus on how the observed morphological features can help us to understand palaeo-conditions better, in this case the direction of palaeo-ice flow and palaeo-ice sheet extent.

Data and methods

This study is mainly based on 2870 km² of industry 3-D migrated seismic data acquired in 1998 (Lillevik & Lyngnes 1998), made available to the University of Tromsø by the former Saga Petroleum ASA (now Norsk Hydro ASA). 2-D seismic data, industry well logs and shallow borehole logs have also been used with the purpose of stratigraphic correlation to previous seismic studies in the southwestern Barents Sea. Constant velocities of 1500 m s⁻¹ and 1750 m s⁻¹ have been applied for the water and the glacigenic sediment layer above the URU when converting time to depth (Sættem *et al.* 1992b). This gives a depth range for the zone of interest of 290–540 m below sea level (bsl).

The three conventional 3-D seismic surveys SG9803, SG9804 and SG9810 (Fig. 1a) were acquired with a depth of source and receiver in the range of 5–8 m. The acquisitions consist of dual sources (flip-flop shooting) with 75 m separation and 6 streamers with 100 m separation (Lillevik & Lyngnes 1998).

With the 3000 m long streamers consisting of 240 channels and a shot point interval of 25 m, the recorded fold is 30 and the recorded bin size is 12.5 m by 37.5 m (inline by cross-line, inlines are orientated parallel to the survey vessel while cross-lines are orientated perpendicular to the inlines). As the applied Hale-McClellan migration algorithm carried out post stack requires input bins to be quadratic, the data were interpolated to 12.5 by 12.5 m. The bin size was finally converted to 25 by 25 m.

The final fold of the data is 20, after spatial decimation from 1440 to 720 channels (30 fold), by common receiver interpolation (60 fold), production of super common mid-point (CMP) input to radon transform demultiple (120 fold) and application of partial stack for the dip-moveout (DMO) and pre stack time migration processors (20 fold). With the applied mute, the processing-generated fold in the glacigenic sediments varies from 1 to maximum 3.

The 3-D seismic surveys SG9803 and SG9804 differ from SG9810 by having a shot point interval of 18.75 m, thus giving a recorded fold of 40. However, the end product is the same; a final fold of 20 and a 25 by 25 m bin size.

Pre-migration Fresnel zone

$$= V(T/F)^{0.5} = 195.66 \text{ m}$$

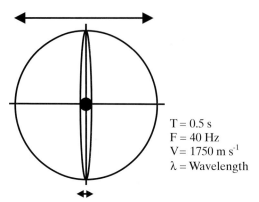

T = 0.5 s
F = 40 Hz
V = 1750 m s^{-1}
λ = Wavelength

Post-migrationFresnelzone

$$= \frac{\lambda}{4} = \frac{V}{4F} = 10.94 \text{ m}$$

Fig. 2. Effect of 2-D and 3-D migration on Fresnel zone size and shape. The Fresnel zone before migration (large circle), after 2-D migration (elliptic shape) and after 3-D migration (small black circle). Modified from Brown (1999).

Early in the processing, the number of samples of the data was reduced by resampling from 2 to 4 ms using a bandpass filter in front with a low cut-off of 6 Hz (24 dB/octave) and a high cut-off of 90 Hz (72 dB/octave) to attenuate the swell noise and to avoid anti-aliasing, respectively. From filter-tests, the highest frequency with geological information in the shallow part is limited up to 80–85 Hz, resulting in a maximum vertical resolution (the ability to distinguish individual reflecting surfaces) of approximately 5 m (=1/4 wavelength). For a dominating frequency of around 40 Hz the corresponding resolution is approximately 10 m.

Velocity analyses were run for every 14 3-D inlines and for every 42 3-D cross-lines to give a square velocity grid of 525 by 525 m. Based on an early phase interpretation of fast-track data, the processing contractor was required to pick velocities on selected target and regional horizons, including the shallow part.

The horizontal resolution (the ability to distinguish separate features on a horizon, given by the post-migrated Fresnel zone; Fig. 2; Brown 1999; Sheriff 1999) is approximately the same as the vertical resolution. Including the relatively high-density grid in the data and the application of the

3-D migration, the total horizontal resolution is good (Fig. 2). By comparison, conventional 2-D migrated seismic data would have an elliptic Fresnel zone (Fig. 2) with improved horizontal resolution only along the line.

In 3-D seismic interpretation, the density of data-points combined with a snap interpretation technique makes it possible to detect sub-sample features. The snap interpretation technique (when searching on maximum amplitude) searches until a true maximum value is found and calculates a sub-sample time value by fitting a quadratic polynomial to the nearest three samples. The peak value of this function is used as a maximum, giving an accuracy of 0.1 ms instead of the sample rate (which is 4 ms). By means of the snap interpretation technique the program can detect the depth of features below the sample interval, explaining why plough marks less than 3 ms (two-way traveltime, TWT) of depth are detectable (Fig. 3). Further, this accuracy is far below the vertical resolution.

The studied 3-D surveys are 16-bit datasets, giving an amplitude range of –32768 to +32767, compared with –127 to +126 in 8-bit datasets. Thus 16-bit datasets give the Automatic Seismic Area Picker (ASAP) more sensitivity in amplitude variation, making the automatic interpretation easier and subtle amplitude variations detectable.

The GeoQuest software GeoFrame 3.8 Charisma IMain, running on a Unix work-station, has been used to interpret the seismic data. Mapping of horizons was achieved mainly by tracing seismic reflections with the ASAP. The ASAP tool provides a more objective way of mapping reflections than hand-drawn interpretations and is able to detect and accurately visualize details on the interpreted horizons. Subtle changes in reflection character that may look like minor disturbances or noise in the seismic section are generally easily recognized on time slices and horizon maps, and may turn out to be interesting geological features that can be traced over significant areas. A good reference is plough marks and lineations (Figs 3 & 4); features which, in the worst case, would not be detected by manual mapping or, at best, would take several days to interpret in detail.

Comparison of maps generated from different specific amplitude extremes, such as maximum amplitude (peak), minimum amplitude (trough) or upper/lower zero crossing (where the amplitude changes from positive to negative, or the opposite), shows that upper and lower zero crossings give the most detailed maps of the shallow palaeo-surfaces. In order to preserve the details from the sub-sampled snap

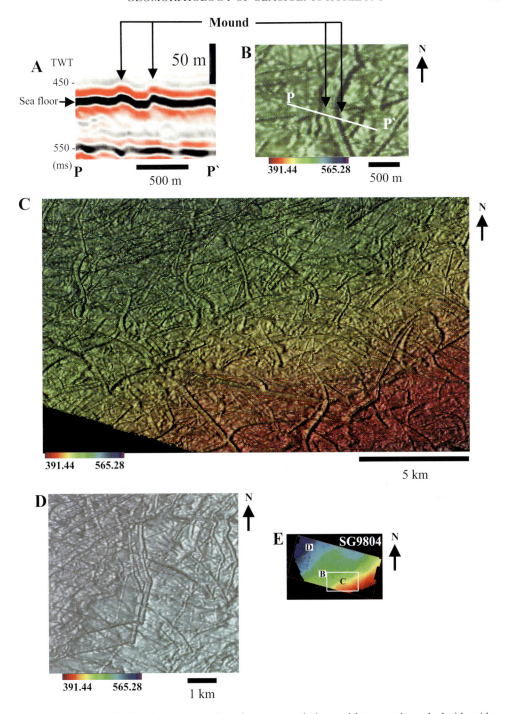

Fig. 3. (**a**) Seismic profile showing a cross-section of an asymmetrical curved furrow at the seabed with a ridge of ploughed sediments on each side. (**b**) An illuminated time–structure map showing the furrow in (a). (**c**) An illuminated time–structure map with curved furrows in a chaotic pattern. (**d**) An illuminated time–structure map showing twinned furrows. (**e**) The approximate location of the illuminated time–structure maps (b), (c) and (d) in 3-D survey SG9804. In all illuminated time–structure maps the light source is located east of the horizons. Colours show depth in ms (two-way traveltime) below sea level.

interpretation, no smoothing has been applied to the final maps.

In addition to application of amplitude information, different attribute maps like instantaneous frequency, reflection strength, cosine of phase, apparent polarity, instantaneous phase, dip and azimuth have been generated in an attempt to extract more information from the dataset regarding specific features.

In the search for geological features not

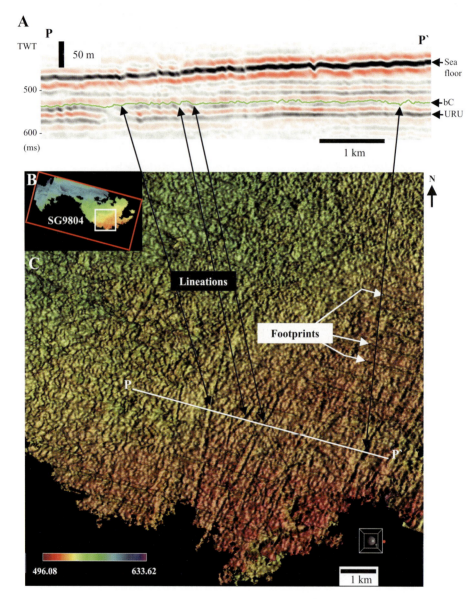

Fig. 4. (**a**) Seismic profile showing irregularities on horizon bC in SG9804. The green line represents the interpreted horizon bC. (**b**) Map with time-structure contours indicating the location of (c) within 3-D survey SG9804. (**c**) An illuminated time-structure map of horizon bC showing lineations interpreted to be formed subglacially. The 'footprints' aligned parallel to the course of the survey vessel (inlines) are artefacts related to the acquisition of the 3-D seismic data. Depth in ms (two-way travel-time) below sea level is shown on the colour bar in (c), but applies for (b) as well. The light source is located east of the horizon and vertical exaggeration (*z*-axis) is 8×.

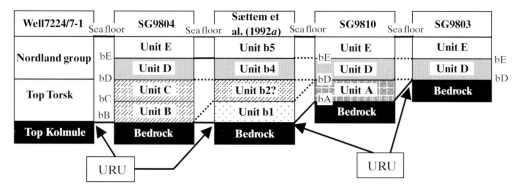

Fig. 5. Seismo-stratigraphic units and horizons from the 3-D seismic surveys correlated with the units of Sættem *et al.* (1992*b*). Unit b2? is inferred to be present only in SG9804 while unit b1 is inferred to be present only in SG9810. Units b4 and b5 (Sættem *et al.* 1992*b*) correlate to units 4E and 5E (Vorren *et al.* 1989, 1990), respectively (Table 1). Broken lines indicate uncertain correlations. Not to scale. Bedrock is mainly of Triassic, Cretaceous and Palaeocene age.

related to horizons, a set of volumetric attributes and horizon slices has been generated. For visually enhancing the subtle features, a program called GeoViz has been used (allowing the user to manipulate colours, select the position and aspect of the light source, the vertical exaggeration (*z*-axis) and the angle of the horizon).

So-called 'footprints' are artefacts occurring in the 3-D seismic data. They resemble lineations parallel to the inline direction (Fig. 4). The footprints have the same spacing throughout the dataset (with a separation equal to every 12 common mid point lines) and should not be regarded as geological information, but rather as a result of the pattern of movement of the survey vessel.

Stratigraphy of the glacigenic sequence

In the study area, the glacigenic sequence is up to 146 m thick. We have identified five seismic horizons (plus seabed) bounding five glacigenic units, termed A (oldest) to E (youngest) (Fig. 5). The oldest seismic unit, A, is only observed in 3-D area SG9810 and is so thin that the lineations observed at its basal reflection are interpreted to be dominated by interference from the seismic response of the top of unit A. The other, overlying younger seismic units, B, C, D and E are separated by pronounced seismic reflectors, here termed horizon bB, bC, bD and bE (bB = base unit B, etc.). The seismic horizons (bA, bB, bC, bD, bE and the sea floor) are here interpreted to represent unconformities formed during several glaciations. The upper regional unconformity is a diachronous surface and is interpreted to be

represented by horizon bD in SG9803, bB in SG9804 and bA in SG9810. The seismic units (A, B, C, D & E) probably represent deposition of glacimarine sediments or basal till (Solheim & Kristoffersen 1984). Units A–C in the 3-D surveys are probably of mid-Pleistocene age, whereas units D and E are of upper Pleistocene age (Sættem *et al.* 1992*b*).

The glacigenic sequence of the southwestern Barents Sea, called the Barents Sea Synthem (Vorren *et al.* 1989), has been mapped and divided into several seismostratigraphic units (Solheim & Kristoffersen 1984; Vorren *et al.* 1990; Sættem *et al.* 1992*b*). A shallow borehole (7222/09-U-01), located close to the 3-D surveys of our study area (Fig. 1), has been correlated to 2-D seismic data by Hald *et al.* (1990) and Sættem *et al.* (1992*a*). Their work is the basis for the correlation of our 3-D seismic data to this shallow borehole and to 2-D seismic data from the same area (Fig. 5; Table 1).

Unit A corresponds with unit b1 in Sættem *et al.* (1992*b*) and is younger than the upper regional unconformity but older than unit B (Table 1). Units B and C (Fig. 5) probably correlate with the lower and upper part of unit b2 of Sættem *et al.* (1992*b*), respectively. Unit b2 is described by Sættem *et al.* (1992*b*) as a soft, dark claystone, where the lower part contains structures that may have been formed by iceberg turbation or glaciotectonics. Unit b2 was probably deposited 200–130 ka BP, and the lower part of unit b2 has been interpreted to be deposited in an ice-proximal glacimarine environment whereas the upper part has been interpreted to be a till (Sættem *et al.* 1992*a*).

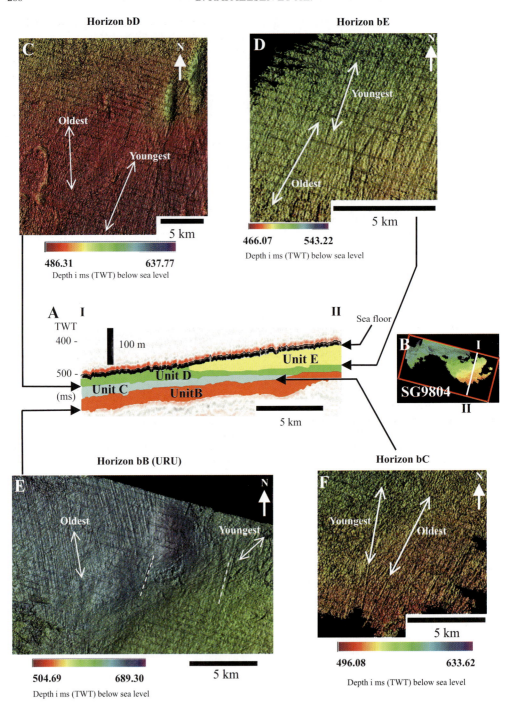

Fig. 6. (**a**) Stratigraphy from the 3-D survey SG9804. (**b**) Horizon bC from SG9804 shown with time–structure contours, indicating the location of the seismic section in (a). (**c–f**) Sections of illuminated time–structure maps from the interpreted horizons in 3-D survey SG9804, showing subglacial depressions and different generations of subglacial lineations. A summary of the dominant orientations is shown in Fig. 8 and Table 2. Depth in ms (two-way travel-time) below sea level is given on the colour bars. The light source is located to the east of the horizon and vertical exaggeration (*z*-axis) is 8×.

A

B

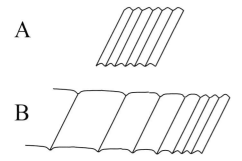

Fig. 7. Sketches illustrating the difficulty in determining whether a feature is positive or negative. (**a**) Closely spaced features prevent the user in giving a positive or negative determination. (**b**) Negative features may be identified when they occur as isolated lineations.

According to Vorren & Laberg (1996), most of the southern Barents Sea was deglaciated during the Arnøya Interstadial (29–24 ka BP; Andreassen et al. 1985), when our unit D was probably deposited. A glacier advance just before this interstadial probably formed horizon bD (this paper). Sediments near the base of unit b4 of Sættem et al. (1992b), which probably correspond to our unit D (Fig. 5; Table 1), are radiocarbon dated to be younger than 27 320 ± 735 years BP (unit b4 of Hald et al. 1990). This corresponds well with the time frame of the Arnøya Interstadial. Unit b4 of Sættem et al. (1992b) is over-consolidated

(Sættem et al. 1992b) and corresponds to unit 4E from Vorren et al. (1989, 1990), which is interpreted to be a glaciofluvial sandy sediment, probably deposited by meltwater (Vorren et al. 1989). In addition, unit D (this paper) correlates with sequence I in Solheim & Kristoffersen (1984).

Lineations on horizon bE (Fig. 6) indicate that our study area was again ice-covered, probably in the beginning of LGM I (Late Glacial Maximum I, 24–22 ka BP; Vorren & Laberg 1996). After LGM I, an ice-free period called the Andøya Interstadial (22–19 ka BP) took place, and unit E was probably deposited in this period. Our unit E corresponds with unit 5E from Vorren et al. (1989, 1990; Table 1), which is interpreted to be deposited in a distal glacimarine environment. The sediments may be derived from meltwater rivers in the south and deposited from suspension together with some ice-rafted debris during the ice-free period between LGM I and LGM II (Vorren & Laberg 1996).

Our 3-D seismic data are acquired from an area where other authors (Sættem et al. 1992b; Vorren & Laberg 1996; Lebesbye 2000) have had difficulty in correlating glacigenic seismic units between the eastern and western Barents Sea. The correlation problem is due mainly to the fact that most of the units belong either to a western or an eastern province and wedge out in and around our study area (Figs 5 & 6a). Additional 2-D and 3-D seismic data are required to resolve this problem.

Table 2. *Size and predominant orientation of subglacial lineations from the four subsurfaces in 3-D surveys SG9804 and SG9810 (youngest generation first, oldest generation last). The dominant orientations are 349°–50°/169°–230°. No lineations were found on the sea floor in the 3-D surveys. Depth calculated using 1750 m s⁻¹ as the velocity of sound in sediments (Sættem* et al. *1992b)*

3-D survey	Horizon (Fig. 5)	Length	Width	Relief	Dominant orientation		N↑	
					Y	O	Y	O
SG 9804	bE	1.2–20 km	100–170 m	2.5–8 m	22° / 202° & 30° / 210°		⤢	⤢
	bD	2.5–12 km	60–150 m	2.5–8 m	26° / 206° & 176° / 356°		⤢	↕
	bC	0.8–7 km	60–180 m	3.5–6 m	10° / 190° & 28° / 208°		↕	⤢
	bB	2–7 km	60–150 m	2.5–7 m	50° / 230° & 169° / 349°		⤡	↕
SG 9810	bE	7–8 km	50–60 m	2.5–3.5 m	48° / 228°		⤢	
	bD	0.5–20 km	50–120 m	2.5–6 m	45° / 225° & 5° / 185°		⤢	↕
	bA	0.5–20 km	50–120 m	3.5–6 m	45° / 225° & 5° / 185°		⤢	↕

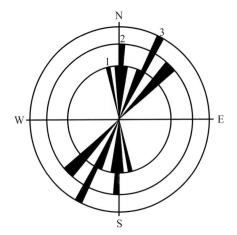

Fig. 8. Rose diagram showing the 13 dominant orientations of subglacial lineations on the horizons in the 3-D surveys. The main trends are 349°–50°/169°–230°.

Glacial erosion surfaces

The units defined here are separated by four horizons interpreted to have been formed by subglacial erosion. Features observed on the horizons in the 3-D seismic surveys are mainly negative features (although it is not clear whether the lineations are positive or negative) and have been divided into three main classes: cross-cutting furrows, straight lineations and depressions. The depressions are further subdivided into type I (smooth edges), type II (indented edges) and type III (formed in bedrock) depressions.

Cross-cutting furrows are mainly observed on the sea floor, but a few occur on one of the buried horizons (bE in SG9810). The cross-cutting furrows observed in the study area are interpreted to be iceberg plough marks, and indicate a glacimarine environment.

Straight lineations and depressions are interpreted as indications of palaeo-ice flow directions which, in turn, may be related to palaeo-ice sheet form and extent. The straight lineations are observed on mapped horizons bB to bE and are interpreted to be subglacial bedforms. This implies that warm-based, or at least periodically warm-based, grounded ice sheets have advanced over the study area at least four times. Oblong depressions with similar orientation to the lineations occur on horizon bD in 3-D survey SG9804. The depressions are interpreted as having been formed by subglacial erosion and to record palaeo-ice flow direction, thus supporting

the indications from the lineations of dominant palaeo-ice flows towards the north or south.

Cross-cutting furrows (iceberg plough marks)

Description. Curved furrows of negative relief that cross-cut each other in a chaotic pattern are the dominant morphological features on the sea floor (Fig. 3). Similar curved furrows also occur on buried horizon bE in SG9810, but not on any other buried horizon in the study area. The sea floor furrows are 30–500 m wide, 1–20 km long and 2.5–25 m (3–28 ms TWT) deep. In profile, their cross-section is either U- or V-shaped (and sometimes asymmetrical), and some have a small rise on each side (Fig. 3a, b). The relief of the furrows is measured from the bottom of the feature to the top of the flanking rise, or to the surrounding sea floor level if there is no rise on either side of the feature. The furrows usually have a constant depth along the entire feature, and on the sea floor they show a slightly east–west dominant orientation. In some cases, the cross-cutting furrows occur in pairs spaced apart up to 700 m (e.g. Fig. 3d).

Interpretation. The cross-cutting curved furrows are interpreted as representing ploughmarks formed in a submarine environment by drifting icebergs, a phenomenon well known in the literature (Lien 1983; Vorren *et al.* 1983; Barnes *et al.* 1984; Stoker & Long 1984; Solheim *et al.* 1988; Longva & Bakkejord 1990; Dowdeswell *et al.* 1992; Vogt *et al.* 1994; Crane *et al.* 1997; Long & Praeg 1997; Polyak 1997; Solheim 1997). The plough-marks in our study area have a maximum relief of 25 m and are up to 500 m wide and 20 km long. Lien (1983) described plough-marks from the Norwegian continental shelf that are up to 27 m deep, and Crane *et al.* (1997) mention plough-marks 450–850 m bsl from the southern Yermak Plateau that are up to 10 m deep, 1000 m wide and, according to Polyak (1997), 15 km long. The largest icebergs may extend deep underwater and have sufficient mass and momentum to force the keel down into the sea floor (Lien 1983). Thus, provided other factors are equal, icebergs reaching the sea floor in deep waters may plough to much greater depths into the underlying sediment than those that run aground in shallower waters. The plough-marks typically have a U- or V-shaped cross-section with a ridge of ploughed sediments on each side (Lien 1983).

The size and shape of the plough-marks

varies greatly. Buried plough-marks 246 ms below the sea floor have been described from the central North Sea and the Norwegian continental shelf (Long & Praeg 1997). Plough-marks have been observed on the Yermak Plateau NW of Svalbard at depths of up to 850 m below the present sea level (Crane *et al.* 1997). In the Barents Sea, plough-marks have been found at all depths down to 450 m, although the keels of present-day icebergs rarely exceed 100 m (Solheim 1997).

Paired furrows observed in our study area (Fig. 3d) are interpreted as having been formed by icebergs with two keels, each forming separate plough-marks, a phenomenon described previously by Barnes *et al.* (1984) and Longva & Bakkejord (1990). Our observations imply that the iceberg forming the largest paired plough marks in the 3-D surveys was at least 700 m wide at the sea floor.

Plough-marks commonly occur on the sea floor in the study area, but only a few were found on buried horizon bE in SG9810. The reason why plough-marks are not found on other buried horizons is probably that subsequent glaciations have removed previously formed plough-marks.

Straight lineations (subglacial bedforms)

Description. Sub-parallel lineations are observed on most of the buried horizons of the glacigenic sequence (Figs 4 & 6) and can be traced over long distances (up to 20 km). Most have a straight linear outline, although a few are curvilinear. When aligned parallel to each other with close spacing (130–250 m) between each lineation, it cannot be determined with certainty whether they are positive or negative features (Fig. 7a). However, detailed studies of isolated lineations and of lineations lying relatively far apart indicate that some of them may be negative rather than positive features (Fig. 7b).

The lineations on buried surfaces in the study area are mostly U-shaped in cross-section and of constant depth along most of the lineation, becoming shallower at both ends before they gradually disappear. These lineations may occur on all parts of some buried horizons, or only in particular areas. They appear on both bedrock horizons (URU) of early Cretaceous age (Sigmond 1992) and horizons within the glacigenic sequence. Usually a surface is dominated by lineations with one or two main orientations, although a few isolated lineations with other orientations may also be present (Fig. 6a). Most lineations range in size from 0.5 to 20 km long, 50–180 m wide and have a relief of 2.5–8 m

(Table 2). The orientations of the lineations are in the range 349°–50° and 169°–230° (Table 2 & Fig. 8). When determining the age relationship between lineations of different orientations on a single horizon in a 3-D survey, the set that clearly cross-cuts another is interpreted to be the younger. When lineations cross-cut one another, the older must be removed or replaced by the younger to make the age relation unequivocal.

Interpretation. The long, straight and relatively narrow features on the buried horizons are interpreted as subglacial bedforms indicating the direction of palaeo-ice flow. They closely resemble large-scale lineations and megaflutes described in the literature (Boulton & Clark 1990; Clark 1993). Glacial flutes are defined as narrow ridges of sediment aligned parallel to glacier flow (Benn 1994) and flutes on the sea floor have been described from the north–central Barents Sea at water depths of 150–340 m (Solheim *et al.* 1990). They have a positive relief of less than 1 m, widths of 4–8 m, lengths typically 100–500 m, and are thus much smaller lineations than those observed on our buried horizons. From the Ross Sea continental shelf, Antarctica, lineations described to be 100–200 m wide, up to several tens of kilometres long and with a relief of the order of several metres are interpreted as mega-flutes (Shipp & Anderson 1997).

Some of the straight lineations in the 3-D seismic areas occur in bedrock and may resemble grooves. Pudsey *et al.* (1997) describe grooves from the Antarctic Peninsula shelf with relief of 2–3 m offshore and 10–20 m inshore at 200–700 m water depth. Bennett & Glasser (1996) interpret glacial grooves, ranging from a few metres to several kilometres long, to be streamlined depressions formed by areal ice flow. Other authors refer to grooves that were probably formed by subglacial erosion with meltwater as the active erosive agent (Kor *et al.* 1991; Rains *et al.* 1993), but the features they describe are not as straight and persistent as the lineations in our 3-D datasets. Morphologically, glacial grooves are comparable to glacial striations, except that grooves are larger and deeper (Bennett & Glasser 1996). Both glacial striations and grooves indicate the direction of local glacier movement and indicate a warm-based glacier. Based on the above discussion, we cannot exclude the possibility that some of the lineations on the upper regional unconformity are grooves.

In recent years, continental lineations composed of glacial sediment have been recognized on satellite images (Punkari 1993, 1997; Boulton

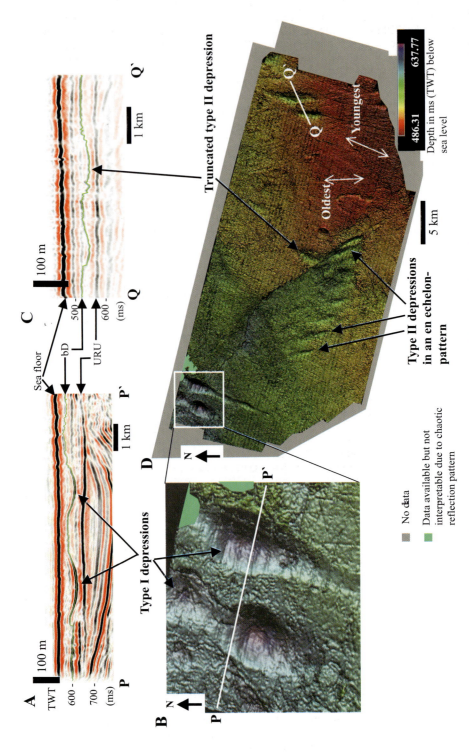

Fig. 9. (**a**) Seismic profile showing buried type I depressions in the northwestern part of 3-D survey SG9804. (**b**) An enlarged area of (d), showing type I depressions. (**c**) Seismic profile showing buried type II depressions in the northeastern part of 3-D survey SG9804. (**d**) Illuminated time–structure map of horizon bD. White arrows indicate the main orientations of the lineations and their relative age (26°/206° = youngest, 176°/356° = oldest). Depth in ms (two-way travel-time) below sea level is shown on the colour bar. The light source is located east of the horizon and vertical exaggeration (z-axis) is 8×.

et al. 2001) and classified as mega-scale glacial lineations (Bennett & Glasser 1996). Typical lengths range between 8 and 70 km, widths between 200 and 1300 m and the spacing between lineations may vary between 300 and 5000 m. On the ground, their morphology is often difficult to detect. According to Clark (1993), mega-scale glacial lineations are formed by differential subglacial sediment deformation caused by variations in bed characteristics, and their long length reflects rapid ice flow and/or long periods of time for development. Using late Quaternary chronologies for the James Bay lowlands, where mega-scale glacial lineations were generated by the last Laurentide Ice Sheet, Clark (1993) concluded that these large stream-lined landforms were produced by a fast-flowing glacier at velocities typical of modern ice streams (400–1600 m a^{-1}). On land in northern Europe and the northwestern part of Russia, 0.1–10 km long erosional forms have been observed on satellite imagery and aerial photographs (Punkari 1993). In the same region, flow-parallel sediment ridges, 10s–1000s of metres in length, metres to 10s of metres in height and 10s to 100s of metres in width have been observed (Boulton *et al.* 2001).

Three-dimensional marine reflection seismic time–structure maps (Figs 4 & 6) provide similar information about the sea floor and buried surfaces that satellite images do for continental surfaces. The lineations observed on 3-D seismic images of buried surfaces in our study area bear many similarities to subglacial lineations described from satellite images (Boulton & Clark 1990; Bennett & Glasser 1996). Differential compaction at burial or the nature of the sediments in and between the lineations might have affected the observed lineations. For the interpretation of their origin, it may be significant that the large-scale lineations are observed on buried horizons whereas the flutes and mega-scale flutes are described from sea floor and continental surfaces. Flutes have a lower preservation potential than negative features because they are readily degraded by wind and water and, according to Boulton & Dent (1974), they are far more common on modern forelands than on older terrain. We suggest that the lineations observed on buried horizons of the glacigenic section are large-scale lineations formed beneath a warm-based ice sheet, and thus indicate dominant ice-flow directions.

Deformation of subglacial bedforms is strongest near the margin of warm-based glaciers where old lineations are obliterated (Bennett & Glasser 1996). Ice streams can effectively erode the orientated elements representing earlier flow phases, but in areas of weak marginal erosion indications of older flow directions of the last ice sheet may remain (Punkari 1993). Several generations of lineations from the last glaciation have been observed in northern Europe and northwestern Russia, where the older ones are suggested to have formed during melting-bed phases of ice-sheet growth and preserved by frozen bed conditions during the glacial maximum (Boulton *et al.* 2001). A change in the ice margin may explain why there are two dominant directions of ice flow on several horizons in our study area. As an ice sheet retreats, the shape of the ice margin and the nature of the calving front may change, causing the ice flow to change accordingly. The lineations trending north–south and NNE–SSW in the study area may, therefore, have formed during deglaciation phases. If this is so, possible lineations from earlier phases may have been removed as the glacier retreated and subjected the study area to strong deformation. Another possibility, although less likely, is that the oldest directions are preserved features from an earlier phase, such as from the ice sheet advance or at its maximum position, and that only the youngest were formed close to the ice margin during ice sheet retreat.

Depressions (subglacial erosion)

Description. Two kinds of incised semi-circular to oblong depressions are observed on horizon bD in 3-D survey SG9804, and a third kind is observed on the horizon bB (URU) in the same 3-D survey. Type I depressions (Fig. 9a & 9b) are 35–53 m (40–60 ms TWT) deep, 1560–1875 m wide and 1875 to more than 9000 m long. Their long axes are orientated 26°/206°. These depressions have a concentric shape, are widest in the middle and become narrower as they shallow. Their deepest point is slightly NNE of their centre. A common feature of the depressions is that they have a smooth surface and even edges, which are not marked by the straight lineations that are observed on the surface surrounding the depressions.

Type II depressions (Fig. 9c & 9d) are relatively shallow, 3750–5625 m in length, 1250–1700 m in width and 9–32 m (10–37 ms TWT) in depth. Their longest axes are orientated 8°/188° and 169°/349°. They are elliptical in shape, deepest in the middle and become shallower towards the edges. The edges are rough and indented and some of the depressions seem to occur in series, where the depressions are mostly aligned parallel to each other, resembling an *en echelon* pattern (Fig. 9d). All of the depressions in such a series are about the same size.

Type III depressions are observed on the horizon bB (URU) in SG9804, in early Cretaceous bedrock (Fig. 6e). They are semicircular with a slight north–south trending elongation, ranging from 3125–3200 m in width, 3500–3850 m in length and 9–13 m (10–15 ms TWT) deep. Their edges seem to be a combination of type I and II depressions, in that they have relatively smooth edges even though the lineations on the horizon reach their edge (Fig. 6e). This is caused by the fact that the depth of the lineations gradually decreases as they reach the edge of the depressions.

Interpretation. The depressions (Figs 6e & 9) are thought to have been formed by subglacial erosion. After the depressions were formed, they may have been modified by meltwater. Type I depressions (Fig. 9a & 9b) may have formed either coeval with or after the formation of the straight lineations, since the depressions have smooth edges. We believe that it is most likely that the type I depressions formed after the nearby lineations, mainly because their orientations differ significantly. Depressions with their deepest point NNE of their centre may indicate that the glacier moved towards the NNE, but ice flow in the opposite direction cannot be ruled out from the morphological observations alone.

Type II depressions (Fig. 9c & 9d) may have been formed in both directions along their longest axis as their deepest point is in their centre. The depressions were probably formed before the straight lineations because the straight lineations extend all the way to the edge of the depressions, making the edges rough and indented. If these interpretations are correct, type II depressions are probably older than the lineations and type I depressions are younger than both lineations and type II depressions. Another possibility is that both type I and type II depressions formed before the lineations, but later only the eastern part of the 3-D survey (type I depressions) was modified by meltwater.

Type III depressions (Fig. 6e) may indicate an ice flow towards the north or south, but because they are so rounded this is speculative. They may have been formed in areas where the bed was easier to erode.

Depressions formed in bedrock by subglacial erosion vary in size from a few metres to several hundred metres in diameter (Bennett & Glasser 1996). Subglacial meltwater erosion in unconsolidated sediments is also thought to have formed depressions of about 200 m in diameter (Rampton 2000); such depressions formed in bedrock are called S-forms and are about 0.5 by 1 km (Beaney & Shaw 2000). Negative, longitudinal forms formed by meltwater erosion are called spindle flutes in the classification of Kor *et al.* (1991), regardless of size. They are narrow, shallow negative forms that are longer than they are wide. They become slightly wider downstream and may be asymmetrical. The shape of these spindle flutes is similar to type II depressions on horizon bD in SG9804 (Fig. 9c & 9d). Examples of eroded depressions (S-forms) formed by meltwater are described from Alberta (Beaney & Shaw 2000) and from the northeastern coast of Georgian Bay, Canada (Kor *et al.* 1991). The depressions observed in this study are interpreted as having been formed by erosion beneath a warm-based ice sheet, and some (especially type I depressions) are suspected to have been modified by meltwater.

Discussion

The morphology of the buried surfaces of late Cenozoic age, including the upper regional unconformity (URU) that separates the upper unconsolidated sediments from underlying consolidated rocks (bedrock), suggests that extensive glacial erosion and deposition has taken place in the study area.

Palaeo-ice flow and extent

Interpretation of 3-D seismic morphological features of the four buried horizons of the study area provides important information about glaciations and deglaciations in the southwestern Barents Sea.

Glacial advances. The existence of subglacial lineations on the four main buried horizons of our study area (horizon bB, bC, bD & bE; Fig. 6; Table 2) indicates at least four glacial advances in the southern Barents Sea. In addition, there may have been a glacial advance prior to these four, as horizon bA is interpreted to represent the URU. Unfortunately, the morphological features on horizon bA are questionable. On the horizons bB, bC, bD and bE the large size and persistence over several tens of kilometres of the lineations suggests that they were formed subglacially. Their orientation indicates that ice flow directions mainly towards north and NNE (or south and SSW) were present in all four glacial advances. The slightly different orientations of the two sets of lineations on all four buried horizons of 3-D

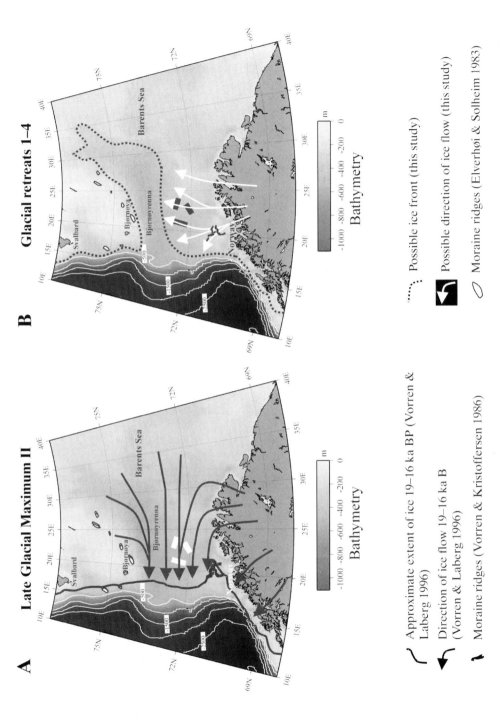

Fig. 10. (a) Extent of ice sheet during advance 5 (LGM II; Vorren & Laberg 1996). (b) Hypothetical extent of ice sheets during retreats 1 to 4 proposed in this study.

areas SG9804 and SG9810 (Table 2) may reflect different ice-flow patterns.

An ice-flow direction towards the west has previously been suggested for the study area during glacial maxima when the ice sheet reached the shelf edge (Fig. 10a; Landvik *et al.* 1998; Vorren *et al.* 1990; Vorren & Laberg 1996). Most of the features in our study area are orientated north–south, and both bathymetry, sub-aerial topography and ice-sheet modelling (Dowdeswell & Siegert 1999) suggests an ice-flow direction towards north. The Bjørnøyrenna is relatively deep (up to 500 m below present sea level) and may have acted as a calving bay during a glacial retreat. The observed lineations and depressions may have formed during an ice-sheet retreat (Fig. 10b), when Bjørnøyrenna was ice free, obliterating features formed during the earlier stages of glaciation. Though it is less likely, the possibility that the oldest generation of lineations on a horizon have formed during a glacial advance and been preserved beneath the ice sheet, while the youngest generation of lineations were formed during deglaciation, cannot be excluded. A third possibility is that these features were formed while the ice sheet reached the shelf edge, and that ice flow turned from trending north–south in our study are to a more east–west orientation when entering the Bjørnøyrenna. Lineations in unlithified glacigenic sediments beneath a warm-based ice sheet are probably degraded relatively rapidly, and we believe that it is more likely that older generations of lineations and depressions have been removed and that just the last couple of generations descending from deglaciation phases (Fig. 10b) have been preserved.

Plough-marks

The plough marks on the sea floor indicate a glacimarine environment. Their chaotic cross-cutting pattern shows a slightly east–west dominant orientation, suggesting that the prevailing wind and ocean-current direction in the study area were towards the east or west. When the study area was ice free, but still proximal to the ice, katabatic winds and melt-water discharge may have affected the direction of iceberg drift. Vorren *et al.* (1990) indicate that Bjørnøyrenna, and part of the area just south of Bjørnøyrenna, may be among the first parts of the southwestern Barents Sea to become ice free during ice retreats, thus forming a large ice-free bay. In this bay, the current direction is likely to have been towards the east in the southern part and towards west in the northern part of the bay (Vorren *et al.* 1990). In the early stages of the deglaciation, the dominant direction of iceberg drift may have been east–west trending, but as larger parts of the southwestern Barents Sea became ice-free the direction of iceberg drift became more chaotic.

Conclusions

1. The present study illustrates that conventional industry 3-D seismic data, acquired for the purpose of studying deeper targets, can provide detailed morphological information on shallow buried glacigenic sequences.

2. The morphology of the buried horizons, together with the repeated glacigenic sequence, suggests that extensive erosion and deposition has taken place.

3. Long straight, negative lineations observed on four buried horizons, are interpreted as having formed beneath a warm-based ice sheet, reflecting the dominant ice-flow directions. The lineations provide conclusive evidence that grounded ice has reached the southern part of Bjørnøyrenna at least four times since the formation of the upper regional unconformity.

4. A dominantly north–south orientation of glacial lineations observed on all buried horizons suggests that they were formed during ice sheet retreats.

5. Plough-marks occur mainly on the sea floor and indicate that icebergs have reached depths of 456 m below present sea level, and that some icebergs were up to 700 m wide at the sea floor.

Norsk Hydro ASA and the European Communities project TriTex (IST–1999–20500) are acknowledged for funding the research project. Norsk Hydro ASA, Statoil ASA, Norsk Agip A/S, Fortum Petroleum A/S and former Saga Petroleum ASA are acknowledged for providing the seismic and well data. The University of Tromsø acknowledges support from GeoQuest via computer software and help on technical issues. We offer our sincere thanks to D. Praeg and A. Solheim who critically reviewed the manuscript and to P. E. B. Armitage who corrected the English language.

References

ANDREASSEN, K., VORREN, T. O. & JOHANSEN, K. B. 1985. Pre-Late Weichselian glacimarine sediments at Arnøy, North Norway. *Geologiska Føreningens i Stockholm Førhandlingar*, **107**, 63–70.

BARNES, P. W., REARIC, D. M. & REIMNITZ, E. 1984. Ice gouging characteristics and processes. *In:* BARNES, W., SCHELL, D. M. & REIMNITZ, E. (eds). *The Alaskan Beaufort Sea: Ecosystems and*

Environments. Academic Press, Orlando, 184–212.

BEANEY, C. L. & SHAW, J. 2000. The subglacial geomorphology of southeast Alberta: evidence for subglacial meltwater erosion. *Canadian Journal of Earth Sciences*, **37**, 51–61.

BENN, D. I 1994. Fluted moraine formation and till genesis below a temperate valley glacier: Slettmarkbreen, Jotunheimen, southern Norway. *Sedimentology*, **41**, 279–292.

BENNET, M. R. & GLASSER, N. F. 1996. *Glacial Geology – Ice Sheets and Landforms*. John Wiley & Sons Ltd., West Sussex.

BOULTON, G. S. & CLARK, C. D. 1990. A highly mobile Laurentide ice sheet revealed by satellite images of glacial lineations. *Nature*, **346**, 813–817.

BOULTON, G. S. & DENT, D. L. 1974. The nature and rates of post-depositional changes in recently deposited till from south-east Iceland. *Geografiska Annaler*, **56A**, 121–134.

BOULTON, G. S., DONGELMANS, P., PUNKARI, M. & BROADGATE, M. 2001. Palaeoglaciology of an ice sheet through a glacial cycle: the European ice sheet through the Weichselian. *Quaternary Science Reviews*, **20**, 591–625.

BROWN, A. R. 1999. *Interpretation of Three-Dimensional Seismic Data*, 5th edition. American Association of Petroleum Geologists, Tulsa, Oklahoma, Memoir, **42**.

CLARK, C. D. 1993. Mega-scale lineations and cross-cutting ice-flow landforms. *Earth Surface Processes and Landforms*, **18**, 1–29.

CRANE, K., VOGT, P. R. & SUNDVOR, E. 1997. Deep Pleistocene iceberg plowmarks on the Yermak Plateau. *In:* DAVIES, T. A., BELL, T., COOPER, A. K., JOSENHANS, H., POLYAK, L., SOLHEIM, A., STOKER, M. S. & STRAVERS, J. A. (eds) *Glaciated Continental Margins: An Atlas of Acoustic Images*. Chapman & Hall, London, 140–141.

DOWDESWELL, J. A. & SIEGERT, M. J. 1999. Ice-sheet numerical modeling and marine geophysical measurements of glacier-derived sedimentation on the Eurasian Arctic continental margins. *Geological Society of America Bulletin*, **111**, 1080–1097.

DOWDESWELL, J. A., WHITTINGTON, R. J. & HODGKINS, R. 1992. The sizes, frequencies and freeboards of East Greenland icebergs observed using ship radar and sextant. *Journal of Geophysical Research*, **97**, 3515–3528.

EIDVIN, T., JANSEN, E. & RIIS, F. 1993. Chronology of Tertiary fan deposits off western Barents Sea: implications for the uplift and erosion history of the Barents Sea shelf. *Marine Geology*, **112**, 109–131.

EIDVIN, T., GOLL, R. M., GROGAN, P., SMELROR, M. & ULLEBERG, K. 1998. The Pleistocene to Middle Eocene stratigraphy and geological evolution of the western Barents Sea continental margin at well site 7316/5-1 (Bjørnøya West area*). Norsk Geologisk Tiddskrift*, **78**, 99–123.

HALD, M., SÆTTEM, J. & NESSE, E. 1990. Middle and late Weichselian stratigraphy in shallow drillings from the southwestern Barents Sea: foraminiferal, amino acid and radiocarbon evidence. *Norsk Geologisk Tidsskrift*, **70**, 241–257.

KNUTSEN, S.-M., RICHARDSEN, G. & VORREN, T. O. 1992. Late Miocene-Pleistocene sequence stratigraphy and mass-movements on the western Barents Sea margin. *In:* VORREN, T. O., BERGSAKER, E., DAHL-STAMNES Ø. A., HOLTER, E., JOHANSEN, B., LIE, E. & LUND, T. B. (eds) *Arctic Geology and Petroleum Potential*. NPF Special Publication, Elsevier, Amsterdam, **2**, 573–606.

KOR, P. S. G., SHAW, J. & SHARPE, D. R. 1991. Erosion of bedrock by subglacial meltwater, Georgian Bay, Ontario: a regional view. *Canadian Journal of Earth Sciences*, **28**, 623–642.

LANDVIK, J. Y., BONDEVIK, S., ELVERHØI, A., FJELDSKAAR, W., MANGERUD, J., SALVIGSEN, O., SIEGERT, M. J., SVENDSEN, J. I. & VORREN, T. O. 1998. The last glacial maximum of Svalbard and the Barents Sea area: Ice sheet extent and configuration. *Quaternary Science Reviews*, **17**, 43–75.

LEBESBYE, E. 2000. *Late Cenozoic Glacial History of the Southwestern Barents Sea*. PhD thesis, University of Tromsø.

LIEN, R. 1983. Pløyemerker etter isfjell på norsk kontinentalsokkel. PhD thesis, *Institutt for kontinentalsokkelundersøkelser*, **109**, 147 pp.

LILLEVIK, A. & LYNGNES, B. 1998. Processing report, Saga Petroleum ASA, Loppa East/South, Seismic 3D processing on SG9803, Geco-Prakla Norway.

LONG, D. & PRAEG, D. 1997. Buried ice-scours: 2D vs 3D-seismic geomorphology. *In:* DAVIES, T. A., BELL, T., COOPER, A. K., JOSENHANS, H., POLYAK, L., SOLHEIM, A., STOKER, M. S. & STRAVERS, J. A. (eds) *Glaciated Continental Margins: An Atlas of Acoustic Images*. Chapman & Hall, London, 142–143.

LONGVA, O. & BAKKEJORD, K. J. 1990. Iceberg deformation and erosion in soft sediments, Southeast Norway. *Marine Geology*, **92**, 87–104.

POLYAK, L. 1997. Overview. *In:* DAVIES, T. A., BELL, T., COOPER, A. K., JOSENHANS, H., POLYAK, L., SOLHEIM, A., STOKER, M. S. & STRAVERS, J. A. (eds) *Glaciated Continental Margins: An Atlas of Acoustic Images*. Chapman & Hall, London, 136–137.

PUDSEY, C. J., BARKER, P. F. & LARTER, R. D. 1997. Glacial flutes and iceberg furrows, Antarctic Peninsula. *In:* DAVIES, T. A., BELL, T., COOPER, A. K., JOSENHANS, H., POLYAK, L., SOLHEIM, A., STOKER, M. S. & STRAVERS, J. A. (eds) *Glaciated Continental Margins: An Atlas of Acoustic Images*. Chapman & Hall, London, 58–59.

PUNKARI, M. 1993. Modelling of the dynamics of the Scandinavian ice sheet using remote sensing and GIS methods. *In:* ABER, J. (ed.). *Glaciotectonics and Mapping Glacial Deposits, Proceedings of the INQUA Comission on Formation and Properties of Glacial Deposits*. Canadian Plains Research Center, University of Regina, 232–250.

PUNKARI, M. 1997. Subglacial processes of the Scandinavian Ice Sheet in Fennoscandia inferred from flow-parallel features and lithostratigraphy. *Sedimentary Geology*, **111**, 263–283.

RAFAELSEN, B., MIDTBØ, M., KULIMAN, L. W., LEBESBYE, E., HOGSTAD, K. & ANDREASSEN, K. 2000. 3-D seismic data used to investigate the Late Cenozoic sediments of the southwestern Barents Sea. *Geonytt*, **1**, 139.

RAINS, B., SHAW, J., SKOYE, R., SJOGREN, D. & KVILL, D. 1993. Late Wisconsin subglacial megaflood paths in Alberta. *Geology*, **21**, 323–326.

RAMPTON, V. N. 2000. Large-scale effects of subglacial meltwater flow in the southern Slave Province, Northwest Territories, Canada. *Canadian Journal of Earth Sciences*, **37**, 81–93.

RICHARDSEN, G., HENRIKSEN, E. & VORREN, T. O. 1991. Evolution of the Cenozoic sedimentary wedge during rifting and seafloor spreading west of the Stappen High, western Barents Sea. *Marine Geology*, **101**, 11–30.

SHERIFF, R. E. 1999. *Encyclopedic Dictionary of Exploration Geophysics*, 3rd edition. Society of Exploration Geophysicists, Tulsa.

SHIPP, S. & ANDERSON, J. B. 1997. Lineations on the Ross Sea Continental Shelf, Antarctica. *In:* DAVIES, T. A., BELL, T., COOPER, A. K., JOSEN-HANS, H., POLYAK, L., SOLHEIM, A., STOKER, M. S. & STRAVERS, J. A. (eds) *Glaciated Continental Margins: An Atlas of Acoustic Images*. Chapman & Hall, London, 54–55.

SIGMOND, E. M. O. 1992, Berggrunnskart, Norge med havområder. Målestokk 1: 3 millioner. *Norges Geologiske Undersøkelse*. Emil Moestue A/S, Oslo.

SOLHEIM, A. 1997. Depth-dependent iceberg plough marks in the Barents Sea. *In:* DAVIES, T. A., BELL, T., COOPER, A. K., JOSENHANS, H., POLYAK, L., SOLHEIM, A., STOKER, M. S. & STRAVERS, J. A. (eds). *Glaciated Continental Margins: An Atlas of Acoustic Images*. Chapman & Hall, London, 138–139.

SOLHEIM, A. & KRISTOFFERSEN, Y. 1984. Sediments above the upper regional unconformity: Thickness, seismic stratigraphy and outline of the glacial history. *Norsk Polarinstitutt Skrifter*, **179B**, 26 pp.

SOLHEIM, A., RUSSWURM, L., ELVERHØI, A. & BERG, M. N. 1990. Glacial geomorphic features in the northern Barents Sea: direct evidence for grounded ice and implications for the pattern of deglaciation and late glacial sedimentation. *In:* DOWDESWELL, J. A. & SCOURSE, J. D. (eds). *Glacimarine Environments: Processes and Sediments*. Geological Society, London, Special Publication, **53**, 253–268.

SOLHEIM, A., MILLIMAN, J. D. & ELVERHØI, A. 1988. Sediment distribution and sea-floor morphology of Storbanken: implications for the glacial history of the northern Barents Sea. *Canadian Journal of Earth Sciences*, **25**, 547–556.

STOKER, M. S. & LONG, D. 1984. A relict ice-scoured erosion surface in the central North Sea. *Marine Geology*, **61**, 85–93.

SÆTTEM, J. 1994. Glaciotectonic structures along the Barents shelf margin. *In:* WARREN, W. P. & CROOT, D. G. (eds) *Formation and Deformation of Glacial Deposits*. Balkerna, Rotterdam, 95–113.

SÆTTEM, J., POOLE, D. A. R., ELLINGSEN, L. & SEJRUP, H. P. 1992*a*. Glacial geology of outer Bjørnøyrenna, southwestern Barents Sea. *Marine Geology*, **103**, 15–51.

SÆTTEM, J., RISE, L. & WESTGAARD, D. A. 1992*b*. Composition and properties of glacigenic sediments in the southwestern Barents Sea. *Marine Geotechnology*, **10**, 229–255.

SÆTTEM, J., BUGGE, T., FANAVOLL, S., GOLL, R. M., MØRK, A., MØRK, M. B. E., SMELROR, M. & VERDENIUS, J. G. 1994. Cenozoic margin development and erosion of the Barents Sea: Core evidence from southwest of Bjørnøya. *Marine Geology*, **118**, 257–281.

VOGT, R. P., CRANE, K. & SUNDVOR, E. 1994. Deep Pleistocene iceberg plowmarks on the Yermak Plateau: Sidescan and 3.5 kHz evidence for thick calving ice fronts and a possible marine ice sheet in the Arctic Ocean. *Geology*, **22**, 403–406.

VORREN, T. O. & KRISTOFFERSEN, Y. 1986. Late Quaternary glaciation in the south-western Barents Sea. *Boreas*, **15**, 51–59.

VORREN, T. O. & LABERG, J. S. 1996. Late glacial air temperature, oceanographic and ice sheet interactions in the southern Barents Sea region. *In:* ANDREWS, J. T., AUSTIN, W. E. N., BERGSTEN, H. & JENNINGS, A. E. (eds) *Late Quaternary Palaeoceanography of the North Atlantic Margins*. Geological Society, London, Special Publications, **111**, 303–321.

VORREN, T. O., HALD, M. EDVARDSEN, M. & LIND-HANSEN, O-W. 1983. Glaciogenic sediments and sedimentary environments on continental shelves: General principles with a case study from the Norwegian shelf. *In:* EHLERS, J. (ed.) *Glacial Deposits in North-west Europe*. Balkema, Rotterdam, 61–73.

VORREN, T. O., KRISTOFFERSEN, Y. & ANDREASSEN, K. 1986. Geology of the inner shelf west of North Cape, Norway. *Norsk Geologisk Tidsskrift*, **66**, 99–105.

VORREN, T. O., HALD, M. & LEBESBYE, E. 1988. Late Cenozoic environments in the Barents sea. *Paleoceanography*, **3**, 601–612.

VORREN, T. O., LEBESBYE, E., ANDREASSEN, K. & LARSEN, K. B. 1989. Glacigenic sediments on a passive continental margin as exemplified by the Barents Sea. *Marine Geology*, **85**, 251–272.

VORREN, T. O., LEBESBYE, E. & LARSEN, K. B. 1990. Geometry and genesis of the glacigenic sediments in the southern Barents Sea. *In:* DOWDESWELL, J. A. & SCOURSE, J. D. (eds) *Glacimarine Environments: Processes and Sediments*. Geological Society, London, Special Publications, **53**, 269–288.

VORREN, T. O., RICHARDSEN, G., KNUTSEN, S.-M. & HENRIKSEN, E. 1991. Cenozoic erosion and sedimentation in the western Barents Sea. *Marine and Petroleum Geology*, **8**, 317–340.

Retreat signature of a polar ice stream: sub-glacial geomorphic features and sediments from the Ross Sea, Antarctica

STEPHANIE S. SHIPP, JULIA S. WELLNER & JOHN B. ANDERSON

Department of Earth Science, MS-126, Rice University, 6100 Main Street, Houston, Texas, 77005, USA (e-mail: shippst@rice.edu)

Abstract: Three research expeditions to the Ross Sea, Antarctica resulted in collection of a dataset of more than 270 km of side-scan and chirp-sonar data, more than 330 km of swath bathymetry and 3.5 kHz data, and 24 cores within a glacially-carved trough. The former ice-stream flow path is divided into six zones, covering a distance of approximately 370 km, distinguished by unique stratigraphic signatures and geomorphic features. An erosional surface with thin, patchy lodgement till characterizes Zone 1. This region is interpreted as having experienced relatively high basal shear stress conditions. Zones 2, 3, and 4 are characterized by an erosional surface and thin, time-transgressive subglacial and grounding-line proximal deposits that include back-stepping moraines, flutes, transverse moraines, and corrugation moraines. These zones represent the transition between erosional and depositional regimes under the expanded LGM ice sheet; material eroded from the inner shelf was transported toward the outer shelf, possibly as a thin deforming till layer. The two outer zones are depositional and include maximum grounding-line (Zone 5) and pro-glacial deposits that were overridden subsequently by the ice sheet (Zone 6). Surface features include mega-scale glacial lineations, corrugation moraines, and iceberg furrows. Ice in these zones is interpreted as having experienced relatively lower basal shear stress, an extensional regime, and faster flow. This advance may have destabilized the ice sheet, initiating local draw-down and production of icebergs that furrowed the sea floor. Corrugation moraines are thought to represent annual retreat moraines, constraining the retreat rate of the ice sheet across the continental shelf to a consistent 40 to 100 m a^{-1}.

A long-standing goal of the glacial geological community is to understand the processes taking place beneath ice sheets (e.g. Boulton 1976; Boulton & Jones 1979; Boulton & Hindmarsh 1987; Clark, C. 1994; Piotrowski & Tulaczyk 1999). Key questions about subglacial conditions concern the roles of meltwater and substrate (e.g. Alley 1989; Alley *et al.* 1989; Fowler & Walder 1993; Clark & Walder 1994; Blankenship *et al.* 2001). For example, is the ice moving over a mobile bed or sliding over pockets or a film of water? Further, what is the distribution of soft versus hard substrate? How might substrate distribution over a broad region influence the movement of an ice sheet, in particular a marine-based ice sheet? Answers to these questions will help to constrain and model the behaviour of past ice sheets and predict the behaviour of modern ice sheets.

Unfortunately, direct observations of the subglacial environment are rare (e.g. Boulton & Hindmarsh 1987; Powell & Molnia 1989), and investigations rely on linking features exposed by the retreat of an ice sheet with the processes responsible for their formation. Many studies that examine features and processes associated with ice sheets focus on either a broad region with a general view (e.g. Boulton & Clark 1990; Alley 1991; Clark, P. 1994; Boulton 1996*b*), specific features (e.g. Lundqvist 1989; Menzies 1989), or regional histories where processes may not be the focus of interest (e.g. Vorren *et al.* 1989; Josenhans & Zevenhuizen 1990; Solheim *et al.* 1990; Shipp *et al.* 1999). Wellner *et al.* (2001) describe systematic changes to large-scale bedforms along the axes of West Antarctic Ice Sheet (WAIS) ice streams based on swath bathymetry data. Recession of the ice sheet is marked by a variety of relatively small-scale geomorphic features that are resolved only in high resolution side-scan sonar records. Characterization of these recessional bedforms and deposits is important for understanding retreat of the ice sheet. Two questions remain unanswered: 1) was the ice sheet in contact with the bed during retreat?; and 2) what is the relationship between bedforms and sediment lithology?

This investigation provides detailed documentation of the features and sediments deposited beneath a previously expanded polar ice sheet along the extent of a single glacial trough, and uses this information to interpret the subglacial processes associated with the deposits. Additionally, an attempt is made to

From: DOWDESWELL, J. A. & Ó COFAIGH, C. (eds) 2002. *Glacier-Influenced Sedimentation on High-Latitude Continental Margins.* Geological Society, London, Special Publications, **203**, 277–304. 0305-8719/02/$15.00
© The Geological Society of London 2002.

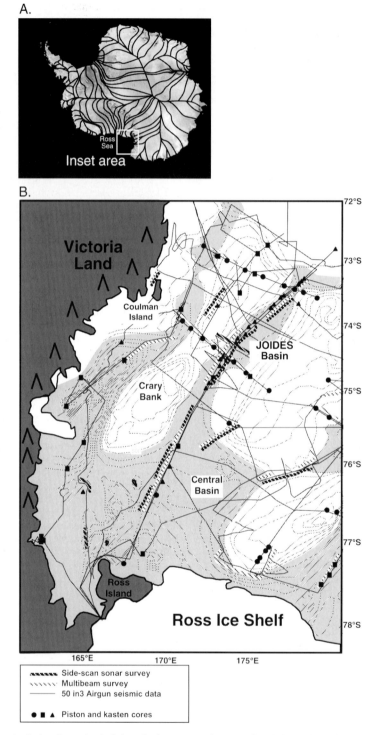

Fig. 1. The geological and geophysical data, bathymetry and geography of the study region. Depths greater than 500 m are marked by light grey stipple. The data were collected during 1994, 1995, and 1998 cruises aboard RV/IB *Nathaniel B. Palmer*.

determine which part of a glacial cycle is represented by the preserved features (e.g. Clark, C. 1999; Stokes & Clark 2001). Toward these goals, a swath bathymetry, high-resolution chirp-sonar, and side-scan sonar dataset was collected in 1998 along what is now one of the most extensively sampled trough axes on the Ross Sea continental shelf (Fig. 1). This study also included sedimentological analyses of cores collected along the trough axis and adjacent banks. The objective was to determine if there was an association between geomorphic features and sedimentary properties that might provide insight into the conditions at the base of the former ice stream. These data augment a pre-existing dataset of cores, lower-resolution seismic profiles, and swath bathymetry.

Investigation setting

The study area occurs within western Ross Sea. The Ross Sea is bounded on the west by the Transantarctic Mountains (TAM) of the Victoria Land coast, to the east by the 180° parallel, to the south by the Ross Ice Shelf edge and to the north by the continental shelf break (Fig. 1). Shelf depths in the Ross Sea range from less than 250 m to greater than 1200 m and average more than 500 m (Barnes & Lien 1988; Davey 1994). NE–SW trending banks, ridges and troughs dominate the bathymetry. The troughs are foredeepened; the inner shelf is deeper than the outer shelf due to repeated periods of glacial erosion during which material was removed from the inner shelf and transported to the outer shelf (Johnson et al. 1982; ten Brink et al. 1995). The inner shelf is characterized by more consolidated strata relative to the outer shelf (Anderson & Bartek 1992). The dataset is concentrated in a trough extending from Central Basin on the inner shelf to JOIDES Basin on the outer shelf (Fig. 1).

During the last glacial maximum (LGM), ice from the East Antarctic Ice Sheet expanded across and through the TAM, grounding on the western Ross Sea continental shelf to the vicinity of Coulman Island (Domack et al. 1995a, b; Licht et al. 1996; Domack et al. 1999; Shipp et al. 1999). The WAIS was grounded across the central and eastern Ross Sea to the continental shelf edge (Domack et al. 1999; Shipp et al. 1999). The expanded ice sheets left a history of their activity in their deposits. The Ross Sea is ideal for observing well-preserved subglacial features. The shelf is over-deepened and not subject to reworking by icebergs, tides, or currents, barring the outer shelf and shallow banks. In addition, the thin cover of the recent glacial-marine sediment does not obscure the subglacial features.

Previous work

There has been considerable debate as to the extent of glaciation in the Ross Sea during the LGM (e.g. Hughes 1977; Drewry 1979; Stuiver et al. 1981; Denton et al. 1989, 1991; Kellogg et al. 1996). Hughes (1977) first suggested that the troughs on the western Ross Sea continental shelf reflected extension of West Antarctic ice streams across the continental shelf during glacial advances. Recent geophysical and sedimentological investigations have established the presence of grounded ice on the continental shelf in the western Ross Sea region to the vicinity of Coulman Island (Domack et al. 1995a, b; Hilfinger et al. 1995; Licht 1995; Licht et al. 1996, 1999; Domack et al. 1999; Shipp et al. 1999). The results from these marine geological investigations are in agreement with the findings of terrestrial studies (Hall & Denton 1998). Grounded ice occupied the troughs and banks. The inner shelf shows signs of glacial erosion; deposition occurred primarily on the outer shelf and in the troughs. The glacial maximum is marked by a grounding-zone wedge that exhibits mega-scale glacial lineations at the surface (Shipp et al. 1999). Ice retreated from the troughs, with each trough exhibiting an independent retreat history (Domack et al. 1999). Ice remained longer on the banks, possibly forming ice shelves over the troughs, as reflected by the off-bank deposits observed in some locations (Shipp et al. 1999).

Methods

The geophysical dataset for this study includes 3.5 kHz (chirp) transects, 12 kHz swath-bathymetry data, 2–7 kHz deep-tow side-scan and 90–100 kHz chirp-sonar records (Fig. 1). Geological data include piston and kasten cores. The 3.5 kHz data delineate the bathymetry and permit characterization of the upper tens of metres of the strata. Chirp-sonar data provide detail to 1 m, complementing the 3.5 kHz dataset. Unit thicknesses were acquired from these datasets. Piston and kasten core information was used to help constrain the sedimentary composition of the features observed on the seismic data.

Plan-view swath bathymetry data display large-scale features of the sea-floor. Side-scan sonar data record features at a finer scale (Shipp et al. 1999). The width of the side-scan footprint matched the width of the larger features

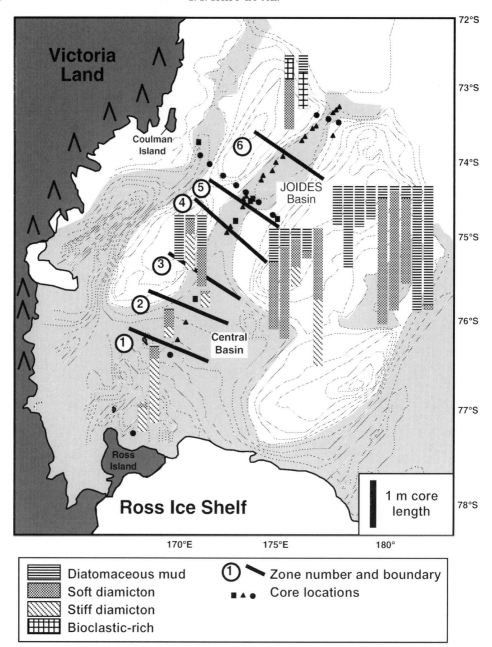

Fig. 2. Lithologies of the cores used in this investigation grouped by zone. Interpretations are generalized. Zonal designations are discussed in the paper and illustrated in subsequent images.

observed on the swath bathymetry. The port channel record from the side-scan system often was weaker than that of the starboard channel; this results in a 'washed-out' appearance of the port display of the records. Where needed, the

swath bathymetry data were relied upon for directional information as the side-scan sonar fish occasionally shifted to being towed at an angle to the path of the ship.

Most of the piston and kasten cores were

collected during a 1998 cruise aboard the RV/IB *Nathaniel B. Palmer*. Based on repeated core attempts, it became apparent that much of the upper section could be lost in a piston core. Kasten cores, however, preserve the upper sediment. At times, cores collected at the same site cannot be correlated completely; this is attributed to loss of the upper section in the piston cores. Piston cores were analysed for magnetic susceptibility and stored, unopened, in a chilled room. Kasten cores were described, photographed, and sampled on board. Piston cores and archive kasten cores were chilled and shipped to the Florida State University's Antarctic Research Facility, where they were opened, X-radiographed, photographed, described, and sampled for shear strength, sediment grain size, microfossils, and radiocarbon age analyses.

Obtaining radiocarbon ages on marine materials from the Antarctic continental shelf is problematic (e.g. Berkman & Forman 1996; Andrews *et al.* 1999; Domack *et al.* 1999; Anderson *et al.* 2002). Most of the dates in this paper are reported uncorrected. Corrected radiocarbon dates are from the work of Domack *et al.* (1999) in which corrections are based on subtracting the surface age date from the dates obtained in the subsurface.

Results

Lithology

The cores used in this study are subdivided into three general lithologies; stiff diamictons, soft diamictons, and diatomaceous mud (Fig. 2).

The stiff diamictons exhibit shear strengths in excess of 0.5 kg cm^{-2}. They are composed of massive, gravelly sandy mud with low water content and sparse, reworked fossils. In X-radiographs, the gravel component rarely displays a preferred alignment. Pebble density and lithology varies little down unit; indeed, lithology varies little over the span of the trough. Magnetic susceptibility shows little variability down core (Licht *et al.* 1999). Where stiff diamicton exists, it occurs below the soft diamicton and diatomaceous mud. A single total organic carbon date in a stiff diamicton yielded an uncorrected age of 31 450 ± 475 years BP (Hilfinger *et al.* 1995).

The soft diamictons exhibit shear strengths below 0.5 kg cm^{-2}. They, too, are gravelly sandy mud. The water content is higher than that of the stiff diamictons. Fossils, including foraminifera, sponge spicules, diatom frustules, and occasional gastropods, bivalves, and corals,

may be present. The fossils may be reworked. Pebble lithology varies little down unit. This unit was sampled in all piston cores. It thickens from the inner shelf to the outer shelf (Fig. 2). Soft diamicton always occurs between the stiff diamicton unit and the diatomaceous mud. The soft diamictons in cores from the outer shelf have been studied most extensively (Fig. 2) (Licht *et al.* 1996, 1999; Domack *et al.* 1999). Soft diamictons from the mid-outer shelf display no stratification, contain reworked microfossils, and yield ages up to 14 000 years BP (Licht *et al.* 1996; Domack *et al.* 1999). Cores from the outer-most shelf may exhibit weak stratification and contain non-reworked fossils. Dates from the deposits in this zone yield ages over 14 000 years BP (Licht *et al.* 1996; Domack *et al.* 1999).

Diatomaceous mud/ooze is distinguished by its olive green colour. This unit caps all other lithologies and is correlated with a transparent drape observed on chirp-sonar images (Shipp *et al.* 1999). Diatomaceous mud contains 10–30% diatoms; ooze contains more than 30%. The unit may be sandy, exhibiting a range of compositions of sand- and gravel-sized debris. Water contents commonly are high (Anderson 1999; Domack *et al.* 1999).

Geomorphic features

The geomorphic features vary along the trough axes and have been broken into six zones. They are described here starting at the landward end of the trough and moving seaward.

Zone 1. The inner shelf, with an average water depth of 725 m, is characterized by a very thin to absent sediment layer overlying a smooth erosional surface. Lineations are observed on swath bathymetry and side-scan chirp sonar data (Fig. 3). The longest measured lineation was more than 4.5 km long. The crest to trough height of the feature is believed to be less than 1 m, based on the available data. The lineations trend approximately N35°E to N55°E. An abrupt change in the basement is observed where a different lithology crops out at the surface; the lineations become more clearly defined and continuous (Fig. 3).

Zone 2. Farther seaward is an area with average water depth of 625 m that is characterized by thin, asymmetric sediment wedges (Fig. 4). The wedges overlie a smooth erosional surface and are capped by a transparent sediment drape. From south (landward) to north (seaward)

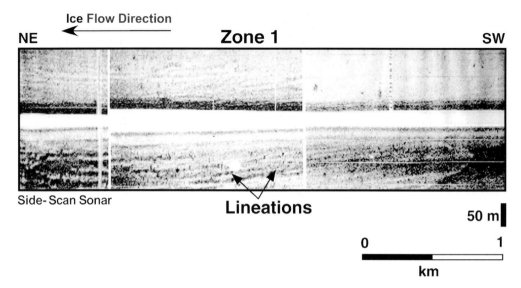

Fig. 3. Side-scan and chirp-sonar data displaying the lineations in Zone 1. This zone is characterized by thin, patchy sediments. Scattered, dark patches are interpreted to be boulders.

through this area, four distinct sub-zones are distinguished based on wedge geometry, and are labelled from A–D. Small (1 m to 2 m high, with crest to crest spacing of less than 100 m), closely spaced, straight-crested symmetrical moraines, referred to as corrugation moraine, occur in each of the sub-areas of Zone 2. They are rare, but where they occur, the spacing between moraines is approximately 60 m.

The southern sub-zone, Zone 2A, extends approximately 8.5 km and contains a single, asymmetric wedge, approximately 12 m high (Fig. 4a). The wedge downlaps onto the lower erosional surface and slopes and thins to the south. The stoss side displays a wavy, chaotic pattern at the surface. The wedge surface is characterized by high frequency, low amplitude corrugations (about 30 m crest to crest; less than 1 m high) and by narrow lineations that occur in association with the corrugations. The lineations are 5 m to 25 m wide and in excess of 7 km long and display a slight sinuosity.

Zone 2B encompasses an 8.5 km long region. It includes 19 asymmetric wedges overlying a smooth erosional surface. The wedge height is approximately 3 m at the crest, although the sediment thickness nowhere exceeds 5 m in this subzone (Fig. 4b). The wedges average 450 m crest to crest. The crests are highly sinuous. Lineations, 5 m to 15 m wide and trending approximately N60°E, occur on the lee side of the wedge, in some locations extending across

the entire wedge back. These are shallower and not as well developed as in the surrounding subzones. The lineations are perpendicular to, and may terminate at, the wedge crests. Small corrugations (up to 1 m high and approximately 50 m crest to crest) occur rarely on the wedge backs.

Zone 2C extends for 9.25 km. It contains 24 asymmetric wedges with sinuous crests. The wedges measure an average of 385 m from crest to crest and range from 6 to 8 m in height (Fig. 4c). Heights are more variable than in Zone 2C. Numerous, well-defined lineations occur on the wedge backs, roughly perpendicular to the wedge crests. The lineations trend from N50°E to N°55 and exhibit approximately 20 m crest to crest spacing. A single lineation may be observed to cross more than one wedge with no degradation in signature over the variable sediment thickness. Lineation terminations appear to occur at wedge crests.

Zone 2D, a 5.4 km long zone, includes 30 asymmetric wedges with an average spacing of 180 m (Fig. 4d). The crests are slightly sinuous. The ridges are 2–3 m high. Occasional, faint lineations, roughly perpendicular to the wedge crests, are recorded on the side-scan sonar data. Lineations trend from N45°E to N°50E.

Overall, the wedges in Zone 2 display increasing number, decreasing sinuosity of the wedge front, decreasing crest to crest distance, and decreasing height from south to north.

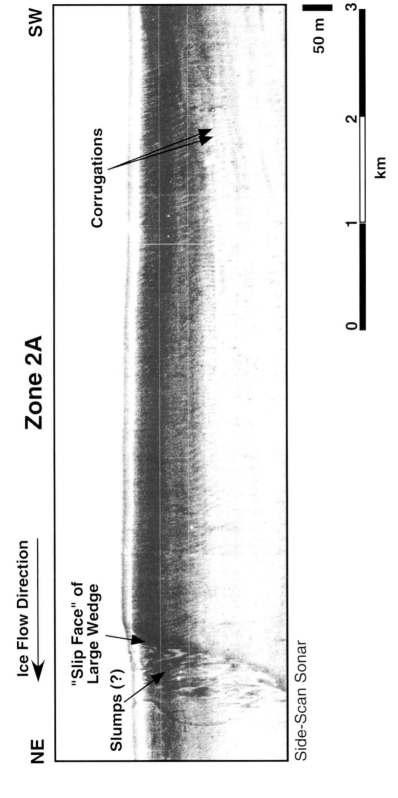

Fig. 4. (**a**) Side-scan sonar data displaying the wedge geometry and surface features for Zone 2A. (**b**) Side-scan sonar image of the back-stepping grounding-zone wedges of Zone 2B. (**c**) Asymmetrical wedges of Zone 2C. (**d**) Sonar image of wedges from Zone 2D. These are smaller and more closely spaced than those in Zone 2C.

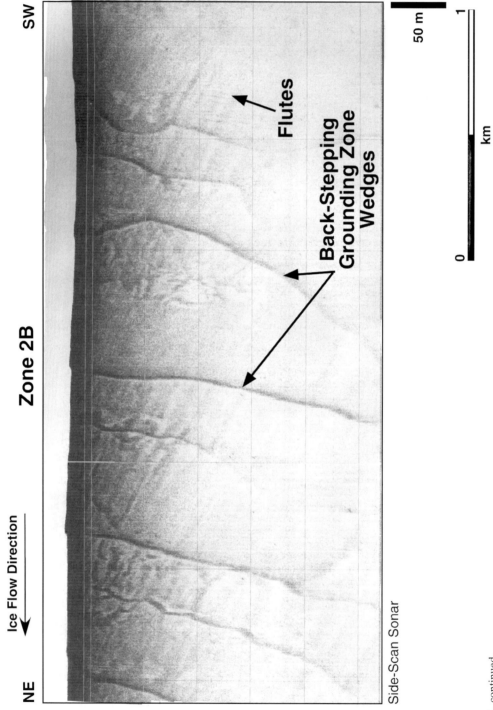

Side-Scan Sonar

Fig. 4. continued.

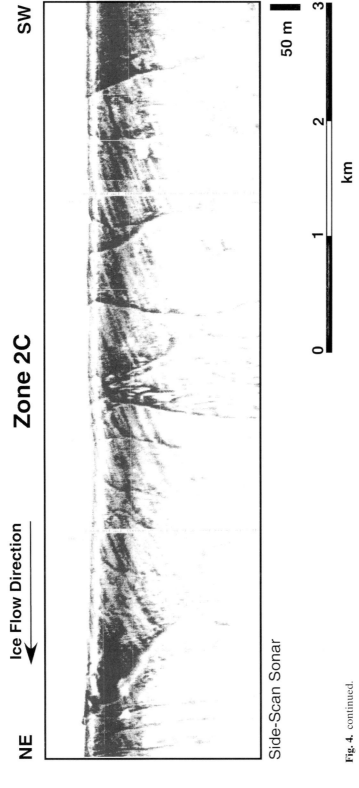

Fig. 4. continued.

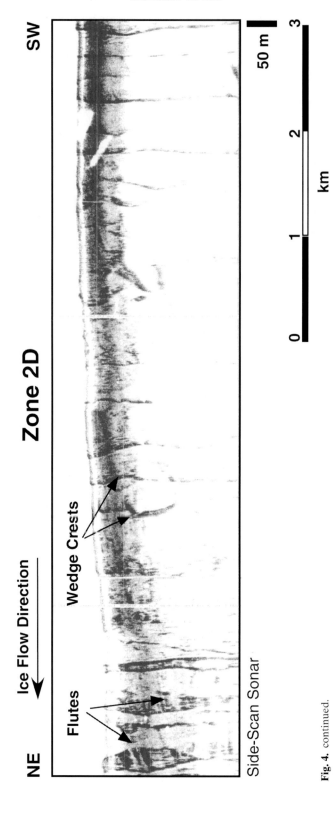

Side-Scan Sonar

Fig. 4. continued.

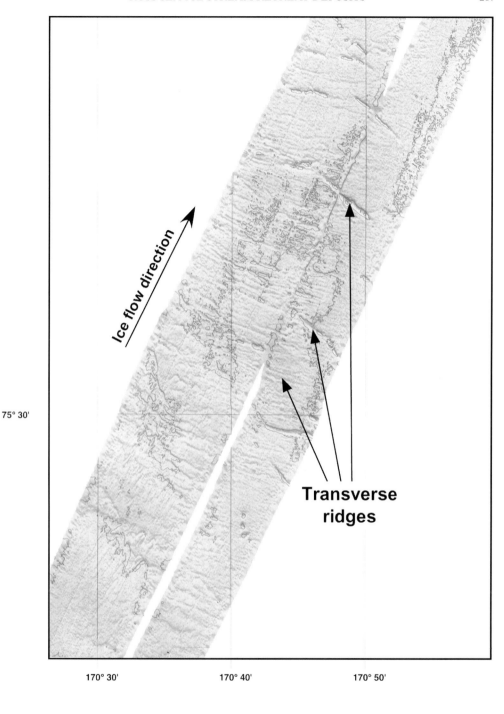

Fig. 5. Swath bathymetry data across zone of ridges in Zone 3; only the largest ridges are visible on the data. The ridges record varying directions of curvature. Note the faint flutes in the southern portion of the display.

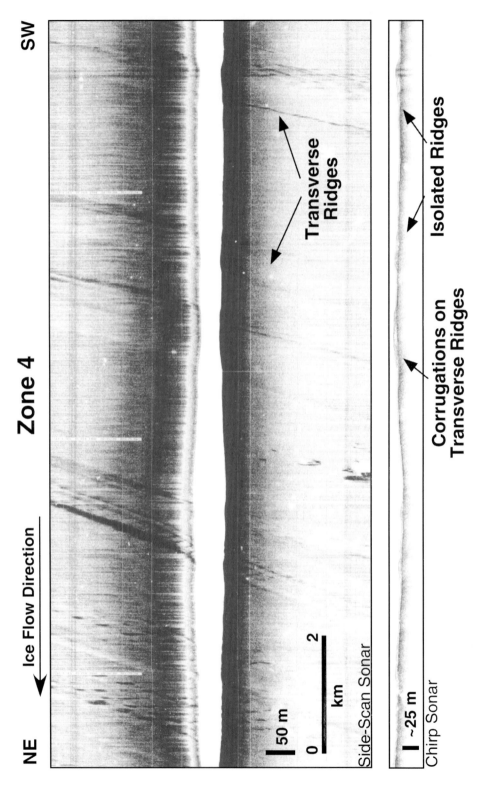

Fig. 6. (a) Side-scan and chirp-sonar data across the straight-crested ridges in Zone 4. Note the corrugations at the surface and the thickening of deposits to the NE. **(b)** Side-scan sonar across the northern edge of Zone 4, near the grounding-zone wedge of Zone 5. Note the corrugations that persist even where cross-cut by iceberg furrows.

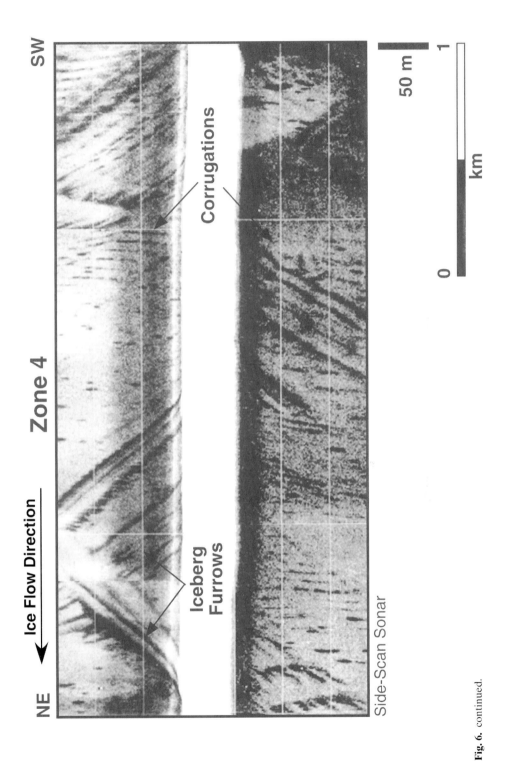

Fig. 6. continued.

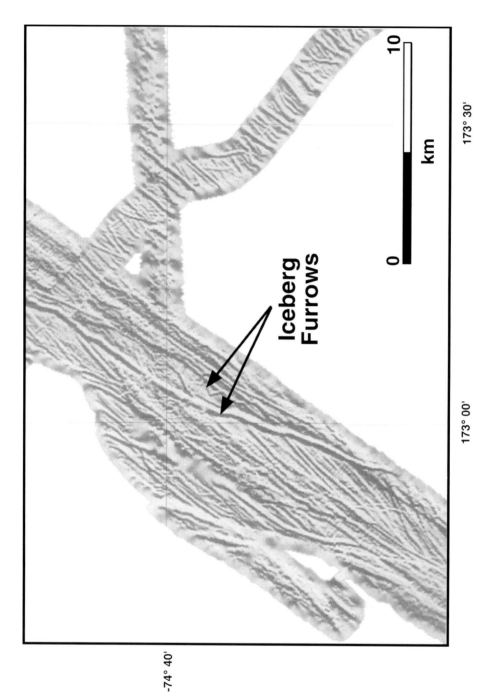

Fig. 7. (a) Swath bathymetry data across the grounding-zone wedge in Zone 5 (Licht *et al.* 1996; Domack *et al.* 1999; Shipp *et al.* 1999). (b) Side-scan sonar data across the wedge display intense iceberg furrowing. Corrugations are apparent in areas that are unfurrowed. Note the larger furrows are cross-cut by smaller furrows. Many of the furrows are 'ribbed'.

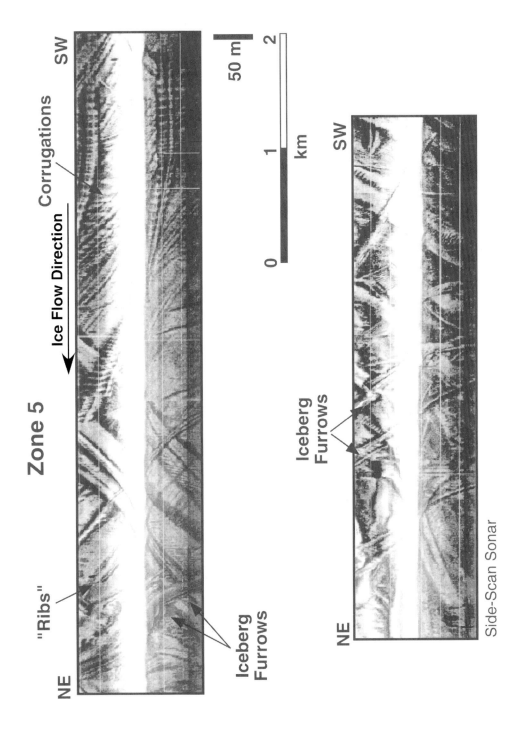

Fig. 7. continued.

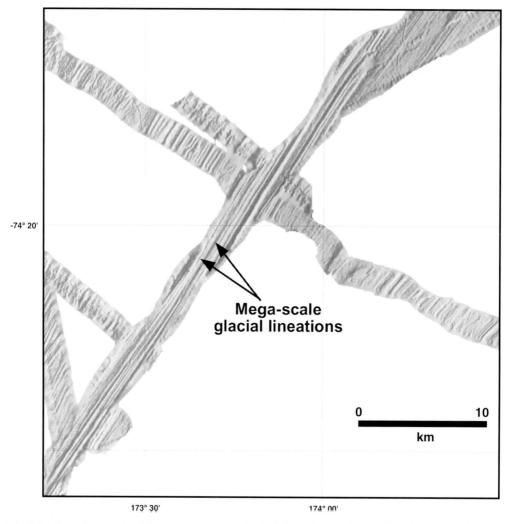

Fig. 8. (a) Swath bathymetry data recording mega-scale glacial lineations in Zone 6. (b) Side-scan and chirp-sonar data across corrugations in Zone 6.

Sediment drape thickness increases to the north. The wedges 'back-step.' In other words, individual ridges downlap the tail of the ridge immediately to the north, indicating that the ridges become increasingly young to the south.

Zone 3. Zone 3 extends 53 km and is distinguished by the presence of several distinct ridges. Side-scan sonar coverage of this region is limited; feature descriptions rely predominately on swath bathymetry and 3.5 kHz data. Side-scan data record numerous, closely spaced, straight-crested ridges with a spacing of 60 to 75 m and height of 2 m (Fig. 5). Rare lineations

occur at the surface. Several larger ridges, ranging from 5 to 19 m in height and measuring a maximum of 300 m in length are observed on the swath bathymetry and 3.5 kHz data (Fig. 5). The ridges have slopes up to 15°. They are at least 7.5 km wide; beyond this distance, they are out of range of the data. In plan view, individual ridges may display either a concave or a convex geometry in a seaward direction. Smaller symmetric wedges of material occur between the ridges (Fig. 5). The wedges are between 2 and 5 m in height. The northern portion of Zone 3 is characterized by smooth sea floor between individual ridges.

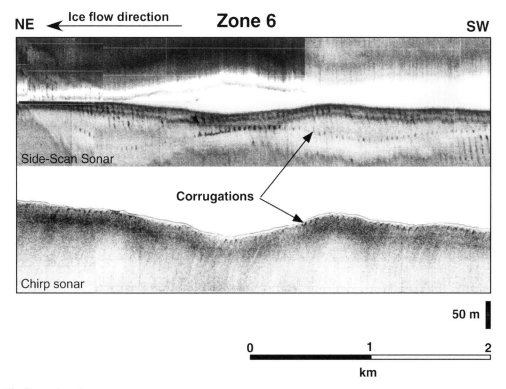

Fig. 8. continued.

Zone 4. Zone 4 covers 81 km and contains isolated ridges of thin sediment and occasional cross-cutting arcuate grooves (Fig. 6). Lineations are not observed in the region. The ridges are straight-crested, 5 m high at the maximum, and less than 770 m wide (Fig. 6a). The crests trend N15°W to N40°W. The ridges are widely spaced in the southern region of Zone 4. A few corrugations are observed on the sea floor and on the ridge crests (Fig. 6b). The crests trend approximately N55°W. The corrugations are less than 1 m high and crest to crest spacing ranges from 40 to 75 m. The sediment drape may obscure their presence; the hummocky signature is observed on cross-sections where corrugations do not appear at the surface. The northern portion of Zone 4 is characterized by more continuous corrugations on the surface (Fig. 6b). The abundance of arcuate grooves increases to the north, forming a pattern of bi-directional grooves on the sea floor.

Zone 5. A distinct wedge overlies an erosional unconformity in Zone 5. It extends 40 km along the trough axis, reaches a maximum thickness of 80 m, and thins in the landward and seaward directions (Shipp *et al.* 1999). Intermediate-resolution seismic data display rare, stacked concave-down reflectors within the predominantly acoustically massive interior. The crest and landward flank of the wedge displays a hummocky surficial character. The hummocks correlate with trough-parallel lineations and later-stage arcuate furrows on swath bathymetry records. Disorganized, slightly sinuous, shallow grooves, the largest of which is 1 km wide, 20 m deep, and in excess of 28 km long, cut the lineations and truncate each other (Fig. 7a). Where corrugations appear, they exhibit a spacing of 70 to 100 m (Fig. 7b).

Zone 6. Zone 6 contains a sheet of massive material, less than 50 m thick, with a subdued signature, that rests on the grounding-zone deposits of Zone 5 or on an older, seaward deposit (Fig. 8) (Shipp *et al.* 1999). Strong, uni-directional lineations exceeding 20 km in length and well-defined corrugations characterize the surface of Zone 6 (Fig. 8). The lineations, aligned roughly parallel to the trough axis, trend an average of N28°E and are 200 m wide (Fig. 8a). The corrugations are regular, with a 50 to 65 m spacing from crest to crest, and a height of up to 2 m (Fig. 8b). The crests trend approximately N50°W. Rare iceberg furrows cross-cut the features; these are seldom apparent on the swath bathymetry data.

Discussion

The geomorphic features and sedimentary facies described above allow discussion of the possible formation processes associated with the features, and placement of these features and deposits in the context of the glacial system.

Sediment associations

The stiff diamicton is interpreted as having been deposited by lodgement processes, based primarily on its high shear strength. Lodgement till develops as sediment is plastered onto the substrate due to pressure melting as the glacier base slides over the substrate (Lundqvist 1988; Dreimanis 1988). The single age date of 31 450 ± 475 years BP (uncorrected) is thought to reflect the presence of open or sub-ice shelf conditions at that time in this location or up-flow line (Hilfinger *et al.* 1995). The lodgement till may have been emplaced by ice advance since about 31 000 years BP. This negates the possibility that the core sampled a lower, older till associated with a previous advance.

A variety of settings, described below, are proposed for deposition of the soft diamicton. The deposits may include grounding-line proximal diamicton, subglacially deformed diamicton, and/or iceberg-turbated diamicton.

Grounding-line proximal diamictons have been described by Licht *et al.* (1996, 1999) and Domack *et al.* (1999). This soft, poorly-sorted sediment is extruded into the marine environment close to the grounding line. The material acquires a marine signature with a non-reworked fossil assemblage and may exhibit faint stratification reflecting subaqueous deposition (Licht *et al.* 1996, 1999). Petrographic distribution exhibits low variability reflecting a single ice-sheet flow-line source (Anderson 1999).

Deforming subglacial till is formed when shear stress at the base of an ice sheet overcomes the strength of the underlying sediment, causing the sediment to move as a deforming bed. Deformation of this material could play a significant role in the rate of glacier movement (Boulton & Jones 1979). A 6 m thick layer of deforming till has been proposed to underlie Ice Stream B in Antarctica and to accelerate its rate of flow (Blankenship *et al.* 1986; Alley *et al.* 1986, 1987, 1989; Engelhardt *et al.* 1990). Deposition from the deforming layer occurs when ice flux or velocity decreases. Hence, distribution of deformation till can be used to infer the dynamic behaviour of former ice sheets (Boulton 1996*a*). This approach has been used in a variety of investigations examining the deposits associated with the Laurentide ice sheet of the LGM (e.g. Boulton & Clark 1990; Alley 1991; Clark, P. 1994).

If the material is extruded into the marine environment and subsequently overridden by an advancing ice sheet, it will acquire a marine signature. Deposits produced in this manner would resemble the soft diamictons observed in the study region. Some of the soft diamictons, in regions where iceberg furrows may be prevalent, may be subglacial diamictons that have been reworked by iceberg activity (Dowdeswell *et al.* 1994). Deformation till can be deposited during advance or retreat of the ice sheet; proximal- and iceberg-turbate glacial-marine deposits would be emplaced during the time of the maximum ice configuration and subsequent retreat.

Diatomaceous mud is deposited in an open-marine environment. Diatoms require light to survive, precluding an ice-shelf, or sub-ice environment. The high variability of petrography reflects ice-rafted debris, deposited from icebergs originating in a variety of settings (Anderson *et al.* 1980, 1984; Anderson 1999).

Of importance to the interpretation are the shelf-wide depositional trends (Fig. 2). Lodgement tills, capped by thin soft diamictons, occur on the inner shelf. The soft diamicton increases in thickness toward the outer shelf. In the central portions of the survey region the deposits are less than 10 m thick, commonly 5 m, based on chirp-sonar data. Outer shelf deposits reach 75 m in thickness. This trend in increasing soft-diamicton thickness is thought to reflect the shift from erosion/little deposition in the interior of the expanded ice sheet to deposition at the outer edges. Material is interpreted as having moved in a conveyor-belt fashion, perhaps as a deforming bed (5 to 10 m thick) under the ice sheet (Powell 1984; Alley *et al.* 1989; Boulton 1996*a*). The materials may have been reworked into the retreat features described below.

Geomorphic-feature associations

On a regional scale, ice sheets erode the substrate underlying the interior portion of the ice sheet. The material is deposited toward the edges in an organized series of features (Sugden & John 1976; Aylesworth & Shilts 1989; Hughes 1995; Van der Wateren 1995; Hart 1999; Wellner *et al.* 2001). The distribution of features can be used to reconstruct ice-sheet conditions.

Zone 1. The erosional surface is close to the sea floor in Zone 1; no deposits can be distinguished on the chirp-sonar or 3.5 kHz data. A single piston core from this region acquired lodgement till, capped by thin soft till and diatomaceous mud. Erosion and patchy lodgement deposition dominate this zone and, thus, little material was available to be reworked during ice-sheet retreat.

Based on the presence of lodgement till, and the thin nature of this material above the erosional surface, ice flow of the expanded ice sheet in Zone 1 may have been slow, with high basal shear stresses. Alternatively, Zone 1 may have been characterized by rapid ice flow by deformation of a very thin till layer. Kamb (2001) argued that a till layer less than 0.5 m thick is all that is required to allow fast flow. A thin layer of till may exist in Zone 1 but was unsampled. Such an interpretation is consistent with the models of the upstream portions of the ice streams that call for little thickening of the ice during the LGM (Ackert *et al.* 1999; Steig *et al.* 2001). This interpretation is supported as well by the work of Wellner *et al.* (2001), which suggests that ice flowing across sedimentary substrate of the continental shelf flows by the mechanism of substrate deformation.

Zone 2. The Zone 2 wedges vary in size from crest to crest (8 km to 180 m) and in height (3 to 12 m) (Fig. 4). The wedge heights match those of the features described in other work; however, they are wider than any of the features described in the literature (e.g. Aylesworth & Shilts 1989; Lundqvist 1989; Zilliacus 1989; Solheim *et al.* 1990). The wedges display an asymmetrical cross-section and are stacked on the backs of each other. Many of them have flutes on the surface, interpreted to have formed simultaneously. The flutes occasionally terminate at a ridge crest, indicating simultaneous formation (Fig. 4a). The flutes do not appear to vary in depth or freshness across the wedge surface. Hence, they are not below the wedges, where thicker sediment might be expected to obscure the flute signature at the crest, nor are they a feature of subsequent re-advance, where the base of the ice would be required to conform exactly to the entire wedge surface.

The wedges in Zone 2 are interpreted as grounding-line features, specifically as grounding-zone wedges deposited as the ice sheet retreated (Powell & Domack 1995). In spite of the intra-zonal variation, all of the wedges are thought to result from the same process. Any interpreted mechanism of formation requires production of an asymmetric wedge with surficial flutes.

The wedges are not considered to be Rogen moraines because they are not known to occur in close proximity to drumlins, nor do they display the classic down-glacier 'horns' of Rogen moraines (Lundqvist 1989). If they were formed by a subglacial process, it may be similar to that proposed by Aylesworth & Shilts (1989) in which material is sheared into the base of the ice and emplaced by meltout upon retreat, although this still requires deposition at the ice-sheet margin. It is also unclear how the flutes could be emplaced and/or preserved on the surface of the features without invoking re-advance of an ice sheet with active basal movement. A mechanism of formation in association with crevasse squeeze (Zilliacus 1989) is discounted because the wedges are strongly asymmetrical, stacked, and display flutes at the surface indicating active ice flow. The crevasse-squeeze mechanism implies extending or stagnant ice; however, preservation of features associated with crevasse squeeze appears to require stagnant ice.

The wedges cannot be classified as classic De Geer moraines, as an annual timing of deposition cannot be ascertained. Although a few cores are available from this region, no features associated with meltwater (e.g. channels, canals) are observed, negating designation as morainal banks (Powell & Domack 1995). The wedges may be associated with glacier push (i.e. push moraines), but no internal structure is observed on chirp-sonar data.

The wedges are classified as grounding-zone wedges deposited as the ice sheet retreated. These are smaller than the grounding-zone wedges (till deltas) described by Alley *et al.* (1987) or the till tongues of King *et al.* (1991), or the wedge observed in Zone 5. Unlike the Alley *et al.* (1987) wedge, no evidence of sediment gravity flow is observed on the available data.

The wedges in Zone 2 were probably emplaced during a pause in grounding-line retreat. As the ice paused, a wedge formed at its outer limit. The shape of the wedge reflects the underside geometry of the ice sheet, based on the presence of flutes. The feature forms either by a combination of lodgement and extrusion processes, or by extrusion of a deforming till into the grounding zone. The features are reworked from material previously in transport under the expanded ice, possibly as a deformation till. Based on the presence of a fairly continuous layer of acoustically massive material in this zone, the deforming layer may have been approximately 5 m thick. The wedges occur above the layer.

Zone 3. Zone 3 deposits are characterized by symmetrical ridges of two different sizes. Rare flutes were observed in a limited area of the smaller ridges. The ridges are orientated transverse to ice flow, and overlie an erosional surface that is close to the sea floor. The large ridges show curvature, although the curves open both up- and down-glacier. Two possible mechanisms of formation are proposed. The ridges formed either in crevasses in the base of the ice or they formed as ice-edge retreat moraines.

A Rogen moraine origin is eliminated because the features do not display a consistent orientation of curvature down-glacier, or an association with drumlins, although the survey area is limited (Figs 1 and 5). Nor are the features considered to be ice-marginal morainal banks or grounding-zone wedges because they exhibit a symmetrical form, apparently are not associated with flutes, and, in the case of morainal banks, do not exhibit association with meltwater-derived features. The large ridges display slopes and geometries similar to the crevasse-fill features interpreted by Zilliacus (1989). The smaller ridges are similar in scale and morphology to the straight-crested De Geer moraines described by Solheim *et al.* (1990) and to the spacing of corrugations observed in other parts of the study.

The deposits in Zone 3 are composed of soft diamicton, and excluding the larger ridges, they are typically thin. A subglacial setting similar to that for Zone 2 is proposed; thin deposits (up to 5 m) were being transported toward the ice margin, probably as a deforming layer. These materials were reworked into the ridge features either just prior to ice retreat or as the ice margin moved over the region. It is possible that the ice underwent a period of extension during deglaciation. Crevasses, formed at the ice-sheet base, were infilled with deforming till, resulting in the formation of the ridges (Zilliacus 1989). A requirement of this model is that ice remains stagnant so that the features are not reworked. In this setting, lift-off of the ice would occur. Alternatively, the ridges could mark ice-edge deposition, although the symmetrical form for the larger ridges would remain problematic. The timing of either setting would be similar;

although the mechanism differs, both occur close to a retreating margin.

Zone 4. Zone 4 is characterized by isolated, slightly asymmetrical, straight-crested wedges, composed of soft diamicton, resting on an erosional surface and covered by a transparent sediment drape of diatomaceous mud (Fig. 6a). The wedges display corrugation moraines at the surface (Fig. 6b). No flutes are observed, possibly because they are obscured by the drape. Large-scale lineations are observed on the swath bathymetry data. The wedges occur at an oblique angle to the lineations and may reflect two stages of flow. The relative ages of the features cannot be determined from the dataset. Arcuate grooves, interpreted as iceberg furrows, cross-cut the features in the seaward region of the zone and are observed on the swath bathymetry data cross-cutting the lineations.

The deposits are interpreted as subglacial transverse moraines, similar to the moraines described by Solheim *et al.* (1990), although considerably larger in width and extent and without surface flutes. The relationship with the lineations is unclear; the wedges are believed to occur over the lineations. The wedges are not considered to be crevasse-squeeze deposits based on geometry and scale. These deposits could be ice-marginal retreat moraines, but the straight edge does not seem compatible with this mode of formation. The presence of corrugation moraine on the surface of the ridges also places these features behind the margin (Fig. 6b). The features of Zone 4 are interpreted as remnant wedges that reflect the subglacial transport of material in isolated sediment patches. These expand in thickness toward the grounding zone of Zone 5 as more material accumulates.

Zone 5. The deposits observed in Zone 5 have been interpreted as a grounding-zone wedge (Shipp & Anderson 1995; Licht *et al.* 1996; Domack *et al.* 1999; Shipp *et al.* 1999). Piston cores sampled lodgement till, overlain by a thick soft diamicton, capped by diatomaceous mud. The soft diamicton was interpreted as subglacial in origin (Licht *et al.* 1996; Domack *et al.* 1999; Licht *et al.* 1999). Radiocarbon dates from the unit overlying the subglacial deposits yield uncorrected total organic carbon ages of approximately 14 290 years BP (corrected value of 11 150 years BP; see Domack *et al.* 1999). The wedge is interpreted as marking a pause in the ice edge during retreat (Domack *et al.* 1999; Shipp *et al.* 1999).

The wedge is much larger than other features described in the literature, barring till tongues (King *et al.* 1991) and till deltas (Alley *et al.* 1987), which are grouped and classified as grounding-zone wedges by Powell & Domack (1995). In plan view, the stoss side of the wedge displays a chaotic folded character, interpreted as slumps down the front slope (Shipp *et al.* 1999).

Based on seismic data, Alley *et al.* (1989) proposed that a several-metre thick layer of unconsolidated sediment, overlying a unit tens of metres thick with beds dipping downstream at approximately 1/2°, occurred at the mouth of Ice Stream B. The feature, tens of kilometres long, is similar in external geometry and size to the grounding-zone wedge identified in this investigation. The grounding-zone wedge also is similar in geometry and scale to a single till tongue of King & Fader (1986) and King *et al.* (1991). The regional extent of the lower erosional surface and its amalgamation with other erosional surfaces on the shelf landward of the zone of deposition also are comparable to the King *et al.* (1991) model. Neither the wedge nor the till tongues show evidence for sediment gravity flow processes. In contrast to the till tongues of King *et al.* (1991), the Ross Sea grounding-zone wedges do not inter-finger with thick glacial-marine deposits. Rather, the deposits are dominated by subglacial till and grounding-zone proximal diamictons.

An abundance of arcuate, cross-cutting grooves, interpreted as iceberg furrows occur on the landward side and crest of the grounding-zone wedge (Fig. 7a). The most recent stage of ice-keel ploughing is composed of furrows that are smaller in width than the older furrows they cross-cut.

Material is thought to have been transported to the ice-sheet edge in Zone 5 as a mobile till layer on top of the lodgement till, accumulating to build the wedge. Following deposition of the grounding-zone wedge and retreat of grounded ice from this position, icebergs ploughed the surface. The earlier, larger furrows may reflect the break-up of the ice sheet and the production of large, deep-keeled, icebergs in a setting proximal to the grounding-zone wedge. Later, smaller, farther-travelled icebergs periodically ploughed the grounding-zone wedge surface, leaving narrower, shallower keel marks (Fig. 7a).

Zone 6. These deposits have been interpreted to contain proglacial material that was over-ridden subsequently by the ice sheet (Domack *et al.* 1999; Shipp *et al.* 1999). Lodgement till was not sampled in Zone 6. Radiocarbon dates from the soft diamicton provide an uncorrected age of approximately 14 000 years BP (NBP95-KC-39; Hilfinger *et al.* 1995; Domack *et al.* 1995*a, b,* 1999; Licht 1995; Licht *et al.* 1996), placing the LGM extent at the northern edge of these deposits.

The prominent lineations are interpreted to be mega-scale glacial lineations of subglacial origin (Fig. 8a). Lineations occur in most of the troughs of West Antarctica (e.g. Canals *et al.* 2000; Anderson *et al.* 2001; Wellner *et al.* 2001). A similar linear fabric is reported from other glaciated environments, although typically the features are smaller in scale (e.g. Barnes 1987; Solheim *et al.* 1990; Josenhans & Zevenhuizen 1990). C. Clark (1993, 1994) and Stokes & Clark (1999) used the term 'mega-scale glacial lineations' to describe lineations with lengths typically from 8 to 70 km, widths of 200–1300 m, and spacing of 0.3 to 5 km. Mega-scale glacial lineations are presumed to form under actively flowing ice. C. Clark (1994) examined the great length, straight form, and parallel arrangement of the lineations and proposed that the features are initiated at a point source, such as that associated with till inhomogeneities and are formed under conditions of rapid basal motion, possibly in response to convergence of ice, draw-down in a marine environment, or encountering of topographic lows or areas of soft sediment. The presence of these features is thought to indicate a deforming substrate (e.g. Menzies 1989; Clark, C. 1993, 1994); soft sediment is squeezed into the lineations as the ice moves forward rapidly. Work by Boulton & Hindmarsh (1987) suggests that parallel-sided flutes, such as those seen in the study region, are composed of weak till that can deform more readily than the ice.

The ice paused in the vicinity of the grounding-zone wedge of Zone 5, and deposited proximal glacial-marine sediment in Zone 6. The ice then advanced over the soft glacial-marine proximal sediment, deforming the upper surface into mega-scale glacial lineations (Fig. 8a). Advance was rapid, and the deformation minimal, based on the lack of additional grounding-zone deposits and intact radiocarbon stratigraphies (Licht *et al.* 1996; Domack *et al.* 1999). This interpretation is supported by the reasoning put forward by C. Clark (1994) for rapid basal flow of ice in association with mega-scale glacial lineation formation. Ice retreated, pausing at the original grounding-zone, depositing additional subglacial till onto the fluted surface.

Iceberg furrows. The arcuate grooves observed in Zones 4, 5, and 6 are interpreted as iceberg furrows, formed by iceberg keels ploughing the sea floor (Figs 6b and 7). Similar features occur on other glaciated shelves (e.g. Josenhans & Zevenhuizen 1990; Solheim *et al.* 1990; Dowdeswell *et al.* 1993), although the Ross Sea features appear to be larger than those described elsewhere. Modern icebergs with drafts of 330 m have been observed in the Ross Sea (Orheim 1980) and modern iceberg furrows have seldom been observed below 400 m. The iceberg furrows described here occur in present water depths up to 575 m (LGM water depths of approximately 450 m). These large icebergs probably were produced at the ice margin during the break up of the ice sheet (Shipp *et al.* 1999).

Meltwater features. Features such as channels, tunnel valleys, outwash plains, or grounding-line fans, are not observed along the Ross Sea transect examined here. Recent research suggests that shallow braided channels or canals could exist in conjunction with a deforming bed (Alley 1989; Alley *et al.* 1989; Walder & Fowler 1994; Catania & Paola 1998), but these are below the scale of resolution of the dataset. The pore water chemistry of subglacial deposits yields a marine signature (Rayne & Domack 1996), supporting the conclusion that little melt-water existed at the base of the ice sheet and that the ice sheet cannibalized pre-existing glacial and glacial-marine sediment and/or that deposition of the material was into the marine environment.

Corrugation Moraine. Small (1 m to 2 m high), closely spaced (40 m to 100 m crest to crest spacing), straight-crested symmetrical corrugations occur at the surface of the soft till in all zones, except for Zone 1 (Figs 4a, 6, 7b and 8b). These features, capped by a transparent sediment drape of diatomaceous mud, are considered to be corrugation moraine. Hypothesized mechanisms of formation include: subglacial deformation of the surface unit; tidal pumping close to the grounding line causing undulations at the base of the ice sheet; crevasse-fill features associated with retreat; or ice-margin retreat features.

The corrugations are thought to be associated with the retreat of the ice margin because they occur on top of all other subglacial features. The corrugations have a complex relationship with the wedges in Zone 2; they occur on the backs of the wedges and in front of them in some locations. They are believed to

be time-transgressive and related to the retreat of the margin. It is difficult to imagine a process that would continue under the entire length of the flow line of an expanded ice sheet and would result in the formation of corrugations at the sea floor.

The hypothesis that the corrugations formed as folds in the surface of a deforming sediment, produced as the flowing ice dragged the upper layer, is discarded because the expected bedforms would be asymmetrical (Van der Wateren 1995). Tidal pumping may be a possible mechanism. The fluctuation of the margin of the ice sheet with daily tides might cause flexure. The 'lift-and-settle' motion may produce hummocks. However, this mechanism does not account for the straightness of the crests or the paucity of current-reworked sediment. It also is difficult to imagine a tidal motion consistently influencing a margin, independent of changing water depth. The corrugations could be crevasse-fill features, such as those described by Zilliacus (1989). This mechanism, however, does not explain the pattern of low and consistent height (1 to 2 m) of the features, or the constant spacing between features, and implies that the ice sheet is in a constant state of extension during retreat.

Similar features are observed in Prydz Bay, East Antarctica (O'Brien *et al.* 1999) and are interpreted as sediment waves formed in a subglacial cavity at the ice margin. This mechanism is ruled out for the Ross Sea ribs because, unlike sand waves, the ribs are symmetrical and cores sampled diamicton for the most part.

Formation of the corrugations is attributed to deposition at the grounding-zone margin as the ice edge retreated across the sea floor. A mechanism of formation similar to the emplacement of De Geer moraines is proposed, in which subglacial material is extruded at the ice margin to produce a small ridge in front of the grounding line. The margin retreats and the process is repeated. This mechanism accounts for the straight-crested nature of the forms, as well as for the relationships observed with other features. The retreating margin moved south across the study area. Only in Zone 2 do wedges occur over the corrugations. Based on the hypothesized mode of formation, this relationship is anticipated. Corrugations are also expected on the lee side of the wedges. Based on the presence of flutes, the ice was in contact with the wedge surface. The ice margin retreated down the lee side, depositing retreat moraines.

De Geer moraines imply annual deposition of ridges. The details of the timing of glacial

Table 1. *Total sediment volume and sediment per unit area for each zone shown in Figure 9*

Zone	1	2	3	4	5	6
Sediment volume (m³)	5.7×10^9	14.7×10^9	2.5×10^9	9.8×10^9	69.4×10^9	39.9×10^9
Sediment volume per unit area (m²)	1	17.8	1	6.5	42.5	19

retreat are not well constrained in the Ross Sea. However, radiocarbon dates from subglacial deposits of Zone 5 yield uncorrected total organic carbon ages of approximately 14 290 years BP (corrected value of 11 150 years BP; see Domack *et al.* 1999). Other work indicates the ice margin passed Ross Island approximately 6430 ±70 years BP (uncorrected; Denton *et al.* 1989, 1991) (Fig. 1). These data suggest grounded ice retreated from the Coulman Island region grounding-zone wedge (Zone 5) approximately 14 000 years BP across a distance of approximately 370 km to Ross Island by about 6400 years BP. This is a rate of approximately 50 m a⁻¹ and agrees well with the corrugation moraine spacing of approximately 40 to 100 m. Conway *et al.* (1999) report the ice sheet to have been in its LGM position about 12 880 years BP (corrected) with its retreat to the vicinity of Ross Island by approximately 7600 years BP (corrected), yielding a retreat rate of approximately 84 m a⁻¹ across 440 km (or about 70 m a⁻¹ across 370 km to compare with the work of Domack *et al.*). These retreat rates are not significantly different from the ice-shelf edge retreat rate of approximately 100 m a⁻¹ calculated by Domack *et al.* (1999) for a site in the adjacent trough. The rates are similar to terrestrial ice-sheet margin retreat rates in the Northern Hemisphere for the LGM (Andrews 1973; Dyke & Prest 1987).

If this interpretation is correct, the data imply that ice retreated consistently at 40 to 100 m a⁻¹ across the continental shelf; the moraines are roughly equally spaced. This negates a collapse of the ice sheet. Instead, the data suggest a gradual retreat of ice as it was continually in contact with the bed. The availability of 'annual' features, if they prove to be such, also opens the possibility of a method of dating features within Ross Sea at a higher resolution than previously possible.

The consistent rate of retreat suggested by the surface corrugations contrasts with expectations that the ice-sheet retreat would accelerate as it retreated into deeper water. It also contrasts with proposed scenarios of collapse of the ice sheet. The features suggest that ice

retreated in a gradual manner across the study region. Perhaps the ice on the deeper inner continental shelf was stabilized to some degree by sticky spots and by adjacent banks that provided pinning points.

Sediment volumes

Sediment volumes for the different zones were estimated based on the thickness of the observed sediment along deep-tow chirp-sonar and 3.5 kHz transects, augmented by high-resolution seismic datasets from earlier investigations (Shipp *et al.* 1999). This information was then extrapolated across the area of the zone within the trough (Table 1). Zones 5 and 6, the grounding zone and over-ridden pro-glacial deposits, contain the thickest sediment accumulations (combined Zones 5 and 6: 62 m³ m⁻²). Zone 1, the inner shelf erosional zone, contained the smallest volume of sediment; spread in a thickness across the area of approximately 1 m. Zones 2, 3, and 4 contained approximately 25 m³ m⁻² of sediment above the basal erosional surface.

Using available age constraints of ice-sheet activity, the estimated sediment volumes provide some constraints on sediment-flux rates. Alley *et al.* (1989) suggested that the grounding-zone wedge at Ice Stream B was deposited over a period of 5000 to 10 000 years. Based on the sediment accumulation in Zones 5 and 6, the present study suggests a sediment-flux rate of 2.2×10^7 m³ a⁻¹ for deposition of the materials in Zones 5 and 6 in 5000 years, or 1.1×10^7 m³ a⁻¹ for Zone 5 and 6 deposits to have accumulated in 10 000 years. These are comparable to the estimate of approximately 6×10^7 m³ a⁻¹ for the sediment flux at the present Ice Stream B terminus (Alley *et al.* 1987). Conway *et al.* (1999) suggested that the expanded ice sheet remained at the LGM position for approximately 15 000 years. Based on the observed sediment distribution in the grounding-zone region (Zones 5 and 6), this suggests a sediment-flux rate of 0.7×10^7 m³ a⁻¹. Using the LGM position of Domack *et al.* (1999; 14 290 years BP) and the timing of retreat to

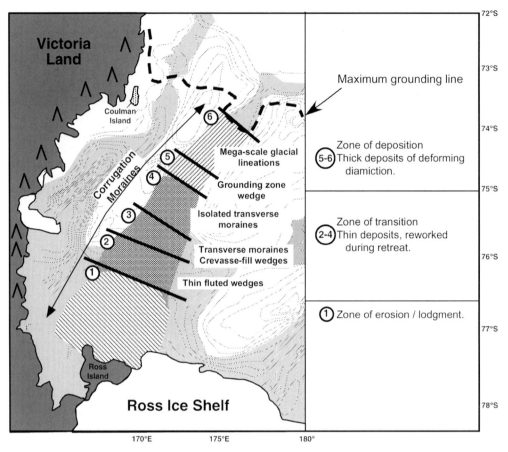

Fig. 9. Summary of the distribution of features in the study region and proposed subglacial conditions. The inner shelf was characterized by high basal shear stress and erosion. The outer shelf was characterized by rapid ice flow and deposition of thick deformation till.

Ross Island (Denton *et al.* 1989, 1991; 6430 ±70 years BP), sediment-flux rates for the duration of retreat across Zones 2, 3, and 4, are approximately 4.2×10^6 m³ a⁻¹. The duration of retreat from Conway *et al.* (1999) yields a sediment-flux rate of approximately 6.1×10^6 m³ a⁻¹.

Summary of interpretations

Ice advanced during the LGM to the grounding-zone wedge of Zone 5 (Fig. 9). The present day inner shelf, Zone 1, is located on the landward edge of an erosional region. The substrate was eroded by the flowing ice sheet; patchy regions of sediment lodgement existed. The ice may have moved over a very thin layer of deforming substrate that was erosional at its base. Farther to the north, in Zones 2, 3 and 4,

sediment was transported actively under the ice, probably as a deforming layer. The layer was not thick (<10 m, commonly <5 m), based on the present thickness of sediment on the shelf. Zone 4 contains straight-crested transverse moraines, indicating little reworking (Fig. 6a). The material moved from Zones 1–4 accumulated in Zone 5 as a grounding-line wedge and in the pro-glacial deposits of Zone 6 (Figs 7a and 8a). The equilibrium line in the polar setting of Antarctica occurs at the ice margin at sea level. Relative to other portions of the ice sheet, the fastest rates of flow would occur in this region.

Ice extended across the soft pro-glacial deposits of Zone 6, based on the presence of surficial lineations and corrugations (Figs 8 and 9). The movement across sediment with a high water content may have caused the initial acceleration, low basal shear stress, and accompanying thinning of the ice sheet, initiating retreat. Corrugations on the surface of Zone 6 record a consistent retreat rate of 50 m a[-1]. The corrugations are larger than those of Zones 2, 3, and 4, perhaps as a result of the softer substrate. The ice paused at the grounding-zone wedge in Zone 5 for a sufficient time to deposit the upper hummocky sediment layer. This pause may have been caused by the higher bathymetric level of the grounding zone (pinning point) and/or the change in substrate conditions. Ice then retreated toward its present-day position with a few, short, pauses. Only infrequent, small moraines mark the pauses, indicating retreat proceeded largely unchecked (Fig. 4). The thin sediment cover on the middle zones (2 to 4) was reworked during retreat into ice-margin deposits.

As the ice retreated from its maximum position, it may have calved large icebergs, which furrowed the sea floor in Zones 4 and 5. These icebergs were in contact with the substrate, based on the presence of ribs in the furrows. Subsequently, smaller bergs were produced, which produced thinner, non-ribbed furrows into the sea floor. An ice shelf may have formed over the trough at some time following break-up (Licht *et al.* 1996; Domack *et al.* 1999; Shipp *et al.* 1999). Open-marine conditions then resulted in the deposition of the thin blanket of diatomaceous mud observed over most of the shelf (Fig. 2).

Conclusions

1) During the last glacial expansion across the Ross Sea continental shelf, the grounded ice sheet acted as a conveyor belt, eroding and transporting sediment from the inner regions toward the outer shelf where material was deposited as a grounding-zone wedge.

2) The inner shelf was a zone of erosion with limited deposition predominantly by lodgement processes. Ice flow may have been slow, with high basal shear stresses, or ice may have moved over a very thin layer of deforming substrate that eroded the material beneath it. A transition zone existed across the central shelf. A thin (<10 m; commonly <5 m) subglacial sediment layer covered this region. The sediment may have been deforming, eroding the smooth surface of the substrate beneath it. Ice-movement velocities and basal shear stresses were intermediate. The deposits are thickest in the depositional zones of the central–outer

shelf. Ice flow was faster in this region, associated with low basal shear stresses. Mega-scale glacial lineations formed on soft pro-glacial diamicton with high water contents. These zones were controlled by the geology of the trough. Consolidated strata occur on the inner shelf; softer strata occur on the outer shelf.

3) The accelerated ice flow interpreted for the outer zones may have led to a thinning of the ice and initiation of retreat. Formation of large icebergs may correlate with break-up of the ice sheet in the initial stages.

4) During retreat, the grounded ice reworked the thin sediment of the transitional zone into retreat features, including back-stepping grounding-zone wedges and corrugation moraine. Features interpreted as crevasse-fill ridges may mark restricted regions of ice extension and lift-off.

5) Corrugation moraines (low amplitude, high frequency, symmetrical, straight-crested features) mark annual pauses in the retreat of the ice-sheet margin. The moraines exhibit a spacing of the order of 40 to 100 m, consistent with retreat rates proposed in other investigations (Conway *et al.* 1999; Domack *et al.* 1999; Shipp *et al.* 1999). If this interpretation is correct, the consistent retreat rate across the Ross Sea floor negates possible collapse scenarios. Discarding the interpretation that these features formed at the ice edge, other features, such as the grounding-zone wedges of Zone 2, necessitate that the ice remained in contact with the substrate throughout retreat.

6) No features associated with actively flowing meltwater are observed; no tunnel valleys, channels, canals, or braid streams. Any water available to the subglacial system was incorporated into a deforming unit at the base of the ice sheet.

The authors extend their appreciation to the crew of the RV/IB *Nathaniel B. Palmer* and to the Antarctic Support Associates for three pleasant and productive cruises. Many thanks to the research teams who worked so hard and so long during these cruises to collect the data. S. O'Hara of Lamont Doherty Earth Observatory kindly assisted in the processing of the multibeam data. We thank T. Janecek and M. Curren of the Florida State University Antarctic Research Facility for curating the cores and for amiably providing assistance in describing and sampling. We are grateful to J. A. Dowdeswell, who provided thorough reviews of two earlier versions of this manuscript, and to an anonymous reviewer. This research was funded by the National Science Foundation, Office of Polar Programs grant numbers DPP-9119683 and OPP-9527876 to J. B. Anderson.

References

ACKERT, R. P., JR., BARCLAY, D. J., BORNS, H. W., JR., CALKIN, E., KURZ, M. D., FASTOOK, J. L. & STEIG, E. J. 1999. Measurements of past ice sheet elevations in interior West Antarctica. *Science*, **286**, 276–280.

ALLEY, R. B. 1989. Water-pressure coupling of sliding and bed deformation. I. Water system. *Journal of Glaciology*, **35**, 108–118.

ALLEY, R. B. 1991. Deforming-bed origin for southern Laurentide till sheets? *Journal of Glaciology*, **37**, 67–76.

ALLEY, R. B., BLANKENSHIP, D. D., BENTLEY, C. R. & ROONEY, S. T. 1986. Deformation of till beneath Ice Stream B, West Antarctica. *Nature*, **322**, 57–59.

ALLEY, R. B., BLANKENSHIP, D. D., BENTLEY, C. R. & ROONEY, S. T. 1987. Till beneath Ice Stream B. 3. Till deformation: Evidence and implications. *Journal of Geophysical Research*, **92**, 8921–8929.

ALLEY, R. B., BLANKENSHIP, D. D., ROONEY, S. T. & BENTLEY, C. R. 1989. Sedimentation beneath ice shelves – The view from Ice Stream B. *Marine Geology*, **85**, 101–120.

ANDERSON, J. B. 1999. *Antarctic Marine Geology.* Cambridge University Press, Cambridge.

ANDERSON, J. B. & BARTEK, L. R. 1992. Ross Sea glacial history revealed by high resolution seismic reflection data combined with drill site information. *In:* KENNETT, J. & WARNKE, D. A. (eds) *The Antarctic Paleoenvironment: A Perspective on Global Change.* American Geophysical Union, Washington, D. C., Antarctic Research Series, **56**, 231–263.

ANDERSON, J. B., KURTZ, D. D., DOMACK, E. W. & BALSHAW, K. M. 1980. Glacial and glacial-marine sediments of the Antarctic continental shelf. *Journal of Geology*, **88**, 399–414.

ANDERSON, J. B., BRAKE, C. F. & MYERS, N. C. 1984. Sedimentation in the Ross Sea, Antarctica. *Marine Geology*, **57**, 295–333.

ANDERSON, J. B., WELLNER, J. S., LOWE, A. L., MOSOLA, A. B. & SHIPP, S. S. 2001. Footprint of the expanded West Antarctic ice sheet: Ice stream history and behavior. *GSA Today*, **11**, 4–9.

ANDERSON, J. B., SHIPP, S. S., LOWE, A. L., WELLNER, J. S. & MOSOLA, A. B. 2002. The Antarctic ice sheet during the last glacial maximum and its subsequent retreat history: A review. *Quaternary Science Reviews*, **21**, 49–70.

ANDREWS, J. T. 1973. The Wisconsin Laurentide ice sheet: dispersal centers, problems of rates of retreat, and climatic implications. *Arctic and Alpine Research*, **5**, 185–199.

ANDREWS, J. T., DOMACK, E. G., CUNNINGHAM, W. L., LEVENTER, A., LICHT, K. J., JULL, A. J. T., DEMASTER, D. J. & JENNINGS, A. E. 1999. Problems and possible solutions concerning radiocarbon dating of surface marine sediments, Ross Sea, Antarctica. *Quaternary Research*, **52**, 206–216.

AYLESWORTH, J. M. AND SHILTS, W. W. 1989. Bedforms of the Keewatin Ice Sheet, Canada. *Sedimentary Geology*, **62**, 407–428.

BARNES, P. W. 1987. Morphologic studies of the Wilkes Land continental shelf, Antarctica–glacial and iceberg effects. *In:* EITTREIM, S. L. & HAMPTON, M. A. (eds) *Geology and Geophysics of Offshore Wilkes Land.* Circum-Pacific Council for Energy and Mineral Resources, Houston, The Antarctic Continental Margin, Earth Science Series, 175–194.

BARNES, P. W. & LIEN, R. 1988. Icebergs rework shelf sediments to 500 m off Antarctica. *Geology*, **16**, 1130–1133.

BERKMAN, P. A. & FORMAN, S. L. 1996. Pre–bomb radiocarbon and the reservoir correction for calcareous marine species in the Southern Ocean. *Geophysical Research Letters*, **23**, 363–366.

BLANKENSHIP, D. D., BENTLEY, C. R., ROONEY, S. T. & ALLEY, R. B. 1986. Seismic measurements reveal a saturated porous layer beneath an active Antarctic ice stream. *Nature*, **322**, 54–57.

BLANKENSHIP, D. D., MORSE, D. L., FINN, C. A., BELL, R. E., PETERS, M. E., KEMPF, S. D., HODGE, S. M., STUDINGER, M., BEHRENDT, J. C. & BROZENA, J. M. 2001. Geological controls on the initiation of rapid basal motion for West Antarctic ice streams; A geophysical perspective including new airborne radar sounding and laser altimetry results. *In:* ALLEY, R. B. & BINDSCHADLER, R. A. (eds) *The West Antarctic Ice Sheet, Behavior and Environment.* American Geophysical Union, Washington, D. C., Antarctic Research Series, **77**, 105–121.

BOULTON, G. S. 1976. The origin of glacially fluted surfaces, observations and theory. *Journal of Glaciology*, **17**, 287–309.

BOULTON, G. S. 1996a. Theory of glacial erosion, transport, and deposition as a consequence of subglacial sediment deformation. *Journal of Glaciology*, **42**, 43–62.

BOULTON, G. S. 1996b. The origin of till sequences by subglacial sediment deformation beneath mid-latitude ice sheets. *Annals of Glaciology*, **22**, 75–84.

BOULTON, G. S. & CLARK, C. D. 1990. A highly mobile Laurentide ice sheet revealed by satellite images of glacial lineations. *Nature*, **346**, 813–817.

BOULTON, G. S. & HINDMARSH, R. C. A. 1987. Sediment deformation beneath glaciers: rheology and geological consequences. *Journal of Geophysical Research*, **92**, 9059–9082.

BOULTON, G. S. & JONES, A. S. 1979. Stability of temperate ice caps and ice sheets resting on beds of deformable sediment. *Journal of Glaciology*, **24**, 29–43.

CANALS, M., R. URGELES & CALAFAT, A. M. 2000. Deep sea-floor evidence of past ice streams off the Antarctic Peninsula. *Geology*, **23**, 31–34.

CATANIA, G. & PAOLA, C. 1998. A physical model of pressurized flow over an unconsolidated bed: Implications for subglacial braided channels. *Scientific Program and Abstracts, West Antarctic Ice Sheet*, American Geophysical Union, Chapman Conference, Orono, Maine, 119.

CLARK, C. D. 1993. Mega-scale glacial lineations and cross-cutting ice-flow landforms. *Earth Science Processes and Landforms*, **18**, 1–29.

CLARK, C. D. 1994. Large-scale ice-moulding: a discussion of genesis and glaciological significance. *Sedimentary Geology*, **91**, 253–268.

CLARK, C. D. 1999. Glaciodynamic context of subglacial bedform generation and preservation. *Annals of Glaciology*, **28**, 23–32.

CLARK, P. U. 1994. Unstable behavior of the Laurentide Ice Sheet over deforming sediment and its implications for climate change. *Quaternary Research*, **41**, 19–25.

CLARK, P. U. & WALDER, J. S. 1994. Subglacial drainage, eskers, and deforming beds beneath the Laurentide and Eurasian ice sheets. *Geological Society of America Bulletin*, **106**, 304–314.

CONWAY, H., HALL, B. L., DENTON, G. H., GADES, A. M. & WADDINGTON, E. D. 1999. Past and future grounding-line retreat of the West Antarctic Ice Sheet. *Science*, **286**, 280–283.

DAVEY, F. J. 1994. Bathymetry and gravity of the Ross Sea, Antarctica. *Terra Antarctica*, **1**, 357–358.

DENTON, G. H., BOCKHEIM, J. G., WILSON, S. C. & STUIVER, M. 1989. Late Wisconsin and Early Holocene glacial history, inner Ross Embayment, Antarctica. *Quaternary Research*, **3**, 151–182.

DENTON, G. H., PRENTICE, M. L. & BURCKLE, L. H. 1991. Cainozoic history of the Antarctic Ice Sheet. *In:* TINGEY, R. J. (ed.) *The Geology of Antarctica*. Clarendon Press, Oxford, 365–433.

DOMACK, E. W., HILFINGER, M. & FRANCESCHINI, J. 1995a. Stratigraphic and facies relationships in the Ross Sea related to Late Pleistocene fluctuation of the West Antarctic Ice Sheet system. *Third Annual Science Workshop, West Antarctic Ice Sheet Initiative*, Arlington, Virginia, 12.

DOMACK, E. W., HILFINGER, M., FRANCESCHINI, J., LICHT, K., JENNINGS, A., ANDREWS, J., SHIPP, S. & ANDERSON, J. B. 1995b. New stratigraphic evidence from the Ross Sea continental shelf for instability of the West Antarctic Ice Sheet during the last glacial maximum. *First Congress on Sedimentary Geology*, Society of Economic Paleontologists and Paleontologists, St Pete Beach, Florida, 46–47.

DOMACK, E. W., JACOBSON, E. K., SHIPP, S. & ANDERSON, J. B. 1999. Sedimentological and stratigraphic signature of the Late Pleistocene/Holocene Fluctuation of the West Antarctic Ice Sheet in the Ross Sea: A New Perspective Part 2. *Geological Society of America Bulletin*, **111**, 1486–1516.

DOWDESWELL, J. A., VILLINGER, H., WHITTINGTON, R. J. & MARIENFELD, P. 1993. Iceberg scouring in Scorsby Sund and on the East Greenland continental shelf. *Marine Geology*, **111**, 37–53.

DOWDESWELL, J. A., WHITTINGTON, R. J. & MARIENFELD, P. 1994. The origin of massive diamicton facies by iceberg rafting and scouring, Scorsby Sund, East Greenland. *Sedimentology*, **41**, 21–35.

DREWRY, D. J. 1979. Late Wisconsin reconstruction for the Ross Sea region, Antarctica. *Journal of Glaciology*, **24**, 231–244.

DREIMANIS 1988. Tills: their genetic terminology and classification. *In:* GOLDTHWAIT, R. P. & MATSCH, C. L. (eds) *Genetic Classification of Glacigenic Deposits*. Balkem, Rotterdam, 17–83.

DYKE, A. S. & PREST, V. K. 1987. Late Wisconsinan and Holocene history of the Laurentide ice sheet. *Geographie Physique et Quaternaire*, **16**, 237–263.

ENGELHARDT, H., HUMPHREY, N., KAMB, B. & FAHNESTOCK, M. 1990. Physical conditions at the base of a fast moving Antarctic ice stream. *Science*, **248**, 57–59.

FOWLER, A. C. & WALDER, J. S. 1993. Creep closure of channels in deforming subglacial till. *Proceedings of the Royal Society of London, Series A*, **441**, 17–31.

HALL, B. L. & DENTON, G. H. 1998. Deglacial chronology of the Western Ross Sea from terrestrial data. *Scientific Program and Abstracts, West Antarctic Ice Sheet*, American Geophysical Union, Chapman Conference, Orono, Maine, 27–28.

HART, J. K. 1999. Identifying fast ice flow from landform assemblages in the geological record: a discussion. *Annals of Glaciology*, **28**, 59–66.

HILFINGER, M., FRANCESCHINI, J. & DOMACK, E. W. 1995. Chronology of glacial marine lithofacies related to the recession of the West Antarctic Ice Sheet in the Ross Sea. *Antarctic Journal of the United States*, **30**, 82–84.

HUGHES, T. J. 1977. West Antarctic ice streams. *Reviews of Geophysics and Space Physics*, **15**, 1–46.

HUGHES, T. J. 1995. Ice sheet modeling and the reconstruction of former ice sheets from glacial geo(morpho)logical field data. *In:* MENZIES, J. (ed.) *Modern Glacial Environments: Processes, Dynamics, and Sediments*. Glacial Environments, Volume 1. Butterworth and Heinemann, Oxford, 77–99.

JOHNSON, G. L., VANNEY, J. R. & HAYES, P. 1982. The Antarctic continental shelf-a review. *In:* CRADDOCK, C. (ed.) *Antarctic Geosciences*. University of Wisconsin Press, Madison, 22–27.

JOSENHANS, H. W. & ZEVENHUIZEN, J. 1990. Dynamics of the Laurentide Ice Sheet in Hudson Bay, Canada. *Marine Geology*, **92**, 1–26.

KAMB, B. 2001. Basal zone of the West Antarctic ice streams and its role in lubrication of their rapid motion. *In:* ALLEY, R. B. & BINDSCHADLER, R. A. (eds) *The West Antarctic Ice Sheet, Behavior and Environment*. American Geophysical Union, Washington, D. C., Antarctic Research Series, **77**, 157–199.

KELLOGG, T. B., HUGHES, T. & KELLOGG, D. E. 1996. Late Pleistocene interactions of East and West Antarctic ice-flow regimes: evidence from the McMurdo Ice Shelf. *Journal of Glaciology*, **42**, 486–499.

KING, L. H. & FADER, G. B. J. 1986. Wisconsinan glaciation of the Atlantic continental shelf of southeast Canada. *Geological Survey of Canada, Bulletin*, **363**, 72 p.

KING, L. H., ROKOENGEN, K., FADER, G. B. J. & GUNLEIKSRUD, T. 1991. Till-tongue stratigraphy. *Geological Society of America Bulletin*, **103**, 637–659.

LICHT, K. J. 1995. *Marine Sedimentary Record of Ice Extent and Late Wisconsin Deglaciation in the Western Ross Sea, Antarctica.* MSc thesis, University of Colorado, Boulder.

LICHT, K. J., JENNINGS, A. E., ANDREWS, J. T. & WILLIAMS, K. M. 1996. Chronology of late Wisconsin ice retreat from the western Ross Sea, Antarctica. *Geology*, **24**, 223–226.

LICHT, K. J., DUNBAR, N. W., ANDREWS, J. T. & JENNINGS, A. E. 1999. Distinguishing subglacial till and glacial marine diamictons in the western Ross Sea, Antarctica: Implications for last glacial maximum grounding line. *Geological Society of America Bulletin*, **111**, 91–103.

LUNDQVIST, J. 1988. Glacigenic processes, deposits, and landforms. *In:* GOLDTHWAIT, R. P. & MATSCH, C. L. (eds) *Genetic Classification of Glacigenic Deposits.* Balkem, Rotterdam, 3–16.

LUNDQVIST, J. 1989. Rogen (ribbed) moraine – identification and possible origin. *Sedimentary Geology*, **62**, 281–292.

MENZIES, J. 1989. Subglacial hydraulic conditions and their possible impact upon subglacial bed formation. *Sedimentary Geology*, **62**, 125–150.

O'BRIEN, P. E., DE SANTIS, L., HARRIS, P. T., DOMACK, E. & QUILTY, P. G. 1999. Palaeo ice shelf grounding zones of western Prydz Bay, Antarctica – Sedimentary processes from seismic and sidescan images. *Antarctic Science*, **11**, 78–91.

ORHEIM, O. 1980. Physical characteristics and life-expectancy of tabular Antarctic icebergs. *Annals of Glaciology*, **1**, 11–18.

PIOTROWSKI, J. A. & TULACZYK, S. 1999. Subglacial conditions under the last ice sheet in northwest Germany: ice-bed separation and enhanced basal sliding? *Quaternary Science Reviews*, **18**, 737–751.

POWELL, R. D. 1984. Glacimarine processes and inductive lithofacies modeling of ice shelf and tidewater glacier sediments based on Quaternary examples. *Marine Geology*, **57**, 1–52.

POWELL, R. D. & DOMACK, E. W. 1995. Modern glacimarine environments. *In:* MENZIES, J. (ed.) *Modern Glacial Environments: Processes, Dynamics, and Sediments.* Glacial Environments, Volume 1, Butterworth-Heinemann, Oxford, 445–486.

POWELL, R. D. & MOLNIA, B. F. 1989. Glacimarine sedimentary processes, facies, and morphology of the south-southeast Alaska shelf and fjords. *Marine Geology*, **85**, 359–390.

RAYNE, T. & DOMACK, E. W. 1996. Pore water geochemistry of Ross Sea diamictons, a till or not a till? *Antarctic Journal of the United States*, **31**, 97–98.

SHIPP, S. & ANDERSON, J. B. 1995. Late Quaternary deglacial history of Ross Sea, Antarctica: Results from recent seismic investigation. *VII International Symposium on Antarctic Earth Sciences*, Siena, Italy, 347.

SHIPP, S., ANDERSON, J. B. & DOMACK, E. W. 1999. Late Pleistocene/Holocene Retreat of the West Antarctic Ice-Sheet System in the Ross Sea: Part 1 – Geophysical results. *Geological Society of America Bulletin*, **111**, 1486–1516.

SOLHEIM, A., RUSSWURM, L., ELVERHØI, A. & BERG, M. N. 1990. Glacial geomorphic features in the northern Barents Sea: direct evidence for grounded ice and implications for the pattern of deglaciation and late glacial sedimentation. *In:* DOWDESWELL, J. A. & SCOURSE, J. D. (eds) *Glacimarine Environments: Processes and Sediments.* Geological Society, London, Special Publications, **53**, 253–268.

STEIG, E. J., FASTOOK, J. L., ZWECK, C., GOODWIN, I. D., LICHT, K. J., WHITE, J. W. C. & ACKERT, R. P., JR. 2001. West Antarctic Ice Sheet elevation changes. *In:* ALLEY, R. B. & BINDSCHADLER, R. A. (eds) *The West Antarctic ice sheet, behavior and environment.* American Geophysical Union, Washington, D. C., Antarctic Research Series, **77**, 75–90.

STOKES, C. R. & CLARK, C. D. 1999. Geomorphological criteria for identifying Pleistocene ice streams. *Annals of Glaciology*, **28**, 67–75.

STOKES, C. R. & CLARK, C. D. 2001. Palaeo-ice streams. *Quaternary Science Reviews*, **20**, 1437–1457.

STUIVER, M., DENTON, G. H., HUGHES, T. J. & FASTOOK, J. L. 1981. History of the marine ice sheet in West Antarctica during the last glaciation: a working hypothesis. *In:* DENTON, G. H. & HUGHES, T. J. (eds) *The last great ice sheets.* Wiley, New York, 319–439.

SUGDEN, D. E. & JOHN, B. 1976. *Glaciers and Landscape: Landforms of Glacial Deposition.* Edward Arnold, London, 235–257.

TEN BRINK, U. S., SCHNEIDER, C. & JOHNSON, A. H. 1995. Morphology and stratal geometry of the Antarctic continental shelf: Insights from models. *In:* COOPER, A. K. & DAVEY, F. J. (eds) *Geology and Geophysics of the Western Ross Sea.* Circum-Pacific Council for Energy and Mineral Resources, Houston, The Antarctic Continental Margin, Earth Science Series, **5B**, 1–24.

VAN DER WATEREN, F. M. 1995. Processes of Glaciotectonism. *In:* MENZIES, J. (ed.) *Modern Glacial Environments: Processes, Dynamics, and Sediments.* Glacial Environments, Volume 1, Butterworth-Heinemann, Oxford, 309–335.

VORREN, T. O., LEBESBYE, E., ANDREASSEN, K. & LARSEN, K.-B. 1989. Glacigenic sediments on a passive margin as exemplified by the Barents Sea. *Marine Geology*, **85**, 251–272.

WALDER, J. S. & FOWLER, A. 1994. Channelized subglacial drainage over a deformable bed. *Journal of Glaciology*, **40**, 3–15.

WELLNER, J. S., LOWE, A. L., SHIPP, S. S. & ANDERSON, J. B. 2001. Distribution of glacial geomorphic features on the Antarctic continental shelf and correlation with substrate: implications for ice behavior. *Journal of Glaciology*, **47**, 397–411.

ZILLIACUS, H. 1989. Genesis of De Geer moraines in Finland. *Sedimentary Geology*, **62**, 309–317.

Grain-size characteristics and provenance of ice-proximal glacial marine sediments

JOHN T. ANDREWS & SARAH M. PRINCIPATO

INSTAAR and Department of Geological Sciences, University of Colorado, Box 450,
Boulder, CO 80309, USA (e-mail: andrewsj@colorado.edu)

Abstract: This paper addresses the issue of the characteristic grain-size spectra of glacial and glacial marine sediments, an important topic because of the increasing emphasis on the use of specific sand-size fractions as indicators of iceberg rafting. Different methods of IRD identification can lead to different palaeoclimatic interpretations. We use three methods of grain-size analyses, with examples from the Labrador Sea, East Greenland, North Iceland, and the Ross Sea, Antarctica. The first method illustrates the 'real' grain size of glacial marine sediments by an examination of X-radiographs by image analysis and the sizing of clasts larger than 2 mm. Typically, grain-size analyses only apply to the matrix fraction of the sediment (smaller than 2 mm), and ignore the larger size clasts. The mean grain size appears to be between 8 to 10 mm. The second method includes counting the number of clasts larger than 2 mm from X-radiographs, and counts of particles (106–1000 μm). These latter counts show that diamictons from the North Iceland shelf indicate a dominance of glacially abraded basaltic grains, but they also contain a small, consistent proportion of angular volcanic glass shards of various compositions. The third method is to examine grain-size spectra of surface samples from East Greenland and the Ross Sea region of Antarctica and compare these data with down-core data from neoglacial-age glacial marine sediments from Nansen Fjord, East Greenland, and with late glacial diamictons from the North Iceland shelf. These sediments have a mode in the silt fraction, but they frequently have secondary peaks in the coarse sand and fine sand/coarse silt areas, and a trough in the range of 100 to 500 μm (medium to coarse sand). Most of the sediment samples analysed in this study contain 20–50% in the below 1 μm grain size, which reinforces the importance of examining this fraction in provenance studies in glacial marine sediments.

In the past decade there has been a significant renewed interest in the late Quaternary palaeoceanographic and palaeoclimatic signals associated with the presence of sand-size mineral grains in deep sea sediments. This fraction is frequently interpreted as iceberg rafted detritus or debris (IRD) (e.g. Ruddiman 1977; Andrews & Matsch 1983; Heinrich 1988; McManus *et al.* 1999; van Kreveld *et al.* 2000). Generally, the investigators are interested in millennial scale variability. Marine core records are commonly compared to the Dansgaard–Oeschger (D–O) cycles in the Greenland Summit isotopic records and to the North Atlantic Heinrich (H-) events (Bond *et al.* 1993; Bond & Lotti 1995; Elliott *et al.* 1998; van Kreveld *et al.* 2000). This recent renewed interest in IRD followed an earlier period when the time-scales of interest were at the orbital scale (Ruddiman 1977).

Parallel to this renewed interest in IRD and use of the sand-size fraction by the deep-sea palaeoceanographic community, there has been a substantial increase in research on present glacial marine environments from locations, such as Greenland and Antarctica. This research typically discusses the seismic stratigraphy, lithofacies, and sedimentology of cores from ice-proximal areas (up to 100 km from glacier margins). The study of ice sheet dynamics, iceberg production and transport, and the introduction of 'glacial sediments' into ocean basins has become increasingly intense in the last decade as ice sheets are important contributors to deep-sea sediment budgets, and interpretations of marine cores are used as indicators of abrupt climate change (Matsumoto 1996, 1997; Seidov 1997; Clarke *et al.* 1999; Schaefer-Neth & Stattegger 1999; Clarke & Prairie in press).

The dynamics of tidewater and calving glacier margins are explicitly linked to the transport of sediments beyond ice sheet and glacier margins. The glaciological community is currently studying the issues of glacial transport of sediments to marine depocentres, both on the glaciated continental margins and in more distal deep sea basins (Powell & Molnia 1989; Dowdeswell & Scourse 1990; Dowdeswell *et al.* 1994; Syvitski *et al.* 1996; Domack *et al.* 1998;

From: DOWDESWELL, J. A. & Ó COFAIGH, C. (eds) 2002. *Glacier-Influenced Sedimentation on High-Latitude Continental Margins.* Geological Society, London, Special Publications, **203**, 305–324. 0305-8719/02/$15.00
© The Geological Society of London 2002.

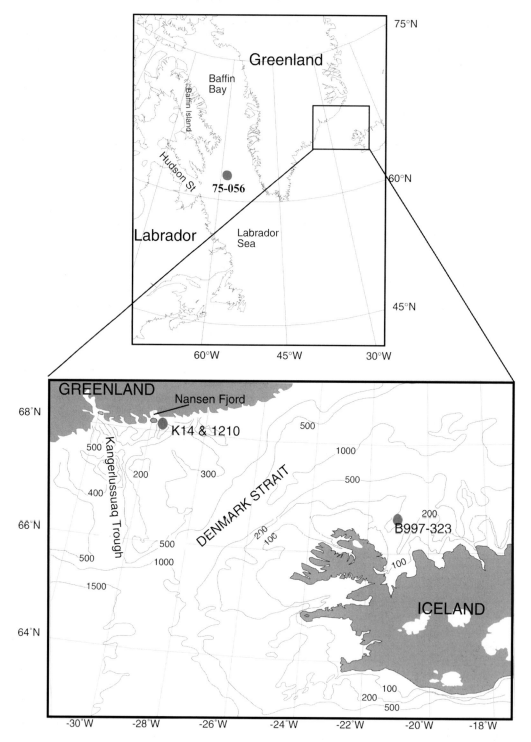

Fig. 1. **(Upper)** Location of HU75-056 in the Labrador Sea. **(Lower)** Bathymetry of the margins on either side of the Denmark Strait, and location of the cores from East Greenland and North Iceland that are used in the paper.

Dowdeswell & Siegert 1999; Licht *et al.* 1999; Ashley & Smith 2000). There are important arguments as to whether these events represent a glacial response to a climate signal or are a function of local ice sheet dynamics (Lowell *et al.* 1995; Dowdeswell *et al.* 1999).

In a recent paper (Andrews 2000), some of the issues that explicitly link the research findings and interests of glacial marine sedimentologists with the deep-sea palaeoceanographers were discussed. In this paper we extend that discussion and develop an argument raised by Fillon and Full (1984) that differences in the grain-size spectra of ice transported sediments can be interpreted in terms of provenance and depositional process. The main focus of our study is on data from the East Greenland and North Iceland margins (Fig. 1), although some data from Antarctica are introduced. This paper addresses the following basic questions:

(1) What is the real (i.e. total) grain-size distribution of glacial marine sediments?
(2) What are the grain-size spectra of the matrix (less than 2 mm, i.e. sand and smaller sediment fractions) of glacial and glacial marine sediments?
(3) Can grain-size spectra be used to distinguish different glacial sources (e.g. bedrock erodibility and/or subglacial regime)?
(4) What differences, if any, exist between glacial marine sediments deposited on continental margins versus deposition in adjacent deep sea basins?

The goals of this study are primarily descriptive and address rather simple questions. For example, do different glaciological and geological (bedrock) environments result in different grain-size distributions? The notion of a terminal grade might imply that bedrock geology exerts some influence on the final grain-size distribution, but this needs to be tested. A paper by Fillon and Full (1984) partly addressed these issues regarding the influence of basal regime and underlying bedrock type on the grain size spectra of glacial marine sediments from the Labrador Sea. Resolution of these questions would enable the ice-proximal glacial marine community to communicate better and compare data with palaeoceanographic colleagues whose primary focus is on sites located several kilometres from the glaciated continental margins.

Many authors use some fraction of the sand spectra (e.g. > 105, >250 µm) to characterize the presence of icebergs in the North Atlantic during the Quaternary, but the limits on the fraction varies from study to study. Is there a magic grain-size limit that uniquely and unambiguously represents iceberg transport? Possibly the most severe problem associated with using different sand-size fraction cut-offs is the effect that it has on palaeoclimate interpretations (Andrews 2000). It is possible that down-core plots of different grain-size fractions might not lead to a unique palaeoclimate signal. Thus different interpretations of IRD events might be determined from the same set of samples purely on the basis of the choice of the sand-size cut-off.

What is the 'real' grain-size distribution of glacial marine sediments?

There are limitations in the collection and description of diamictons from the marine environment due to common marine coring strategies. Diamictons retrieved in a 7 or 11 cm diameter core do not provide a complete nor adequate representation of the grain size spectra of the sediments. Most reports of grain-size, including those from our own laboratory, usually only describe the matrix of the sediment sample, specifically the fraction less than 2 mm. However, X-radiographs of cores that penetrate diamicton units (Licht *et al.* 1999) commonly show that the sediment includes large clasts with diameters measured in centimetres (Fig. 2). Even with a complete analysis of clasts of all sizes in marine core sediments, there is a sampling bias because of the core operation itself, as stated above. Sediments greater than the core diameter cannot be recovered, and the coring is probably halted on many occasions when large clasts are encountered (Fillon & Harmes 1982). In the following sections we will examine grain-size data on the very coarse fraction (> 2 mm) of diamicton units and issues pertaining to the study of grain size and provenance on the sand fraction (2 mm to 0.63 mm).

Clasts over 2 mm

Figure 2 shows tracings of the clast content from X-radiographs from parts of two of the marine cores, B91-K14 and B997-323 (Fig. 1). Clasts were traced manually to have greater control in identifying actual clasts rather than non-geological shadings on the X-radiographs. B91-K14 is from an ice-proximal site close to a calving tidewater glacier in Nansen Fjord, East Greenland, and B997-323 is from a trough on the North Iceland shelf. The tracings of the clasts from X-radiographs were outlined and filled in Adobe Photoshop. The resulting images

North Iceland
B997-323pc1,
(core depth: 190-242cm)

Greenland
K14
(core depth: 140-165cm)

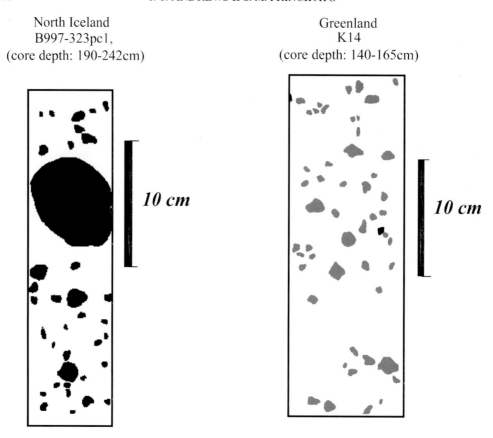

Fig. 2. Outline of the distribution of clasts over 2 mm from X-radiographs of cores from off East Greenland and North Iceland (see Fig. 1B for location).

of clasts were exported as binary files, and the clast distribution, grain size, and orientations were analysed using the public domain NIH Image program (developed at the US National Institutes of Health and available on the Internet at http://rsb.info.nih.gov/nih-image/). Figure 2 indicates that in terms of the surface area, the matrix fraction constitutes between 20 and 50% of the sediment. In some cores from Denmark Strait clast-supported diamictons are observed, and the matrix contribution can be less than 20% (Cartee-Schoolfield 2000). If IRD transport to the deep sea really does consist of basal glacial, englacial and supraglacial sediments, then an explicit indication of IRD transport might be the presence of large clasts (Fig. 2) in ice-distal deep-sea sediments.

The grain-size spectra of these scanned images may be presented as either a fraction of the area or as equivalent diameters. A selection of grain-size data from our X-radiograph work

(Principato and Andrews 2001) is illustrated in Figure 3, and it shows the spectra of total clast counts for two cores. The size distributions are generally normal to log-normal in distribution with a modal size around 8 mm. Maximum pebble sizes are between 40 and 60 mm.

IRD counts from X-radiographs

An intermediate approach to determining the extent of iceberg rafting was developed by Grobe (1987). This method has been used extensively, especially on cores from the East Greenland margin (Marienfeld 1992; Stein *et al.* 1996; Andrews *et al.* 1997) and the Labrador Sea (Andrews and Barber in press). The original method consisted of counting the gravel fraction visible in 1 cm slab X-radiographs. We use a similar method, but instead of slab X-radiographs, we use X-radiographs of the entire archive half of the core and count the number

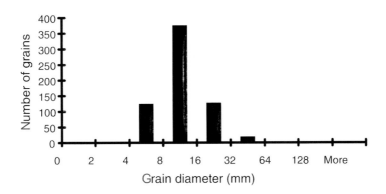

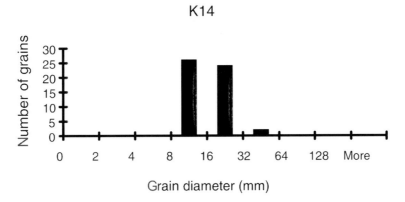

Fig. 3. Histograms of the diameters of clasts larger than 2 mm in terms of equivalent diameters from cores K14 and B997-323 (Fig. 1B).

of clasts over 2 mm in size. Although our method is less exact than the method of Grobe, our experiments (Andrews *et al.* 1997) indicated that our method has excellent reproducibility. Unfortunately, deep-sea cores are not always X-rayed, and many of the 'type' cores for the North Atlantic were collected decades ago and have been extensively sampled, hence such an approach is not feasible.

Figure 4 shows three examples of IRD counts using this approach. The first example, HU75-056 from the Labrador Sea (Fig. 1), shows the ambivalence of the relationship between Heinrich (H-) events and IRD events in this area which is directly east of the Hudson Strait shelf break (Andrews 2000). The second example, B997-323 from the North Iceland shelf (Fig. 1), contains a diamicton with foraminifera; post-glacial mud overlies this unit with an

abrupt contact at 80 cm. The third example, 91-K14 from Nansen Fjord, East Greenland, contains diamictons and fine-grained mud (Jennings & Weiner 1996; Smith & Andrews 2000). These examples are from a variety of glacial marine settings, and they show that the apparent number of clasts larger than 2 mm varies from 0 to *c.* 3 clasts per square centimetre. A flux estimate is possible if the rate of sediment accumulation is known.

One drawback to our approach of counting IRD is that there is no explicit measurement of size, other than being above the limit of 2 mm. Sometimes a low count merely represents a series of large pebbles, but other times it represents an actual lack of clasts (e.g. Figs 2 & 3). It would be instructive to see whether Heinrich events in the deep-sea basins of the North Atlantic are characterized by this

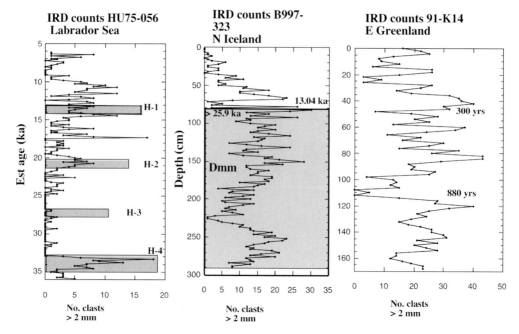

Fig. 4. Counts of visible clasts on X-radiographs of split cores (Fig. 1A & B for location). Cores HU75-056 and B997-323 are 7 cm in diameter; K14 is 10 cm in diameter. The counts represent successive 2 cm (vertical) measurements.

criterion. Icebergs transported to the deep sea carry and deposit clasts all along their trajectories. Icebergs that reach the deep sea have travelled much longer distances than icebergs proximal to the glaciated margin. Is the apparent lack of clasts due to decreased probability of retaining clasts over large distances? Our methods of describing sediment by weight or volume % might be more appropriately recalculated to the number of grains or clasts per size-fraction per unit volume (e.g. Fig. 3).

The sand fraction (Between 63 μm and 2 mm)

Down-core variations of different sand-size fractions and the correlation between these different fractions show that there is not a consistent relationship between the sand fractions (Fig. 5). The scatter plots indicate that the explained variance (r^2) is generally below 0.50, but this correlation increases in adjacent size intervals. An example of the increased correlation (r^2 of 0.95) is present in K14 when the data between 500 and 250 μm are compared to the weight per cent in the 250–125 μm interval. These results explicitly show that different interpretations of IRD events might be determined from the same

set of samples purely on the basis of the choice of the sand-size cut-off.

Provenance of lithic grains and glass

A common approach in deep-sea palaeoceanographic provenance studies is to determine the composition of a certain sand-size fraction (Heinrich 1988; Bond *et al.* 1993). This is rarely attempted in the ice-proximal environments on continental shelves probably because the sand fraction is so abundant, and samples would have to be split innumerable times to reduce the fraction to a countable amount. In addition, the source(s) of the glacial and glacial marine sediments is usually known, as the sites lie immediately offshore from glaciated landscapes. However, there is now a need to understand the provenance of ice-proximal sources because the provenance of ice distal sediments is directly related to the provenance of ice proximal sediments (Barber 2001; Hemming *et al.* 2000a,b; Dowdeswell *et al.* 2001).

On the North Iceland shelf (Fig. 1), a thin cover of fine-grained muds (postglacial marine sediments) overlie diamicton units (Fig. 4B & 6). The origin of the diamictons could be basal tills, primary glacial marine diamictons, or

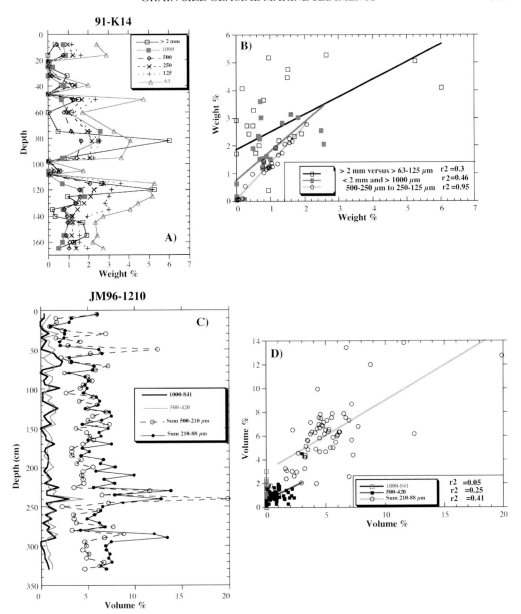

Fig. 5. (**A**) Down-core plot of different sand fractions from K14, East Greenland margin. (**B**) Scatter plots of the weight% of different sand fractions and the explained variances (r²). (**C**) Down-core plot of different sand fractions from JM96-1210, East Greenland margin. (**D**) Scatter plots of the volume% of different sand fractions and the explained variances (r²).

reworked glacial marine diamictons (Principato and Andrews 2001; Andrews and Helgadottir 1999). Core B997-323PC1 (Fig. 1) consists of a lower diamicton facies with an abrupt transition into a finer grained mud facies with varying amounts of IRD (Fig. 4B). We counted the

composition of 200 to 300 sand-size particles (105–1000 μm) under a binocular microscope and express the results as ratios to avoid the closed array problem (Aitcheson 1986). The compositions that we identify in our counts include: quartz, basalts, light (silica-rich) glass

B997-323PC1

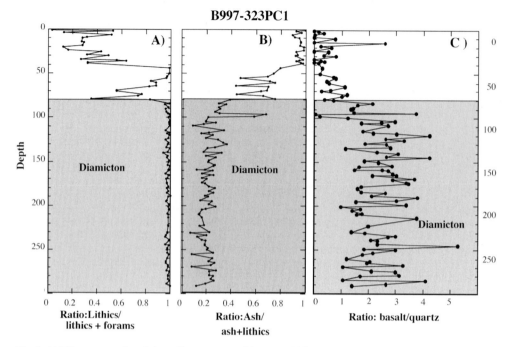

Fig. 6. (**A**) Down core plot of the sediment composition over 105 μm of the ratio of lithic fragments (not including clear and basaltic glass shards) over the total of lithic fragments and foraminifera. (**B**) Plot of the ratio of sediment over 105 μm of volcanic ash (tephra) shards of all compositions over the volcanic ash plus lithic counts. (**C**) Ratio of basaltic sand-size particles (not 'glassy') over quartz grains in the larger than 105 μm sand-size fraction.

shards, basaltic shards, foraminifera, 'red' lithics, sedimentary 'siltstones', and 'other' (mainly feldspars). We might expect *a priori* that an Iceland glacial source would be rich in basaltic grains and low in quartz or other lithics.

Interpreting the counts of volcanic glass shards in the sand size fraction is not straightforward because they may represent primary or secondary deposits. On the Iceland shelf, the presence of volcanic glass shards may be deposited by primary volcanic eruptions, or they may be transported to the local ice sheet by wind. In the latter case, they are carried supraglacially or englacially in the ice until it reaches the terminus where it is deposited. The work of Jennings *et al.* (pers. comm. 2001) and Bond *et al.* (2001), indicates that tephra deposits do not always constitute simple isochrons.

A commonly used ratio in deep-sea settings is the ratio of lithics/(lithics + foraminifera) (Bond *et al.* 1993). This ratio expresses the competing contributions of IRD versus calcareous marine productivity. We use a modification of this ratio and two others to describe our counts. The first ratio that we derive is (quartz

+ red + basalt + others) / (quartz + red + basalt + others + foraminifera) because the volcanic glass might be contributed to the sediment from airfall, fluvial discharge, or glacial meltout. The results (Fig. 6A) indicate that the relative proportion of foraminifera is low but constant throughout the diamicton. Above the basal diamicton there is a rapid increase in foraminiferal abundance, with the exception of a foraminiferal barren zone deposited at about the time of the Younger Dryas cold event (Helgadottir & Andrews 1999). During the Holocene, net sediment accumulation has been low and foraminifera dominate over the lithic fraction; the ratio rarely drops below 0.2 (Fig. 6A). This may indicate some sediment rafting on sea-ice and far-travelled icebergs, plus erosion and transport from the shallow banks. In order to examine the relative changes in lithics and volcanic glass, a second ratio of volcanic ash shards/lithics+volcanic ash shards is computed. Throughout the diamicton, irregular, sharply edged basaltic and rhyolitic glass constitute about 0.2 of the ash/lithics+ash ratio (Fig. 6B). This ratio increases rapidly above the 80 cm transition zone and in the last 10 000 ^{14}C years, i.e. above 50 cm, the ash shards dominate. The third ratio used shows variations in the percentages of quartz and basalt, which is an index of possible allochthonous sources on the North Iceland margin (Fig. 6C). The ratio of abraded basaltic particles to quartz sand (Fig. 6C) indicates that within the diamicton the ratio averages 2.5/1, but during the postglacial interval, the ratio decreases, indicating that the quartz becomes rare in the sand fraction. Quartz is an important indicator of Holocene ice-rafting off North Iceland (Eiriksson *et al.* 2000*a*, *b*).

Deep-sea cores from the North Atlantic, south of Iceland, contain basaltic Icelandic materials, which occur in association with IRD intervals (Bond & Lotti 1995; Elliott *et al.* 1998). In some cases, basaltic sediments may also be derived from the glacial erosion of extensive outcrops of flood basalts along the East Greenland coast (Larsen 1983). It is important to discriminate between particles that originate from the East Greenland coast and Iceland (Larsen 1983). We are investigating isotopic ways of distinguishing between Icelandic and East Greenland basalts in order to specify the provenance of sand-size basaltic particles.

What is the grain-size spectrum of ice transported glacial sediments in the matrix fraction (below 2 mm)?

Glacial marine sedimentologists and palaeoceanographers investigate the grain-size of IRD in two distinct ways. For the first group, the investigation usually consists of a lithofacies description of the sediment followed by an analysis of the matrix fraction (below 2mm). Some glacial marine sedimentologists count lithics and describe the petrological composition of the matrix, but in general, such analyses are not presented. The source of the sediment is generally not an important issue. In contrast, palaeoceanographers rarely X-ray their cores, and lithofacies descriptions are uncommon. A cut-off within the sand fraction is selected for counting foraminifera and determining the composition of this fraction. The palaeoceanographic community is increasingly working with geochemists and geochronologists to constrain the provenance of the IRD sediments of all size fractions better. This is especially useful for studies of massive iceberg outbreak events that are referred to as Heinrich events (Grousset *et al.* 1993, 2000; Revel *et al.* 1996; Hemming *et al.* 1998, 2000*b*; Barber 2001).

Matrix grain-size attributes of ice-proximal surface sediments

Are there characteristic grain-size distributions associated with glacial erosion and subsequent transport and deposition? Fillon and Full (1984) made the most serious effort to address the problem of discriminating between glacial sediments deposited in deep-sea basins. Their work has largely been ignored. They used a set of surface samples from areas SW of Iceland, the Labrador Sea, and eastern Baffin Bay (Fig. 1). The samples were decalcified, and particle size measurements were made with an Elzone particle-size analyser. The upper size limit of their samples was 212 µm (2–3 phi), and the lower limit was fine silt, 4 µm (8 phi). They used statistical methods to extract five 'principal components', which were used to explain provenance and depositional processes.

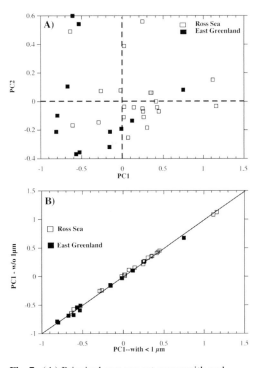

Fig. 7. (**A**) Principal component scores with and without the below 1 µm fraction from surface sediments in currently glaciated environments of East Greenland and the Ross Sea. (**B**) Correlation of the scores on the first principal component axis for the sediments on Figure 5A with and without the under 1 µm fraction.

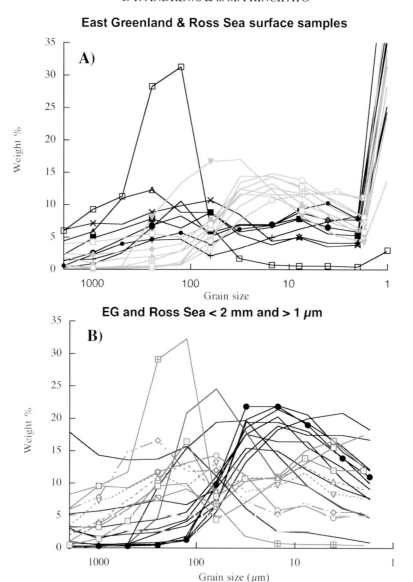

Fig. 8. (**A**) Sedigraph data for surface samples less than 2 mm from East Greenland and for the Ross Sea area, Antarctica (note that the fraction under 1 μm is truncated; typically this fraction constitutes more than 20% of the sediment. (**B**) Samples recalculated to 100% with the fraction under 1 μm removed.

Our data (Figs 7 & 8) are generated from particles in the range of 2000 μm to 1 μm. Because the erosion of detrital carbonate from Palaeozoic basins around the North Atlantic is a fundamental feature of Heinrich events (Andrews & Tedesco 1992; Bond *et al.* 1992; Kirby & Andrews 1999) and glacial marine transport, we did not decalcify the sediments. Issues pertaining to different methods of

grain-size determination are addressed in papers in Syvitski (1991). We present data obtained from a Sedigraph and a Malvern long-bed laser sizing system (Agrawal *et al.* 1991; Konert 1997). Our comparisons between these two instruments (Andrews *et al.* 2002) indicate that for sediments abundant in sand and silt the results are comparable.

We use principal component analysis (PCA)

Table 1. *Principal component analysis: Ross & EG grain size*

(1) Eigenvalues

	Axis 1	Axis 2	Axis 3
Eigenvalues	8.313	2.125	1.074
Cumulative percentage	69.278	86.983	95.932

(2) Ross and EG less 1micron and recalculated to 100% Eigenvalues

	Axis 1	Axis 2
Eigenvalues	8.091	2.196
Cumulative percentage	73.550	93.516

PCA variable loadings

Size μm	(1) Axis 1	Axis 2	Axis 3	(2) Axis 1	Axis 2
4000–2000	0.275	−0.286	−0.349	0.28	−0.301
2000–1000	0.309	−0.257	−0.102	0.314	−0.254
500	0.330	−0.128	0.081	0.333	−0.093
250	**0.335**	0.036	0.132	0.335	0.084
125	0.232	0.469	0.108	0.208	0.511
63	0.051	**0.662**	−0.073	0.007	0.655
32	−0.274	0.350	−0.221	−0.299	0.295
16	−0.331	−0.009	−0.266	−0.346	−0.039
8	**−0.340**	−0.097	−0.072	−0.345	−0.106
4	**−0.338**	−0.125	−0.048	−0.342	−0.129
2	−0.334	−0.160	0.059	−0.342	−0.129
1	−0.161	−0.062	**0.835**		

Bold numbers refer to significant loadings.
Data standardized.

to evaluate the grain-size data employing a similar methodology to Fillon and Full (1984). Because grain-size data sum to 100% by weight or volume, we use the log-ratio transformation (Aitchison 1986; Reyment & Savazzi 1999) to avoid the problems associated with analysis of closed arrays. As Fillon and Full (1984) correctly note, the grain-size samples that we obtain are probably derived from a combination of depositional processes. An original 'pure' glacial sediment, till, may be deposited when ice was grounded on the shelf and sea level was lower, or it may be transported seaward in an iceberg. During the release of the sediment from an iceberg, it will mix with concurrent sedimentary processes. For example, Syvitksi *et al.* (1996) showed that in Kangerlussuaq Fjord, East Greenland, sediment in meltwater plumes mixes with IRD sediments. In glaciated continental margin settings, the boundary between diamicton units directly associated with glacial erosion, reworking and remolding of previously deposited ice-proximal or ice-distal glacial marine and marine sediments is unclear (Fig. 6). It is possible that a 'pure' glacial end member is illusory.

We compare Sedigraph generated records of surface sediments from the western and central Ross Sea with those from the East Greenland margin around Kangerlussuaq Fjord and Trough (Licht 1999; Smith 1997). Both areas are classified as 'glacial marine', but with radically different glaciological settings. However, glaciological processes at the base of the fast-flowing ice streams heading into the Ross Sea may be similar to those at the bottom of Kangerlussuaq Glacier, which is flowing at an estimated rate of more than 1000 m a^{-1} (Andrews *et al.* 1994). Log-ratio PCA analysis of centred and standardized grain-size data (all intervals given equal weight) indicated that 69.3% of the variance was explained on the first PC and an additional 18% is associated with the second PC axis (Table 1, Fig. 7A). These two axes are positively associated with the medium sand (125–250 μm) and negatively with two fine silt fractions (PC1) and the fine-sand fraction respectively. Samples from these two glaciological regimes have substantial overlap when plotted in the 2-D PCA space (Fig. 7A). This can also be seen in the individual grain-size spectra (Fig. 8A).

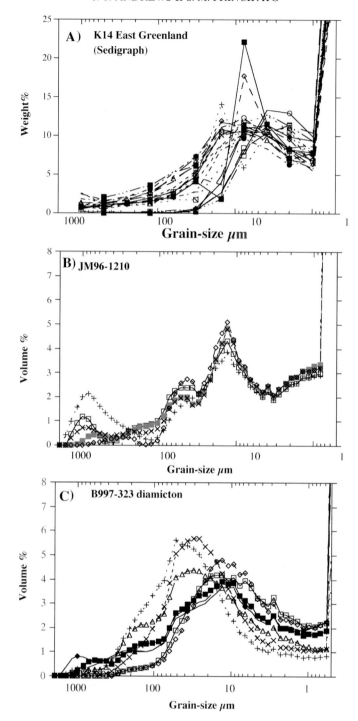

Fig. 9. (A) Plot of grain size data from present-day ice-proximal settings in Nansen Fjord, East Greenland (K14) (Fig. 1A). **(B)** Data from JM96-1210, also in Nansen Fjord. **(C)** Grain-size distribution in the diamicton at the base of B997-323 North Iceland (Fig. 1A). The difference in percentages between the Sedigraph (K14) and the Malvern data (JM96-1210 and B997-323) are caused by the fact that there are 12 grain-size intervals in our Sedigraph data compared to 49 for the Malvern laser system.

Both sets of samples are marked by a substantial fraction of fines, less than 1 µm in size (Fig. 8A). In the samples from the Ross Sea, a substantial proportion of this fine fraction is composed of biogenic silica because of the high rates of diatom production (DeMaster *et al.* 1996; Andrews *et al.* 1999). Shelf sediments from East Greenland contain substantially lower amounts of biogenic silica (Licht *et al.* 1998). Might such differences in the biogenic inputs bias the grain-size spectra? We excluded the very fine fraction and recalculated the spectra to sum to 100%. The resulting graphs (Fig. 8B) indicate that the main mode in these sediments that exclude the fraction below 1 µm is in the medium to fine silt (30–10 µm). A PCA of these recalculated samples resulted in a new set of PCA scores. Fig. 7B shows that there is very little difference between individual PCA scores on the first and second axes whether or not the very fine fraction was removed, suggesting that this process of removing the fraction below 1 µm did not change the distributions materially.

These data indicate that these glacial marine sediments are principally silty-clays and that the fraction of sand ranges from a few percent up to 50% of the sample. However, only in three samples was the modal grain size in the sand-size range (Fig. 8A). These results indicate that the sand fraction is not usually the modal fraction; a more representative fraction of glacial marine and glacial sediments is within the silt range.

Matrix grain-size of ice-proximal subsurface glacial marine sediments

Our analysis of present seafloor sediments (Figs 7 & 8) might not be typical of deposition on continental margins during intervals of more extensive glaciation. Thus, we extend our analysis of glacial marine sediments to those recovered in cores from East Greenland and North Iceland.

Ice-proximal glacial marine sediments reflect a mixture of iceberg rafted sediments, sediment transported in meltwater plumes, and biogenic components — a pure glacial marine ice-proximal signal might be difficult to define. Two cores (B91-K14 and JM96-1210) were taken from Nansen Fjord, East Greenland, within a few kilometres of the actively calving Christian IV Gletcher (Fig. 1; Fig. 5). This glacier has an estimated calving output of around 2 km³ per year (Andrews *et al.* 1994). The lithofacies, sediment accumulation rates, and palaeo-environmental changes have been presented for K14 (Jennings & Weiner 1996; Smith &

Andrews 2000) and some data have been presented from 1210 (Eastman & Andrews 1997). Grain-size data from K14 were determined with a Sedigraph, whereas the 1210 data were determined from the Malvern laser system and represent volume per cent data. The grain-size attributes of sediments over 2 mm from K14 are illustrated in Fig. 4.

The grain-size distributions of these ice-proximal sediments are multi-modal (Fig. 9A & B). Although these data were collected within 5 km of the ice margin, they are quite similar to the regional array of surface sediments illustrated in Figure 8. This is especially true in terms of the silt peak between 10 and 15 µm at the fine end of the 'sortable silt' fraction (10–15 µm) (McCave *et al.* 1995). In 1210 this peak is noticeable, but in K14 there is another mode at about 3 µm. The Malvern data from 1210 indicates a small mode at 1000 µm followed by a decrease in the per cent volume to 500 µm. Both grain-size methods indicate that a substantial fraction of the sediment is finer than 1 µm (Fig. 9A & B). The multi-modality of the grain-size spectra might indicate a combination of different erosional processes, different bedrock lithologies and associated mineral grain sizes, and different transport and depositional mechanisms (Drewry 1986).

Cores and seismic data from the North Iceland shelf shows that diamictons underlie a thin (less than 1 m) postglacial marine mud (Helgadottir & Andrews 1999) (e.g. Fig. 4B & 6). The origin(s) of the diamictons is under investigation (Principato & Andrews 2001). They probably represent glacially overrun glacial marine sediments, or less likely, original glacial marine sediments. They are interpreted as 'glacial marine' and not subglacial till because they possess a consistent proportion of benthic foraminifera throughout the diamicton unit (Helgadottir & Andrews 1999) (Fig. 6A). Grain-size analysis (Fig. 9C) of the diamicton matrix shows two main modes. One mode is in the coarse sand fraction, around 1000 µm, and the other is a dominant peak in the silt fraction from 30 to 10 µm. These data compared with the glacial marine sediments in 1210 (Fig. 9B), show a slightly coarser silt mode and lack the small peak around 63 µm. It would be interesting to compare the mineralogies of the silt-size fractions at these two sites to ascertain whether the differences in grain size are associated with changes in the source bedrock (cf. Andrews *et al.* 1989).

Based on the 49 grain-size intervals of the Malvern laser system, log-ratio transformed PCA analysis of the 1210 and 323 data (Fig. 9B

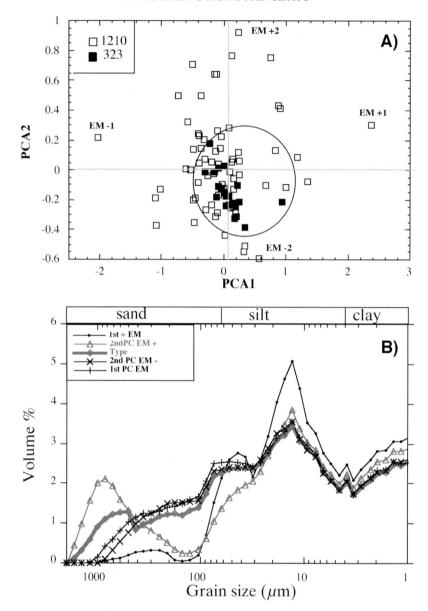

Fig. 10. (**A**) Plot of principal component scores from grain-size data (centred and standardized) from JM96-1210 and B997-323. Note the clustered nature of the North Iceland diamicton samples compared with the more dispersed glacial marine diamicton samples from Nansen Fjord. (**B**) Plot of end member grain-size distributions from PCA 1 and 2 (labelled relative to their PCA and sign, hence EM 1+ versus EM 2–) and a 'type core' (defined by PCA scores of close to 0,0 on PCA 1 and 2).

& C) enable us to determine grain-size end-members (cf. Wang & Hesse 1996; Fillon & Full 1984). The first two axes explain seventy-two per cent of the variance. When the two sets of data are plotted together (Fig. 10A), the data from the Iceland diamicton (323) are significantly less dispersed than the spectra from 1210. On Figure 10B we plot the four labelled end members (EM), all from 1210, and a type spectrum which includes a significant number of the samples from 323. The first end member (EM 1+) has limited sand but a strong secondary peak in fine sand and coarse silt. This is followed by a dominant mode in the silt fraction at about 15 μm. In contrast, EM2 + shows a very strong mode in the coarse sand, no fine sand peak, and a peak in the sortable silt fraction. EM 1- and EM 2- are very similar (Fig. 10B). They are both typified by a consistent increase in the sand fraction from coarse to fine with a plateau in the fine sand/coarse silt area and a peak similar in volume per cent to EM 2+. The 'type' curve is only marginally different from EM 1- with the exception of a small peak in the medium to coarse sand. From the viewpoint of the deep-sea palaeoceanographer, some end members have very low percentages in the sand fractions between 100 and 500 μm.

Clay-size mineralogy and provenance

Ideally, provenance indicators should be obtained on all fractions of glacially derived sediments. However, provenance studies frequently focus only on the sand and clay fractions, although as shown in Figures 8 and 10, the most diagnostic fraction may be silt-sized material, which is commonly considered to be the product of abrasion and crushing at the base of ice sheets and glaciers (Drewry 1986). Significant differences in the mineralogy of silt versus clay-size minerals from glacial marine offshore sediments has been demonstrated (Andrews *et al.* 1989) and the under 63 μm fraction is preferred by some for isotopic identification of source areas (Groussett *et al.* 1993, 2000; Barber 2001).

Clay-size particles are transported seaward in meltwater plumes, and they are probably also released from the meltout of sediment-rich basal glacial sediments. X-ray diffraction (XRD) analyses of the below 1 μm fraction from core 1210 indicate abundant smectite (Fig. 11A). This suggests that a substantial proportion of the fine fraction is associated with the erosion of the weathering products of the Tertiary basalts of East Greenland *c.* 68° N (Larsen 1983). There is little feldspar or quartz, which we would expect from the erosion of the Precambrian shield rocks that crop out a few kilometres from the coast of Greenland. In contrast with Greenland, clay mineralogy on sediments from Baffin Island fjords is dominated by illite (mica). This is due to erosion of Precambrian granites and gneisses of the Canadian Shield (Andrews *et al.* 1989; Andrews 1990).

In the XRD analyses of core 323 from North Iceland (Fig. 1), mica is absent within the diamicton (Dmm), and it only occurs in the deglacial and postglacial marine sediments (above 80 cm) (Fig. 11B). On average, smectite makes up about 90% of the clay-size mineral composition, compared with approximately 70% smectite in Nansen Fjord, East Greenland. XRD analysis of the clay-size fraction appears to distinguish between the East Greenland and Icelandic source areas, but these observations are too spatially limited to make an explicit argument. More analyses, including XRD, isotopic analyses, and other provenance indicators are needed to characterize and discriminate between Greenland and Iceland sourced sediments. This is especially true if the goal is to link modern day glacial sources with deep-sea palaeo-environmental indicators of provenance.

Discussion

There is a need to communicate the differences in approaches in grain size analyses between the glacial marine community, which is commonly focused on sedimentary processes, and the palaeoceanographic community, which is governed by a goal of reconstructing past climatic scenarios. Heinrich events (Heinrich 1988; Andrews 1998) have provided a catalyst to develop a better understanding of the physics underlying ice sheet behaviour on 6–7 ka time scales (Clarke *et al.* 1999; Dowdeswell & Siegert 1999). Progress is also being made on understanding the controls on iceberg trajectories and melt rates (Matsumoto 1997; Schaefer-Neth & Stattegger 1999; Clarke & Prairie in press). Despite this progress, however, these two communities should recognize that IRD is defined in a variety of ways, and this could lead to different palaeoclimate interpretations of the same data.

There are three restrictions on the successes of future modelling exercises on glacial sediment source to sink trajectories:

(1) we have extremely limited empirical knowledge of the volume of sediment entrained in icebergs (Anderson *et al.* 1980);
(2) we need unique tracers for the main iceberg sources so we can unmix the relative contributions of the different ice sheets (Groussett *et al.* 1993; Hemming *et al.* 2000*b*; Barber 2001); and
(3) we have not explicitly considered how best to define an IRD event in terms of sediment grain size.

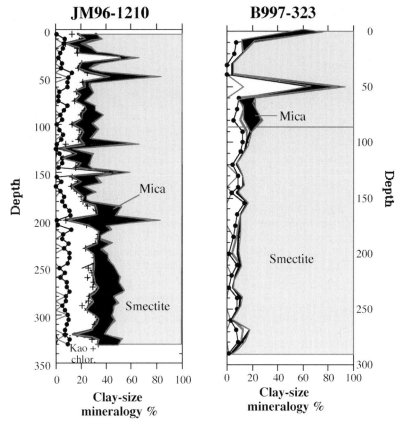

Fig. 11. (**A**) Cumulative percentage X-ray diffraction data on the fraction below 4 μm from JM96-1210 from the East Greenland margin (Fig. 1). (**B**) Cumulative percentage X-ray diffraction data on the fraction less than 4 μm from B996-323 from the North Iceland margin (Fig. 1). The bedrock geology of the source areas are somewhat different although both have a significant Tertiary basalt component, but East Greenland has a substantial outcrop of Precambrian granites and granite gneisses. The cumulative fraction marked by the solid black dots is composed of carbonate minerals, feldspars, and quartz. In B997-323 quartz is virtually absent.

The focus of tracking IRD events in the palaeoceanographic community is usually to determine their climatic significance, but provenance studies are increasing and undoubtedly will result in significant progress in our understanding of the temporal interactions between ice sheets. Ultimately, we need to understand the physics of glacial erosion and transportation better, and to model iceberg trajectories and melt rates in order to understand glacial sediment source to sink paths better (Schaefer-Neth and Stattegger 1999).

This paper has attempted to document some properties of sediments deposited in ice proximal settings and to provide examples of what sediment distributions might look like when deposition from far-travelled icebergs

occurs in deep-sea basins (Ruddiman 1977). Little is known about the grain-size spectra of sediments in icebergs, and we can only infer grain-size distributions from deposited sediments. However, is there any reason to expect an evolution or change in the grain size of iceberg rafted sediments from glaciated margins to the deep-sea basins? This might occur if there is some consistent vertical change in sediment characteristics within an ice sheet. These sediment variations might also be due to a change in grain size deposited by icebergs as they travel to deeper water. This iceberg trajectory explanation is not compelling because icebergs overturn as they travel. Thus, we maintain that the grain-size attributes illustrated on Figures 2–11 might form an initial basis for comparison with sediments deposited in deep-sea basins adjacent to glaciated continental margins. However, as noted earlier the probability of observing large clasts (Figs 2–4) depends on their numbers per unit volume within the ice, hence we imagine that the probability of clasts over 2 mm being observed in ice-distal sediments (i.e. in the temperate North Atlantic) is low.

The focus on the provenance of sand-size particles in deep-sea sediments ignores the dominance of the silt and clay fraction in the total ice-proximal sediment spectra (Figs 8 & 9). Isotopic analysis of the sand-fraction in Heinrich events yields a distinct identification of provenance (Gwiazda *et al.* 1996; Hemming *et al.* 1998, 2000*a*), but this approach needs to be combined with an examination of the fraction below 63 µm. Studies that have analysed the under 63 µm fraction alone (Grousset *et al.* 1993) are possibly suspect because the authors lack a database of the critical isotopic signature of ice-proximal sediments from point sources around the North Atlantic (Barber *et al.* 1995; Barber 2001). Recent and current research at the University of Colorado and our associates at Lamont-Doherty (Hemming) is focused on defining the isotopic composition of source end-members from around the North Atlantic, from Iceland to the eastern Canadian margin (Barber 2001; Farmer and Hemming unpubl. data). More studies, using isotopic analyses and other provenance indicators from all size fractions, are urgently needed to improve the resolution of ice sheet–ocean interactions (Dowdeswell *et al.* 2001)

Our research on the glacial marine sediments on the NE Canadian, East Greenland and Iceland margins has been supported by a number of grants from the National Science Foundation, especially the Arctic Natural Science Program and the Antarctic Geology Program of the Office of Polar Programs, and the PARCS/ESH programs on Earth History. We appreciate the critical comments of C. Ó Cofaigh, R. Gilbert, and C. Pudsey on an earlier draft. Grain-size analyses were performed by R. Kihl and in the INSTAAR Sedimentological Laboratory.

References

AGRAWAL, Y. C., McCAVE, I. N. & RILEY, J. B. 1991. Laser diffraction size analysis. *In:* SYVITSKI, J. P. M. (ed.) *Principles, Methods and Application of Particle Size Analysis.* Cambridge University Press, Cambridge, 119–128.

AITCHISON, J. 1986. *The Statistical Analysis of Compositional Data.* Chapman and Hall, London.

ANDERSON, J. B. 1999. *Antarctic Marine Geology.* Cambridge University Press, Cambridge.

ANDERSON, J. B., DOMACK, E. W. & KURTZ, D. D. 1980. Observations of sediment-laden icebergs in Antarctic waters: Implications to glacial erosion & transport. *Journal of Glaciology*, **25**, 387–396.

ANDREWS, J. T. 1990. Fiord to Deep-Sea sediment transfers along the northeastern Canadian continental margin: Models and data. *Géographie Physique et Quaternaire*, **44**, 55–70.

ANDREWS, J. T. 1998. Abrupt changes (Heinrich events) in late Quaternary North Atlantic marine envrionments: a history and review of data and concepts. *Journal of Quaternary Science*, **13**, 3–16.

ANDREWS, J. T. 2000. Icebergs and iceberg rafted detritus (IRD) in the North Atlantic: Facts and assumptions. *Oceanography*, **13**, 100–108.

ANDREWS, J. T. & BARBER, D. C. 2002. Dansgaard-Oescher events: Is there a signal off the Hudson Strait Ice Stream? *Quaternary Science Reviews*, **21**, 443–454.

ANDREWS, J. T. & HELGADOTTIR, G. 1999. Late Quaternary glacial and marine environments off North Iceland: Where was the LGM? *Abstract volume, Geological Society of America*, **31**, A314.

ANDREWS, J. T. & MATSCH, C. L. 1983. *Glacial marine sediments and sedimentation: An annotated bibliography.* Geo Abstracts Ltd., Norwich, UK.

ANDREWS, J. T. & TEDESCO, K. 1992. Detrital carbonate-rich sediments, northwestern Labrador Sea: Implications for ice-sheet dynamics and iceberg rafting (Heinrich) events in the North Atlantic. *Geology*, **20**, 1087–1090.

ANDREWS, J. T., GEIRSDOTTIR, A. & JENNINGS, A. E. 1989. Spatial and temporal variations in clay- and silt-size mineralogies of shelf and fiord cores, Baffin Island. *Continental Shelf Research*, **9**, 445–463.

ANDREWS, J. T., MILLIMAN, J. D., JENNINGS, A. E., RYNES, N. & DWYER, J. 1994. Sediment thicknesses and Holocene glacial marine sedimentation rates in three east Greenland fjords (ca. 68°N). *Journal of Geology*, **102**, 669–683.

ANDREWS, J. T., SMITH, L. M., PRESTON, R., COOPER, T. & JENNINGS, A. E. 1997. Spatial and temporal patterns of iceberg rafting (IRD) along the East Greenland margin, ca. 68 N, over the last 14 cal.ka. *Journal of Quaternary Science*, **12**, 1–13.

ANDREWS, J. T., DOMACK, E. W., CUNNINGHAM, W. L., LEVENTER, A., LICHT, K. J., JULL, A. J. T., DeMASTERS, D. J. & JENNINGS, A. E. 1999. Problems and possible solutions concerning radiocarbon dating of surface marine sediments, Ross Sea, Antarctica. *Quaternary Research*, **52**, 206–216.

ANDREWS, J. T., KIHL, R., KRISTJÁNSDÓTTIR, G. B., SMITH, L. M., HELGADÓTTIR, G., GEIRSDÓTTIR, Á. & JENNINGS, A. E. 2002. Holocene sediment properties of the East Greenland and Iceland continental shelves bordering Denmark Strait (64°–68°N), North Atlantic. *Sedimentology*, **49**, 5–24.

ASHLEY, G. M. & SMITH, N. D. 2000. Marine sedimentation at a calving glacier margin. *GSA Bulletin of the Geological Society of America*, **112**, 657–667.

BARBER, D. C. 2001. *Laurentide ice sheet dynamics*

from 35 to 7 ka: Sr-Nd-Pb isotopic provenance of northwest Atlantic margin sediments. PhD dissertation, University of Colorado, Boulder, CO.

BARBER, D. C., FARMER, L., ANDREWS, J. T. & KIRBY, M. E. 1995. Mineralogic and istopic tracers in late Quaternary sediments in the Labrador Sea: Implications for iceberg sources during Heinrich events. *EOS*, **76**, F296.

BOND, G. C. & LOTTI, R. 1995. Iceberg discharges into the North Atlantic on millennial time scales during the last glaciation. *Science*, **267**, 1005–1009.

BOND, G., HEINRICH, H., BROECKER, W. S., LABEYRIE, L., MCMANUS, J., ANDREWS, J. T., HUON, S., JANTSCHIK, R., CLASEN, S., SIMET, C., TEDESCO, K., KLAS, M., BONANI, G. & IVY, S. 1992. Evidence for massive discharges of icebergs into the glacial Northern Atlantic. *Nature*, **360**, 245–249.

BOND, G., BROECKER, W. S., JOHNSEN, S., MCMANUS, J., LABEYRIE, L., JOUZEL, J. & BONANI, G. 1993. Correlations between climate records from North Atlantic sediments and Greenland ice. *Nature*, **365**, 143–147.

BOND, G., MANDEVILLE, C. & HOFFMAN, S. 2001. Were rhyolitic glasses in the Vedde ash and in the North Atlantic Ash Zone 1 produced by the same volcanic eruption. *Quaternary Science Reviews*, **20** (11), 1189–1199.

CARTEE-SCHOOLFIELD, S. 2000. *Late Pleistocene Sedimentation in the Denmark Strait Region.* MSc thesis, University of Colorado.

CLARKE, G. K. & PRAIRIE, I. L. in press. Modelling iceberg drift and ice-rafted sedimentation. *In:* GREVE, R. (ed.) *Continuum Mechanics and Applications in Geophysics and the Environment*, Springer-Verlag, Berlin, 1–18.

CLARKE, G. K. C., MARSHALL, S. J., HILLAIRE-MARCEL, C., BILODEAU, G. & VEIGA-PIRES, C. 1999. A glaciological perspective on Heinrich events. *In:* CLARK, P. U., WEBB, R. S. & KEIGWIN, L. D. (eds) *Mechanisms of Global Climate Change at Millennial Time Scales*, American Geophysical Union, Washington, DC, 243–262.

DEMASTER, D. J., RAGUENEAU, O. & NITTROUER, C. A. 1996. Preservation efficiencies and accumulation rates for biogenic silica and organic C, N, and P in high-latitude sediments: The Ross Sea. *Journal of Geophysical Research*, **101** (C8), 18,501–18,518.

DOMACK, E., O'BRIEN, P. E., HARRIS, P. T., TAYLOR, F., QUILTY, P. G., DESANTIS, L. & RAKER, B. 1998. Late Quaternary sediment facies in Prydz Bay, East Antarctica and their relationship to glacial advance onto the continental shelf. *Antarctic Science*, **10**, 236–246.

DOWDESWELL, J. A. & SCOURSE, J. D. 1990. Glacimarine Environments: Processes and sediments. Geological Society, London, Special Publications, **53**.

DOWDESWELL, J. A. & SIEGERT, M. J. 1999. Ice-sheet numerical modeling and marine geophysical measurements of glacier-derived sedimentation on the Eurasion Arctic continental margins. *GSA Bulletin of the Geological Society of America*, **111**, 1080–1097.

DOWDESWELL, J. A., WHITTINGTON, R. J. & MARIENFELD, P. 1994. The origin of massive diamicton facies by iceberg rafting and scouring, Scorsby Sund, East Greenland. *Sedimentology*, **41**, 21–35.

DOWDESWELL, J. A., ELVERHØI, A., ANDREWS, J. T. & HEBBELN, D. 1999. Asynchronous deposition of ice-rafted layers in the Nordic seas and North Atlantic Ocean. *Nature*, **400**, 348–351.

DOWDESWELL, J. A., Ó COFAIGH, C., ANDREWS, J. T. & SCOURSE, J. D. 2001. Workshop explores debris transported by icebergs and its paleoenvironmental implications. *EOS*, **82** (35), 382, 386.

DREWRY, D. 1986. *Glacial geologic processes.* Edward Arnold, London, 276 pp.

EASTMAN, K. & ANDREWS, J. T. 1997 Changes in inshore marine and glacial conditions over the last 3,000 years, East Greenland (68°N) based on high-resolution marine sediment cores. *Geological Society of America, Abstracts* 29, A38.

EIRIKSSON, J., KNUDSEN, K. L., HAFLIDASON, H. & HEINEMEIER, H. 2000a. Chronology of the late Holocene climatic events in the northern North Atlantic based on AMS 14C dates and tephra markers from the volcano, Hekla, Iceland. *Journal of Quaternary Science*, **15**, 573–580.

EIRIKSSON, J., KNUDSEN, K. L., HAFLIDASON, H. & HENRIKSEN, P. 2000b. Late-glacial and Holocene paleoceanography of the North Iceland Shelf. *Journal of Quaternary Science*, **15**, 23–42.

ELLIOTT, M., LABEYRIE, L., BOND, G., CORTIJO, E., TURON, J. L., TISNERAT, N. & DUPLESSY, J. C. 1998. Millennial-scale iceberg discharges in the Irminger Basin during the last glacial period: Relationship with the Heinrich events and environmental settings. *Paleoceanography*, **13**, 433–446.

FILLON, R. H. & FULL, W. E. 1984. Grain-size variations in North Atlantic non-carbonate sediments and sources of terrigenous components. *Marine Geology*, **59**, 13–50.

FILLON, R. H. & HARMES, R. A. 1982. Northern Labrador Shelf glacial chronology and depositional environments. *Canadian Journal of Earth Sciences*, **19**, 162–192.

GROBE, H. 1987. A simple method for the determination of ice-rafted debris in sediment cores. *Polarforschung*, **57**, 123–126.

GROUSSET, F. E., LABEYRIE, L., SINKO, J. A., CREMER, M., BOND, G., DUPRAT, J., CORTIJO, E. & HUON, S. 1993. Patterns of ice-rafted detritus in the glacial North Atlantic (40–55°N). *Paleoceanography*, **8**, 175–192.

GROUSSET, F. E., PUJOL, C., LABEYRIE, L., AUFFRET, G. & BOELAERT, A. 2000. Were the North Atlantic Heinrich events triggered by the behavior of the European ice sheets. *Geology*, **28**, 123–126.

GWIAZDA, R. H., HEMMING, S. R., BROECKER, W. S., ONSTTOT, T. & MUELLER, C. 1996. Evidence from Ar/Ar ages for a Churchill province source of ice-rafted amphiboles in Heinrich layer 2. *Journal of Glaciology*, **42**, 440–446.

HEMMING, S. R., BROECKER, W. S., SHARP, W. D., BOND, G. C., GWIAZDA, R. H., MCMANUS, J. F., KLAS, M. & HAJDAS, I. 1998. Provenance of Heinrich layers in core V28–82, northeastern Atlantic: ^{40}Ar/^{39}Ar ages of ice-rafted hornblende, Pb isotopes in feldspar grains, and Nd-Sr-Pb isotopes in the fine sediment fraction. *Earth and Planetary Science Letters*, **164**, 317–333.

HEMMING, S. R., BOND, G. C., BROECKER, W. S., SHARP, W. D. & KLAS-MENDELSON, M. 2000a. Evidence from ^{40}Ar/^{39}Ar ages of individual hornblende grains for varying Laurentide sources of iceberg discharges 22,000 to 10,500 yr B.P. *Quaternary Research*, **54**, 372–383.

HEMMING, S. R., GWIADZA, R. H., ANDREWS, J. T., BROECKER, W. S., JENNINGS, A. E. & ONSTOTT, T. C. 2000b. ^{40}Ar/^{39}Ar and Pb-Pb study of individual hornblende and feldspar grains from southeastern Baffin Island glacial sediments: implications for the provenance of the Heinrich layers. *Canadian Journal of Earth Sciences*, **37**, 879–890.

HEINRICH, H. 1988. Origin and consequences of cyclic ice rafting in the Northeast Atlantic Ocean during the past 130,000 years. *Quaternary Research*, **29**, 143–152.

HELGADOTTIR, G. & ANDREWS, J. T. 1999. Late Quaternary shifts in sediment north of Iceland: Foraminiferal data and stable oxygen isotope analyses. *Geological Society of America Abstract*, **31**, A314.

JENNINGS, A. E. & WEINER, N. J. 1996. Environmental change in eastern Greenland during the last 1300 years: Evidence from foraminifera and lithofacies in Nansen Fjord, 68°N. *The Holocene*, **6**, 179–191.

KIRBY, M. E. & ANDREWS, J. T. 1999. Mid-Wisconsin Laurentide ice sheet growth and decay: Implications for Heinrich events-3 and -4. *Paleoceanography*, **14**, 211–223.

KONERT, M. & VANDENBERGHE, J. 1997. Comparison of laser grain size analysis with pipette and sieve analysis: a solution for the underestimation of the clay fraction. *Sedimentology*, **44**, 523–535.

LARSEN, B. 1983. Geology of the Greenland–Iceland Ridge in the Denmark Strait. *In*: BOTT, M. H. P., SAXOV, S., TALWANI, M. & THIEDE, J. (eds) *Structure and Development of the Greenland–Scotland Ridge*. Plenum Publishing Corp., London, 425–444.

LICHT, K. J. 1999. *Investigations into the Late Quaternary History of the Ross Sea, Antarctica*. PhD thesis, University of Colorado.

LICHT, K. J., CUNNINGHAM, W. L., ANDREWS, J. T., DOMACK, E. W. & JENNINGS, A. E. 1998. Establishing chronologies from acid-insoluble organic ^{14}C dates on Antarctic (Ross Sea) and Arctic (North Atlantic) marine sediments. *Polar Research*, **17**, 202–216.

LICHT, K. J., DUNBAR, N. W., ANDREWS, J. T. & JENNINGS, A. E. 1999. Distinguishing subglacial till and glacial marine diamictions in the western Ross Sea, Antarctica: Implications for a last glacial maximum grounding line. *Bulletin of the Geological Society of America*, **111**, 91–103.

LOWELL, T. V., HEUSSER, C. J., ANDERSEN, B. G., MORENO, P. I., HAUSER, A., HEUSSER, L. E., SCHLUCHTER, C., MARCHANT, D. R. & DENTON, G. H. 1995. Interhemispheric correlation of late Pleistocene glacial events. *Science*, **269**, 1541–1549.

MARIENFELD, P. 1992. Recent sedimentary processes in Scoresby Sund, East Greenland. *Boreas*, **21**, 169–186.

MATSUMOTO, K. 1996. An iceberg drift and decay model to compute the ice-rafted debris and iceberg meltwater flux: Application to the interglacial North Atlantic. *Paleoceanography*, **11**, 729–742.

MATSUMOTO, K. 1997. Modeled glacial North Atlantic ice-rafted debris pattern and its sensitivity to various boundary conditions. *Paleoceanography*, **12**, 271–280.

McCAVE, I. N., MANIGHETTI, B. & ROBINSON, S. G. (1995). Sortable silt and fine sediment size/composition slicing: Parameters for palaeocurrent speed and palaeoceanography. *Paleoceanography*, **10**, 593–610.

McMANUS, J. R., OPPO, D. W. & CULLEN, J. L. 1999. A 0.5 million-year record of millennial-scale climate variability in the North Atlantic. *Science*, **283**, 971–975.

POWELL, R. D. & MOLNIA, B. F. 1989. Glacimarine sedimentary processes, facies and morphology of the south-southeast Alaska shelf and fjords. *Marine Geology*, **85**, 359–390.

PRINCIPATO, S. M. & ANDREWS, J. T. 2001. Interpreting the origin of diamicton units in marine cores from the Iceland Shelf. *The 31st International Arctic Workshop, Program and Abstracts*, 89–90.

REVEL, M., SINKO, J. A., GROUSSET, F. E. & BISCAYE, P. E. 1996. Sr and Nd isotopes as tracers of North Atlantic lithic particles: Paleoclimatic implications. *Paleoceaonography*, **11**, 95–113.

REYMENT, R. A. & SAVAZZI, E. 1999. *Aspects of Multivariate Statistical Analysis in Geology*. Elsevier, New York.

RUDDIMAN, W. F. 1977. Late Quaternary deposition of ice-rafted sand in the sub-polar North Atlantic (40–60 N). *Bulletin of the Geological Society of America*, **88**, 1813–1827.

SCHAEFER-NETH, C. & STATTEGGER, K. 1999. Icebergs in the North Atlantic: Modelling circulation changes and glacio-marine deposition. *In*: HARFF, U. A. (ed.) *Computerized Modelling of Sedimentary Systems*. Springer-Verlag, Berlin, 63–78.

SEIDOV, D. & HAUPT, B. J. 1997. Simulated ocean circulation and sediment transport in the North Atlantic during the last glacial maximum and today. *Paleoceanography*, **12**, 281–305.

SMITH, L. M. 1997. *Late Quaternary glacial marine sedimentation in the Kangerlussuaq Region, East Greenland, 68° N*. MSc thesis, University of Colorado, Boulder.

SMITH, L. M. & ANDREWS, J. T. 2000. Sediment characteristics in iceberg dominated fjords, Kangerlussuaq region, East Greenland. *Sedimentary Geology*, **130**, 11–25.

STEIN, R., NAM, S.-I., GROBE, H. & HUBBERTEN, H. 1996. Late Quaternary glacial history and short-term ice-rafted debris fluctuations along the East Greenland continental margin. *In*: ANDREWS, J. T., AUSTEN, W. A., BERGSETN, H. & JENNINGS, A. E. (eds) *Late Quaternary Paleoceanography of North Atlantic Margins*. Geological Society, London, Special Publications, **111**, 135–151.

Syvitski, J. P. M. (ed.) 1991. *Principles, Methods, and Applications of Particle Size Analysis*. Cambridge University Press, London.

Syvitski, J. P. M., Andrews, J. T. & Dowdeswell, J. A. 1996. Sediment deposition in an iceberg-dominated glacimarine environment, East Greenland: Basin fill implications. *Global and Planetary Change*, **12**, 251–270.

van Kreveld, S., Sarnthein, M., Erlenkeuser, H., Grootes, P., Jung, S., Nadeau, M. J.,

Pflaumann, U. & Voelker, A. 2000. Potential links between surging ice sheets, circulation changes, and the Dansgaard-Oeschger cycles in the Irminger Sea, 60–18 ka. *Paleoceanography*, **15**, 425–442.

Wang, D. & Hesse, R. 1996. Continental slope sedimentation adjacent to an ice-margin. II. Glaciomarine depositional facies on Labrador Slope and glacial cycles. *Marine Geology*, **135**, 65–96.

Sediment reworking on high-latitude continental margins and its implications for palaeoceanographic studies: insights from the Norwegian–Greenland Sea

COLM Ó COFAIGH[1], JUSTIN TAYLOR[2], JULIAN A. DOWDESWELL[1],
ANTONI ROSELL-MELÉ[3], NEIL H. KENYON[4], JEFFREY EVANS[5]
& JÜRGEN MIENERT[6]

[1]Scott Polar Research Institute and Department of Geography, University of Cambridge, Cambridge CB2 1ER, UK (e-mail: co232@cam.ac.uk)
[2]Bristol Glaciology Centre, School of Geographical Sciences, University of Bristol, Bristol BS8 1SS, UK
[3]ICREA, Centre of Environmental Studies, Universitat Autonoma de Barcalona, 018193 Bellaterra, Catalonia, Spain
[4]Southampton Oceanography Centre, European Way, Southampton SO14 3ZH, UK
[5]British Antarctic Survey, High Cross, Madingley Road, Cambridge CB3 0ET, UK
[6]Department of Geology, University of Tromsø, Dramsveien 201, 9037 Tromsø, Norway

Abstract: Geological evidence indicates that sediment reworking is common around the continental margins and abyssal depths of the Norwegian–Greenland Sea, a high-latitude setting with glacier-influenced margins. Detailed analysis of 22 cores up to 5 m long, placed in context by accompanying geophysical data including high resolution sub-bottom profiles, swath bathymetry and backscatter maps, indicates that reworking is variable and ranges from debris flows and turbidity currents, to bottom-current activity, as well as iceberg scouring. Reworking by debris flows appears to be restricted mainly to the main trough-mouth fans and sediment slides. Elsewhere, turbidity-current activity frequently dominates, although iceberg ploughing down to 600 m depth and current winnowing assume increasing significance on continental shelves. Reworking in the Norwegian–Greenland Sea reflects variations in ice-sheet dynamics that, in turn, influence the rate of sediment delivery and location of depocentres. Spatial variations in the style of reworking may also reflect the influence of continental slope gradient and bedrock geology on continental shelves. The widespread nature of sediment reworking has important implications for palaeoceanographic investigations in the region, as reworking can result in erosion and disturbance of the sediment column. It is estimated that less than 7% of material delivered to the Norwegian–Greenland Sea since the Late Weichselian is derived from hemipelagic and pelagic sedimentation. This problem is significant where continuous, high-resolution records of hemipelagic and pelagic sedimentation are required, and attempts are made to correlate with other high-resolution proxy records, such as ice cores, at sub-millennial scales. Bioturbation results in the smoothing of high-resolution records and imposes a maximum resolution for sediment-core time-slices of generally 400 years or more. In the Norwegian–Greenland Sea, areas of high sedimentation such as trough-mouth fans or contourite drifts are commonly associated with extensive reworking. Identification of reworking is particularly important where attempts are made to link records of iceberg-rafted debris to past ice-sheet dynamics, as bottom-current winnowing and mass-flow processes can increase the concentration of coarse-grained iceberg-rafted debris. Such localized accentuation of the iceberg-rafted debris signal may lead to erroneous palaeo-environmental interpretations. It is therefore critical that palaeoceanographic interpretations are firmly underpinned by an explicit sedimentological assessment of reworking.

Sediment reworking on high-latitude continental margins is of two general styles: (1) erosion and deposition by sediment-gravity flow processes and contour/bottom currents, and (2) disturbance or 'smoothing' of the sediment record. The latter may be due to bioturbation, which disturbs the upper part of the sedimentary sequence, and iceberg scouring, which

From: DOWDESWELL, J. A. & Ó COFAIGH, C. (eds) 2002. *Glacier-Influenced Sedimentation on High-Latitude Continental Margins.* Geological Society, London, Special Publications, **203**, 325–348. 0305-8719/02/$15.00

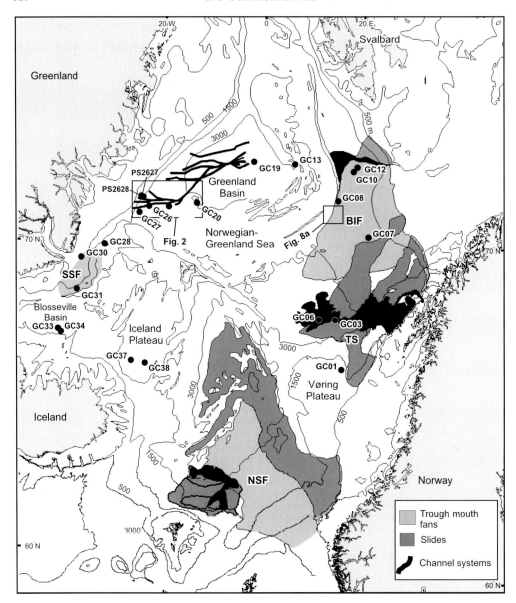

Fig. 1. Map of Norwegian–Greenland Sea showing main sediment slides, trough-mouth fans, submarine channel systems and location of cores discussed in text. BIF, Bear Island Fan; SSF, Scoresby Sund Fan; NSF, North Sea Fan; TS, Trænadjupet Slide. Bathymetric contours are in metres. The inset box shows the area covered by Figure 2.

turbates and homogenizes the substrate. The nature of reworking and its resulting sedimentary signature reflects temporal and spatial variations in the interplay of climate, ice (ice sheets and sea-ice) and oceanography. For example, where ice sheets expand across continental shelves and ice streams develop during full-glacial periods, large volumes of glacigenic sediment may be delivered to the continental slope and reworked by sediment gravity flow processes (e.g. Hesse *et al.* 1999; King *et al.* 1998).

Recognition of sediment reworking in the geological record is important because it can

Table 1. *Site information on sediment cores from the Norwegian–Greenland Sea. Locations of cores are shown in Figure 1*

Core No.	Location	Date acquired	Water depth (m)	Recovery (m)
JR51-GC01	67°48.00'N, 07°19.96'E	29/07/00	1557	3.24
JR51-GC03	69°12.50'N, 07°16.94'E	03/08/00	3150	0.41
JR51-GC06	69°19.54'N, 05°38.79'E	04/08/00	3234	1.34
JR51-GC07	71°56.00'N, 12°27.00'E	04/08/00	1875	3.70
JR51-GC08	73°10.00'N, 09°40.00'E	06/08/00	2289	3.20
JR51-GC10	74°29.30'N, 10°52.00'E	08/08/00	2409	2.16
JR51-GC12	74°10.98'N, 10°44.04'E	08/08/00	2320	1.96
JR51-GC13	74°33.23'N, 05°03.64'E	08/08/00	3272	3.50
JR51-GC19	74°26.73'N, 02°21.12'W	13/08/00	3676	4.66
JR51-GC20	73°14.99'N, 09°00.02'W	14/08/00	3039	3.68
JR51-GC25	73°06.93'N, 11°52.98'W	24/08/00	2735	1.46
JR51-GC26	72°58.49'N, 13°47.54'W	24/08/00	2567	1.40
JR51-GC27	72°47.16'N, 14°44.27'W	24/08/00	2300	3.04
JR51-GC28	71°12.03'N, 18°28.37'W	25/08/00	1600	4.45
JR51-GC30	70°22.01'N, 19°04.36'W	25/08/00	460	0.60
JR51-GC31	69°22.47'N, 19°36.90'W	25/08/00	1072	2.14
JR51-GC33	67°59.97'N, 21°44.97'W	26/08/00	820	4.70
JR51-GC34	67°50.05'N, 21°10.00'W	26/08/00	855	5.16
JR51-GC37	67°56.32'N, 12°02.02'W	27/08/00	1856	1.42
JR51-GC38	67°41.00'N, 12°26.00'W	27/08/00	1794	3.98
PS2627	73°07.40'N, 15° 40.85'W	09/94	2009	4.14
PS2628	73°09.79'N, 15° 57.98'W	09/94	1694	2.35

exert a fundamental limitation on the resolution and interpretation of palaeoceanographic data obtained from marine sediment cores. Palaeoceanographic investigations use a variety of proxies, such as percentage carbonate, geochemical markers (biomarkers), and $\delta^{18}O$ studies of foraminifera, to reconstruct palaeoceanographic changes through time. A fundamental premise is that variability in these proxies reflects changes in wider environmental controls such as ocean-surface temperature or ice-sheet volume. In order for this premise to be valid and high-resolution sediment records to be obtained, cores are best selected from areas where sedimentation has been continuous through time and has not been interrupted or disturbed significantly, or emplaced by 'instantaneous' mass-wasting events.

In this paper, we describe geological evidence for sediment reworking in marine cores from high-latitude margins that are influenced strongly by glacier ice; the continental margins and deep-marine environment of the Norwegian–Greenland Sea. Cores were recovered from a variety of locations, including: the Greenland Basin, the East Greenland margin, the Scoresby Sund Fan, Blosseville Basin, the Iceland and Vøring plateaux, and the Bear Island Fan (Fig. 1). Our objective is to document the contrasting geological signatures of sediment reworking in this region and relate these to wider environmental controls such as ice-sheet extent and bottom-current activity. We then discuss the implications of these results for palaeoceanographic investigations in the Norwegian–Greenland Sea and for reconstructing palaeoenvironmental change from the ice-rafted debris (IRD) record.

Methods

The principal data source for this paper is a suite of gravity cores collected during cruise JR51 of the RRS *James Clark Ross* to the Norwegian–Greenland Sea in 2000. Two gravity cores collected during cruise ARK X/1 of the RV *Polarstern* in 1994 are also discussed. Cores range from 0.60 to 5.16 m in length (Table 1). The cores were logged visually, and information regarding sedimentary structures, stratigraphic relationships, bedding and laminae contacts, sediment texture and colour was recorded and lithofacies identified. In addition, the paper uses geophysical data collected during cruises of the RRS *James Clark Ross* in 1994 and 2000, the RV *Livonia* in 1992, and RV *Polarstern* in 1994. These data comprise 6.5 kHz GLORIA long-range side-scan sonar imagery and sub-bottom

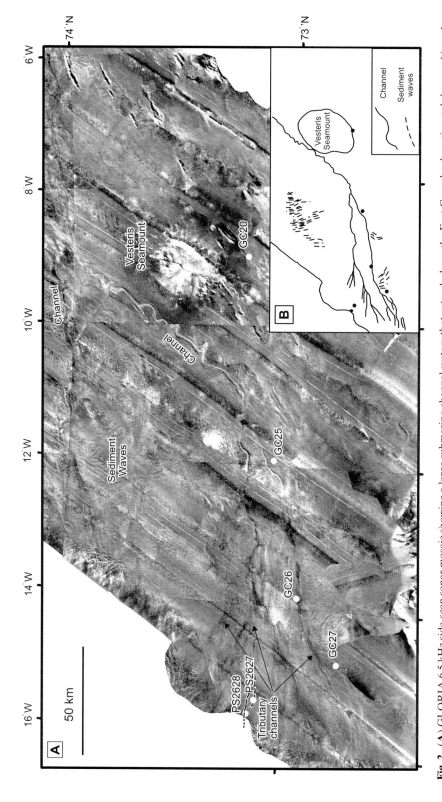

Fig. 2. (**A**) GLORIA 6.5 kHz side-scan sonar mosaic showing a large submarine channel system that extends down the East Greenland continental slope and into the Greenland Basin (Fig. 1). Areas of high backscatter are shown as light tones. Locations of cores discussed in the text are shown. The locations of the Parasound profiles of Figures 11a and 11c are shown by the dashed lines through the core sites of PS2627 and PS2628; (**B**) Interpretation of GLORIA image shown in a.

profiler records (3.5 kHz, TOPAS and Parasound systems). Acoustic backscatter contrasts on the GLORIA imagery allow the identification of the modes and extent of reworking on a scale of more than 1 km². AMS radiocarbon dates were obtained on samples of 2000–30 000 planktonic foraminifera (predominantly *Neogloboquadrina pachyderma*) from selected horizons in gravity cores.

Results

Sedimentary evidence for down-slope and cross-slope reworking

Greenland Basin. The Greenland Basin descends from the 500 m-deep edge of the East Greenland shelf to water depths of more than 3500 m in the abyssal plain, with an intervening continental slope of up to approximately 5°. The topographic boundaries of the 250 000 km² basin are defined by the Jan Mayen and Greenland fracture zones to the SW and NE, and by the mid-ocean spreading centre of the Mohns Ridge to the SE. Geophysical data demonstrate that the basin is dominated by well-developed networks of submarine channels (Mienert *et al.* 1993) (Fig. 2). These channels extend about 300 km in length from the upper continental slope to the deepest parts of the basin, are up to about 80 m deep, several kilometres wide, and have braided, anastomosing and sinuous reaches. We describe six cores from the Greenland Basin to illustrate the styles of sediment reworking (Figs 2 and 3; Table 1). Cores JR51-GC27, JR51-GC26 and JR51-GC25 are from the middle and lower reaches of the most southerly channel system. Cores JR51-GC19 and JR51-GC13 are from the abyssal plain, and core JR51-GC20 is from the lower flanks of a volcanic seamount within the basin.

Core JR51-GC27 was collected 6 km south of a tributary channel on the mid-slope in 2300 m of water. Sediment waves characterize this region of the slope. The core is dominated by grey to grey-brown (10YR5/1 to 10YR5/2) laminated mud (Fig. 3 and Table 1). Four principal types of laminae or laminae sequence are recognized in this core: (1) a lowermost sharp-based lamina of silt-fine sand that is 0.3–0.5 cm thick and is gradationally overlain by bioturbated clayey mud; (2) a basal silt–fine sand lamina that is overlain by convoluted silty laminae; (3) a basal silt–fine sand lamina that is overlain by more laterally continuous, lenticular, irregularly-spaced silty laminae which are overlain in turn by more wispy silty laminae; and (4) a basal silt–fine sand lamina that is overlain by

convoluted silty laminae, overlain in turn by more continuous, irregularly-spaced, lenticular silty laminae. Internally, the coarse-grained basal lamina ranges from massive to normally graded in structure. The thickness of individual laminated units ranges from 0.3–5 cm. These features are consistent with deposition from fine-grained turbidity currents (Stow & Shanmugam 1980; Stow *et al.* 1996; Wang & Hesse 1996). Prominent massive and normally graded sandy turbidites occur from 286–280 cm and 23–20.5 cm depth (Fig. 3).

JR51-GC26 was recovered from a bar situated between braided thalwegs in the channel system and it contains 0.43 m of massive, pebbly coarse sand from 79–122.5 cm depth that rests directly on massive sandy mud (Fig. 3 and Table 1). The pebbly sand contains occasional units of muddy coarse sand or mud as well as mud rip-ups. It is interpreted as the product of high-concentration, turbidity-current sedimentation. A waning flow regime up-core is suggested by the presence of sharp-based beds of medium to fine-grained massive and normally graded sandy turbidites resting on top of the pebbly sand unit (Fig. 3). A final transition to hemipelagic sedimentation is indicated by the presence of brown to yellowish brown (10YR5/4) massive mud with dispersed foraminifera in the uppermost 0.5 m.

Core JR51-GC25 was recovered from a braid bar situated adjacent to the main channel thalweg. The core contains numerous sharp-based laminae and beds of massive or normally-graded fine sand (Fig. 3). The bases of individual beds exhibit subtle internal scour surfaces. These sandy laminae units are separated by massive or weakly laminated mud. The sequence suggests episodic turbidity current events that deposited the sands, interspersed with more quiescent deposition of hemipelagic mud. An AMS radiocarbon date of 12 990±60 years BP (CAMS-77203)(corrected for an ocean reservoir effect of 550 years in East Greenland) was obtained on planktonic foraminifera (predominantly *N. pachyderma*) from 39–41 cm depth. This provides a date on the most recent transition from turbiditic to hemipelagic sedimentation.

Core JR51-GC19 was collected from an area of acoustically stratified sediment beyond the mouth of the channel system, in 3676 m of water in the Greenland Basin. Sedimentologically, the core comprises three sections (Fig. 3). The lower part of the core, from 395–466 cm, consists of a single bed of dark grey (10YR4/1), massive to weakly laminated silty mud grading upwards to bioturbated clayey mud. It records

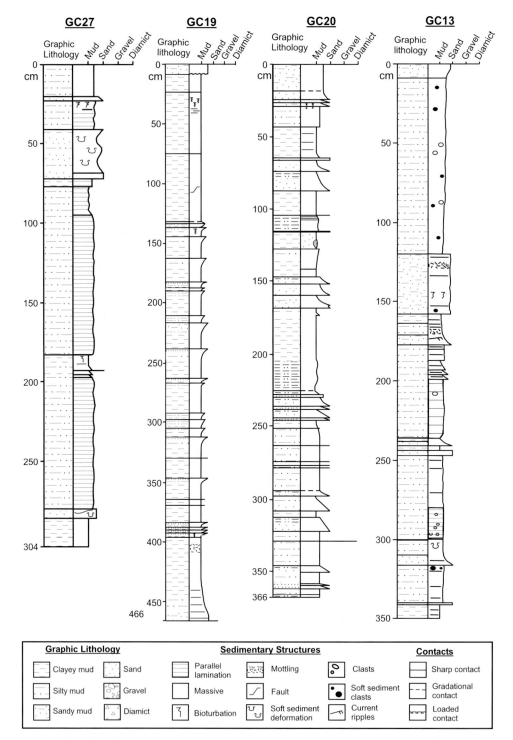

Fig. 3. Sediment cores from the Greenland Basin and abyssal plain of the Norwegian–Greenland Sea. Core locations are shown in Figures 1 and 2. A key to the graphic lithology, sedimentary structures and bedding contacts is provided.

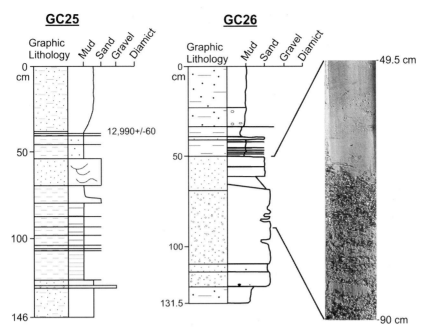

Fig. 3. continued.

deposition as a single thick turbidite (e.g. Stow *et al.* 1996). From 133–395 cm, the core comprises a series of coarsening-upward units. Individual units range in thickness from 2.5–26 cm and consist of grey and greyish brown, massive clayey mud that coarsens upwards gradationally into a thinner cap of silty or sandy mud (Fig. 3). The coarse unit is terminated by a sharp, commonly planar, contact that is overlain directly by massive clayey mud, marking the start of the next coarsening-upward unit. Twenty-four such inversely graded units, stacked directly on top of each other, were observed in this core. The repetitive, stacked nature of these units, inverse grading, and fine grain-size suggest deposition by bottom currents, either emanating from the channel network or flowing across slope (contourites). Sharp tops reflect either abrupt termination of the flows or erosion at the base of the subsequent flow. The upper part of the core comprises massive, locally bioturbated, hemipelagic mud.

Core JR51-GC20 was recovered from the south side of the Vesteris Seamount, an isolated volcanic edifice at 73° 30' N, 9° 10' W (Fig. 3; Table 1). The flanks of the seamount are rugged and give way to a smooth, sedimentary sea floor at about 3100 m water depth, where there is a marked break of slope. The core comprises homogeneous clayey mud that was deposited under low-energy conditions by hemipelagic sedimentation, interspersed with normally graded black (5Y2.5/1) sandy turbidites. The turbidites are composed of medium to coarse-grained basaltic sand with glass shards, overlying a sharp erosional contact and fining upwards into dark greyish brown (2.5Y4/2) and olive brown (2.5Y5/4) silty mud. They typically occur in a stacked fashion, with up to three turbidites together in sequence (Fig. 3). These black turbidites record episodic downslope sediment transport and deposition by turbidity currents moving off the seamount into deeper water. Antonow *et al.* (1997) also discuss cores from the vicinity of the Vesteris Seamount that contain prominent packages of black ash turbidites similar to those described above, which those authors term volcaniclastic turbidites.

A final core, JR51-GC13, was recovered from the abyssal plain on the Greenland side of the mid-Atlantic Ridge, in 3272 m water depth (Figs 1 and 3; Table 1). This core contains sharp-based, normally graded and massive turbidites of sand and sandy mud in its lower half (Fig. 3). The upper 120 cm of the core is characterized by yellowish brown (10YR5/4) massive silty mud

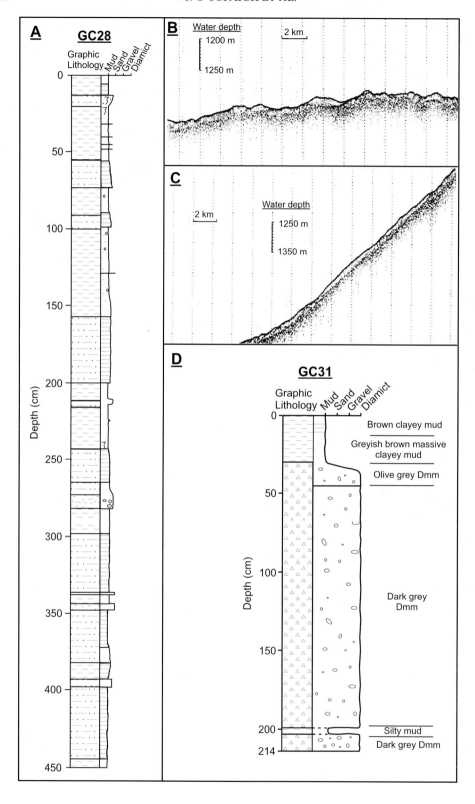

with dispersed foraminifera, interpreted to record hemipelagic sedimentation. The hemipelagic muds are separated from the underlying turbidites by 38 cm of yellowish brown (10YR5/4-10YR5/6), bioturbated sandy mud containing abundant foraminifera. Zones of weakly developed lamination are occasionally present within this muddy unit. Laminae are characterized by subtle textural changes and are diffuse and wispy. The unit has a sharp and slightly undulating lower contact. These characteristics are consistent with formation by bottom-current activity (Stow et al. 1998).

East Greenland margin and Scoresby Sund Fan. Two representative cores are shown in Figure 4 to illustrate contrasting styles of sediment reworking along the East Greenland margin south of the Greenland Basin (Fig. 1). Core JR51-GC28 was recovered from an area of acoustically stratified sediment on the continental slope, in 1600 m of water between the Greenland Basin and the Scoresby Sund Fan (Fig. 4A; Table 1). This core is 4.5 m in length and is dominated by grey (10YR5/1 to 10YR6/1) laminated mud. The laminated facies comprises a lower sharp-based silt or fine sand lamina that fines upwards into thicker clayey laminae. The upper contact of the coarser lamina ranges from sharp to gradational. Laminae range from lenticular and irregularly spaced, to more laterally continuous and parallel. Bioturbation is restricted to the upper parts of the clayey mud laminae. Such packages of laminated mud occur repeatedly throughout the core and are interpreted as muddy turbidites (cf. Stow & Shanmugam 1980; Stow et al. 1996). Occasional units of massive and normally graded sand and sandy mud with sharp contacts indicate deposition of coarser-grained turbidites.

In the southern part of the Scoresby Sund trough-mouth fan, 3.5 kHz acoustic data show a series of acoustically transparent lenses, up to about 15 m thick and 0.5–2 km in width, that are elongate downslope and have an irregular surface morphology (Fig. 4B, C)

(Dowdeswell et al. 1997). These have been interpreted as debris flows by previous workers (Dowdeswell et al. 1997). Core JR51-GC31 was collected from a debris flow lobe in this region of the fan and is dominated by massive diamict facies (Fig. 4D; Table 1). The base of the core comprises 11 cm of dark grey (7.5YRN4/), massive muddy diamict. The diamict is structureless and unsorted, and contains frequent clasts up to pebble size in a sandy mud matrix. It is gradationally overlain by a thin (4 cm) unit of light grey (10YR5/1-10YR6/1), massive silty mud which is overlain in turn by a second unit of dark grey (7.5YRN4/) massive diamict, 1.54 m in thickness. Occasional broken and whole marine shells were noted within this upper dark grey diamict. The lower contact of the diamict with the underlying silty mud facies is abrupt but appears to be non-erosive. Sharply-overlying the upper and thicker of the two dark grey diamict units is an olive grey (5Y4/2) muddy diamict from 45.5–13 cm depth (Fig. 4D). This diamict is also massive, contains small pebbles and can be differentiated easily from the underlying dark grey diamict facies by its olive grey colour. The diamict fines gradationally upwards into greyish brown (2.5Y5/2) massive clayey mud. The core is capped by 13 cm of brown (10YR5/3) clayey mud with occasional wispy, irregularly spaced laminations.

Based on their poorly-sorted nature, massive structure, stacked stratigraphic context and geophysical signature (acoustically-transparent elongate lenses; Fig. 4), the dark grey diamict units are interpreted as debris flows composed of reworked glacigenic sediment deposited when glacier ice was positioned at, or close to, the shelf break. The absence of an erosive contact between the upper dark grey diamict and underlying silty mud suggests that erosion was absent or minimal at the base of the flow. The overlying olive grey diamict facies is also interpreted as a debris-flow deposit on the basis of its sharp contact with the underlying dark grey diamict, massive structure and poor

Fig. 4. Geological and geophysical evidence of contrasting styles of sediment reworking along the East Greenland margin, south of the Greenland Basin. Core locations are shown on Figure 1. (**A**) Core JR51-GC28 from the continental slope north of the Scoresby Sund Fan. The core was recovered from an area of acoustically stratified sediment and contains repetitive packages of laminated muddy turbidites; (**B**) and (**C**) 3.5 kHz profiles of strike (**B**) and dip (**C**) lines from the southern part of the Scoresby Sund trough–mouth fan illustrating the acoustic character of debris flow deposits. Note the hummocky surface relief of the acoustically transparent debris flow lenses in the strike section, and the elongate, lenticular nature of the debris flows in the dip section; (**D**) Core JR51-GC31 from the Scoresby Sund trough–mouth fan. The core is dominated by massive diamict facies that records debris flows on the East Greenland continental slope. Key to core logs is provided in Figure 3.

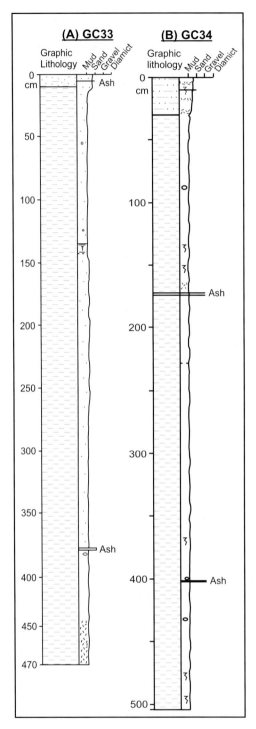

Fig. 5. Sediment cores from Blosseville Basin (Fig.
1). Both cores are predominantly composed of
massive bioturbated silty-clayey mud with occasional
ash layers. Key to core logs is provided in Figure 3.

sorting. The capping clayey mud represents
postglacial hemipelagic sedimentation.

Blosseville Basin. Cores JR51-GC33 and JR51-
GC34 were collected from acoustically stratified
sediments, at 800–860 m water depth, on the
southern margin of the Blosseville Basin,
between East Greenland and Iceland (Figs 1
and 5; Table 1). The cores are composed of
homogeneous greyish brown and grey silty and
clayey mud, apart from several thin (0.1–1 cm
thick) ash layers. The cores do not appear to
contain any sedimentary evidence for reworking, apart from zones of bioturbation and a thin
(8 cm) unit of diffusely laminated, bioturbated
mud in JR51-GC33 which may record minor
bottom-current activity. Bedding is poorly
developed and distinct boundaries within the
muds are difficult to define. These characteristics are consistent with a hemipelagic origin
(Stow & Piper 1984).

Iceland Plateau. Two cores were obtained from
the Iceland Plateau, an area situated to the NW
of Iceland with water depths of 1200–1600 m
(Fig. 1; Table 1). Core JR51-GC38 recovered
almost 4 m of sediment (Fig. 6A). This core
contains numerous massive and normally
graded units of silty mud and fine sand, ranging
from 0.5–10 cm in thickness. Bounding contacts
are typically sharp, although locally, basal
contacts may exhibit loading into underlying
units and upper contacts may be gradational
with bioturbated clayey mud. These characteristics are indicative of a turbidite origin. Core
JR51-GC37 contains two sandy units interpreted as turbidites (Fig. 6B). The lower unit
occurs between 109.5–95 cm depth and
comprises a dark grey (5Y3/1) massive to
diffusely laminated silty sand with a loaded
basal contact. A second prominent unit of
normally graded fine sand (5Y3/1) with a sharp
basal contact occurs from 87.5–77.5 cm depth.

*Vøring Plateau, distal Trænadjupet Slide and
Lofoten Channel.* Core JR51-GC01 was recovered from the Vøring Plateau, an area off

Fig. 6. Sediment cores from the Icelandic and
Norwegian continental margins of the
Norwegian–Greenland Sea. (**A** and **B**) Iceland
Plateau. Note the numerous normally graded and
massive sandy units (turbidites). (**C**) Vøring Plateau.
(**D**) Distal Trænadjupet Slide; (**E**) West of the mouth
of the Lofoten Channel. Core locations are shown on
Figure 1. Key to core logs is provided in Figure 3.

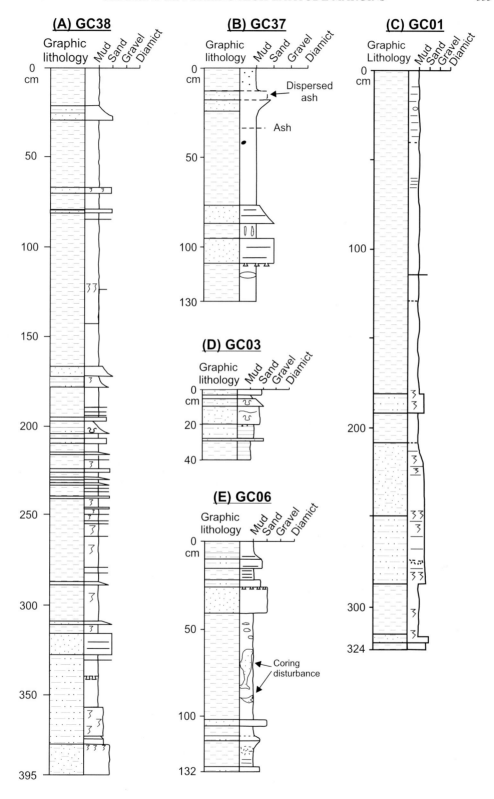

mid-Norway where the continental margin broadens out into the Norwegian–Greenland Sea (Fig. 1). Water depths across the plateau range mainly from 1200–1600 m. In the lower part of the core, thin (5 cm) units of massive, poorly sorted, dark grey to dark greyish brown (7.5YRN4/-2YR4/2) silty mud (287–283 cm; 320–315.5 cm) erosively overlies massive clayey mud, and subtle loading is locally present along lower contacts. Zones of faint lamination with bioturbation and gradational contacts occur between 283–249 cm and 225–212 cm depth. The upper half of the core from 0–181 cm comprises predominantly massive clayey mud (Fig. 6C). Within these muds occasional weakly laminated zones occur at 65–60 cm and 40–10 cm depth. Laminae have a faint 'streaky' appearance and are laterally discontinuous. Contacts between units throughout the upper half of the core are gradational. The silty and diamictic mud units with erosive to locally-loaded basal contacts are interpreted as mass flows. Zones of diffuse streaky lamination that

occur within the clayey muds probably reflect reworking by bottom currents (Stow *et al.* 1998). Bottom current winnowing is also suggested by the presence of a 0.5 cm thick lag of coarse sand with foraminifera at 114 cm depth.

The Trænadjupet Slide is located on the continental slope immediately east and NE of the Vøring Plateau (Fig. 1). The slide has a run-out distance of about 200 km, and extends from the continental shelf break to over 3000 m water depth in the Lofoten Basin (Laberg & Vorren 2000a). The slide is terminated to the north by the Lofoten Channel. Core JR51-GC03 was recovered from the distal 'toe' of the slide, while JR51-GC06 was recovered from immediately west of the mouth of the Lofoten Channel. The cores contain massive and normally graded turbidites of fine-medium sand (Figs 1 and 6D, E). In JR51-GC06 these units have been disturbed by the coring process; however, they maintain sharp contacts with bounding muds. Interbedded with the sandy turbidites are clayey and silty muds, which include laminated facies

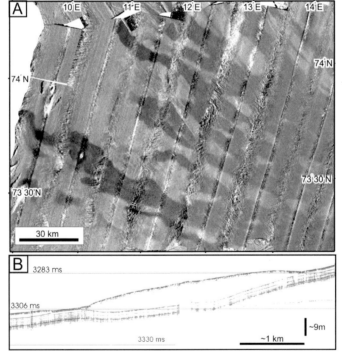

Fig. 7. Debris flows on the Bear Island Fan. (**A**) GLORIA 6.5 kHz side-scan sonar image of debris-flow lobes orientated east–west (located in Figure 1). The dark lines running approximately north–south across the image are ship tracks; (**B**) TOPAS parametric echo sounder record of a debris flow lens on the Bear Island Fan. Note elongate, lenticular geometry and acoustically transparent internal structure. The constant thickness of the underlying layer indicates that it may be predominantly hemipelagic sediment.

characterized by sharp-based, micro-graded laminae, and are interpreted as muddy turbidites. These cores therefore record the passage of turbidity currents through the Lofoten Channel and across the distal part of the Trænadjupet Slide into the Lofoten Basin.

Bear Island Fan. Debris flows comprise the main building blocks of trough-mouth fans around the margins of the Norwegian–Greenland Sea (Vogt *et al.* 1993; Laberg & Vorren 1995; Dowdeswell *et al.* 1996; Vorren & Laberg 1997). Flows are clearly visible on GLORIA side-scan sonar imagery and sub-bottom profiler records from the Bear Island Fan, where they consist of well-defined lobes with distinct terminations, ranging from 30–200 km in length, 2–10 km in width and 10–50 m in thickness (Fig. 7) (Dowdeswell *et al.* 1998; Taylor *et al.* 2002*a*). Three cores were recovered from distal debris-flow lobes on the Bear Island Fan (JR51-GC08, JR51-GC10, JR51-GC12) (Figs 1 and 8). The cores are dominated by dark grey (10YR4/1), massive diamict facies containing dispersed

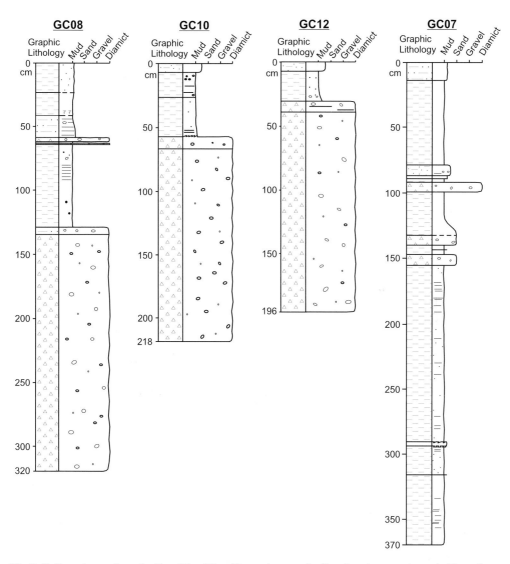

Fig. 8. Sediment cores from the Bear Island Fan, Norwegian margin. Core locations are shown in Figure 1. Key to core logs is provided in Figure 3.

clasts in a poorly sorted, sandy mud matrix (Fig. 8). This diamict facies has been widely described in previous investigations from the Bear Island Fan (Vorren *et al.* 1998; Laberg & Vorren 1995, 2000*b*) and is interpreted as having been deposited from glacigenic debris flows that were delivered to the continental slope when glacier ice formerly extended to the shelf break under full glacial conditions. The dark grey diamict is sharply overlain by massive or weakly-stratified olive grey (5Y4/2), silty

mud/muddy diamict facies. The upper parts of the three cores comprise brown (10YR5/3 to 10YR5/4), massive to diffusely laminated mud with dispersed foraminifera that is predominantly the result of hemipelagic sedimentation.

Core JR51-GC07 was recovered from the southern part of the Bear Island Fan (Figs 1 and 8; Table 1). 3.5 kHz profiles from this region show at least 20 m of acoustically stratified sediment (Taylor *et al.* 2002*a*). Buried debris-flow lenses are visible on the 3.5 kHz

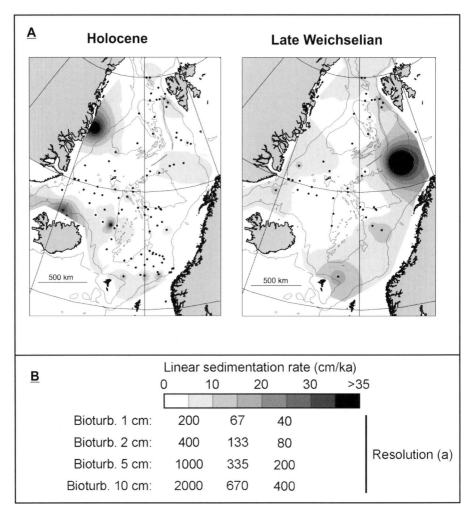

Fig. 9. (**A**) Linear sedimentation rates (cm ka^{-1}) based on radiocarbon-dated sediment core records for the Norwegian–Greenland Sea during the Holocene (*n* = 125 cores) and Late Weichselian (*n* = 61 cores). For details about cores refer to Taylor *et al.* 2002*b*. Core locations are marked by dots. Note that most of the deep sea floor of the Norwegian–Greenland Sea receives less than 5 cm of sediment per thousand years during the Holocene. (**B**) Likely temporal resolution of core records from the Norwegian–Greenland Sea based on the relationship between the depth of bioturbation and the linear sedimentation rate. With increasing bioturbation depths, the temporal resolution of the core record decreases.

profiles beneath the stratified sediments and indicate debris-flow delivery to this region of the Bear Island Fan. Most of core JR51-GC07 consists of massive to weakly laminated clayey mud (Fig. 8). However, from 155–79 cm, thin beds of massive muddy diamict and poorly-sorted sandy mud separated by sharp contacts occur. These units record the last interval of mass-flow delivery to this region of the fan. The maximum thickness of these units is 30 cm. Below 155 cm depth, the core comprises a thick

sequence of massive grey (5Y5/2) clayey mud with dispersed sand grains and weak lamination. Laminae appear to form weakly developed rhythmic cycles within the massive clayey mud. Zones of sandier mud (2–3 cm thick), that are gradational with bounding sediments and exhibit bioturbation-induced mottling, also occur in this unit. These sediments are collectively interpreted as the product of glacimarine sedimentation by suspension settling and contour current activity.

Mixing and disturbance of sea-floor sediments

Bioturbation. The maximum resolution of geological time slices from core records is limited by the depth of bioturbation, such that it is not possible to differentiate between signals which were produced within the period of time needed to form an individual time-slice. By calculating linear sedimentation rates in the Norwegian–Greenland Sea from published and publicly available core records, and then mapping these out across the region (Taylor *et al.* 2002*b*), it is possible to assess the probable limits of temporal resolution based on likely bioturbation depths (Fig. 9). According to Graf *et al.* (1995), the upper 2 cm of sediments in the Norwegian–Greenland Sea are strongly mixed by bioturbation and hence will represent the maximum resolution for core time-slices in this region. As most of the deep sea-floor in the Norwegian–Greenland Sea receives less than 5 cm of sediment every thousand years (Fig. 9), this will generally limit the temporal resolution of core records to approximately 400 years. Indeed, there are some indications that mixing can occur up to about 10 cm (Thomson *et al.* 2000), in which case the temporal resolution decreases to 2000 years. Thus, only in areas where linear sedimentation rates exceed 15 cm ka^{-1} are sediment cores suitable for centennial-scale investigations.

Iceberg scouring. In fjords, on continental shelves, and on the upper continental slope, typically down to 400–600 m depth, scouring of the sea-floor by grounded iceberg keels results in the disturbance of the upper part of the sediment column and formation of iceberg plough marks (e.g. Belderson *et al.* 1973; Dowdeswell *et al.* 1993; Solheim 1997; Syvitski *et al.* 2001). Associated sediments comprise structureless diamicts formed from homogenization of pre-existing marine sediments and iceberg rain-out (Vorren *et al.* 1983; Dowdeswell

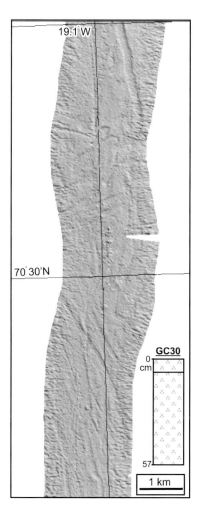

Fig. 10. Geophysical and geological evidence of iceberg reworking on the East Greenland continental shelf immediately east of the mouth of the Scoresby Sund fjord system (Fig. 1). EM120 shaded relief swath bathymetry image of iceberg plough marks in about 400 m of water. Cores recovered from this region (e.g. JR51-GC30) are composed of massive diamict facies formed by iceberg rafting and scouring.

et al. 1994*a*). Figure 10 shows an example of iceberg reworking from the East Greenland continental shelf, east of the mouth of the Scoresby Sund fjord system (Fig. 1). Swath bathymetric data from this area show widespread iceberg plough marks across the shelf (Fig. 10). Sediment cores recovered from this region are predominantly composed of massive, coarse-grained, diamict facies (e.g. core JR51-GC30; Fig. 10) that is interpreted as the product of iceberg rafting and scouring (cf. Dowdeswell *et al.* 1994*a*).

Discussion

Sediment reworking in the Norwegian–Greenland Sea: spatial variations and controls

The cores discussed above were collected from contrasting physiographic settings around the Norwegian–Greenland Sea, ranging from continental shelf to abyssal plain (Fig. 1). Sediment reworking is common in the cores, with the exception of those from Blosseville Basin (Fig. 5). A variety of processes are responsible for this reworking. On continental slopes and in the deep sea, downslope mass transport processes, in the form of turbidity currents and debris flows, dominate. Turbidite facies recovered range from massive pebbly sands that record the passage of high-concentration turbidity currents (JR51-GC26), to finer-grained muddy facies (clay to silt), that are characteristic of distal-turbidite deposition from low concentration turbidity currents (e.g. JR51-GC27, JR51-GC28). The influence of bottom currents and/or contour currents was also noted in some cores (e.g. JR51-GC01 and JR51-GC13). On shallower (less than 500 m) continental shelves, such as along the East Greenland margin, ploughing of iceberg-keels through the substrate, in conjunction with rain-out of ice-rafted debris, results in the production of massive diamict facies (JR51-GC30). Winnowing by bottom currents is also a significant process on continental shelves and the upper continental slopes around the Norwegian–Greenland Sea (Vorren *et al.* 1984; Kenyon 1986; Mienert *et al.* 1992; Cadman 1996; Evans *et al.* 2002).

Marked spatial variations in the style of sediment reworking can be seen across the region. Reworking by debris flow appears to be restricted mainly to the large trough-mouth fans and sediment slides along the East Greenland and Norwegian margins. Elsewhere, the core sedimentology indicates that reworking is primarily associated with turbidity currents. Trough-mouth fans are absent along the East Greenland margin, north of the Scoresby Sund Fan and debris flows appear to play a relatively minor role in sediment reworking in this region compared to turbidity currents (Mienert *et al.* 1993; Dowdeswell *et al.* 1996; Evans *et al.* 2002). A variety of sand and mud facies in cores from the Greenland Basin demonstrate the passage of turbidity currents moving downslope though a large channel system that descends from the upper continental slope to the abyssal plain (Fig. 2). Turbidite reworking is not just confined to the environs of this channel system, however, as core JR51-GC20 from the Vesteris Seamount contains numerous black sandy layers that record deposition from turbidity currents moving downslope off the seamount into deeper water. Turbidites dominate reworking further south along the East Greenland margin where packages of laminated mud in core JR51-GC28 record repeated deposition from low-concentration, muddy turbidity currents. Turbidity current reworking is also common on the Iceland Plateau.

These spatial variations in sediment reworking reflect past ice-sheet dynamics around the margins of the Norwegian–Greenland Sea, as well as the gradient of the continental slope and, possibly, bedrock lithology on the shelf. Collectively, these factors have influenced the rate of sediment delivery and the location of depocentres. Major intervals of progradation on the North Sea, Bear Island and Scoresby Sund fans are related to episodes when glacier ice extended to the continental shelf break via cross-shelf troughs. Numerical modelling indicates that ice flow in these cross-shelf troughs occurred as fast-flowing ice streams (Dowdeswell & Siegert 1999). Large volumes of sediment were delivered directly to the upper continental slope from the subglacial deforming layer at the base of these ice streams (cf. Alley *et al.* 1989) and reworked downslope into deeper water as debris flows (Laberg & Vorren 1995, 2000*b*; Dowdeswell *et al.* 1996; Taylor *et al.* 2002*a*). By contrast, in areas where slower-moving glacier ice extended to the shelf break between ice streams and trough-mouth fans, rates of sediment delivery were lower and sediment was reworked more slowly. Sediment accumulation at the shelf break in these inter-fan areas was thus greater than in areas of the upper slope fed directly by ice streams (Taylor *et al.* 2002*c*). Sediment was therefore able to build up incrementally prior to downslope reworking by, typically, large-scale sliding (e.g.

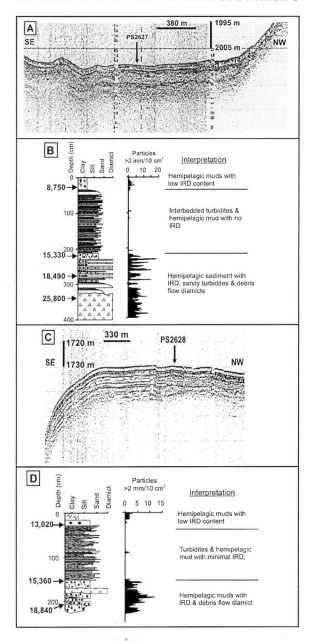

Fig. 11. Examples of cores recovered from areas of acoustically stratified sediment, illustrating how such stratified packages may contain ubiquitous reworked elements that cannot be resolved by the geophysical records. Both cores are from the middle-continental slope on the East Greenland margin (Figs 1 and 2). (A) Parasound sub-bottom profiler record of acoustically stratified sediments on the middle-continental slope. The location of the Parasound profile is shown on Figure 2. (B) Lithofacies log and coarse particle count (particles over 0.2cm/10cm³) of sediment core PS2627 recovered from the acoustically stratified sediments shown in (A). Debris flow diamicts and turbidites dominate the core. Note the distinct spikes in the coarse particle plot below 200 cm depth, which record individual debris-flow pulses rather than discrete iceberg-rafting events. (C) Parasound sub-bottom profiler record of acoustically stratified sediments on the middle-continental slope of East Greenland. The location of the Parasound profile is shown on Figure 2. (D) Lithofacies log and coarse particle count of sediment core PS2628 recovered from the acoustically stratified sediments shown in (C). Note the ubiquitous thin turbidites.

Laberg *et al.* 1999, 2000; Laberg & Vorren 2000*a*).

The well developed network of submarine channels that extends from the upper continental slope to the abyssal plain of the Greenland Basin records a different set of formative processes than that of the trough-mouth fans and large sediment slides. Although debris flows are present in cores and on acoustic records from upper-mid slope (Fig. 11), turbidity currents are the principal style of sediment reworking associated with the mid-lower channel system. The Greenland Ice Sheet appears to have exhibited relatively minor fluctuations between recent glacial and interglacial periods, and current evidence suggests that the ice sheet was restricted to the inner-middle continental shelf during the LGM (Dowdeswell *et al.* 1994*b*; Funder & Hansen 1996; Funder *et al.* 1998; Evans *et al.* 2002). Under this scenario, limited quantities of sediment would be transported across the shelf to the continental slope and, in comparison with the Norwegian margin, this would be an area of low sediment delivery. Formation of the channels is inferred to result from erosion by turbidity currents and cascading dense cold water formed by brine rejection during sea-ice formation on the shelf (Mienert *et al.* 1993; Dowdeswell *et al.* 1996; Vorren *et al.* 1998; Dowdeswell *et al.* 2002).

Alternatively, however, formation of the channels may have been influenced by more direct glacial activity, whereby the Greenland Ice Sheet expanded to the shelf break (cf. Wang & Hesse 1996; Hesse *et al.* 1999), but sediment supply and continental slope gradient mitigated against fan development. Limited bathymetric data from this region show troughs extending across the continental shelf from the mouths of the large fjords and these are likely to be the product of successive episodes of glacial erosion during ice-sheet expansion to the shelf break.

The gradient of the continental slope in this region is ≤5° and is, therefore, relatively steep compared to other areas of the margin characterized by trough-mouth fans, which are commonly less than 2° (Vorren *et al.* 1998). Hence, debris delivered to the upper continental slope would be relatively quickly reworked into turbidity currents. Cores from the upper channel system contain diamict units interpreted as debris flow deposits (Fig. 11), but a lack of similar sediments in cores from further downslope (Fig. 3) suggests the rapid evolution of debris flows into turbidity currents. Furthermore, bedrock across much of the continental shelf in this region is volcanic (Escher & Pulvertaft 1995). This would not be as conducive as sedimentary strata to the formation of a well-developed subglacial deforming layer (Marshall *et al.* 1996; Anandakrishnan *et al.* 1998), and sediment erosion and delivery to the shelf break would be correspondingly lower. Thus, the combination of steep slope and low (inferred) sediment delivery could have acted together to inhibit trough-mouth fan development.

Implications for palaeoceanographic investigations in the Norwegian–Greenland Sea

The geophysical and core data discussed above indicate that sediment reworking is widespread around the continental margins (shelf to slope) of the Norwegian–Greenland Sea. The style of this reworking is highly variable, and ranges from downslope mass-transport processes such as debris flows, turbidity currents and sediment slides, to the effects of bottom-currents and iceberg scour. The scale of reworking ranges from millimetres (e.g. thin turbidite laminae in cores) to tens or hundreds of metres in thickness (e.g. debris flows on the Bear Island Fan and large sediment slides along the Norwegian margin).

Reworking also occurs in the abyssal plain of the Norwegian–Greenland Sea as indicated by core JR51-GC13, which was recovered from the vicinity of the mid-Atlantic Ridge (Fig. 1) and contains coarse-grained turbidites. This demonstrates the significance of the ridge in generating downslope mass-flows and, therefore, cores collected from areas in the vicinity of tectonically-formed ridges across the region may also contain reworked elements. Similarly, the presence of turbidite facies in cores recovered from the distal Trænadjupet Slide and beyond the mouth of the Lofoten Channel indicate the passage of turbidity currents into the Lofoten Basin that may be recorded in cores recovered from the more distal parts of the basin. Furthermore, cores from the Greenland Basin and Iceland Plateau also contain extensive reworked elements in the form of turbidites and bottom-current deposits. These are areas from which cores have routinely been used for palaeoceanographic purposes. Of those areas investigated in this study, Blosseville Basin contains the most complete record of hemipelagic sedimentation (cores JR51-GC33 and JR51-GC34; Voelker *et al.* 1998), and is thus one of the prime areas for the recovery of undisturbed sediment core records in the Norwegian–Greenland Sea.

Identification of reworking in marine

sediment cores is important where such cores are used for reconstructing changes in palaeo-oceanography (e.g. Sarnthein *et al.* 1995; Antonow *et al.* 1997; Hebbeln *et al.* 1998), because reworking places a fundamental constraint on the resolution of these records. Incorrect genetic interpretation of core litho-facies (e.g. iceberg-rafted and hemipelagic sediments v. debris flow material) may result in erroneous palaeoenvironmental reconstruction. This is particularly significant where high-resolution (millennial to centennial scale) records are required and attempts are made to correlate marine sediment cores to other high-resolution climatic records such as ice cores (e.g. matching Dansgaard–Oeschger events and Bond cycles). Where cores contain sandy turbidites or bottom-current deposits, it may be difficult or impossible to quantify the amount of erosion resulting from such reworking, and an uninterrupted record is therefore not available. In addition, bioturbation-induced smoothing will also mask such high-resolution records (Berger & Heath 1968; Anderson 2001; Evans *et al.* 2002). The widespread nature of sediment reworking around the continental margins of the Norwegian–Greenland Sea demonstrates that these areas are likely to contain reworked elements (e.g. Antonow *et al.* 1996; Stein *et al.* 1996; Henrich 1998), and thus the assumption of a continuous uninterrupted sediment record will be negated.

To obtain high-resolution records for palaeo-oceanographic investigations, cores should be collected from areas that are, or have been, characterized by high sedimentation rates (Fig. 9). However, such areas are spatially restricted in the Norwegian–Greenland Sea and are commonly associated with extensive reworking. During the Holocene, they have been limited to the Greenland and north Iceland continental shelves, and one section of the Norwegian basin, adjacent to the Iceland Plateau (Fig. 9), although contourite drifts on the Norwegian margin are also areas of high sedimentation (Van Weering *et al.* 1998; Laberg *et al.* 1999). Many of these areas have been subject to Holocene reworking by iceberg scour (e.g. Dowdeswell *et al.* 1993; Syvitski *et al.* 2001) and bottom-current winnowing (Mienert *et al.* 1992; Evans *et al.* 2002). During the Late Weichselian, maximum sedimentation rates are associated with the large trough-mouth fans (Fig. 9), areas where downslope reworking is common. Even away from the main areas of debris-flow activity on fans, sedimentological data indicate that deposition of fine-grained sediments from suspension settling and bottom-current activity

is common (Taylor *et al.* 2002*a*), and hence reworking would be expected. Indeed, recent empirical and ice-sheet modelling data indicate that, in terms of the relative importance of sediment accumulation, less than 7% of material delivered to the Norwegian–Greenland Sea since the Late Weichselian is derived from hemipelagic and pelagic sedimentation (Taylor *et al.* in press *c*). The remainder is subject to large-scale reworking in the form of debris flows, slides, slumps, turbidity currents and channel deposition.

Ice-rafted debris layers: genesis and palaeoenvironmental significance

Over the last decade an increasingly important direction in palaeoceanographic research has involved the investigation of ice-rafted debris in deep-sea sediments (e.g. Heinrich 1988; Bond *et al.* 1992; Henrich *et al.* 1995; Bischoff 2000; Grousset *et al.* 2001). A key premise of this work is that the IRD layers result from episodes of intense rain-out of iceberg-rafted debris coupled to background hemipelagic sedimentation and are linked to large-scale ice-sheet dynamics. This clearly requires accurate discrimination between those relatively coarse-grained units composed of IRD that were emplaced by iceberg rafting, and those units resulting from debris flow delivery or bottom-current winnowing. This is critical because reworking by sediment gravity flow or current winnowing can accentuate the coarse-grained sediment fraction, resulting in apparent enhancement of the IRD signal and erroneous interpretations *vis-à-vis* ice-sheet dynamics. Cores recovered from the East Greenland continental slope (Fig. 11) illustrate this problem. Core PS2627 was recovered from acoustically-stratified sediments and contains numerous diamict layers of variable thickness deposited by debris flows, that correlate to distinct spikes in the associated coarse particle plot (Fig. 11B). In the absence of detailed sedimentological analysis, such spikes could be mistaken for discrete episodes of enhanced IRD delivery to the slope.

Stein *et al.* (1996) discuss a transect of cores extending from middle shelf to deep sea on the East Greenland margin. This transect includes several cores from the continental slope of the Scoresby Sund Fan. IRD peaks are reconstructed by counting of the fraction over 0.2 cm size from X-radiographs (cf. Grobe 1987). Shelf cores comprise consolidated sandy and silty muds overlying stiff diamicts (Nam *et al.* 1995). The authors state that the cores contain low to moderate amounts of IRD. However, the

IRD signal in these cores is unlikely to be a direct record of iceberg rafting given that iceberg scour is common across the shelf and produces massive diamicts (Dowdeswell *et al.* 1993, 1994*a*) (Fig. 10). In addition, the East Greenland Current may act to winnow finer-grained material and produce coarser sediment lags in this region (Mienert *et al.* 1992; Evans *et al.* 2002). The main focus of attention, however, was on the cores from the slope and deep-sea (Stein *et al.* 1996). These cores contain peaks in particles more than 0.2 cm in diameter that are interpreted as IRD, indicating repeated fluctuations of the Greenland Ice Sheet. The authors note that important IRD events occur at intervals of 1000–3000 years, suggesting short-term collapses of the Greenland Ice Sheet at a higher frequency than that of the Heinrich Events. A potential problem with this interpretation, and the attempt to correlate the IRD peaks with the Greenland ice cores, is that the role of reworking is not addressed and is difficult to assess due to the lack of sedimentological description of the IRD layers. Given that several of the cores were from the Scoresby Sund Fan, it is possible that reworking by downslope mass-transport may have affected the coarse particle signal in the cores (cf. Hesse & Khodabakhsh 1998; Andrews *et al.* 1998).

Reworking by bottom-current winnowing can also accentuate the coarse-grained fraction in sediment cores, resulting in the apparent enhancement of the IRD signal. A recent study by Anderson & Andrews (1999) highlights the role of current winnowing in the production of coarse-grained IRD lags in cores from the continental slope of the Weddell Sea, Antarctica. Grain-size distributions of the IRD layers are negatively skewed, indicating that winnowing of the fine-grained fraction has taken place, thus increasing the IRD concentration in these cores. This problem will assume greater significance where the age of IRD layers is not constrained tightly by chronological control.

Distinguishing between iceberg-rafted, debris-flow and current-winnowed sediments

Iceberg-rafted sediments are poorly sorted, commonly massive with low to absent bioturbation and they have sharp boundaries where the onset and termination of iceberg rafting is both abrupt and intense (Hesse & Khodabakhsh 1998). In the case of the Heinrich layers, included microfaunal contents are low (Bond *et al.* 1992). Iceberg-rafted sediments have been

deposited under zero or low stress conditions and, hence, at a micromorphological scale, finer particles should have a random orientation and may show fine lamination that is deformed by larger particles (dropstones). The plasmic (silt-clay) fabric of iceberg-rafted diamicts would be expected to be weak and would probably also record a vertical component. Rain out of IRD is a passive sedimentary process and on acoustic records would be expected to result in the deposition of a blanket of sediment with a draped geometry. However, such sequences can contain significant reworked elements in the form of turbidites or debris flows that are too thin to be resolved by geophysical records (e.g. Fig. 11C and D) and thus thick packages of acoustically-stratified sediment may contain reworked elements. In continental slope settings, preservation of iceberg-rafted diamict facies might be expected to be low, given the possibility for downslope remobilization of iceberg-rafted material, particularly where sedimentation rates are high.

Diamict facies deposited by debris flows may contain flow-banding, rip-ups of underlying units, larger (freighted) clasts towards bed tops, and often have abrupt upper and lower contacts. The presence of grading, either inverse or normal, is also diagnostic of a mass-flow origin, and relates to the development of turbulence during downslope flow. Important indirect evidence of a mass-flow origin is also provided by the facies sequence context, where diamicts are interbedded with turbidites or slump deposits. On acoustic records, debris flows are characteristically lenticular in geometry, have an irregular, hummocky upper surface and are internally acoustically transparent and homogeneous.

Cross-slope and down-slope bottom currents can winnow fine-grained material, resulting in the production of coarse-grained sediments and concentration of IRD. Such coarser material can be differentiated from iceberg-rafted deposits by their generally better sorting with negatively skewed size distributions, in contrast to the unsorted distribution of IRD. Included microfauna may be concentrated in layers. Faint scour structures may be present at the base of such deposits and they may also contain discontinuous internal stratification.

Provenance data such as IRD lithology and mineralogy can be used to constrain the sediment sources and pathways (Scourse *et al.* 2000; Grousset *et al.* 2001). They could also be used to assist in the genetic interpretation of the sediments themselves. For example, mineralogically, sediments derived from downslope mass failures might be more locally constrained than far-travelled material brought in by iceberg rafting. However, if the debris flow is sourced from remobilized IRD, this discriminator is negated and, on continental slopes, the possibility of remobilization of IRD by debris flow is probably high. However, the presence of shelf benthic foraminifera in slope sediments may imply marked dislocations from the usual depth habitats and support a mass-flow or bottom-current origin for the surrounding sediments.

Conclusions

(1) Geological evidence for sediment reworking in marine cores from around the Norwegian–Greenland Sea has been presented. Cores were collected from a variety of contrasting physiographic settings, from shelf to abyssal plain. These areas include the Greenland Basin, the Bear Island and Scoresby Sund trough mouth fans, the Vøring and Iceland plateaux, and Blosseville Basin (Fig. 1).

(2) Reworking is common in cores from the continental margins (shelf-slope) of the Norwegian–Greenland Sea, and also occurs in cores from abyssal depths. The style of reworking is variable and ranges from downslope mass-transport processes, such as debris flows and turbidity currents, to the action of along-slope and down-slope bottom currents, as well as iceberg scour. Reworking by debris flows appears to be restricted mainly to the large trough-mouth fans and giant sediment slides along the Norwegian and East Greenland margins. Elsewhere, reworking by turbidity currents frequently dominates, although iceberg ploughing and current winnowing are also significant on continental shelves and upper slopes.

(3) Cores from Blosseville Basin contain the most complete, undisturbed record of hemipelagic sedimentation obtained during this investigation. Reworked elements indicative of downslope and cross-slope processes are generally absent. Blosseville Basin is thus a key area for the recovery of undisturbed sediment-core records in the Norwegian–Greenland Sea.

(4) Sediment reworking across the region is linked strongly to ice-sheet dynamics that, in turn, influence directly the rate of sediment delivery and location of depocentres around the continental margins of the Norwegian–Greenland Sea. Spatial contrasts in the style of reworking may also reflect the influence of continental slope gradient and bedrock geology on the continental shelf.

(5) Widespread sediment reworking in the Norwegian–Greenland Sea has important implications for palaeoceanographic investigations in this region. Reworking can result in the erosion or disturbance of the marine sediment column. This problem is significant where continuous high-resolution records of hemipelagic and pelagic sedimentation are required and attempts are made to correlate with other high-resolution climate records such as ice cores. Bioturbation acts to smooth such high-resolution records and will impose a maximum resolution on core time-slices which is directly related to the sedimentation rate. Thus, much of the continental margin of the Norwegian–Greenland Sea is unlikely to contain undisturbed records of sedimentation suitable for high-resolution palaeoeanographic investigation.

(6) It is important to obtain cores from areas that are, or have been, characterized by high sedimentation rates. However, in the Norwegian–Greenland Sea, such areas of high sedimentation are commonly associated with extensive reworking (Fig. 9).

(7) Reworking is particularly significant in the case of investigations of iceberg-rafted debris, where attempts are made to link such sediments to past ice-sheet dynamics. Bottom-current winnowing and downslope mass-flow processes can increase the concentration of coarse-grained IRD in marine sediments. Such reworked IRD layers or lags often simply record the operation of *local* processes and may have little direct relationship to wider changes in ice-sheet volume and dynamics.

(8) Finally, the significance of reworking will obviously depend on the nature of the palaeoceanographic investigation. For example, where temporal changes in bottom-current formation are the focus of investigation, contourites and bottom current reworking will play a key role in the interpretation and reconstruction of the geological record. However, where the focus is on iceberg-rafting events and their relationship to ice-sheet dynamics, failure to identify which elements of the IRD record are related to current reworking (IRD lags), and which record direct iceberg-rafting, may lead to incorrect palaeoenvironmental interpretations. The key point is that interpretations of palaeoceanographic proxies should be underpinned by a thorough sedimentological assessment of the effects of reworking.

This work was supported by the Natural Environment Research Council, UK (grants GST/02/2198 – ARCICE thematic programme, and GR3/JIF/02). The cores discussed in this paper were collected during cruise JR51 of the RRS *James Clark Ross* in July–August, 2000. We thank Neil Campbell (British Geological Survey) for assistance with core collection and handling. Geophysical data were also collected during cruises of the RRS *James Clark Ross* in 1994 (NERC grant GR3/8508), the RV *Livonia* in 1992 and the RV *Polarstern* in 1994. The NERC British Ocean Sediment Core Repository (BOSCOR) provided facilities for core analysis and storage. Radiocarbon dates were obtained through the Natural Environment Research Council Radiocarbon Laboratory, and we thank C. Bryant and M. Garnett for their assistance. We are grateful to J. T. Andrews and H. Bauch for constructive reviews, which improved the final manuscript.

References

ALLEY, R. B., BLANKENSHIP, D. D., ROONEY, S. T. & BENTLEY, C. R. 1989. Sedimentation beneath ice shelves: the view from Ice Stream B. *Marine Geology*, **85**, 101–120.

ANANDAKRISHNAN, S., BLANKENSHIP, D. D., ALLEY, R. B. & STOFFA, P. L. 1998. Influence of subglacial geology on the position of a West Antarctic ice stream from seismic observations. *Nature*, **394**, 62–65.

ANDERSON, D. M. 2001. Attenuation of millennial-scale events by bioturbation in marine sediments. *Paleoceanography*, **16**, 353–357.

ANDERSON, J. B. & ANDREWS, J. T. 1999. Radiocarbon constraints on ice sheet advance and retreat in the Weddell Sea, Antarctica. *Geology*, **27**, 179–182.

ANTONOW, M., GOLDSCHMIDT, P. M. & ERLENKEUSER, H. 1997. The climate-sensitive Vesterisbanken area (central Greenland Sea): depositional environment and paleoceanography during the past 250,000 years. *In*: HASS, H. C. & KAMINSKI, M. A. (eds) *Contributions to the Micropaleontology and Paleoceanography of the Northern North Atlantic*. Grzybowski Foundation Special Publication, **5**, 101–118.

ANDREWS, J. T., KIRBY, M., JENNINGS, A. E. & BARBER, D. C. 1998. Late Quaternary stratigraphy, chronology, and depositional processes on the slope of S.E. Baffin Island, detrital carbonate and Heinrich events: Implications for onshore glacial history. *Géographie Physique et Quaternaire*, **52**, 1–15.

BELDERSON, R. H., KENYON, N. H. & WILSON, J. B. 1973. Iceberg plough marks in the northeast Atlantic. *Palaeogeography, Palaeoclimatology and Palaeoecology*, **13**, 215–224.

BERGER, W. H. & HEATH, G. R. 1968. Vertical mixing in pelagic sediments. *Journal of Marine Research*, **26**, 134–143.

BISCHOF, J. 2000. *Ice Drift, Ocean Circulation and Climate Change*. Springer-Praxis, Chichester.

BOND, G., HEINRICH, H., BROECKER, W., LABEYRIE, L., MCMANUS, J., ANDREWS, J., HUON, S., JANTSCHIK, R., CLASEN, S., SIMIET, C., TEDESCO, K., KLAS, M., BONANI, G. & IVY, S. 1992. Evidence for massive discharges of icebergs into

the North Atlantic during the last glacial period. *Nature*, **360**, 245–249.

CADMAN, V. M. 1996. *Glacimarine sedimentation and environments during the Late Weichselian and Holocene in the Bellsund Trough and Van Keulenfjorden, Svalbard.* PhD thesis, University of Cambridge, Cambridge.

DOWDESWELL, J. A. & SIEGERT, M. J. 1999. Ice-sheet numerical modelling and marine geophysical measurements of glacier-derived sedimentation on the Eurasian Arctic continental margins. *Geological Society of America Bulletin*, **111**, 1080–1097.

DOWDESWELL, J. A., VILLINGER, H., WHITTINGTON, R. J. & MARIENFELD, P. 1993. Iceberg scouring in Scoresby Sund and on the East Greenland continental shelf. *Marine Geology*, **111**, 37–53.

DOWDESWELL, J. A., WHITTINGTON, R. J. & MARIENFELD, P. 1994a. The origin of massive diamicton facies by iceberg rafting and scouring, Scoresby Sund, East Greenland. *Sedimentology*, **41**, 21–35.

DOWDESWELL, J. A., UENZELMANN-NEBEN, G., WHITTINGTON, R. J. & MARIENFELD, P. 1994b. The Late Quaternary sedimentary record in Scoresby Sund, East Greenland. *Boreas*, **23**, 294–311.

DOWDESWELL, J. A., KENYON, N., ELVERHØI, A., LABERG, J. S., MIENERT, J. & SIEGERT, M. J. 1996. Large-scale sedimentation on the glacier-influenced Polar North Atlantic margins: long-range side-scan sonar evidence. *Geophysical Research Letters*, **23**, 3535–3538.

DOWDESWELL, J. A., KENYON, N. H. & LABERG, J. S. 1997. The glacier-influenced Scoresby Sund Fan, East Greenland continental margin: evidence from GLORIA and 3.5kHz records. *Marine Geology*, **143**, 207–221.

DOWDESWELL, J. A., ELVERHØI, A. & SPIELHAGEN, R. 1998. Glacimarine sedimentary processes and facies on the Polar North Atlantic Margins. *Quaternary Science Reviews*, **17**, 243–272.

DOWDESWELL, J. A., Ó COFAIGH, C., TAYLOR, J., KENYON, N. H. & MIENERT, J. 2002. On the architecture of high-latitude continental margins: the influence of ice-sheet and sea-ice processes in the Polar North Atlantic. *In*: DOWDESWELL, J. A. & Ó COFAIGH, C. (eds) *Glacier-Influenced Sedimentation on High-Latitude Continental Margins*. Geological Society, London, Special Publications, **203**, 33–54.

ESCHER, J. C. & PULVERTAFT, T. C. R. 1995. *Geological map of Greenland, 1: 2,500,000.* Geological Survey of Greenland, Copenhagen.

EVANS, J., DOWDESWELL, J. A. GROBE, H., NIESSEN, F., STEIN, R., HUBBERTEN, H.-W. & WHITTINGTON, R. J. 2002. Late Quaternary sedimentation in Kejser Franz Joseph Fjord and the continental margin of East Greenland. *In*: DOWDESWELL, J. A. & Ó COFAIGH, C. (eds) *Glacier-Influenced Sedimentation on High-Latitude Continental Margins*. Geological Society, London, Special Publications, **203**, 149–179.

FUNDER, S. & HANSEN, L. 1996. The Greenland ice

sheet – a model for its culmination and decay during and after the Last Glacial Maximum. *Bulletin of the Geological Society of Denmark*, **42**, 137–152.

FUNDER, S., HJORT, C., LANDVIK, J. Y., NAM, S-I., REEH, N. & STEIN, R. 1998. History of a stable ice margin – East Greenland during the middle and upper Pleistocene. *Quaternary Science Reviews*, **17**, 77–123.

GRAF, G., GERLACH, S. A., LINKE, P., QUEISSER, W., RITZRAU, W., SCHELTZ, A., THOMSEN, L. & WITTE, U. 1995. Benthic-pelagic coupling in the Greenland-Norwegian Sea and its effect on the geological record. *Geologische Rundschau*, **84**, 49–58.

GROBE, H. 1987. A simple method for the determination of ice-rafted debris in sediment cores. *Polarforschung*, **57**, 123–126.

GROUSSET, F. E., CORTIJO, E., HUON, S., HERVÉ, L., RICHTER, T., BURDLOFF, D., DUPRAT, J. & WEBER, O. 2001. Zooming in on Heinrich layers. *Paleoceanography*, **16**, 240–259.

HEBBELN, D., HENRICH, R. & BAUMANN, K.-H. 1998. Paleoceanography of the last interglacial/glacial cycle in the Polar North Atlantic. *Quaternary Science Reviews*, **17**, 125–153.

HEINRICH, H. 1988. Origin and consequences of cyclic ice rafting in the Northeast Atlantic Ocean during the past 130,000 years. *Quaternary Research*, **29**, 143–152.

HENRICH, R. 1998. Dynamics of Atlantic water advection to the Norwegian-Greenland Sea – a time-slice record of carbonate distribution in the last 300 ky. *Marine Geology*, **145**, 95–131.

HENRICH, R., WAGNER, T., GOLDSCHMIDT, P. & MICHELS, K. 1995. Depositional regimes in the Norwegian-Greenland Sea: the last two glacial to interglacial transitions. *Geologische Rundschau*, **84**, 28–48.

HESSE, R. & KHODABAKHSH, S. 1998. Depositional facies of late Pleistocene Heinrich Events in the Labrador Sea. *Geology*, **26**, 103–106.

HESSE, R., KLAUCKE, I., KHODABAKHSH, S. & PIPER, D. 1999. Continental slope sedimentation adjacent to an ice margin. III. The upper Labrador Slope. *Marine Geology*, **155**, 249–276.

KENYON, N. H. 1986. Evidence from bedforms for a strong polewards current along the upper continental slope of Northwest Europe. *Marine Geology*, **72**, 187–198.

KING, E. L., HAFLIDASON, H., SEJRUP, H. P. & LØVLIE, R. 1998. Glacigenic debris flows on the North Sea Trough Mouth Fan during ice-stream maxima. *Marine Geology*, **152**, 217–246.

LABERG, J. S. & VORREN, T. O. 1995. Late Weichselian submarine debris flow deposits on the Bear Island Trough Mouth Fan. *Marine Geology*, **127**, 45–72.

LABERG, J. S. & VORREN, T. O. 2000a. The Trænadjupet Slide, offshore Norway – morphology, evacuation and triggering mechanisms. *Marine Geology*, **171**, 95–114.

LABERG, J. S. & VORREN, T. O. 2000b. Flow behaviour of the submarine glacigenic debris flows on the

Bear Island Trough Mouth Fan, western Barents Sea. *Sedimentology*, **47**, 1105–1117.

LABERG, J. S., VORREN, T. O. & KNUTSEN, S.-M. 1999. The Lofoten contourite drift off Norway. *Marine Geology*, **159**, 1–6.

LABERG, J. S., VORREN, T. O., DOWDESWELL, J. A., KENYON, N. H. & TAYLOR, J. 2000. The Andøya Slide and the Andøya Canyon, north-eastern Norwegian-Greenland Sea. *Marine Geology*, **162**, 259–275.

MARSHALL, S. J., CLARKE, G. K. C., DYKE, A. S. & FISHER, D. A. 1996. Geologic and topographic controls on fast flow in the Laurentide and Cordilleran Ice Sheets. *Journal of Geophysical Research*, **101**, 17827–17839.

MIENERT, J., ANDREWS, J. T. & MILLIMAN, J. D. 1992. The East Greenland continental margin (65°N) since the last deglaciation: changes in seafloor properties and ocean circulation. *Marine Geology*, **106**, 217–238.

MIENERT, J., KENYON, N. H., THIEDE, J. & HOLLENDER, F.-J. 1993. Polar continental margins: studies off East Greenland. *EOS, Transactions of the American Geophysical Union*, **74**, 225, 234, 236.

NAM, S. I., STEIN, R., GROBE, H. & HUBBERTEN, H. 1995. Late Quaternary glacial-interglacial changes in sediment composition at the East Greenland continental margin and their paleoceanographic implications. *Marine Geology*, **122**, 243–262.

SARNTHEIN, M., JANSEN, E., WEINELT, M., ET AL. 1995. Variations in the Atlantic surface ocean paleoceanography, 50°–80°N: a time-slice record of the last 30,0000 years. *Paleoceanography*, **10**, 1063–1094.

SCOURSE, J. D., HALL, I. R., MCCAVE, I. N., YOUNG, J. R. & SUGDON, C. 2000. The origin of Heinrich layers: evidence from H2 for European precursor events. *Earth and Planetary Science Letters*, **182**, 187–195.

SOLHEIM, A. 1997. Depth-dependent iceberg plough marks in the Barents Sea. *In*: DAVIES, T. A. ET AL. (eds) *Glaciated Continental Margins: An Atlas of Acoustic Images*. Chapman and Hall, New York, 138–139.

STEIN, R., NAM, S. I., GROBE, H. & HUBBERTEN, H. 1996. Late Quaternary glacial history and short-term ice-rafted debris fluctuations along the East Greenland continental margin. *In*: ANDREWS, J. T., AUSTIN, W. E. N., BERGSTEN, H. & JENNINGS, A. E. (eds) *Late Quaternary Palaeoceanography of the North Atlantic Margins*. Geological Society, London, Special Publications, **111**, 135–151.

STOW, D. V. & SHANMUGAM, G. 1980. Sequence of structures in fine-grained turbidites: comparison of recent deep-sea and ancient flysch sediments. *Sedimentary Geology*, **25**, 23–42.

STOW, D. A. V. & PIPER, D. J. W. 1984. Deep-water fine-grained sediments: facies models. *In*: STOW, D. A. V. & PIPER, D. J. W. (eds) *Fine-Grained Sediments: Deep-Water Processes and Facies*. Geological Society, London, Special Publications, **15**, 611–646.

STOW, D. A. V., READING, H. G. & COLLINSON, J. D. 1996. Deep Seas. *In*: READING, H. G. (ed.)

Sedimentary Environments: Processes, Facies and stratigraphy. Blackwell, Oxford, 395–453.

STOW, D. A. V., FAUGÈRES, J.-C., VIANA, A. & GONTHIER, E. 1998. Fossil contourites: a critical review. *Sedimentary Geology*, **115**, 3–31.

SYVITSKI, J. P. M., STEIN, A. B., ANDREWS, J. T. & MILLIMAN, J. D. 2001. Icebergs and the sea floor of the East Greenland (Kangerlussuaq) continental margin. *Arctic, Antarctic and Alpine Research*, **33**, 52–61.

TAYLOR, J., DOWDESWELL, J. A., KENYON, N. H. & Ó COFAIGH, C. 2002. Late Quaternary architecture of trough–mouth fans: debris flows and suspended sediments on the Norwegian margin. *In*: DOWDESWELL, J. A. & Ó COFAIGH, C. (eds) *Glacier-Influenced Sedimentation on High-Latitude Continental Margins*. Geological Society, London, Special Publications, **203**, 55–71.

TAYLOR, J., TRANTER, M. & MUNHOVEN, G. 2002*b*. Carbon cycling and burial in a glacially influenced polar North Atlantic. *Paleoceanography*, **17**, 10.1029/2001PA000644.

TAYLOR, J., DOWDESWELL, J. A. & SIEGERT, M. J. 2002*c*. Late Weichselian depositional processes, fluxes, and sediment volumes on the margins of the Norwegian Sea (62–75°N). *Marine Geology*, **188**, 61–77.

THOMPSON, J., BROWN, L., NIXON, S., COOK, G. T. & MACKENZIE, A. B. 2000. Bioturbation and Holocene sediment accumulation fluxes in the north-east Atlantic Ocean (Benthic Boundary Layer experiment sites). *Marine Geology*, **169**, 21–39.

VAN WEERING, T. C. E., NIELSEN, T., KENYON, N. H., AKENTIEVA, K. & KUIJPERS, A. H. 1998. Sediments and sedimentation at the NE Faeroe continental margin; contourites and large-scale sliding. *Marine Geology*, **152**, 159–176.

VOELKER, A. H. L., SARNTHEIN, M., GROOTES, P. M., ERLENKEUSER, H., LAJ, C., MAZAUD, A., NADEAU, M.-J. & SCHLEICHER, M. 1998. Correlation of marine ^{14}C ages from the Nordic seas with the GISP2 isotope record: implications for ^{14}C calibration beyond 25 ka BP. *Radiocarbon*, **40**, 517–534.

VOGT, P. R., CRANE, K. & SUNVOR, E. 1993. Glacigenic mudflows on the Bear Island submarine fan. *EOS, Transactions of the American Geophysical Union*, **74**, 449, 452–453.

VORREN, T. O. & LABERG, J. S. 1997. Trough mouth fans – palaeoclimate and ice-sheet monitors. *Quaternary Science Reviews*, **16**, 865–881.

VORREN, T. O., HALD, M., EDVARDSEN, M. & LIND-HANSEN, O.-W. 1983. Glacigenic sediments and sedimentary environments on continental shelves: general principles with a case study from the Norwegian Shelf. *In*: EHLERS, J. (ed.) *Glacial Deposits in North-west Europe*. Balkema, Rotterdam, 61–73.

VORREN, T. O., HALD, M. & THOMSEN, E. 1984. Quaternary sediments and environments on the continental shelf off northern Norway. *Marine Geology*, **57**, 229–257.

VORREN, T. O., LABERG, J. S., BLAUME, F., DOWDESWELL, J. A., KENYON, N. H., MIENERT, J., RUMOHR, J. & WERNER, F. 1998. The Norwegian-Greenland Sea continental margins: morphology and late Quaternary sedimentary processes and environment. *Quaternary Science Reviews*, **17**, 273–302.

WANG, D. & HESSE, R. 1996. Continental slope sedimentation adjacent to an ice margin. II. Glaciomarine depositional facies on the Labrador slope and glacial cycles. *Marine Geology*, **135**, 65–96.

Millennial and sub-millennial-scale variability in sediment colour from the Barra Fan, NW Scotland: implications for British ice sheet dynamics

LINDSAY J. WILSON & WILLIAM E. N. AUSTIN

School of Geography and Geosciences, Irvine Building, University of St Andrews, St Andrews, Fife KY16 9AL, UK
(e-mail: wena@st-andrews.ac.uk)

Abstract: Sediment colour, together with other proxy data, provides a novel, rapid and non-destructive tool in the investigation of glacier-influenced sedimentation on the Barra Fan, NW Scotland. Lightness (L^*) and reflectance (400–700 nm) measurements at this site provide a quantitative estimate of changes in calcium carbonate and clay content. Inter-stadials are carbonate-rich/clay-poor (higher L^* and reflectivity), whereas stadials are carbonate-poor/clay-rich (lower L^* and reflectivity). Detailed sedimentological investigations suggest that the last British Ice Sheet (BIS) extended to the outer continental shelf-break shortly after 30 ka BP. This climatic response of the BIS to global cooling at the Marine Isotope Stage (MIS) 3 – 2 transition marks a significant increase in sediment delivery to the Barra Fan. Prior to 30 ka BP, strong Dansgaard/Oeschger (D/O) cyclicity dominates the record. After 30 ka BP, shorter periodicities prevailed as the BIS reached its maximum extent. Glacier dynamics plays a significant role in the delivery of ice-rafted debris (IRD) across this margin, highlighting the inherent difficulties of correlating millennial-scale IRD events when the IRD is derived from different ice sheets. An event stratigraphy based upon carbonate-rich interstadials provides a more robust means of amphi-Atlantic correlation during this interval.

During much of the late Pleistocene, glacier-influenced sedimentation was the dominant depositional feature of the North East Atlantic margins (e.g. Andrews *et al*. 1996). Marine sediment records recovered from the 'open' North Atlantic provide the basis for a significant body of literature describing glacial variability at millennial time-scales during this period (e.g. Bond *et al*. 1992, 1993, 1999). However, there are relatively few well-dated continental margin records predating the last glacial maximum (LGM), particularly from formerly glaciated regions, which provide this millennial-scale detail.

During the past 60 000 years, periods of increased ice-rafting, the so-called 'Heinrich Events' (after Heinrich 1988), punctuate and dominate the 'open' ocean record (Bond *et al*. 1992). Early provenance studies linked the carbonate-rich ice-rafted debris (IRD) of these Heinrich layers to the Laurentide Ice Sheet (LIS) (e.g. Andrews and Tedesco 1992). However, glacier-influenced sedimentation at these oceanic sites represents an integrated record of IRD from numerous potential source regions (e.g. Dowdeswell *et al*. 1995; Revel *et al*. 1996) and compositional differences may have

climatic implications (e.g. Bond & Lotti 1995). More recently, there has been considerable debate concerning the origin and timing of the emplacement of detrital material within Heinrich Events (e.g. Gwiazda *et al*. 1996; Grousset *et al*. 2000).

In addition to the Heinrich Events, the last glacial period is characterized by rapid warm to cold climate transitions (Dansgaard/Oeschger (D/O) events), occurring with periodicities of 500–2000 years (e.g. Dansgaard *et al*. 1993; Bond *et al*. 1999). There is growing evidence to suggest that these cycles are a persistent feature of the North Atlantic's climate system, albeit with considerably diminished Holocene cycles (Bond *et al*. 1997). The fresh water forcing mechanism of D/O events suggested by Broecker *et al*. (1990) has recently been discussed by Bond *et al*. (1999), who propose that cooling phases caused freshening of surface waters in the glacial North Atlantic through increased rates of iceberg discharge.

There is, as yet, limited understanding of the contribution made by the last British ice sheet (BIS) to North Atlantic IRD (see Richter *et al*. 2001). However, Scourse *et al*. (2000), in a detailed investigation of Heinrich Event 2 at

From: DOWDESWELL, J. A. & Ó COFAIGH, C. (eds) 2002. *Glacier-Influenced Sedimentation on High-Latitude Continental Margins.* Geological Society, London, Special Publications, **203**, 349–365. 0305-8719/02/$15.00
© The Geological Society of London 2002.

the Goban Spur, demonstrated a precursor event of British origin, predating the main Laurentide IRD signature. On the Barra Fan, NW Scotland, Knutz *et al.* (2001) documented the general pattern of ice-rafting during Marine Isotope Stage (MIS) 2–3 and suggested a strong regional IRD signature closely coupled to the D/O climate cycle. Since this region represents the main depo-centre of the last BIS, it is well placed to record the dynamic response of this climatically-sensitive ice margin (Boulton 1990).

Sediment colour

Colour is 'the perception of the visible light-reflectance spectrum of an object' (Deaton 1987). The earliest procedure used for identifying colour variations was the Munsell Renotation System (Munsell 1941). This is based on colour deduction by the naked eye and is therefore dependent on the colour perception of the individual. The development of simple colorimeters and later spectrophotometers, has allowed uniform colour values to be represented by a single quantitative measurement. Measurements are rapid and non-destructive, so that data can be generated at a very high sampling resolution (density) (Chapman & Shackleton 1998).

Late Pleistocene marine sediment cores from the open North Atlantic Ocean commonly contain carbonate-rich (light-coloured) and clay-rich (dark-coloured) sediment layers (e.g. Nagao & Nakashima 1992). Sediment colour and reflectance provide a useful tool in this region, not only in characterizing intervals of enhanced IRD deposition (e.g. Grousset *et al.* 1993), but also in characterizing sediment biogenic carbonate content (e.g. Cortijo *et al.* 1995; Ortiz *et al.* 1999; Helmke & Bauch 2001). Rapid temperature fluctuations (Dansgaard/Oeschger cycles) identified in the ice core records of Greenland (e.g. Dansgaard *et al.* 1993) have clear parallels within the colour variations observed in North Atlantic sediment records (Bond *et al.* 1992; Cortijo *et al.* 1995). Sediment colour has also been used as a proxy for changing North Atlantic circulation intensity (e.g. Chapman & Shackleton 1998). The potential therefore exists for the identification of sediment composition changes and, hence, fluctuations in environmental conditions through the analysis of marine sediment colour variations. Few studies, however, have investigated this methodology in a continental margin setting, partly because of the complexity of the depositional systems involved.

Giant piston core MD95-2006

Giant piston core MD95-2006 (57°01.82 N, 10°03.48 W, water depth 2120 m) was recovered in 1995 by the RV *Marion Dufresne* from the northern limits of the Barra Fan, NW Scotland as part of the IMAGES programme (Fig. 1). The Barra Fan extends from the Hebridean continental slope to the deep-water basin of the NE Rockall Trough (Holmes *et al.* 1998). It is primarily composed of Neogene sands and Pleistocene glacigenic sediment from NW Britain (Armishaw *et al.* 1998). This latter component is thought to have been deposited when locally grounded ice reached the shelf edge. Further evidence of late Pleistocene glacial activity is present on the shelf margin in the form of morainal banks (Selby 1989; Stoker 1995) and iceberg plough marks (Kenyon 1987). A 400–700 m undisturbed shelf-margin wedge forms the bulk of the landward depocentre of the Barra Fan (Stoker 1995; Holmes *et al.* 1998).

Two vibrocores (57/-09/89 and 57/-09/46) from the St Kilda Basin of the Hebridean Shelf have documented the timing of the last deglaciation of NW Scotland and provide an exceptionally high resolution record of the Younger Dryas cold phase (Austin 1991; Peacock *et al.* 1992; Austin *et al.* 1995; Austin & Kroon 1996). In addition, two short cores from the Barra Fan (56/-10/36 and 57/-11/59) have

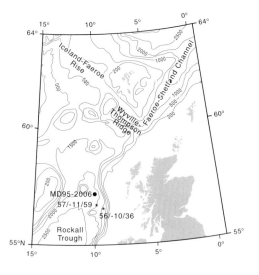

Fig. 1. Location map of core MD95-2006 (57°01.82 N, 10°03.48 W, water depth 2120 m), Barra Fan, NW Scotland. Nearby cores 56/-10/36 (Kroon *et al.* 1997) and 57/-11/59 (Austin and Kroon 2001) are located. Bathymetric contours are shown in metres.

established the detail of deglaciation along this margin within the context of surface (Kroon *et al.* 1997) and deep water circulation changes (Austin & Kroon 2001). Piston core MD95-2006 recovered 30 m of distal glacimarine sediments with average accumulation rates of greater than 0.5m/ka (Kroon *et al.* 2000; Knutz *et al.* 2001). It therefore exhibits the potential for higher resolution marine sediment studies over MIS 2 and 3, from a continental margin setting which is known to record the main fluctuations of the last British ice sheet.

The characterization of millennial-scale events by spectrophotometry is evaluated at this site and the significance of the MD95-2006 record is discussed in the context of amphi-Atlantic climate change. The aims of this paper are to present a new and revised radiocarbon-based age: depth model for core MD95-2006 and to determine the response of the British ice sheet through the MIS3 to MIS2 transition using sediment colour and reflectance measurements.

Methods

Spectrophotometry

The cleaned split-core surfaces of the MD95-2006 archive sections were analysed with a hand-held Minolta CM-2002 spectrophotometer. A calibration and measurement procedure was established based on the Ocean Drilling Program's use of the Minolta CM-2002 (Blum 1997). The illumination system was set at an angle of 2° and spectral reflectance measurements were made on the SCE (secular component excluded) setting (Balsam *et al.* 1997).

Polyethylene film was placed over the sediment surface in accordance with the methodology of Chapman & Shackleton (1998). Measurements were taken on a single track along the centre of the core, with the spectrophotometer held orthogonal to the sediment surface. The sampling interval was set at 5 cm and was increased to 1 or 2 cm where higher resolution records were required, i.e. at a clearly visible transition between two colours, or, at bands highlighted within other proxy studies. The time interval between core cleaning and recording of each reading was kept to a minimum.

The recorded data is displayed using the Commission Internationale de l'Eclairage (CIE) L*a*b* colour space, the most commonly employed colour indicator in palaeoceanography in recent times (Merrill & Beck 1996). This colour definition was established in 1931 as a standard component in aiding international colour matching. In this study $L*$ indicates lightness ($\approx$greyscale reflectance) where, on a scale of zero to one hundred, zero represents black and one hundred is white. In addition to $L*$, colour reflectance data is presented every 100 nm from 400–700 nm (i.e. within the visible spectrum).

Particle size measurements

Particle size distributions were measured from samples of core MD95-2006 on a Coulter LS 230 Particle Size Machine (PSM). Sediment sub-samples were treated for the removal of carbonate through dissolution. The decalcified results obtained reflect the lithogenic (non-carbonate) particle size distribution of the sample. The biogenic opal content of the sediments is generally less than 5% and was estimated from smear slides.

Decalcified data were obtained from approximately 0.55–0.60 g of 'wet' sediment, which were placed in a test tube with distilled water and then centrifuged, before excess fluid was decanted. To remove all traces of salt crystals this procedure was repeated twice. Following decalcification of the sediment overnight in a 20% solution of acetic acid, samples were centrifuged to allow settlement of all fine particles. Excess fluid was decanted, distilled water was added and the procedure repeated twice. A fixed volume of sodium hexametaphosphate was then transferred to the sample and this solution was added to the PSM and the grain size distributions recorded. This methodology is consistent with Austin & Kroon (2001).

Calcium carbonate

Calcium carbonate content was determined downcore at 10 cm intervals using a modified back titration method (Grimaldi *et al.* 1966) as described in Austin & Kroon (1996) and Austin & Evans (2000). 0.5g of dry sediment was accurately weighed and treated with an excess of acid (25 ml of 0.5M HCl solution). 0.5ml of bromophenol blue indicator was added and the solution was back-titrated against 0.35M NaOH solution until the yellow to violet end-point was reached. The calcium carbonate content was calculated (weight in grams, volume in litres) by:

$$\%CaCO_3 = 100\,[(vol.\,HCl \times 1.007225) - \quad (1)$$
$$(vol.\,NaOH \times molarity\,NaOH)] \times (2 \times$$
$$sample\,weight)^{-1} \times 100.$$

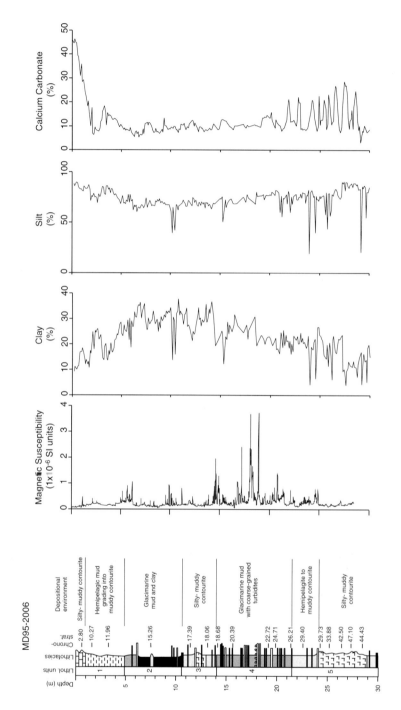

Fig. 2. Lithological summary of core MD95-2006. General log modified after Kroon *et al.* (2000) with additional radiocarbon ages indicated. Clay and silt values (lithogenic) are volume %, calcium carbonate is weight %.

Table 1. *Radiocarbon ages of Barra Fan cores VE 56/-10/36, VE 57/-11/59, and MD95-2006*

Laboratory number	Core depth (cm)	Conventional radiocarbon age (^{14}C years BP $\pm 1\sigma$)	Calendar age (years)	Species
VE 56/-10/36				
AA-13902*	72.5	10 040 ± 80	11062, 10984, 10834	G. bulloides
AA-13899*	75.5	10 585 ± 85	11674	N. pachyderma (sinistral)
AA-13900*	355	12 055 ± 100	13476	G. bulloides
AA-13898*	431.5	13 020 ± 115	15123, 14701, 14393	G. bulloides
VE 57/-11/59				
AA-23932*	2.5	2535 ± 45	2183	G. bulloides
AA-23933*	21.5	5540 ± 60	5910	G. bulloides
AA-23934*	51.5	7175 ± 80	7639	G. bulloides
AA-23935*	71.5	9010 ± 65	9619	G. bulloides
AA-24177*	180.5	11 725 ± 95	13161	G. bulloides
AA-24179*	224	12 720 ± 170	14144	G. bulloides
AA-24178*	282	12 940 ± 90	14355	G. bulloides
MD95-2006				
AA-40438*	0.5	2799 ± 44	2526	G. bulloides
AA-40439*	164.5	10 270 ± 73	11153	G. bulloides
AA-40440*	323	11 960 ± 120	13442	G. bulloides
AA-22347*	770	15 260 ± 140	17664	N. pachyderma (sinistral)
AA-35119*	1175.5	17 390 ± 190	20115	N. pachyderma (sinistral)
AA-22348*	1340	18 060 ± 130	20886	N. pachyderma (sinistral)
AA-35120*	1411	18 680 ± 130	21600	N. pachyderma (sinistral)
AA-35121**	1591.5	20 390 ± 150	23567	N. pachyderma (sinistral)
CAMS-60835**	1941.5	22 720 ± 130	26740	N. pachyderma (sinistral)
AA-22349**	2020.5	24 710 ± 280	29022	N. pachyderma (sinistral)
AA-32312**	2173.5	26 210 ± 270	30726	N. pachyderma (sinistral)
AA-32313**	2288	29 400 ± 370	34305	G. bulloides
AA-32314**	2418.5	29 730 ± 470	34672	G. bulloides
AA-22350**	2539	33 880 ± 610	39229	G. bulloides
AA-35122**	2653.25	42 500 ± 1800	48361	G. bulloides
AA-35123**	2728.5	47 100 ± 3000	53052	G. bulloides
AA-35124**	2860	44 430 ± 2000	50345	G. bulloides

Samples marked * have been calibrated using Calib4.2 (Stuiver & Reimer 1993; Stuiver *et al.* 1998); samples marked ** have been calibrated using U/Th ages and a second-order polynomial equation (Bard *et al.* 1998).

Magnetic susceptibility

The volumetric magnetic susceptibility was measured down-core in SI units in 1cm increments using a Bartington Instruments magnetic susceptibility meter (Model MS2, adapted with a MS2F probe with an operating frequency of 0.58 Hz). The results are reported as 1×10^{-6} SI Units.

Results

Lithostratigraphy

Five lithological units were identified in core MD95-2006 which may be subdivided into six lithofacies, based upon sedimentary structure, texture, colour and carbonate content (Fig. 2) (Kroon *et al.* 2000; Knutz *et al.* 2001). The sequence consists of soft, light grey–brown silts and clays, which are interspersed with occasional clasts and thin sandy horizons, tending to exhibit sharp bases and gradational upper contacts (typically < 1 cm; maximum 5 cm). The latter are interpreted as turbidite layers and are most prevalent through the interval 14–22 m. Well-developed dark blue-black monosulphide streaks were common between 7–13 m. Lighter grey–pale blue horizons, corresponding to higher carbonate contents, are visible between 22–24 m.

Magnetic susceptibility (ms) measurements reveal significant down-core variability. Below 5 m the first in a series of marked peaks in magnetic susceptibility are observed, many coincide with silty to sandy muds containing dropstones. Prior to this, small magnetic susceptibility peaks are observed between 1 and 3 m core depth. The most pronounced magnetic susceptibility events are recorded at approximately 10 m, 15 m, 17–19 m, 21 m and 24.5 m. Between 14 to 24 m core depth, the silty to sandy muds are interspersed with numerous sandy turbidites and occasional gravel layers.

Silt content is typically greater than 60% throughout MD95-2006, as previously shown by Kroon *et al.* (2000). Below 20 m, there appears to be considerable variation in silt content, coinciding with low magnetic susceptibility and variable $CaCO_3$. Above approximately 6–7 m, silt content rises steadily, coinciding with an increase in $CaCO_3$ and general decrease in clay content. Clay content varies between 10% to 35% and exhibits a very clear long-term trend. Lowest values are observed at the base of the core, increase steadily through a series of cycles to a maximum at approximately 6–7 m, and decline in a cyclical manner towards the core top.

Calcium carbonate data were reported for MD95-2006 by Kroon *et al.* (2000), with maximum core top values of approximately 20%. The data presented here suggest that near-surface values range from approximately 10% to 50%, which agree with Holocene data reported by Austin & Kroon (2001) on nearby

core 57/-11/59. The $CaCO_3$ percentage data reported by Kroon *et al.* (2000) are actually Ca weight data since they were measured by sediment X-ray diffraction. Calcium carbonate values in MD95-2006 are generally low, between 6–19 m, coinciding with an interval of prominent magnetic susceptibility peaks and increased clay content. Below 19 m, values exhibit a pronounced cyclicity, ranging between 10–25% calcium carbonate content.

Chronostratigraphy

Seventeen ^{14}C accelerator mass spectrometer (AMS) radiocarbon dates were obtained from core MD95-2006, based upon monospecific foraminiferal samples of sub-polar *Globogerina bulloides* and polar *Neogloboquadrina pachyderma* (sinistral)(Table 1). Samples were prepared to graphite at the NERC Radiocarbon Laboratory, East Kilbride, and ^{14}C analysis was carried out at the University of Arizona NSF-AMS facility. One tephra layer (1 Thol. 2 Ash) was identified to constrain the Younger Dryas period further. This is a constituent of North Atlantic Ash Zone 1 (Kvamme *et al.* 1989) and provides a useful chronostratigraphic marker within an interval of significant radiocarbon dating uncertainty (e.g. Austin *et al.* 1995).

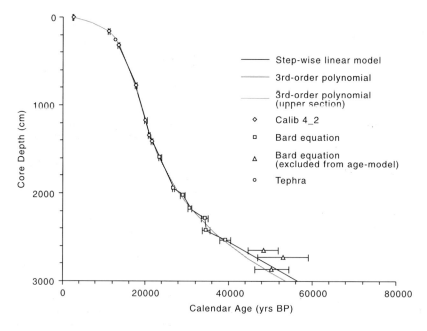

Fig. 3. Age:depth models of core MD95-2006. Dated levels shown are calibrated radiocarbon ages (years BP ± 1σ); see text for explanation of calibration procedure.

Seven ^{14}C ages were calibrated into calendar ages using the Calib4.2 programme (Stuiver & Reimer 1993; Stuiver *et al.* 1998) which incorporates a 400 year correction for the modern surface ocean reservoir effect at this latitude. This calibration of ^{14}C ages was only applied to radiocarbon dates younger than 21 000 ^{14}C years; the calibration data set available beyond this age is limited and has significantly decreased reliability. The nine remaining ^{14}C dates were calibrated using the U/Th ages and second-order polynomial equation of Bard *et al.* (1998) which statistically extends the current coral calibration set of Stuiver *et al.* (1998):

$$[\text{cal age BP}] = -3.0126 \times 10^{-6} \times [^{14}\text{C age BP}]^2 + 1.2896 \times [^{14}\text{C age BP}] - 1005 \quad (2)$$

An age–depth model was constructed for core MD95-2006 using 14 of the calibrated AMS ^{14}C dates and the one tephra date (1 Thol. 2 Ash) (Fig. 3). The oldest three dates obtained (AA35122, AA35123 and AA35124) were not used in the construction of the age–depth model due to the large dating uncertainty beyond 40 000 ^{14}C years BP. These dates lie close to the routine ^{14}C detection limit and are not statistically distinguishable at the 95% confidence level (Table1, Fig. 3). Contrasting age–depth models may yield significant age differences, particularly at the greatest core depths, where dating uncertainties are greater.

The first age–depth modelling approach employs the 14 calibrated AMS ^{14}C dates and the dated tephra horizon to fit two third-order polynomial curves to the data (Fig. 3). From 0–323 cm, the well-dated record of nearby core 57/-11/59 (Austin & Kroon 2001) provides an excellent constraint and test of the following equation (equation 3) for the age–depth relationship ($r = 1$):

$$[\text{cal age BP}] = 2482.2 + 87.78 \times [\text{core depth}] \quad (3) - 0.263 \times [\text{core depth}]^2 + 3.04\text{e}^{-4} \times [\text{core depth}]^3$$

Equation 3 is based upon three calibrated AMS ^{14}C dates and dated tephra horizon at 262.5 cm. From 323 cm to the bottom of the core, the following equation for the age–depth model is employed ($r = 0.995$):

$$[\text{cal age BP}] = 9929.9 + 13.528 \times [\text{core depth}] - 7.0233\text{e}^{-3} \times [\text{core depth}]^2 + 2.4709\text{e}^{-6} \times [\text{core depth}]^3 \quad (4)$$

Using the above relationship (equation 4), the bottom of the core, at a depth of 3000 cm,

yields an age estimate of 53 800 years BP. This differs from the previously published date of 45 000 years BP reported in Knutz *et al.* (2001) and emphasizes the significance of the new dates reported here (Table 1).

The second age–depth modelling approach assumes a constant sedimentation rate between the calibrated AMS ^{14}C ages, enabling the construction of a step-wise linear age–depth relationship (Fig. 3). The age of the core bottom was inferred by fitting a linear regression line through three calibrated ages between 34 000 and 39 000 years BP (AA32313, AA32314 and AA22350), but excluding the oldest three dates obtained (i.e. AA35122, AA35123 and AA35124). This yields an age estimate of 56 661 years BP at the base of the core (i.e. 3000 cm).

There is therefore relatively little age difference between the two modelling approaches (Fig. 3). The third-order polynomial age–depth models yield good statistical fits to the age control points. Large changes in sediment accumulation rates can arise in step-wise linear age–depth models as an artefact of dating uncertainties and sample density. Given the very rapid, sub-millennial changes of sediment accumulation defined during the last deglaciation at this site (Kroon *et al.* 2000), the third-order polynomial age–depth model is preferentially applied in all further age–depth calculations.

Colour variability

The stratigraphic record of lightness (L^*) and reflectance (400–700 nm) is presented in Figure 4. Three main subdivisions of the record are very clearly observed within the sequence. These can be defined as (1) a basal zone of highly variable L^* and reflectance, pre-dating a transition beginning at about 30 000 years BP to (2) a zone of darker, lower amplitude variability in L^* and reflectance, which gives way to (3) a light coloured, high reflectance zone postdating approximately 11 000 years BP. Colour variability pre-dating approximately 30 000 years BP exhibits an apparent cyclicity of about 3000 years. Between 11 000 to 30 000 years, L^* and reflectance show high frequency variability. L^* and reflectance have been less variable in the last 11 000 years and show a steady increase towards the top of the core; the latter is particularly evident at 600 and 700 nm.

Reflectance spectra, L^* and percentage clay results have been plotted against percentage bulk calcium carbonate to investigate relationships between these variables (Fig. 5). A clear positive relationship is observed between L^*

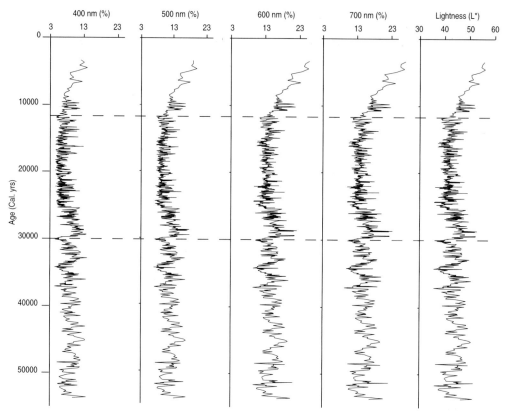

Fig. 4. Stratigraphic summary of sediment reflectance (400–700 nm) and lightness (*L**) of core MD95-2006. Age is calendar years BP.

and calcium carbonate (*r* = 0.66, Fig. 5a), confirming that lighter coloured sediments have a higher calcium carbonate content. However, carbonate contents are generally low throughout much of the sequence and the relationship with *L** is less clearly defined when calcium carbonate values fall below 15–20%. Within the same range of observed *L**, there also exists a negative relationship with sediment clay content (*r* = 0.53, Fig. 5b). Therefore, in general, we observe high *L** when clay content is low and calcium carbonate content is high and *vice versa*.

Specific reflectance wavelengths (400–700 nm) coinciding with the visible spectrum reveal positive relationships with the calcium carbonate content of the sediments (Fig. 5c–f). The strongest correlations (*r* = 0.73) are observed at 600 nm and 700 nm (Fig. 5e–f), corresponding to blue light reflectance. A weaker correlation (*r* = 0.23) is noted for red light reflectance at 400 nm (Fig. 5c).

Spectral analysis

Spectral analysis of the data (*L**, 400–700 nm reflectance, % clay and % $CaCO_3$) were carried out using the Blackman-Tukey method in the Analyseries 1.1 programme (Paillard *et al.* 1996). The results were compiled for two distinct chronostratigraphic intervals, corresponding to 15–22 ka BP and 30–50 ka BP (Fig. 6). These intervals were chosen as periods of relative signal stability and avoid some of the main transitions in depositional pattern, notably after 15 ka BP and between 22–30 ka BP.

In the interval 30–50 ka BP, a strong periodicity is observed at 3077 years, notably in the following proxies: 500 nm, 600 nm, 700 nm, $CaCO_3$. *L** and 400 nm reflectance exhibit a strong periodicity at 3125 years. Longer periodicities, notably in clay and silt content, are also observed through this interval, with greatest power at 4878 and 9524 years, respectively. In addition, a periodicity of 1980 years is observed

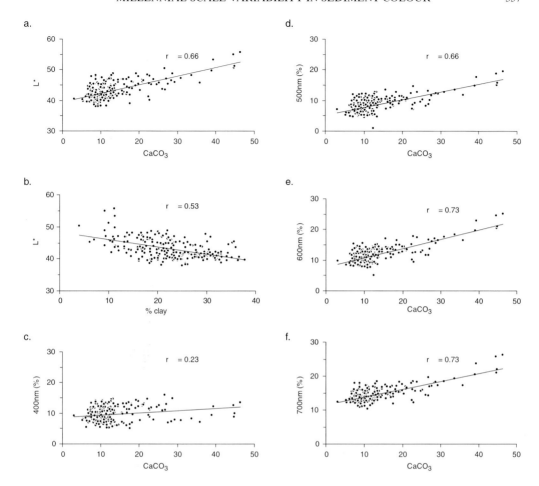

Fig. 5. Summary plots of sediment reflectance and lightness against calcium carbonate (weight %) and clay (volume %) from core MD95-2006. Best-fit linear regression lines shown.

in all the proxies investigated between 30–50 ka BP, although at a reduced power.

In the interval 15–22 ka BP, periodicities generally exhibit reduced power and higher frequency. The most common periodicities observed typically range from approximately 1000 to 2500 years. Many of the proxies investigated, notably L^* and reflectance, exhibit numerous periodicities of similar power. However, clay, silt and $CaCO_3$ content show maximum power at a consistent periodicity of approximately 2500 years.

Discussion

The lithostratigraphy of core MD95-2006 was originally defined by Kroon et al. (2000) and

Knutz et al. (2001). Kroon et al. (2000) describe the general depositional setting at this site, emphasizing the rapidity of sub-millennial scale changes in the depositional cycles observed. Knutz et al. (2001) concentrated upon the IRD signature and its implications for fluctuations along the western limits of the last British ice sheet, postulating a coupling between BIS-derived IRD and D/O cyclicity. Here, the dynamics of this ice sheet are investigated in the context of a refined age: depth model, allowing the response of the depositional pattern along this margin to be quantified through the last glacial period. Of particular note is the marked increase in sediment accumulation rate at approximately 30 ka BP, which suggests the sedimentological response is in-phase with the

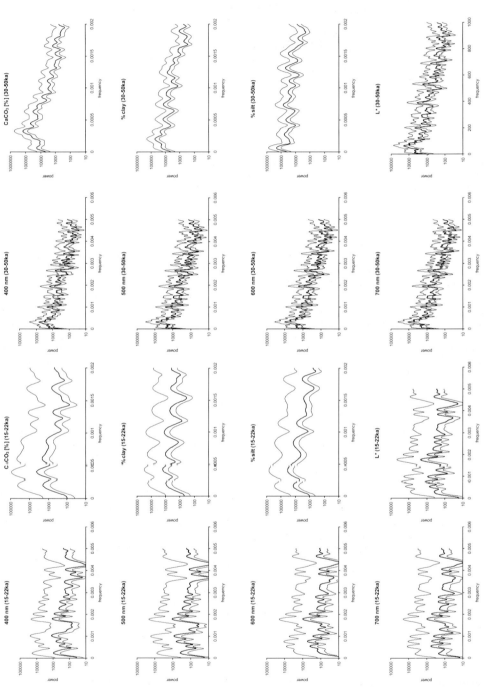

Fig. 6. Summary figures illustrating power spectra of various sedimentological proxies from the intervals 15–22 ka BP and 30–50 ka BP from MD95-2006. Spectral analysis employs the Blackman-Tukey method in the AnalySeries 1.1 programme (Paillard *et al.* 1996). Upper and lower confidence limits (80%) are denoted by thin lines as

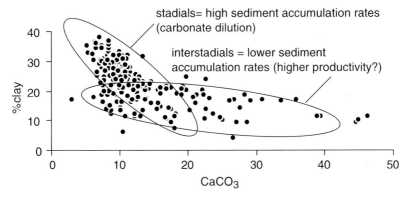

Fig. 7. Summary plot showing clay (volume %) against calcium carbonate (weight %) from core MD95-2006. Stadial intervals are characterized by high sediment accumulation rates; interstadials by lower sediment accumulation rates.

climatological transition of the MIS 3–2 boundary. The timing of this transition agrees with the age estimate from GISP2 and VM19-30 reported by Bond *et al.* (1997, 1999), but is 6000 years older than estimated in the original definition of the MIS 3–2 boundary by SPECMAP (Martinson *et al.* 1987).

The refined age: depth model of core MD95-2006 (Fig. 3) represents a significant improvement in constraining the depositional features at this site on the Barra Fan. In particular, the age at the bottom of core (*c.* 3000 cm) is estimated to be 53.8 ka BP, considerably older than the previous age estimate of 45 ka BP. This finding has implications for some of the previously reported IRD correlations, notably the position of H5, which was not recognized by Knutz *et al.* (2001). The latter has been tentatively assigned here to the quartz-rich IRD interval at approximately 28 m, defined as BF16 by Knutz *et al.* (2001).

The equivalent of the Heinrich Event stratigraphy of the open NE Atlantic (e.g. Bond *et al.* 1992) can be identified along this margin, despite some uncertainty regarding H2, H3 and H5 (Table 2). Only one of the IRD events (H4 at 24.7 m) has the characteristically high concentration of distinct pale yellow dolomitic carbonate (G. Bond, *pers. comm.* 2000). Since the timing of H1 is well constrained in this and other NW European records (e.g. McCabe & Clark 1998; Scourse *et al.* 2000), the interval of intense ice-rafting identified between 15.1–16.6 ka BP is confidently assigned to this event. However, there are numerous additional IRD events in the MD95-2006 record which do not coincide with

Heinrich Events, most of which lack a distinctive Laurentide signature. At this site, these IRD lithologies are dominated by basaltic grains in the lithic sand-sized fraction, supporting the interpretation that they are derived from the Tertiary provinces of the NW British Isles (Austin & Kroon 1996; Knutz *et al.* 2001). Given this strong regional IRD signal, derived from a relatively small ice sheet, amphi-Atlantic correlations based on the IRD events described at this site are extremely difficult. These significant 'non-Heinrich' IRD events highlight the problem of IRD correlation and asynchronous deposition at sites distal to the main Laurentide IRD belt (Dowdeswell *et al.* 1999), yet proximal to, for example, the BIS.

In this study, a more promising approach to amphi-Atlantic correlation is the use of clearly defined cycles of carbonate-rich sediment in defining interstadial events. These cycles have been investigated through the analysis of sediment colour variability. Lightness (equivalent to grey-scale measurements) and blue colour reflectance measurements provide a useful proxy for calcium carbonate content (Fig. 5). In an attempt to provide a quantitative estimate of % $CaCO_3$, we have simplified the multiple regression approach of Ortiz *et al.* (1999) and applied the following regression equation:

$$\text{Proxy \% } CaCO_3 = 1.7733R_{600} - 7.7483 \quad (5)$$

This approach provides a rapid means of estimating calcium carbonate content and has the potential to resolve the fine detail of the interstadial events within MIS 3.

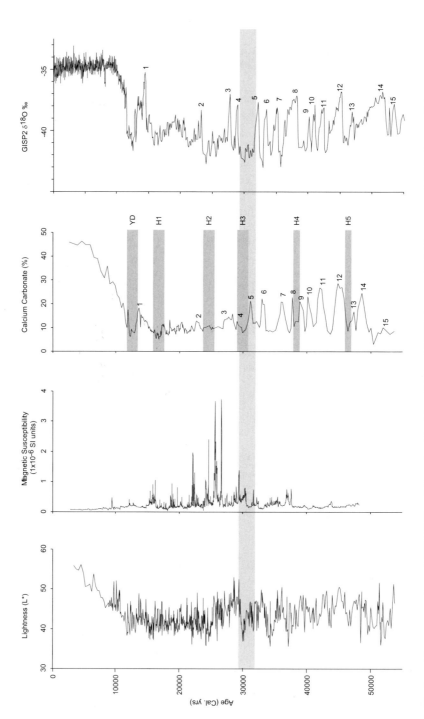

Fig. 8. Stratigraphic summary of core MD95-2006 and Greenland Ice Sheet Project (GISP2) ice core δ18O (Grootes *et al.* 1993). Interstadial events are numbered after Dansgaard *et al.* (1993) and corresponding intervals of increased calcium carbonate content are identified. Light shading represents the transition in depositional environment from MIS3 to MIS2 at approximately 30 000 years BP.

Table 2. *Age estimates of Heinrich events 1 to 5. The GISP 2 data are derived from Grootes & Stuiver (1997)*

Heinrich event	Curry *et al.* 1999	GISP 2	This study
1	16.5	15.8–16.5	15.1–16.6
2	24.3	24.0	24–24.8
3	31.6	30.2	30.8–29.3
4	39.9	38.5–39.6	38.0–36.7
5	46.0	45.5–46.2	46.4

The clay content of the sediment plays a significant role in carbonate dilution, so that interstadial events correspond to lower % clay, associated with lower sediment accumulation rates (Fig. 7). Stadials, which have low calcium carbonate contents, are characterized by higher % clay, reflecting higher sediment accumulation rates. This depositional pattern is consistent with a switching mechanism between hemipelagic and hemiturbiditic sedimentation (e.g. Stow & Tabrez 1998) during stadials, when ice margin sediment supply is high and bottom current activity reduced. During the interstadials sediment accumulation rates are reduced, due to glacier retreat and sediment entrapment on the continental shelf and shelf break. In addition, enhanced bottom current activity may act to winnow fine particles and hence reduce clay content. In response, carbonate content increases. There is some evidence in the literature which suggests enhanced ventilation during the interstadials (e.g. Curry *et al.* 1999) of MIS 3. Alternatively, a change in the actual mechanism of convection between interstadials and stadials (e.g. Dokken & Jansen 1999), would also be consistent with our interpretation of the data.

Spectral analysis of the data reveals a marked difference in periodicities pre- and post-30 ka BP, corresponding to the transition from MIS 3 to 2. Prior to 30 ka BP, there is a strong 3077 year periodicity, which is somewhat longer than the well-known Dansgaard–Oeschger cycles (Dansgaard *et al.* 1993; Bond *et al.* 1993). However, the most persistent cycle observed within this interval has a frequency of 1980 years, which corresponds rather more closely to the Dansgaard–Oeschger frequency (e.g. Bond *et al.* 1997). This quantitative observation compares well with a visual inspection of the data (Fig. 8), where 10 distinct peaks in calcium carbonate are recorded within the interval 30–50 ka BP. These 10 peaks, having an average periodicity of approximately 2000 years, allow us to assign Greenland interstadial events 5–14

(Dansgaard *et al.* 1993) to our stratigraphic record. The correlation of these interstadial events gives us additional confidence in our age: depth model (since the age estimates are independently derived) and suggests that the reported problems of radiocarbon dating this time period (e.g. Voelker *et al.* 1998) have not significantly affected this record.

After 30 ka BP, the observed periodicities in our proxy data do not appear to coincide with those reported from the Greenland ice core record (Grootes & Stuiver 1997) or open North Atlantic (Bond *et al.* 1997). The higher frequency variability recorded, often at considerably lower power, within the interval 15–22 ka BP suggests that changes in the prevailing depositional environment were more dynamic than during MIS 3. This quantitative observation may be anticipated, given the marked transition in numerous sedimentological proxies at 30 ka BP (Fig. 8). During MIS 2, ice sheets surrounding NW Europe reached their maximum extent at or shortly before 22 ka BP (e.g. Sejrup *et al.* 1994). Regional mapping of extensive submarine moraine ridge complexes on the Hebridean shelf suggests that the BIS extended to the outer shelf during this period (Selby 1989; Stoker 1995). At its maximum extent, lying on deformable sediments, the last BIS had the potential to respond rapidly to both internal and external forcing mechanisms. The exact driving mechanism of the observed periodicities at this time remains uncertain, but it is apparent that glacier-influenced sedimentation was not operating with a clear D/O cyclicity.

As the ice sheet extended across the continental shelf during the MIS 3–2 transition, enhanced sediment delivery across the shelf–slope break resulted in rapidly increasing accumulation rates (Fig. 3). During this period small, discrete turbidite horizons appear in the record and are associated with intervals of IRD (Knutz *et al.* 2001) (Fig. 2). This suggests that large volumes of sediment were being deposited on the upper slope of the Barra Fan, providing a source material for the generation of turbidites. One potential trigger mechanism, which may have acted to destabilize the sediment-laden upper slopes of the Barra Fan, would have been iceberg scouring. Iceberg scours, extending to depths of over 400 m, are well known along this margin (e.g. Belderson *et al.* 1972; Kenyon 1987; Armishaw *et al.* 1998). Therefore, at a time of maximum ice-sheet extent and sediment delivery to a proximal site, strong periodicities in sediment proxies are likely to have been weaker. This observation illustrates the difficulty in correlating millennial-scale IRD events deriving from major ice sheets, particularly during the last glacial maximum.

Summary and Conclusions

The combined use of sediment colour with other proxy data provides a powerful tool in the detailed analysis of highly expanded sediment sequences. Sediment colour, although widely used in deep ocean records, has not commonly been used in the investigation of high latitude continental margin records. Despite difficulties associated with low calcium carbonate contents, the potential exists to investigate depositional processes at an extremely high resolution and in a non-destructive manner. The records obtained from MD95-2006 reveal how ice sheet extent influences sediment delivery rates across the shelf–slope break. The response of the BIS across the MIS 3–2 transition is clearly defined in a large number of sedimentological proxies and illustrates the highly responsive nature of this ice sheet to external climate forcing at about 30 ka BP. Sediment colour allows the characterization of rapid, sub-millennial-scale variability, suggesting a shift at about 30 ka BP to lower amplitude, higher frequency in all the proxies investigated.

Numerous IRD intervals within MD95-2006, which do not coincide with Heinrich Events, have a clear BIS provenance. These 'non-Heinrich' IRD events are similar in magnitude to the IRD events correlated with H1 to H5 and appear to have a pacing which matches the Dansgaard/Oeschger (D/O) cyclicity (Knutz *et al.* 2001). Whereas the local, Heinrich-equivalent events may have been triggered by instabilities in the BIS generated as a direct response to massive iceberg discharges from the LIS, this mechanism is unlikely to have generated all the observed IRD throughout the MD95-2006 record. Unfortunately, the absence of a distinctive LIS geochemical signature in the Barra Fan record make amphi-Atlantic correlations

extremely difficult. As such, the recent debate concerning so-called 'precursor' events of European origin (Grousset *et al.* 2000; Scourse *et al.* 2000), predating the major Heinrich Events, may require further consideration of the numerous IRD events characterizing North Atlantic sediment records at the millennial and sub-millennial-scale. In particular, the MD95-2006 record suggests that ice-rafted debris originating from the last BIS during MIS 2 is a pervasive sub-millennial-scale feature, operating at higher frequencies than the D/O cycle. Care should therefore be taken when identifying 'precursor' events for H1 to H3, since these may simply reflect a background signal of sub-millennial ice-rafting.

Despite the difficulties in radiocarbon chronologies and the problems which exist in the determination of lead-lag relationships between the marine, terrestrial and Greenland ice core records, reliable amphi-Atlantic correlations can still be achieved from an event stratigraphy based on interstadials.

We are grateful to D. Gunn for his support while obtaining the spectrophotometry data at the Southampton Oceanography Centre and to C. Byrne for assistance with the measurements. M. Chapman provided useful advice on sediment colour and encouragement at an early stage of the project. Constructive referee comments were provided by J. Scourse and C. Ó Cofaigh. M. Currie and C. Bryant, NERC Radiocarbon Laboratory, supported the AMS dating programme. We are grateful to the captain and crew of the RV *Marion Dufresne*, and to L. Labeyrie in particular. LJW's graduate (Ph.D.) study is funded by the University of St Andrews. This work is supported by the NERC, through grants GST/02/723; GST/02/1174; GR9/01595'A'.

References

ANDREWS, J. T. & TEDESCO, K. 1992. Detrital carbonate-rich sediments, northwestern Labrador Sea; implications for ice-sheet dynamics and iceberg rafting (Heinrich) events in the North Atlantic. *Geology*, **20**, 1087–1090.

ANDREWS, J. T., AUSTIN, W. E. N., BERGSTEN, H. & JENNINGS, A. E. 1996. The Late Quaternary palaeoceanography of North Atlantic margins: an introduction. *In*: ANDREWS, J. T., AUSTIN, W. E. N., BERGSTEN, H. & JENNINGS, A. E. (eds) *Late Quaternary Palaeoceanography of the North Atlantic Margins*. Geological Society, London, Special Publications, **111**, 1–6.

ARMISHAW, J. E., HOLMES, R. & STOW, A. V. 1998. Morphology and sedimentation on the Hebrides Slope and Barra Fan, NW UK continental margin. *In*: STOKER, M. S., EVANS, D. & CRAMP, A. (eds) *Geological Processes on Continental Margins: Sedimentation, Mass-wasting and*

Stability. Geological Society, London, Special Publications, **129**, 81–104.

AUSTIN, W. E. N. 1991. *Late Quaternary benthonic foraminiferal stratigraphy of the western UK continental shelf*. PhD Thesis, University of Wales.

AUSTIN, W. E. N. & EVANS, J. R. 2000. North East Atlantic benthic foraminifera: modern distribution patterns and palaeoecological significance. *Journal of the Geological Society*, **157**, 679–691.

AUSTIN, W. E. N. & KROON, D. 1996. Late glacial sedimentology, foraminifera and stable isotope stratigraphy of the Hebridean continental shelf, northwest Scotland. *In*: ANDREWS, J. T., AUSTIN, W. E. N., BERGSTEN, H. & JENNINGS, A. E. (eds) *Late Quaternary Palaeoceanography of the North Atlantic Margins*. Geological Society, London, Special Publications, **111**, 187–213.

AUSTIN, W. E. N. & KROON, D. 2001. Deep sea ventilation of the northeastern Atlantic during the last 15,000 years. *Global and Planetary Change*, **30**, 13–31.

AUSTIN, W. E. N., BARD, E., HUNT, J. B., KROON, D. & PEACOCK, J. D. 1995. The ^{14}C age of the Icelandic Vedde Ash: implications for Younger Dryas marine reservoir age corrections. *Radiocarbon*, **37**, 53–62.

BALSAM, W. L., DAMUTH, J. E. & SCHNEIDER, R. R. 1997. Comparison of shipboard vs. shore-based spectral data from Amazon Fan cores: implications for interpreting sediment composition. *In*: FLOOD, R. D., PIPER, D. J. W., KLAUS, A. & PETERSON, L. C. (eds) *Proceedings of the Ocean Drilling Program, Scientific Results*. Ocean Drilling Program, College Station, TX, **155**, 193–215.

BARD, E., ARNOLD, M., HAMELIN, B., TISNERAT-LABORDE, N. & CABIOCH, G. 1998. Radiocarbon calibration by means of mass spectrometric Th^{230}/U^{234} and C^{14} ages of corals: an updated database including samples from Barbados, Mururoa and Tahiti. *Radiocarbon*, **40**, 1085–1092.

BELDERSON, R. H., KENYON, N. H., STRIDE, A. H. & STUBBS, A. R. 1972. *Sonographs of the Sea Floor*. Elsevier, Amsterdam.

BLUM, P. 1997. *Physical properties handbook: A guide to shipboard measurement of physical properties of deep-sea cores*. Ocean Drilling Program. Ocean Drilling Program, College Station, TX, Technical note **26**.

BOND, G. C. & LOTTI, R. 1995. Iceberg discharges into the North Atlantic on millennial time scales during the last glaciation. *Science*, **267**, 1005–1010.

BOND, G. C., HEINRICH, H., BROECKER, W. S., LABEYRIE, L., McMANUS, J., ANDREWS, J., HUON, S., JANTSCHIK, R., CLASEN, S., SIMET, C., TEDESCO, K., KLAS, M., BONANI, G. & IVY, S. 1992. Evidence for massive discharges of icebergs into the North Atlantic Ocean during the last glacial period. *Nature*, **360**, 245–249.

BOND, G. C., BROECKER, W., JOHNSEN, S., McMANUS, J., LABEYRIE, L., JOUZEL, J. & BONANI, G. 1993. Correlations between climate records from North Atlantic sediments and Greenland ice. *Nature*, **365**, 143–147.

BOND, G. C., SHOWERS, W. J., CHESEBY, M., LOTTI, R., ALMASI, P., DE MENOCAL, P., PRIORE, P., CULLEN, H., HAJDAS, I. & BONANI, G. 1997. A pervasive millennial-scale cycle in North Atlantic Holocene and glacial climates. *Science*, **278**, 1257–1266.

BOND, G. C., SHOWERS, W., ELLIOT, M., EVANS, M., LOTTI, R., HAJDAS, I., BONANI, G. & JOHNSON, S. 1999. The North Atlantic's 1–2 kyr climate rhythm: Relation to Heinrich Events, Dansgaard/Oeschger cycles and the Little Ice Age. *In*: CLARK, P. U., WEBB, R. S. & KEIGWIN, L. D. (eds) *Mechanisms of Global Climate Change at Millennial Time Scales*, Geophysical Monograph, **112**, 35–58.

BOULTON, G. S. 1990. Sedimentary and sea level changes during glacial cycles and their control on glacimarine facies architecture. *In*: DOWDESWELL, J. A. & SCOURSE, J. D. (eds) *Glacimarine Environments: Processes and Sediments*, Geological Society, London, Special Publications, **53**, 15–52.

BROECKER, W. S., BOND, G., KLAS, M., BONANI, G. & WOLFI, W. 1990. A salt oscillator in the glacial Atlantic? *Paleoceanography*, **5**, 469–477.

CHAPMAN, M. R. & SHACKLETON, N. J. 1998. What level of resolution is attainable in a deep-sea core? Results of a spectrophotometer study. *Paleoceanography*, **13**, 311–315.

CORTIJO, E., YIOU, P., LABEYRIE, L. & CREMER, M. 1995. Sedimentary record of rapid climatic variability in the North Atlantic Ocean during the last glacial cycle. *Paleoceanography*, **10**, 911–926.

CURRY, W. B., MARCHITTO, T. M., McMANUS, J. F., OPPO, D. W. & LAARKAMP, K. L. 1999. Millennial-scale changes in ventilation of the thermocline, intermediate, and deep waters of the glacial North Atlantic. *In*: CLARK, P. U., WEBB, R. S. & KEIGWIN, L. D. (eds) *Mechanisms of Global Climate Change at Millennial Time Scales*, Geophysical Monograph, **112**, 59–76.

DANSGAARD. W., JOHNSEN, S. J., CLAUSEN, H. B., DAHL-JENSEN, D., GUNDESTRUP, N. S., HAMMER, C. U., HVIDBERG, C. S., STEFFENSON, J. P., SVEINJORNSDÖTTIR, A. E., JOUZEL, J. & BOND, G. 1993. Evidence for general instability of past climate from a 250-kyr ice-core record. *Nature*, **364**, 218–220.

DEATON, B. C. 1987. Quantification of rock color from Munsell chips. *Journal of Sedimentary Petrology*, **57**, 774–776.

DOKKEN, T. M. & JANSEN, E. 1999. Rapid changes in the mechanisms of ocean convection during the last glacial period. *Nature*, **401**, 458–461.

DOWDESWELL, J. A., MASLIN, M. A., ANDREWS, J. T. & McCAVE, I. N. 1995. Iceberg production, debris rafting, and the extent and thickness of Heinrich layers (H-1, H-2) in North-Atlantic sediments. *Geology*, **23**, 301–304.

DOWDESWELL, J. A., ELVERHØI, A. & ANDREWS, J. T. 1999. Asynchronous deposition of ice-rafted layers in the Nordic seas and the North Atlantic Ocean. *Nature*, **400**, 348–351.

GRIMALDI, F. S., SHAPIRO, L. & SCHNEPFE, M. 1966. Determination of carbon dioxide in limestone and dolomite by acid-base titration. *US Geological Survey, Professional Paper*, **550**, 186–188.

GROOTES, P. M. & STUIVER, M. 1997. Oxygen 18/16 variability in Greenland snow and ice with 10^{-3}- to 10^5-year time resolution. *Journal of Geophysical Research*, **102**, C12, 26 455–26 470.

GROOTES, P. M., STUIVER, M., WHITE, J. W. C., JOHNSEN, S. & JOUZEL, J. 1993. Comparison of oxygen isotope records from the GISP2 and GRIP Greenland ice cores. *Nature*, **366**, 552–554.

GROUSSET, F. E., LABEYRIE, L., SINKO, J. A., CREMER, M., BOND, G., DUPRAT, J., CORTIJO, E. & HUON, S. 1993. Patterns of ice-rafted detritus in the glacial North Atlantic (40–55°N). *Paleoceanography*, **8**, 175–192.

GROUSSET, F. E., PUJOL, C., LABEYRIE, L., AUFFRET, G. & BOELAERT, A. 2000. Were the North Atlantic Heinrich events triggered by the behaviour of the European ice sheets? *Geology*, **28**, 123–126.

GWIAZDA, R. H., HEMMING, S. R. & BROECKER, W. S. 1996. Tracking iceberg sources with lead isotopes: The provenance of ice-rafted debris in Heinrich layer 2. *Paleoceanography*, **11**, 77–94.

HEINRICH, H. 1988. Origin and consequences of cyclic ice rafting in the northeast Atlantic Ocean during the past 130,000 years. *Quaternary Research*, **29**, 142–152.

HELMKE, J. P. & BAUCH, H. A. 2001. Glacial-interglacial relationship between carbonate components and sediment reflectance in the North Atlantic. *Geo-Marine Letters*, **21**, 16–22.

HOLMES, R., LONG, D. & DODD, L. R. 1998. Large-scale debrites and submarine landslides on the Barra Fan, west of Britain. *In*: STOKER, M. S., EVANS, D. & CRAMP, A. (eds) *Geological Processes on Continental Margins: Sedimentation, Mass-wasting and Stability*. Geological Society, London, Special Publications, **129**, 67–79.

KENYON, N. H. 1987. Mass-wasting features on the continental slope of Northwest Europe. *Marine Geology*, **74**, 57–77.

KNUTZ, P. C., AUSTIN, W. E. N. & JONES, E. J. W. 2001. Millennial-scale depositional cycles related to British ice sheet variability and North Atlantic paleocirculation since 45 kyr BP, Barra Fan, UK margin. *Paleoceanography*, **16**, 53–64.

KROON, D., AUSTIN, W. E. N., CHAPMAN, M. R. & GANSSEN, G. M. 1997. Deglacial surface circulation changes in the Northeastern Atlantic: temperature and salinity records off NW Scotland on a century scale. *Paleoceanography*, **12**, 755–763.

KROON, D., SHIMMIELD, G., AUSTIN, W. E. N., DERRICK, S., KNUYZ, P. & SHIMMIELD, T. 2000. Century- to millennial-scale sedimentological-geochemical records of glacial-Holocene sediment variations from the Barra Fan (NE Atlantic). *Journal of the Geological Society*, **157**, 643–653.

KVAMME, T., MANGERUD, J., FURNES, H. & RUDDIMAN, W. 1989. Geochemistry of Pleistocene ash zones in cores from the North Atlantic. *Norsk Geologisk Tidsskrift*, **69**, 251–272.

McCABE, A. M. & CLARK, P. U. 1998. Ice-sheet variability around the North Atlantic Ocean during the last deglaciation. *Nature*, **392**, 373–377.

MARTINSON, D. G., PISIAS, N. G., HAYS, J. D., IMBRIE, J., MOORE, T. C. & SHACKLETON, N. J. 1987. Age dating and the orbital theory of ice ages: development of a high-resolution 0–300,000 year chronostratigraphy. *Quaternary Research*, **27**, 1–29.

MERRILL, R. B. & BECK, J. W. 1996. The ODP color digital system; color logs of Quaternary sediments from the Santa Barbara Basin, Site 893. *Marine Georesources & Geotechnology*, **14**, 381–408.

MUNSELL, A. H. 1941. *A color notation* (9th edn). Baltimore, Md., Munsell Color Co.

NAGAO, S. & NAKASHIMA, S. 1992. The factors controlling vertical color variations of North Atlantic Madeira abyssal plain sediments. The geochemistry of North Atlantic abyssal plains. *Marine Geology*, **109**, 83–94.

ORTIZ, J. D., O'CONNELL, S. & MIX, A. 1999. Data report; spectral reflectance observations from recovered sediments. Proceedings of the Ocean Drilling Program, scientific results, North Atlantic-Arctic gateways II; covering Leg 162 of the cruises of the drilling vessel JOIDES Resolution, Edinburgh, United Kingdom, to Malaga, Spain, sites 980–987, 7 July–2 September 1995. *Proceedings of the Ocean Drilling Program, Scientific Results*. Ocean Drilling Program, College Station, TX, **162**, 259–264.

PAILLARD, D., LABEYRIE, L. D. & YIOU, P. 1996. AnalySeries 1.0: a Macintosh software for the analysis of geophysical time-series. *EOS*, **77**, 379.

PEACOCK, J. D., AUSTIN, W. E. N., SELBY, I., HARLAND, R., WILKINSON, I. P. & GRAHAM, D. K. 1992. Late Devensian and Holocene palaeoenvironmental changes on the Scottish continental shelf west of the Outer Hebrides. *Journal of Quaternary Science*, **7**, 145–161.

REVEL, M. SINKO, J. A. & GROUSSET, F. E. 1996. Sr and Nd isotopes as tracers of North Atlantic lithic particles: Paleoclimatic implications. *Paleoceanography*, **11**, 95–113.

RICHTER, T. O., LASSEN, S., VAN WEERING, T. C. E. & HAAS, H. DE. 2001. Magnetic susceptibility patterns and provenance of ice-rafted material at Feni Drift, Rockall Trough: implications for the history of the British-Irish ice sheet. *Marine Geology*, **173**, 37–54.

SCOURSE, J. D., HALL, I. R., McCAVE, I. N., YOUNG, J. R. & SUGDON, C. 2000. The origin of Heinrich layers; evidence from H2 for European precursor events. *Earth and Planetary Science Letters*, **182**, 187–195.

SEJRUP, H. P., HAFLIDASON, H., AARSETH, I., KING, E., FORSBERG, C. F., LONG, D. & ROKOENGEN, K. 1994. Late Weichselian glaciation history of the northern North Sea. *Boreas*, **23**, 1–13.

SELBY, I. 1989. *The Quaternary geology of the*

Hebridean continental margin. PhD thesis, University of Nottingham.

STOKER, M. S. 1995. The influence of glacigenic sedimentation on slope–apron development on the continental margin off Northwest Britain. *In*: SCRUTTON, R. A., STOKER, M. S. SHIMMIELD, G. B. & TUDHOPE, A. W. (eds) *The Tectonics, Sedimentation and Palaeoceanography of the North Atlantic Region.* Geological Society, London, Special Publications, **90**, 159–178.

STOW, D. A. & TABREZ, A. R. 1998. Hemipelagites: processes, facies and model. *In*: STOKER, M. S., EVANS, D. & CRAMP, A. (eds) *Geological Processes on Continental Margins: Sedimentation, Mass-wasting and Stability.* Geological Society, London, Special Publications, **129**, 317–337.

STUIVER, M. & REIMER, P. J. 1993. Extended ^{14}C database and revised CALIB 3.0 ^{14}C age calibration program. *Radiocarbon,* **35**, 215–230.

STUIVER, M., REIMER, P. J., BARD, E., BECK, J. W., BURR, G. S., HUGHEN, K. A., KROMER, B., McCORMAC, G., VAN DER PLICHT, J. & SPURK, M. 1998. INTCAL98 Radiocarbon age calibration, 24,000–0 cal BP. *Radiocarbon,* **40**, 1041–1083.

VOELKER, A. H. L., SARNTHEIN, M., GROOTES, P. M., ERLENKEUSER, H., LAJ, C., MAZAUD, A., NADEAU, M.-J. & SCHLEICHER, M. 1998. Correlation of marine ^{14}C ages from the Nordic Seas with the GISP2 isotope record: Implications for ^{14}C calibration beyond 25 ka BP. *Radiocarbon,* **40**, 517–534.

Observations of surge periodicity in East Greenland using molybdenum records from marine sediment cores

JOHN WOODWARD[1], STEVE CARVER[2], HELMAR KUNZENDORF[3] & OLE BENNIKE[4]

[1]Physical Sciences Division, British Antarctic Survey, Natural Environmental Research Council, High Cross, Madingley Road, Cambridge CB3 0ET, UK
(e-mail: jwoo@bas.ac.uk)
[2]School of Geography, University of Leeds, Leeds LS2 9JT, UK
[3]Plant Biology and Biogeochemistry, RISØ National Laboratory, Roskilde, DK-4000
[4]GEUS, Thoravej 8, 2400 København NV, Denmark

Abstract: This paper describes a unique record of glacier flow instability for East Greenland during the Little Ice Age. Trace metal analysis of sediment cores collected during 1998 from the Noret Inlet in the Mesters Vig area of East Greenland shows two peaks in the molybdenum (Mo) record at 495 ± 40 years BP and 95 ± 2 years BP. This is notable as there is no molybdenum mineralization in the geology of the Noret Inlet catchment area. Molybdenum is found, however, in the drainage basin of Mesters Vig Inlet, just to the south of the Noret Inlet. The molybdenum record in the Noret core provides a long-term surge record for the Östre Gletscher, a large surge-type glacier in the Werner Bjerge that drains into Mesters Vig Inlet. The two molybdenum peaks indicate surge termination for the glacier, indicating a surge recurrence interval of around 400 years.

Long-term records of glacier flow instability are extremely rare, particularly in uninhabited regions of the world. Glacier flow instabilities, resulting in glacier advance or retreat, can be attributed to climate change or surging (Meier & Post 1969; Dowdeswell et al. 1995; Raymond 1987). Surge-type glaciers oscillate between periods of extremely rapid, short-term movement, and long periods of relative inactivity or quiescence (Meier & Post 1969). Understanding the periodicity of surge events, and any influence imposed by climate on the surge cycle, has important implications for understanding the response of glaciers and ice sheets to climate change.

The geographical distribution of surge-type glaciers is markedly non-random (Meier & Post 1969; Hamilton & Dowdeswell 1996; Jiskoot et al. 2000). Clusters of surge-type glaciers are found in Alaska, the Yukon, Canada, east and west Greenland, the Pamirs, the Caucasus, the Tien-Shan, parts of the Andes, Iceland and Svalbard (Paterson 1994). Such clusters suggest that environmental conditions may play a controlling role in the distribution of surge-type glaciers (Raymond 1987; Hamilton & Dowdeswell 1996; Jiskoot et al. 2000). As the geographical distribution of surging glaciers varies, so does the duration of quiescent and surge phases. For most glaciers the duration of

the surge is between 1–2 years (Meier & Post 1969). Well documented surge-type glaciers such as Bering Glacier, Variegated Glacier and Medvezhiy Glacier have surge periods of 1–2 years and quiescent periods of about 20, 20 and 15 years, respectively (Dolgoushin & Osipova 1975; Fleisher et al. 1995; Lawson 1996). By contrast, in Svalbard, the surge typically lasts from 3–15 years, with the period of quiescence estimated to last from 50 to 500 years (Dowdeswell et al. 1991, 1995; Hamilton & Dowdeswell 1996).

In Greenland surge-type glaciers are found in three clusters: in the basalt provenance of Disko Island and Nûgssuaq Peninsular on the west coast; between Kangerdlugssuaq and Scoresby Sund on the east coast, also an area of basalt rocks; and in the Staunings Alper of East Greenland, where the provenance is mixed sedimentary rocks with some intruded basalt and gneisses (Weidick 1988). Isolated, possible surge instabilities are also recorded for Harald Moltke Bræ in the NW (Mock 1966), Brikkerne Gletscher in the north (Higgins & Weidick 1988) and Eqalorutsit Kitdlît Sermiat and Søndre Sermilik in the south (Weidick 1984a, b). Surge duration of up to seven years (Rutishauser 1971; Weidick 1988), and quiescent periods in the order of 70–100 years have been inferred. A recorded surge advance of 10 km for

From: DOWDESWELL, J. A. & Ó COFAIGH, C. (eds) 2002. Glacier-Influenced Sedimentation on High-Latitude Continental Margins. Geological Society, London, Special Publications, **203**, 367–373. 0305-8719/02/$15.00
© The Geological Society of London 2002.

Storstrømmen in 1978 reached the same position as the 1913 glacier front (Reeh *et al.* 1994). However, the frontal position in 1913 is the first recorded position for the glacier, so it is questionable as to whether this represents a surge advance or the location of a stable tidewater margin prior to 1913. Work on the looped medial moraines of the Bjørnbo Gletscher suggests a quiescent period of at least 100 years (Rutishauser 1971). Sortebræ is know to have surged at least twice, in 1933–43 and 1992–95, though the advances were the result of surging in the west and main trunk glacier, respectively (Jiskoot 1999).

It is difficult, however, to infer surge conditions or periodicity in Greenland because many glaciated areas are remote and long-term observational records do not exist. In remote areas, alternative methods of identifying surge events in the Quaternary record are required because of the lack historical records from direct observation. In many instances, air photographs show rapid advance or retreat of a glacier together with surface structures, such as looped medial moraines and crevasse patterns, that are indicative of surge conditions (Weidick 1988; Hamilton & Dowdeswell 1996). The study of sedimentary structures such as crevasse-fill features and sediment-rich thrust planes in glacier ice and the sediment deposited from them in the glacier forefield may also be indicative of surge behaviour (e.g. Sharp 1985; Hambrey *et al.* 1999; Woodward 1999). Evidence from sedimentary structures is of

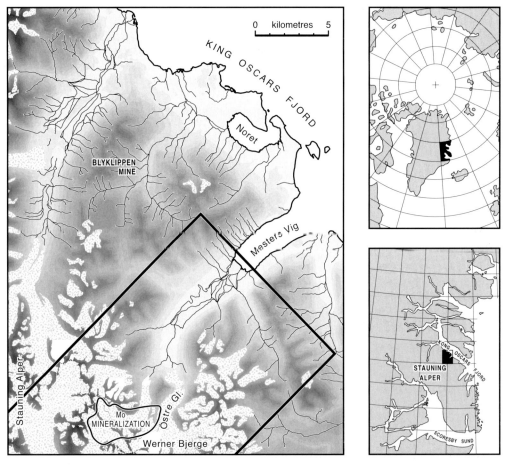

Fig. 1. Map showing the location of Noret Inlet and Mesters Vig on the east coast of Greenland (Original 1: 100 000 DEM data supplied by GEUS). Outcrops of molybdenum are also shown. The box marks the location of the Landsat TM image shown in Figure 3.

limited use, however, in that these structures often record only the latest advance and rarely record a succession of surge advances.

Study site

The Mesters Vig district is located north of the Arctic Circle (72°15' N 24° W) in the fjord region of East Greenland. The Noret Inlet is an almost enclosed inlet of King Oscars Fjord (Fig. 1). The inlet drains the Skeldal catchment, which is bordered on the west by the Stauning Alper, and to the south by Mesters Vig Inlet and the Werner Bjerge. A number of large glaciers flow into the Mesters Vig Inlet, including the Mellem, Vestre and Östre glaciers. A number of glaciers in the Stauning Alper are known to be of surge type, including Løberen, a previously unnamed glacier which surged over 7.5 km into Nordvestfjord between 1950 and 1961 (Henriksen & Watt 1968), Bjørnbo Gletscher which surged around 1890 (Rutishauser 1971), the Dalmore and Roslin gletschers, which may both have surged prior to 1950 (Colvill 1984) and the Sirius Gletscher, which field observations indicate surged immediately before 1998.

The Noret Inlet and Skeldal catchment is underlain by Triassic, Permian and Carboniferous sandstones, shales and conglomerates with intruded basalt dykes and sills (Harpøth *et al.* 1986; Sear & Carver 1996). A veneer of till, emerged deltaic sediments and alluvium covers many of the lower slopes. The Mesters Vig area is underlain by Precambrian sedimentary rocks and gneisses and granites (Harpøth *et al.* 1986). The Östre Gletscher flows north into Deltadal and Mesters Vig inlet from the Werner Bjerge, a massif of intruded alkali granite and syenite rocks bearing molybdenum (Mo) mineralization (Harpøth *et al.* 1986).

Methodology

Cores were originally collected for the analysis of heavy metal pollution reaching the Noret Inlet from the Blyklippen lead and zinc mine. The mine was active from 1956–1963 when tailings were deposited on the tundra surface (Sear & Carver 1996). Tailings have been washed into the Tunnelelv river system which drains the Skeldal catchment into the Noret Inlet.

A single-beam echo sounder was used to obtain a bathymetric profile the Noret Inlet. The deepest area, of 55 m, was selected for the extraction of cores. Cores were collected using a Mackareth corer, which uses a compressed gas source to push a 1 m long, 5 cm diameter plastic core tube into the bottom sediments. The sediments present in the Noret Inlet were ideal for this method of extraction, resulting in 100% core recovery. The sediments were then extruded from the core tubes and the cores cut into 1 and 2 cm sections for analysis.

Core samples were air dried before being crushed through a 150 micron nylon sieve. Samples were digested in *aqua regia* before being analysed in an Atomscan Advantage Sequential ICP-OES for Cu, Fe, Mn, Mo, Pb, Zn and Ag. Multiple digestions and analysis were carried out for a random selection of samples to test for repeatability. Sample results were found to be repeatable within 1% of metal results. The estimated error for Mo concentration is $\pm$ 0.13 mg kg^{-1}. Particle-size analysis was carried out using a Coulter laser particle analyser. 80–90% of the core sediments fall into the silt and clay category. The top 0.09 m of the core was dated using gamma spectroscopy ^{210}Pb dating techniques, giving dates from 2 to 93 $\pm$ 2 years BP (from a CRS model). Background levels of ^{210}Pb were reached below 0.09-m depth in the core.

A radiocarbon date was obtained from a sample of plant material extracted from the core at a depth of 0.45 m. The sample was submitted to Beta Analytic Inc. who converted the sample carbon to graphite before carrying out analysis using an accelerator mass spectrometer (AMS). This resulted in a conventional radiocarbon age of 900 $\pm$ 40 years BP (Beta-132275: C^{13}/C^{12} ratio of –28.1‰). When corrected, this produces a calendar date of 1025–1225 AD at the two standard deviation range (following methodology of Talma & Vogel 1993). This suggests an average annual accumulation rate of around 0.045 cm yr^{-1} for the core, assuming linear interpolation from the base of the ^{210}Pb chronology to the radiocarbon date. This accumulation rate was then extrapolated to the base of the core. Such linear extrapolation ignores any compression effects and assumes a constant sedimentation rate. A constant sedimentation rate for this fjord environment is questionable as sediment supply rates are likely to have fluctuated with climate fluctuations. As organic material was only available in the core at this depth, however, chronology must be estimated based upon this single conventional radiocarbon age.

Results from the cores

The Mo record for the core is shown in Figure 2. There are two sections to the Mo record in the core:

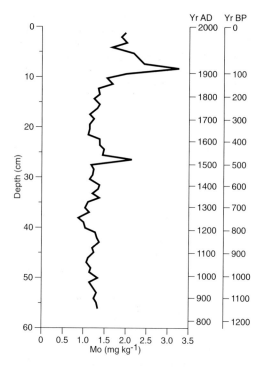

Fig. 2. Molybdenum (Mo) record from a sediment core extracted from the Noret Inlet. The first peak occurs in 1505 ± 40 AD (495 ± 40 BP) and has a Mo concentration of 2.19 mg kg^{-1}. The second peak, in 1905 ± 2 AD (95 ± 2), has a Mo concentration of 3.31 mg kg^{-1}. Both peaks have a steep rising limb and more gradual falling limb. Plant material extracted from the core at a depth of 0.44 m gave a conventional radiocarbon age of 900 ± 40 BP (C^{13}/C^{12} Ratio of –28.1‰). This records the maximum of the Medieval Warm Period for Noret. Mo concentrations were analysed using an Atomscan Advantage Sequential ICP-OES with detection levels of 0.01 mg kg^{-1}.

(1) Before 495 ± 40 years BP, around 1500 AD, there appears to be little variability in the Mo record. The average Mo is 1.21 ± 0.13 mg kg^{-1}, close to the crustal average of approximately 1.5 mg kg^{-1}.
(2) After 495 ± 40 years BP, about 1500 AD, there is an increase in the amount of Mo in the core from 1.25 to 2.02 mg kg^{-1}. In this latter period there are also two dominant peaks. The first peak occurs at 495 ± 40 years BP and has a Mo concentration of 2.19 mg kg^{-1}. The second peak at 95 ± 2 years BP has a Mo concentration of 3.31 mg kg^{-1}. Both peaks have a similar form, with a steep rising limb followed by a more gradual falling limb. These two peaks are superimposed upon the

underlying long-term trend of Mo increase from 1.25 to 2.02 mg kg^{-1}.

The presence of Mo in the cores is notable as there is no Mo mineralization in the geology of the Noret Inlet basin. However, significant Mo mineralization is found in the drainage basin of Mesters Vig Inlet to the south of the Noret Inlet which is underlain by Precambrian sedimentary rocks, gneisses and granites (Harpøth *et al.* 1986).

Discussion

Four issues become apparent in the core record. First, the source of the Mo is unknown. Second, the cause of the Mo peaks at 495 ± 40 years BP and 95 ± 2 years BP must be explained. Third, the reason for the increase in background levels of Mo since 495 ± 40 years BP must be addressed. Finally, there is an issue of how resolvable the Mo peaks are at greater depths in the core, as there is a suggestion of two smeared peaks in Mo concentration around 600 ± 40 years BP and 850 ± 40 years BP.

As previously stated, the Mo is not native to the Noret Inlet catchment, suggesting an external source for the Mo. Mo is found to outcrop in the Stauning Alper to the south and west of the basin, indicating a local source, the most likely being the outcrops around the Werner Bjerge at the head of the Mesters Vig Inlet.

Peaks in the Mo record

There are a number of possible mechanisms for transporting Mo into the inlet that might produce peaks:

(1) Iceberg transport. Ice-rafted debris (IRD) rich in Mo may have been transported into the Noret Inlet from the break-up of a tide-water glacier.
(2) Aeolian transport of Mo. If large areas of fine-grained sediments were exposed either in a glacier forefield or by devegetation, the Mo-rich sediment could be transported from neighbouring catchments, by strong winds, during snow free summer periods.
(3) Glacial retreat. Periods of glacier retreat exposing large areas of sediment in a glacial forefield, along with increased meltwater output, could produce sediment-rich melt-water. This meltwater could be supplied to the Noret Inlet from Mesters Vig Inlet, producing a Mo-rich sediment layer in the Noret Inlet. Periodic outburst floods from glacier dammed or proglacial lakes might

add to this meltwater drainage and produce peaks.

(4) Glacier-surge termination. As a surge terminates, a pulse of sediment-rich water is often released by the glacier (Kamb *et al.* 1985; Humphrey & Raymond 1994; Fleisher *et al.* 1995). If there was a surge-type glacier in the Werner Bjerge area, Mo-rich water released at surge termination could enter the inlet from the Mesters Vig fjord, and then deposit a Mo-rich sediment layer.

No dropstones were identified in any of the cores in the Noret inlet, suggesting little supply of IRD to the Inlet. The morphology of the Noret Inlet is such that it would preclude entry of large icebergs due to its shallow entrance (approximately 4 m at present sea level); thus it is unlikely that large amounts of IRD are transported into the inlet.

Aoelian transport could move molybdenum-rich sediments from the unvegetated, snow-free slopes of the Werner Bjerge. An aeolian source would, however, deposit Mo throughout the Skeldal catchment, not just directly into the Noret Inlet. This general deposition would be unlikely to result in peaks of Mo supply as the Mo deposited in the terrestrial environment would be constantly reworked and transported into the inlet.

Glacial retreat might expose Mo-rich proglacial sediments, which could then be reworked by glacial meltwater. This mechanism is thought unlikely to produce a strongly peaked response, unless glacier retreat was extremely rapid. Also, as Mo is poorly particle-reactive at seawater ionic strengths (Goldberg & Foster 1998), high concentrations of Mo sediment would have to be released into the Mesters Vig inlet in order to enable the Mo to be transported into the Noret Inlet. The gradual retreat and exposure of fore-field sediments is unlikely to produced highly sediment-rich meltwater output. The gradual exposure of forefield sediments could, however, influence the underlying trend of the Mo curve, if background levels of Mo supply increased as a result. No ice-dammed lakes or terraces from previous ice dammed lakes are visible in the Werner Bjerge area. This indicates that only small, proglacial or marginal lakes are likely to have existed, the drainage of which would not cause a significant Mo peak in the sedimentary record in the Noret Inlet core.

Aerial photographs, satellite imagery and field observations of the Östre Gletscher at the head of Mesters Vig Inlet indicate that it is of surge-type (Figures 3 and 4). Indeed, on 1965 maps, Östre Gletscher was labelled as the Retreater Glacier (Washburn 1965), indicating that the glacier has retreated considerably in the

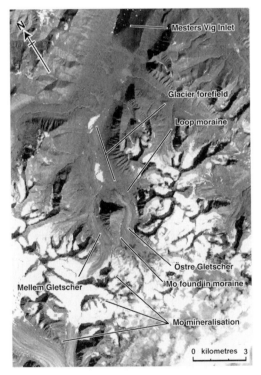

Fig. 3. Landsat TM image (05 August 1989), showing the recently exposed forefield of the Östre Gletscher, that was filled by ice during the last surge advance in 1905 AD. A loop moraine formed during the surge can also be seen on the glacier surface. The location of the image is shown by the box in Figure 1.

Fig. 4. Oblique photograph taken in 1998 showing the snout of Östre Gletscher. A loop formed by the surge is visible in the medial moraine of the glacier.

years previous to this. Such evidence supports a possible retreat from a surge advance just prior to 1905. The peaks of molybdenum in the core are, therefore, thought to indicate surge termination events for Östre Gletscher. As the surge terminated, a pulse of sediment-laden water may have been discharged into the Mesters Vig Inlet and have flowed out into Kong Oscars Fjord. Observations from aerial photographs, satellite imagery and the field suggest sediment plumes from glacial discharge flow both to the west and east into Kong Oscars Fjord, depending presumably on tidal conditions (cf. Dowdeswell & Collin 1990). Some sediment-rich water from this sediment plume may, therefore, have flowed into the backwater of the Noret Inlet. Once trapped in the Noret Inlet, the sediment in the plume may have settled out, producing a Mo-rich layer of sediment. The Mo record from the core, therefore, indicates termination of surge advances for Östre Gletscher around 1505, and again around 1905. This suggests a surge recurrence interval of approximately 400 years.

Background Mo levels

Before 495 ± 40 years BP, background Mo levels appear to show little variability, averaging 1.21 ± 0.13 mg kg^{-1}. After 495 ± 40 years BP, there is an increase in the background amount of Mo in the core from 1.25 to 2.02 mg kg^{-1}. It seems likely that this background Mo originates from the same source as the Mo recorded in the peaks in the core.

The increase in background Mo concentrations from 495 BP suggests an increase in the supply of sediment-rich meltwater and/or an increase the sediment concentrations of the meltwater from the Mesters Vig Inlet. Increased meltwater is likely to result when glaciers are retreating. An increase in sediment concentration could also be expected if water is able to access the basal sediments under the glaciers, possibly due to a change in glacier thermal regime, or if fresh sediments are exposed in the proglacial environment due to glacier retreat. Decreased snow-cover on Mo bearing rocks would also produce more available sediment for transportation. Thus, the increase in background levels of Mo may indicate a decrease in glacier and snow cover in the Mesters Vig region during the last few hundred years.

The Mo record prior to 495 ± 40 years BP shows little variability (Fig. 2). Either there were no peaks of Mo released prior to 495 ± 40 years BP, suggesting the Östre Gletscher has

only developed surge-type flow instabilities since then, or any peaks are lost in the noise of the record. There is the suggestion of two smeared peaks in Mo concentration around 850 BP and around 600 BP, which are close to the noise visible in the record at this time. The peak at 850 BP would give a recurrence interval of 450 ± 40 years, similar to the 400 ± 40 years reported between the other two peaks. Further coring is required to provide larger samples for analysis to investigate the early part of the surge record.

Conclusions

Cores taken from the Noret Inlet, East Greenland, are inferred to record a unique record of surging for the Östre Gletscher. The surge record, represented by peaks in Mo concentrations in the core, and thought to indicate surge-termination events, are visible in the core because Mo is not native to the Noret basin, although it is found in the neighbouring Mesters Vig Inlet. The peaks are thought to have been produced by sediment-rich water being released from the Östre Gletscher at surge-termination, the plumes entering the Noret Inlet, where Mo has been deposited subsequently. The peaks suggest a recurrence interval for surging of approximately 400 years. This is similar in length to the surge periodicity of some large glaciers and ice caps in Svalbard (Hamilton & Dowdeswell 1996).

Further work is required to trace the Mo input to the inlet by studying input from glaciated catchments in the Mesters Vig area, and water entering Noret Inlet both from the fjord outlet and the river catchments within the Noret basin itself. When compared to particulate analysis of the cores, this should provide a quantification of the source of the Mo. Further coring work, particularly in Mesters Vig Inlet is needed to identify surge episodes in the sediment record. Further cores will also be used to examine the effects of bioturbation on the core stratigraphy and to constrain the sedimentation rates for Noret Inlet.

We wish to thank J. Taylor, two anonymous reviewers and P. Collins (Brunel University) for their comments and additions. The University of Leeds and The Royal Society provided funding for field research. The Danish Polar Centre granted permission to carry out research at Mesters Vig in 1998. Andrew Bainbridge, Bruce Carlisle and Ian Heywood assisted with fieldwork. The laboratory staff at the School of Geography, University of Leeds assisted with processing the core data.

References

COLVILL, A. J. 1984. Some observations on glacier surges, with notes on the Roslin glacier, East Greenland. *In*: MILLER, K. J. (ed.) *The International Karakoram Project*. Cambridge University Press, Cambridge, **1**.

DOLGOUSHIN, L. D. & OSIPOVA, G. B. 1975. Glacier surges and the problem of their forecasting. *IAHS-AISH Publ.*, **104**, 292–304.

DOWDESWELL, J. A. & COLLIN, R. L. 1990. Fast-flowing outlet glaciers on Svalbard ice caps. *Geology*, **18**, 778–781.

DOWDESWELL, J. A., HAMILTON, G. S. & HAGEN, J. O. 1991. The duration of the active phase on surge-type glaciers: contrasts between Svalbard and other regions. *Journal of Glaciology*, **37**, 338–400.

DOWDESWELL, J. A., HODGKINS, R., NUTTALL, A.-M., HAGEN, J. O. & HAMILTON, G. S. 1995. Mass balance change as a control on the frequency and occurrence of glacier surges in Svalbard, Norwegian High Arctic. *Geophysical Research Letters*, **22**, 2909–2912.

FLEISCHER, P. J., MULLER, E. H., CADWELL, D. H., ROSENFELD, C. L., BAILEY, P. K., PELTON, J. M. & PUGLESI, M. A. 1995. The surging advance of Bering Glacier, Alaska, USA: a progress report. *Journal of Glaciology*, **41**, 207–213.

GOLDBERG, S. & FORSTER, H. S. 1998. Factors affecting molybdenum adsorption by soils and minerals. *Soil Science*, **163**, 109–114.

HAMBREY, M. J., BENNETT, M. R., DOWDESWELL, J. A., GLASSER, N. F. & HUDDARD, D. 1999. Debris entrainment and transfer in polythermal valley glaciers. *Journal of Glaciology*, **45**, 69–86.

HAMILTON, G. S. & DOWDESWELL, J. A. 1996. Controls on glacier surging in Svalbard. *Journal of Glaciology*, **42**, 157–168.

HARPØTH, O., PEDERSEN, J. L., SCHØNWANDT, H. K. & THOMASSEN, B. 1986. The mineral occurrences of central East Greenland. *Meddelser om Grønland*, **17**, 139.

HENRIKSEN, N. & WATT, W. S. 1968. Geological reconnaisance of the Scoresby Sund fjord complex. *Grønlands Geologiske Undersøgelse Raport*, **15**, 72–77.

HIGGINS, A. K. & WEIDICK, A. 1988. The world's northernmost surging glacier? *Zeitschrift für Gletscherkunde und Glazialgeologie*, **24**, 111–123.

HUMPHREY, N. F. & RAYMOND, C. F. 1994. Hydrology, erosion and sediment production in a surging glacier: Variegated Glacier, Alaska, 1982–83. *Journal of Glaciology*, **40**, 539–552.

JISKOOT, H. 1999. *Characteristics of surge-type glaciers*. Unpublished PhD Thesis, University of Leeds.

JISKOOT, H., MURRAY, T. & BOYLE, P. 2000. Controls on the distribution of surge-type glaciers in Svalbard, *Journal of Glaciology*, **46**, 412–422.

KAMB, B., RAYMOND, R. F., HARRISON, W. D., ENGELHARDT, H., ECHELMEYER, K. A., HUMPHREY, N., BRUGMAN, M. M. & PFEFFER, T. 1985. Glacier surge mechanisms: 1982–1983 surge of Variegated Glacier, Alaska. *Science*, **227**, 469–479.

LAWSON, W. 1996. Structural evolution of Variegated Glacier, Alaska, USA, since 1948. *Journal of Glaciology*, **42**, 261–270.

MEIER, M. F. & POST, A. 1969. What are glacier surges? *Canadian Journal of Earth Sciences*, **6**, 807–817.

MOCK, S. J. 1966. Fluctuations of the terminus of Harold Moltke Bræ, Greenland. *Journal of Glaciology*, **6**, 369–374.

PATERSON, W. S. B. 1994. *The Physics of Glaciers* (3rd edn). Pergamon Press, London.

RAYMOND, C. F. 1987. How do glaciers surge? A review. *Journal of Geophysical Research*, **92**, 9121–9134.

REEH, N., BØGGILD, C. E. & OERTER, H. 1994. Surge of Storstrømmen, a large outlet glacier from the Inland Ice of North-East Greenland. *Grønlands Geologica Understat*, **162**, 201–209.

RUTISHAUSER, H. 1971. Observations on a surging glacier in East Greenland. *Journal of Glaciology*, **10**, 227–236.

SEAR, D. & CARVER, S. 1996. The release and dispersal of Pb and Zn contaminated sediments within an Arctic braided river system. *Applied Geochemistry*, **11**, 187–195.

SHARP, M. J. 1985. 'Crevasse-fill' ridges – a landform characteristic of surge-type glaciers? *Geografiska Annaler*, **67A**, 213–220.

TALMA, A. S. & VOGEL, J. C. 1993. A simplified approach to calibrating C14 dates. *Radiocarbon*, **35**, 317–322.

WASHBURN, A. L. 1965. Geomorphic and vegetational studies in the Mesters Vig District, northeast Greenland. *Meddelelser om Grønland*, **166**, 1–40.

WEIDICK, A. 1984a. Studies of glacier behaviour and glacier mass balance in Greenland – a review. *Geografiska Annaler*, **66A**, 183–195.

WEIDICK, A. 1984b. Location of two glacier surges in West Greenland. *Grønlands Geologica Understat*, **120**, 100–104.

WEIDICK, A. 1988. Surging glaciers in Greenland – a status. *Grønlands Geologiske Undersøgelse Raport*, **140**, 106–110.

WOODWARD, J. 1999. *Structural glaciology of Kongsvegen, Svalbard, using ground-penetrating radar*. Unpublished PhD Thesis, University of Leeds.

Index

Note: Page numbers in **bold** type refer to tables.